Vegetation Dynamics, Changing Ecosystems and Human Responsibility

Edited by Levente Hufnagel and Mohamed A. El-Esawi

Published in London, United Kingdom

Vegetation Dynamics, Changing Ecosystems and Human Responsibility
http://dx.doi.org/10.5772/intechopen.100937
Edited by Levente Hufnagel and Mohamed A. El-Esawi

Contributors
Santiago Alexander Guamán Rivera, Oscar Giovanny Fuentes Quisaguano, Mădălina-Andreea Ciocan, Mihaela Hnatiuc, Simona Ghiță, Victoria Artem, Aurora Ranca, Gyan Shri Kaushal, Rajiv Umrao, Emoke Dalma Kovacs, Melinda-Haydee Kovacs, Veronika Piscová, Juraj Hreško, Michal Ševčík, Terézia Slobodová, Zenebe Mekonnen, Ramesh Prasad Bhatt, Takayoshi Koike, Laiye Qu, Yannan Wang, Noboru Masui, Xiaoke Wang, Thomas Rötzer, Cong Shi, Toshihiro Watanabe, Lizete Stumpf, Maria Bertaso De Garcia Fernandez, Pablo Miguel, Luiz Fernando Spinelli Pinto, Tania Hipolito Montiel, Ryan Noremberg Schubert, Luis Iuñes Carlos de Oliveira Filho, Lucas Da Silva Barbosa, Jeferson Diego Leidemer, Thábata Barbosa Duarte, Ricardo O. Russo, André Johannes J. Pelser, Rujeko Samanthia Chimukuche, Tolulope Olatoye, Oluwayemi Olatoye, Sonwabo Mazinyo, Gbadebo Odularu, Akinwunmi Sunday Odeyemi, Ally K. Kiyenze Nkwabi, Pius Yoram Kavana, Tinyiko Salome Mthombeni, Sérgio Lousada, José Manuel Naranjo Gómez, Patrick O. Ayeku, Nalukui Matakala, Kondwani Kapinga, Jules Christian Zekeng, Lackson Chama, Stanford Siachoono, Obote Shakacite, Concilia Monde, Stephen Syampungani, Aladesanmi Daniel Agbelade, Oğuz Başkan, Tülay Tunçay, Akinbi John Olarewaju John, Ajayi Olalekan Kehinde, Akinbowale Akinlolu Sylvester, Agbeje Abiodun Michael, Moses Z. Zakhele Sithole, Azikiwe I. Agholor, Shalia M. Ndlovu, Abdillahi Maoulida Mohamed, Frédéric Bioret, Vicent Boullet, Ana Daniela Lopes, João Paulo Francisco, Maria Graciela Iecher Faria Nunes, Eveline Henrique dos Santos, Raymond Aabeyir, Kenneth Peprah, Gervase Kuuwaabong, Hayriye Yildiz Dasgan, Tugce Temtek, Semwanga Mohammed, Nakiguli Fatumah, Nasseje Shadia, Kigozi Abas, Ashraf Nkumba, Nakimwanyi Shamirah

First published in London, United Kingdom, 2023 by IntechOpen
IntechOpen is the global imprint of INTECHOPEN LIMITED, registered in England and Wales, registration number: 11086078, 5 Princes Gate Court, London, SW7 2QJ, United Kingdom
Printed in Croatia

British Library Cataloguing-in-Publication Data
A catalogue record for this book is available from the British Library

Additional hard and PDF copies can be obtained from orders@intechopen.com

Vegetation Dynamics, Changing Ecosystems and Human Responsibility
Edited by Levente Hufnagel and Mohamed A. El-Esawi
p. cm.

This title is part of the Environmental Sciences Book Series, Volume 6
Topic: Ecosystems and Biodiversity
Series Editor: J. Kevin Summers
Topic Editors: Salustiano Mato, Josefina Garrido and Francisco Ramil

Print ISBN 978-1-80356-137-0
Online ISBN 978-1-80356-138-7
eBook (PDF) ISBN 978-1-80356-139-4
ISSN 2754-6713

For EU product safety concerns:
IN TECH d.o.o., Prolaz Marije Krucifikse Kozulić 3, 51000 Rijeka, Croatia,
info@intechopen.com or visit our website at intechopen.com.

IntechOpen Book Series

Environmental Sciences

Volume 6

Aims and Scope of the Series

Scientists have long researched to understand the environment and man's place in it. The search for this knowledge grows in importance as rapid increases in population and economic development intensify humans' stresses on ecosystems. Fortunately, rapid increases in multiple scientific areas are advancing our understanding of environmental sciences. Breakthroughs in computing, molecular biology, ecology, and sustainability science are enhancing our ability to utilize environmental sciences to address real-world problems.

The four topics of this book series - Pollution; Environmental Resilience and Management; Ecosystems and Biodiversity; and Water Science - will address important areas of advancement in the environmental sciences. They will represent an excellent initial grouping of published works on these critical topics.

Meet the Series Editor

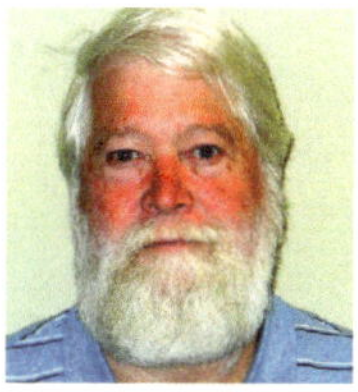

J. Kevin Summers is a Senior Research Ecologist at the Environmental Protection Agency's (EPA) Gulf Ecosystem Measurement and Modeling Division. He is currently working with colleagues in the Sustainable and Healthy Communities Program to develop an index of community resilience to natural hazards, an index of human well-being that can be linked to changes in the ecosystem, social and economic services, and a community sustainability tool for communities with populations under 40,000. He leads research efforts for indicator and indices development. Dr. Summers is a systems ecologist and began his career at the EPA in 1989 and has worked in various programs and capacities. This includes leading the National Coastal Assessment in collaboration with the Office of Water which culminated in the award-winning National Coastal Condition Report series (four volumes between 2001 and 2012), and which integrates water quality, sediment quality, habitat, and biological data to assess the ecosystem condition of the United States estuaries. He was acting National Program Director for Ecology for the EPA between 2004 and 2006. He has authored approximately 150 peer-reviewed journal articles, book chapters, and reports and has received many awards for technical accomplishments from the EPA and from outside of the agency. Dr. Summers holds a BA in Zoology and Psychology, an MA in Ecology, and Ph.D. in Systems Ecology/Biology.

Meet the Volume Editors

Dr. Levente Hufnagel is an associate professor and the head of the Research Institute of Multidisciplinary Ecotheology, John Wesley Theological College, Budapest, Hungary working on ecology, biogeography, ecological research methodology, ecotheology, and sustainability. He obtained a master's degree in Ecology and Evolutionary Biology and a Ph.D. in Hydrobiology from Eötvös Lorand University, Budapest, Hungary. He obtained another Ph.D. in Agricultural Science from Szent István University, Gödöllő, Hungary. Dr. Hufnagel has additional degrees from Corvinus University of Budapest and Adventist Theological College. He has more than 20 years of experience leading Hungarian academic institutions and teaching Ph.D., MSc, and BSc students from various social and cultural backgrounds. He has supervised numerous PhD, BSc, and MSc theses. He has more than 240 scientific publications on aquatic and terrestrial ecological aspects of plants, animals, and microbes at both the community and population levels to his credit. Dr. Hufnagel has participated in several big ecological research and development projects and multidisciplinary collaborations.

Dr. Mohamed A. El-Esawi is a visiting research fellow at the University of Cambridge, United Kingdom, and Associate Professor of Molecular Genetics at Tanta University, Egypt. Dr. El-Esawi received his BSc and MSc from Tanta University, and his Ph.D. in Plant Molecular Biology from Dublin Institute of Technology, TU Dublin, Ireland. Afterward, Dr. El-Esawi joined the University of Warwick, UK, Sorbonne University, France, and KU Leuven, Belgium as a visiting research fellow. His research focuses on plant genetics, genomics, molecular biology, molecular physiology, developmental biology, plant-microbe interaction, and bioinformatics. He has authored more than 140 articles, book chapters, and books, and has participated in more than 60 conferences and workshops worldwide. Dr. El-Esawi is currently involved in several biological science research projects.

Contents

Preface

Humankind, together with its global culture, economy, and society, is an important but small part of our planet's very complex and diverse ecological system known as the biosphere. Maintaining the biodiversity, vegetation, flora, and fauna of our biosphere is vital to our survival. However, it is a difficult task owing to constant changes in vegetation and ecosystems.

Vegetation dynamics is the science of natural, near-natural, and human-influenced changes in vegetation over time and space. These dynamics involve a range of processes, which can vary greatly in spatial scale, from ecophysiological phenomena to changes in communities, ecosystems, biomes, and global biogeographical patterns. The distribution, phenology, and productivity of vegetation are very sensitive to changes in climate, land use, and land cover patterns, and of course agricultural, industrial, and other human activities, all of which affect the Earth's ecosystems.

Vegetation degradation refers to a temporary or permanent loss of biodiversity, biomass, primer production, species richness, and physiognomic structure. Therefore, continuous monitoring, modeling, analysis, and scientific research of this dynamic pattern and process are important tasks from both pure and applied ecological viewpoints.

In the maintenance and conservation of our planet's biodiversity and ecosystem services, knowledge of current vegetation dynamical and biogeographical patterns as well as changes in flora and fauna have outstanding importance.

Vegetation Dynamics, Changing Ecosystems and Human Responsibility presents interesting information about natural and near-natural forests as well as human-influenced ecosystems in agrarian and urban habitats. It is a useful resource for those interested in vegetation research and applied plant ecology.

Levente Hufnagel
Research Institute of Multidisciplinary Ecotheology,
John Wesley Theological College,
Budapest, Hungary

Mohamed A. El-Esawi
Tanta University,
Tanta, Egypt

Section 1

Introduction

Chapter 1

Introductory Chapter: Vegetation Dynamics, Basic Phenomena, and Processes

Levente Hufnagel and Ferenc Mics

1. Introduction

Vegetation dynamics is the science about the concepts, theories, observations, and models that deals with changes in vegetation over time [1–4]. Changes in vegetation are a constant phenomenon on Earth. The simplest example of this is the appearance of weeds on a well-maintained lawn or the appearance of shrubs in an abandoned hayfield. Often the change is not so obvious, because the changes or the rearrangement of vegetation that is difficult to observe with the naked eye are slow compared to human life. Each vegetation patch changes dynamically, with every single plant eventually dying and being replaced by another. When environmental conditions change, including the opportunity of vegetation to influence its own environment, the balance between birth and death is disrupted. As a result, the relative proportion of plant species in the community will also change. The dominant species in the community largely determines succession, productivity, and stability, whereas the less abundant species determine the species richness of the community [5]. If the mortality rate exceeds the birth rate for a long period of time, the species becomes extinct and disappears from the community. New species are constantly being introduced from the area surrounding the vegetation patch, some of which may successfully establish if space is available. New species can also invade from the edges of patches, mainly vegetatively, by shoots and clones, which can also be a source of change in species composition. Largely spontaneously growing populations, evolving in accordance with the conditions of their habitat, form part of the ecosystem, along with external factors and other life forms [6]. Plant communities are assemblages of plant species that have evolved randomly throughout the history of the vegetation cover and then reorganized following the climate [7]. A plant community changes when there is a change in species composition, assuming that the community can be characterized based on the species composition.

Vegetation dynamics involve a range of processes, which can vary greatly in spatial scale, from the closure of stomas to the shift of entire biomes between geographical areas over centuries. Several researches are devoted to understanding and predicting how the physiological functioning and processes of individual plants, combined with each other, determine the structure, functioning, and dynamics of vegetation on large spatial scales. In order to study changes over time, of course, space and the spatial and physical properties of the vegetation have to be taken into account, the fact how vegetation exists in a given area at a given time. No two vegetation patches are exactly alike, the combinations and proportions of species are always changing [8].

2. Succession

Ecological succession has been a focus of research for almost a century. The process of succession can be studied from two perspectives: systems ecology and population ecology. In the latter case, there is a trend in the evolution of the system [9], and succession is a series of communities with a characteristic composition and characteristic ecological factors. In this approach, population dynamics is the dynamics of phytocoenoses. As a result, it is necessary to explain the fate of each phytocoenosis from the beginning to the equilibrium, or the fate of the phytocoenoses succeeding each other. Clements' [10] theory of the alternation of communities during succession is still the basis of several scientific works and theories. In contrast to the holistic approach, the individualistic approach emphasizes the importance of population processes in the biocoenosis dynamics when interpreting successional processes. Also, great emphasis is placed on disturbances, whether human or natural, and any influence that causes instability in vegetation [5]. Thus, succession can be interpreted at the ecosystem level according to one of the concepts. According to the other idea, succession is the consequence of the interaction between species as well as between species and the environment [11]. The latter results in the reproducibility of ecosystem structure and functioning. Taking into account data from population genetics and demography, succession is increasingly understood as a process of species replacement, with the role of individual species in vegetation change being related to life history strategy, growth, and reproduction [12]. In the past, these two theories were thought to be alternatives to each other; however, today they are rather complementary [13]. Long-term studies of the characteristics of certain plants, populations, and communities have led to the conclusion that it is correct to combine theories and seemingly contradictory methods in order to interpret successional processes [14]. In previous decades, there was a debate about whether succession could have only one endpoint, that is, whether it ends in a final climax community, or multiple climax communities could also be the result of a successional process [15, 16]. The concept of a more or less stable climax community was replaced by the idea that the relative frequency of changes decreases toward a supposed climax [17]; however, the climax itself also changes, and only the rate of change slows down but does not become zero [18]. In some cases, the changes may even be in the opposite direction [19]. The theoretically possible climax also changes in terms of species composition as a result of the changing climate, and new invasive species may appear [19]. The factors that influence succession vary in time and space [20]. With these constraints, vegetation approaches an endpoint, where only little change occurs, especially in terms of dominant species [21]. However, due to often irreversible changes caused by humans (e.g., [22, 23]), there may be changes in the species pool and vegetation structure, and the vegetation does not always reach the same hypothetical state [24]. Thus, there are several alternative endpoints.

3. Vegetation and climate change

The distribution, phenology, and productivity of vegetation are highly sensitive to changes in climate, which affects all ecosystems on Earth [25]. Vegetation shifts in terms of altitude and geographical latitude due to rising temperatures, and the vulnerability of many ecosystems increases [26–28]. Higher temperatures cause the growing season to start earlier in spring and last longer in temperate regions [29]. Production increases at higher latitudes, while in arid areas and desert regions, it tends to decrease further from the current low values [30]. Increasingly severe drought and fires also increase the destruction of vegetation [31]. Shifts in phenological phases (e.g., [32]) also result in changes in albedo, vegetation conductance, surface roughness, and the fluxes of water, energy, carbon dioxide, and volatile organic compounds [33].

Global warming causes glaciers to recede more and more and the free surface left behind is eventually covered by plants. Such primary succession means the colonization of previously unvegetated land and is one of the most important concepts [34]. During the process, pioneer plants colonize and stabilize the surface. A specific pattern of colonization and extinction can be observed in the community controlled by biotic and abiotic factors [35]. The structural complexity of the plant community is gradually increasing, and along with this, the biomass, production, species numbers, and the interactions between them are doing so [36]. When Krakatoa erupted in 1883, the entire island was sterilized, leaving no trace of the former soil and vegetation on the completely transformed island (Docters [37]). After the eruption, the area was covered by a nutrient-rich layer of vitric tuff (hypersthene-augite), which provided a suitable medium for plant roots, although no organic matter was present in it yet. The first colonizing species to appear were blue-green algae, forming a coating on the surface. Then the pteridophytes came, as they could reach the island in the easiest way with their spores dispersed by the wind. These were followed by seeds floating on the water. Finally, species spread by animals arrived. Today, there is even forest on the island; however, the vegetation is not as species-rich as in the area not affected by the disaster. The species present are not those typical of the climax community, the succession is ongoing and far from reaching the status of the characteristic tropical forests there, not disturbed by humans [38]. Succession also resumes after the abandonment of agricultural land, however, this is secondary, as these are not sterile areas, plants and other organisms had been present previously as well, but the cessation of human activity causes the process to resume, changing the resilience of the system and its response to external influences [39].

During succession, the microclimate and the physiognomy of the vegetation change [40]. The complexity of vegetation determines the diversity, species composition, and abundance of animal communities [41]. Some animal species are associated with vegetation of a certain complexity, where certain resources occur, such as prey animals, seeds, fruits, and shelter [42]. Forest loss also has the effect of increasing the visibility of animals, both prey and predators. This leads to a change in their behavior, for example, the cohesion of flocks of birds is reduced [43]. The changes also apply, of course, to microorganisms. During vegetation degradation, carbon dioxide efflux increases due to soil respiration, because respiration continues without photosynthesis, that is, carbon dioxide fixation as well, so there is nothing to counteract this [44]. In a healthy ecosystem, carbon accumulation is rather typical, with carbon dioxide getting into vegetation and soil. As a consequence of the activity of microorganisms, carbon is released back into the atmosphere as carbon dioxide. During the degradation of vegetation, the transformation of soil organic matter into carbon dioxide predominates [45]. During succession, the amount of carbon dioxide sequestered increases due to the increasing amount of photosynthesizing plants, while net carbon dioxide efflux decreases [46]. The changing vegetation depends on the soil microbiome network and regulates the community composition and, through this, the productivity of the whole system during succession. A resistant microbiome community also promotes the process of succession [47].

Vegetation degradation refers to a temporary or permanent loss of biomass, production, species richness, cover, and structure [48]. The definition includes not only quantitative but also qualitative change. This also includes the appearance of species that are less favored by grazing animals and less nutritious due to changing species composition in grazed areas [49]. Changes in vegetation are the result of a wide range of variables, from adaptation to changing conditions, through disasters, to human activities [50]. Therefore, it is very important to distinguish natural variation from human-induced changes [51]. Vegetation degradation is a worldwide phenomenon, very often caused by false human management practices or climate change [52]. NPP is the highest in the tropics, accounting for up to one-third of the total global NPP, and its dynamics are therefore very important in the geochemical cycle of carbon [53]. Human activity and climate change are causing the loss and variability of natural vegetation cover and NPP, so any action to combat climate change is crucial to prevent further deterioration of vegetation and desertification [54]. Due to land use and rapid urbanization, significant areas are losing natural vegetation [55]. As urbanization increases, the vegetation index also decreases, however, vegetation is present in the metropolitan area as well [56]. Changes in vegetation dynamics are closely related to climate change and human activities, so continuous monitoring of the dynamics and the prediction of changes are crucial tasks [57]. The extent of desertification in inner Mongolia reached 620.000 km^2 by 2009, which is more than 50% of the area [58]. Fire is also a force that can strongly shape the vegetation pattern. It is an extremely important component of terrestrial ecosystems and has the potential to greatly alter vegetation structure and distribution, carbon, and other element cycling, as well as water and energy budgets [59]. Fires release large amounts of gases and aerosols into the air, which then affect radiation reaching the surface and the climate [60]. The planned lighting of fires in forests and rangelands has been common practice for thousands of years, from hunting and gathering societies to modern-day farmers. Burned forests are converted into pastures, and in the pastures, the emergence of shrubs and trees is prevented. The biomass and species diversity of herbaceous plants increases in regularly burned pastures. After the fire, grazing ungulates influence succession and species composition [61]. Regular fire burning and grazing result in a diverse, mosaic vegetation pattern with different stages of succession, increasing the spatial heterogeneity of the landscape, which in turn leads to greater species richness [62]. In developed countries, however, fire is rarely used as a means of shaping vegetation, and roads and other infrastructure also contribute to a lower incidence of fires than in less developed countries and in the pre-industrial era [63].

Climate plays a central role in the distribution of vegetation and plant species. Climate change could lead to significant changes in the distribution of vegetation across the globe in the coming decades and centuries [64]. Plenty of research focuses on estimating the impact of climate change on vegetation, also in light of research data on past climate changes [65]. In the near future, the impact of climate change will be comparable to changes at the glacial-interglacial boundary, causing significant changes in vegetation properties as well [66]. By the end of the century, average annual temperatures will have increased by 1.8–4°C, and by up to 6.4°C in the case of high emissions, compared to the 1980–1999 average [64]. The temperature increase will be the greatest around the poles, with a range of 5–8°C according to the A1B scenario [64]. The spatial shift of the climate could reach an average of 0.42 km per year, it could be slightly slower in mountainous areas (0.08 km per year) and faster in lowland areas (1.26 km per year) [67]. Associated with this, there will be a number of changes in the environment. Snow cover will decrease, permafrost will thaw, the frequency of weather anomalies will increase in terms of temperature and precipitation, the frequency and intensity of tropical cyclones will increase, and the direction of extratropical storms will shift toward the poles, where precipitation will increase but it will decrease further at the tropics [64]. The rise in carbon dioxide concentrations and sea level are directly linked to this. These changes will have drastic impacts on plant populations as well, both indirectly and directly [68, 69]. According to all scenarios, these changes and their consequences will continue in the coming centuries [64]. Changes in vegetation structure are difficult to predict because changes in climate are followed by changes in vegetation with a lag [70], probably with very complex dynamics.

4. Human effects and conservation efforts

Nowadays, one-third of the human population is already feeling the negative effects of degradation, which include soil erosion, salinization, draining of marshes and bogs, and deforestation [71]. There are already more than 500 million hectares of degraded forests in the tropics and this area is steadily increasing [72]. Deforestation is caused by economic, demographic, technological, and political factors [73]. In total, 52% of the felled timber becomes lumber, 31% firewood and charcoal, 9% is the victim of an uncontrolled fire, and 7% is lost due to grazing [74]. Every biome is losing NPP due to human activities, with a degraded area reaching 2.7 billion hectares worldwide [75]. In abandoned areas, vegetation is able to regenerate. This is often done deliberately as part of a rotation system in order to regenerate soil nutrients [76] or in response to socioeconomic impacts, which alter profitability, access to labor, capital, and markets [77]. Secondary forests appear in the place of previously deforested forests, and their area increased in Brazil from 10 to 17 million hectares between 2004 and 2014, which is very important from the viewpoint of the situation of rainforests in the twenty-first century [78, 79]. Today, there are several programs to restore the original natural vegetation. Theoretical knowledge of the succession process in a given location and consideration of climatic conditions allow decision-making in order to achieve the goal of restoring a given area to a near-natural state or a condition desired by the human community [80]. During the restoration program, vegetation should be monitored continuously. The data obtained as a result of monitoring provide feedback, which can be used to adjust predictions and modify plans if necessary. In addition, monitoring can be the basis for other scientific work, increasing our knowledge of succession and vegetation dynamics. In order to control succession, it is important to know, for example, when it is time for the emergence of desirable species or for their artificial dispersal. What will their mortality be due to the competition? How to reduce harmful abiotic effects? The spread of emerging undesirable adventitious species also needs to be controlled or they have to be eradicated [81]. An adequate response to ongoing degradation is often lacking due to missing adequate knowledge of the causes that trigger it. Even measures that have been initiated are not always successful if the process of degradation itself and the underlying causes are not linked. In the absence of appropriate countermeasures, vegetation cover and soil nutrients may disappear [82].

Author details

Levente Hufnagel[1*] and Ferenc Mics[2]

1 Ecotheology Research Group of John Wesley Theological College, Budapest, Hungary

2 Department of Environmental Security, John Wesley Theological College, Budapest, Hungary

*Address all correspondence to: wjlf.hu@gmail.com

References

[1] Finegan B. Forest succession. Nature. 1984;**312**:109-114

[2] Miles J. Vegetation Dynamics. London: Chapman and Hall; 1979

[3] Tansley AG. The use and abuse of vegetational concepts and terms. Ecology. 1935;**16**:284-307

[4] van Andel J, Bakker JP, Grootjans AP. Mechanisms of vegetation succession: A review of concepts and perspectives. Acta Botanica Neerlandica. 1993;**42**(4):413-433

[5] Grime JP. Dominant and subordinate components of plant communities: Implications for succession, stability and diversity. In: Gray AJ, Crawley MJ, Edwards PJ, editors. Colonization, Succession and Stability. Oxford: Blackwell Scientific Publications; 1987

[6] Westhoff V. Phytosociology in the Netherlands: History, present state, future. In: Werger MJA, editor. Den Haag: Junk Publishers; 1979

[7] Tallis JH. Plant Community History: Long-term Changes in Plant Distribution and Diversity. Dordrecht: Springer; 1991

[8] Gleason HA. Further views on the succession-concept. Ecology. 1927;**8**(3):299-326

[9] Whittaker RH. Communities and Ecosystems. London, UK: Collier MacMillan; 1975

[10] Clements FE. Plant Succession: Analysis of the Development of Vegetation. Washington, DC: Carnegie Institution of Washington Publication Sciences, 242; 1916. pp. 1-512

[11] MacMahon JA. Ecosystems over time: Succession and other types of change. In: Warning R, editor. Forests: Fresh Perspectives from Ecosystem Analyses. Corvallis, Oregon, USA: Oregon State University Biology Colloquium; 1980

[12] van der Maarel S, Orlóci E, Pignatti L. Data-processing in Phytosociology, Retrospect and Anticipation. The Hague: Junk; 1980

[13] Prentice IC. The Design of a Forest Succession Model. Wageningen: PUDOC Publisher; 1986

[14] Falińska K. Plant population processes in the course of forest succession in abandoned meadows. I. Variability and diversity of floristic compositions, and biological mechanisms of species turnover. Acta Societatis Botanicorum Poloniae. 1989;**58**(3):439-465

[15] Clements FE. Plant Succession and Indicators. New York: The H. W. Wilson Company; 1928

[16] Gleason HA. The individualistic concept of the plant succession. American Midland Naturalist. 1926;**21**:92-110

[17] Marleau JN, Jin Y, Bishop JG, Fagan WF, Lewis MA. A stoichiometric model of early plant primary succession. The American Naturalist. 2011;**177**:2

[18] Jackson ST. Vegetation, environment, and time: The origination and termination of ecosystems. Journal of Vegetation Science. 2006;**17**(5):549-557

[19] del Moral R, Walker LR, Bakker JP. Insights gained from succession for the restoration of landscape structure

and function. In: Walker LR, Walker J, Hobbs RJ, editors. Linking Restoration and Ecological Succession. New York: Springer; 2006

[20] Wright JP, Fridley JD. Biogeographic synthesis of secondary succession rates in eastern North America. Journal of Biogeography. 2006;**37**(8):1584-1596

[21] Loidi J, del Arco M, de Paz PLP, Asensi A, Garretas BD, Costa M, et al. Understanding properly the 'potential natural vegetation' concept. Journal of Biogeography. 2010;**37**(11):2209-2211

[22] Cui LF, Wang Z, Deng LH, Qu S. Vegetation dynamics and their relations with climate change at seasonal scales in the Yangtze River Basin, China. Applied Ecology and Environmental Research. 2020;**18**:3543-3556

[23] Gao NN, Li F, Zeng H, Zheng YR. The impact of human activities, natural factors and climate time-lag effects over 33 years in the Heihe River Basin. China – Applied Ecology and Environmental Research. 2021;**19**(3):1589-1606

[24] Prach K, Tichý L, Lencová K, Adámek M, Koutecký T, Sádlo J, et al. Does succession run towards potential natural vegetation? An analysis across seres. Journal of Vegetation Science. 2016;**27**(3):515-523

[25] Richardson AD, Keenan TF, Migliavacca M, Ryu Y, Sonnentag O, Toomey M. Climate change, phenology, and phenological control of vegetation feedbacks to the climate system. Agricultural and Forest Meteorology. 2013;**169**:156-173

[26] Garamvölgyi Á, Hufnagel L. Impacts of climate change on vegetation distribution. No. 1 Climate Change induced vegetation shifts in the Palearctic Region. Applied Ecology and Environmental Research. 2013;**11**(1):79-122

[27] Hufnagel L, Garamvölgyi Á. Impacts of climate change on vegetation distribution. No. 1 Climate Change induced vegetation shifts in the New World. Applied Ecology and Environmental Research. 2014;**12**(2):355-422

[28] Kelly AE, Goulden MI. Rapid shifts in plant distribution with recent climate change. Proceedings of the National Academy of Sciences of the United States of America. 2008;**105**:11823-11826

[29] Jia GJ, Epstein HE, Walker DA. Vegetation greening in the canadian arctic related to decadal warming. Journal of Environmental Monitoring. 2009;**11**:2231-2238

[30] Ma X, Huete A, Moran S, Ponce-Campos G, Eamus D. Abrupt shifts in phenology and vegetation productivity under climate extremes. Journal of Geophysical Research Biogeosciences. 2015;**120**:2036-2052

[31] Mueller RC, Scudder CM, Porter ME, Talbot Trotter R, Gehring CA, Whitham TG. Differential treemortality in response to severe drought: Evidence for long-termvegetation shifts. Journal of Ecology. 2005;**93**:1085-1093

[32] Eppich B, Dede L, Ferenczy A, Garamvölgyi Á, Horváth L, Isépy I, et al. Climatic effects on the phenology of geophytes. Applied Ecology and Environmental Research. 2009;7:253-256

[33] Gonzalez P, Neilson RP, Lenihan JM, Drapek RJ. Global patterns in the vulnerability of ecosystems to vegetation shifts due to climate change. Global Ecology and Biogeography. 2010;**19**:755-768

[34] Begon M, Harper JL, Townsend CR. Ecology. Individuals, Populations and Communities. Oxford: Blackwell Science; 1996

[35] Arróniz-Crespo M, Pérez-Ortega S, De Los Ríos A, Allan Green TG, Ochoa-Hueso R, Casermeiro MA, et al. Bryophyte-Cyanobacteria Associations during Primary Succession in Recently Deglaciated Areas of Tierra del Fuego (Chile). PLOS One. 2014;**9**(5):e96081

[36] Marston RA. Geomorphology and vegetation on hillslopes: Interactions, dependencies, and feedback loops. Geomorphology. 2010;**116**(3-4):206-217

[37] van Leeuwen D. Krakatau, 1883 to 1933. Annales du Jardin botanique de Buitenzorg. 1936;**46-47**:1-506

[38] Whittaker RJ, Bush MB, Richards K. Plant Recolonization and Vegetation Succession on the Krakatau Islands, Indonesia. Ecological Monographs. 1989;**59**(2):59-123

[39] García-Ruiz JM, Regüés D, Alvera B, Lana-Renault N, Serrano-Muela P, Nadal-Romero E, et al. Flood generation and sediment transport in experimental catchment affected by land use changes in the central Pyrennes. Journal of Hydrology. 2008;**356**(1-2):245-260

[40] Guariguata MR, Ostertag R. Neotropical secondary forest succession: Changes in structural and functional characteristics. Forest Ecology and Management. 2001;**148**:185-206

[41] Johnston DW, Odum EP. Breeding bird populations in relation to plant succession on the Piedmont of Georgia. Ecology. 1956;**37**(1):50-62

[42] Baguette M, Deceuninck B, Muller Y. Effect of Spruce Afforestation On Bird Community Dynamics in a Native Broad-leaved Forest Area. Paris: Gauthier-Villars; 1994

[43] Tubelis DP, Cowling A, Donnelly C. Role of mixed-species flocks in the use of adjacent savannas by forest birds in the central Cerrado, Brazil. Biological Conservation. 2006;**116**(1):19-26

[44] Waddington JM, Warner KD, Kennedy GW. Cutover peatlands:a persistent source of atmospheric CO_2. Global Biochemical Cycles. 2002;**16**:1002

[45] Belyea IR, Malmer N. Carbon sequestration in peatland: Patternsand mechanisms of response to climate change. Global Change Biology. 2004;**10**:1043-1052

[46] Waddington JM, Warner KD. Atmospheric CO_2 sequestration inrestored mined peatlands. Ecoscience. 2001;**8**:359-369

[47] de Araujo ASF, Mendes LW, Lemos LN, Antunes JEL, Beserra JEA, de Lyra MCCP, et al. Protist species richness and soil microbiome complexity increase towards climax vegetation in the Brazilian Cerrado. Communications Biology. 2018;**1**:135

[48] Grainger A. The degradation of tropical rain forest in Southeast Asia: Taxonomy and appraisal. In: Eden ME, Parry JT, editors. Land Degradation in the Tropics. London: Mansell Publishers; 1996

[49] Masoudi M. Risk Assessment and Remedial Measures of Land Degradation, in Parts of Southern Iran. Germany: Lambert Academic Publishing; 2010

[50] Ringrose S, Sefe F, Ekose G. Progress towards the evaluation of desertification in Botswana. Desertiffication Control Bulletin. 1995;**27**:62-68

[51] Pickup G, Bastin GN, Chewings VH. Remote sensing based condition assessment for non-equilibrium rangelands under largescale commercial grazing. Ecological Applications. 1994;**4**:497-517

[52] Kharin N. Vegetation Degradation in Central Asia under the Impact of Human Activities. Moscow: Springer Science+Business Media, LLC; 2002

[53] Wu W, De Pauw E, Zucca C. Using remote sensing to assess impacts of land management policies in the Ordos rangelands in China. International Journal of Digital Earth. 2013a;**6**(2):81-102

[54] Wu W, De Pauw E, Helldén U. Assessing woody biomass in African tropical savannahs by multiscale remote sensing. International Journal of Remote Sensing. 2013b;**34**(13):4525-4549

[55] Zhong Q, Ma J, Zhao B, Wang X, Zong J, Xiao X. Assessing spatialtemporal dynamics of urban expansion, vegetation greenness and photosynthesis in megacity Shanghai, China during 2000-2016. Remote Sensing of Environment. 2019;**233**:111374

[56] Zhao S, Liu S, Zhou D. Prevalent vegetation growth enhancement in urban environment. Proceedings of the National Academy of Sciences. 2016;**113**:6313-6331

[57] Fu BJ, Li SG, Yu XB, Yang P, Yu GR, Feng RG, et al. Chinese ecosystem research network: Progress and perspectives. Ecological Complexity. 2010;**7**:225-233

[58] State Forestry Administration, P.R. China. A Bulletin of Status Quo of Desertification and Sandification in China. China: State Forestry Administration; 2011

[59] Lasslop G, Hantson S, Harrison SP, Bachelet D, Burton C, Forkel M, et al. Global ecosystems and fire: Multi-model assessment of fire-induced tree-cover and carbon storage reduction. Global Change Biology. 2020;**26**(9):5027-5041

[60] Knorr W, Jiang L, Arneth A. Climate, CO_2 and human population impacts on global wildfire emissions. Biogeosciences. 2016;**13**(1):267-282

[61] Kerns BK, Day MA. Prescribed fire regimes subtly alter ponderosa pine forest plant community structure. Ecosphere. 2018;**9**:e02529

[62] Fuhlendorf SD, Engle DM, Kerby J, Hamilton R. Pyric Herbivory: Rewilding landscapes through the recoupling of fire and grazing. Conservation Biology. 2009;**23**(3):588-598

[63] van der Werf GR, Randerson JT, Giglio L, van Leeuwen TT, Chen Y, Rogers BM, et al. Global fire emissions estimates during 1997-2016. Earth System Science Data. 2017;**9**:697-720

[64] IPCC. Climate Change 2007: Synthesis report. Contribution of Working Groups 1, II, and III to the Fourth Assessment Report of the Intergovernmental Panel on Climate Change. Geneva, Switzerland: IPCC; 2007

[65] Correa-Metrio A, Bush MB, Cabrera KR, Sully S, Brenner M, Hodell DA, et al. Rapid climate change and no-analog vegetation in lowland Central America during the last 86,000 years. Quaternary Science Reviews. 2012;**38**:63

[66] Bush MB, Silman MR, Urrego DH. 48,000 years of climate and forest change in a biodiversity hot spot. Science. 2004;**303**:827

[67] Loarie SR, Duffy PB, Hamilton H, Asner GP, Field CB, Ackerly DD. The

velocity of climate change. Nature. 2009;**42**:1052-1059

[68] Higgins SI, Scheiter S. Atmospheric CO_2 forces abrupt vegetation shifts locally, but not globally. Nature. 2012;**488**:209-212

[69] Lenoir J, Gégout J, Dupouey J, Bert D, Svenning J-C. Forest plant community changes during 1989-2007 in response to climate warming in the Jura Mountains (France and Switzerland). Journal of Vegetation Science. 2010;**21**:949-964

[70] Ding Y, Li Z, Peng S. Global analysis of time-lag and -accumulation effects of climate on vegetation growth. International Journal of Applied Earth Observation and Geoinformation. 2020;**92**:102179

[71] UNEP. Global Environment Outlook GEO4: Environment for Development. United Nations Environment Programme. 2007;**36**(3):337-338

[72] ITTO. Annual Review and Assessment of the World Tropical Timber Situation 2012. Yokohama: International Timber Trade Organisation; 2012

[73] Stanturf JA, Palik BJ, Dumroese RK. Contemporary forest restoration: A review emphasizing function. Forest Ecology and Management. 2014;**331**:292-323

[74] FAO. Hacia una Definición de Degradación de los Bosques: Análisis Comparativo de las Definiciones Existentes. Roma, Italy: Departamento Forestal, Organización de las Naciones Unidas para la Alimentación y la Agricultura; 2009

[75] Bai ZG, Dent DL, Olsson L, Schaepman ME. Proxy global assessment of land degradation. Soil Use and Management. 2008;**24**(3):223-234

[76] Zarin DJ, Ducey MJ, Tucker JM, Salas WA. Potential biomass accumulation in Amazonian regrowth forests. Ecosystems. 2001;**4**:658-668

[77] Brito B, Barreto P, Brandão A, Baima S, Gomes PH. Stimulus for land grabbing and deforestation in the Brazilian Amazon. Environmental Research Letters. 2019;**14**:064018

[78] Mics F, Rozak AH, Kocsis M, Homoródi R, Hufnagel L. Rainforests at the beginning of the 21st century. Applied Ecology and Environmental Research. 2013;**11**(1):1-20

[79] TerraClass. Projeto TerraClass 2014 [WWW Document]. raClass 2014 Projeto TerraClass 2014 [WWW Document]. 2014. Available from: http://inpe.br/cra/projetos_pesquisas/terraclass2014.php

[80] Kirmer A, Mahn EG. Spontaneous and initiated succession on unvegetated slope sites in the abandoned lignite-mining area of Goitsche, Germany. Applied Vegetation Science. 2001;**4**:19-28

[81] Karel P, Bartha S, Joyce CB, Pyšek P, van Diggelen R, Wiegleb G. The role of spontaneous vegetation succession in ecosystem restoration: A perspective. Applied Vegetation Science. 2001;**4**:111-114

[82] Ayoub AT. Indicators of dryland degradation. In: Squires VR, Sidahmed AE, editors. Drylands—Sustainable Use of Rangelands into the Twenty-first Century. Rome: FAO; 1998

Section 2

Near-Natural Ecosystems and Conservation

Chapter 2

Impact of Revegetation on Ecological Restoration of a Constructed Soil in a Coal Mining in Southern Brazil

Lizete Stumpf, Maria Bertaso De Garcia Fernandez, Pablo Miguel, Luiz Fernando Spinelli Pinto, Ryan Noremberg Schubert, Luís Carlos Iuñes de Oliveira Filho, Tania Hipolito Montiel, Lucas Da Silva Barbosa, Jeferson Diego Leidemer and Thábata Barbosa Duarte

Abstract

The main problems in the constructed soils are the generation of acid mine drainage promoted by the presence of coal debris in the overburden layer and the compaction of the topsoil promoted by the machine traffic when the material used in the overburden cover is more clayey. This book chapter aimed to show an overview of the impact of more than a decade of revegetation with different perennial grasses on the chemical, physical, and biological quality of constructed soil after coal mining. The study was carried out in a coal mining area, located in southern Brazil. The soil was constructed in early 2003 and the perennial grasses, *Hemarthria altissima*; *Paspalum notatum* cv. Pensacola; *Cynodon dactylon* cv Tifton; and *Urochloa brizantha;* were implanted in November/December 2003. In 11.5, 17.6 and 18 years of revegetation soil samples were collected and the chemical, physical, and biological attributes were determined. Our results show that liming is an important practice in the restoration of these strongly anthropized soils because this positively impacts the plants' development, facilitating the roots system expansion. Biological attributes such as soil fauna and the microorganism's population are the attributes that possibly takes longer to establish itself in these areas.

Keywords: *Hemarthria altissima*, *Paspalum notatum* cv. Pensacola, *Cynodon dactylon* cv Tifton, *Urochloa brizantha*, ecological restoration

1. Introduction

Despite the great advancement of renewable energies in recent years, coal still plays an important role in the supply of electricity, as it has important reserves on all continents, contributing to the security of the energy matrix worldwide. It is estimated that in emerging markets by the year 2040, coal will supply 39% of the world's electricity [1]. In Brazil, the main coal reserve, called the Candiota Mine, is located in the state of Rio Grande do Sul, covering 38% of all national coal [2]. The process of coal extraction in the Candiota Mine occurs in the form of open-pit mining, which promotes intense impacts on the environment, such as the suppression of vegetation, soil, and rocks overlapping to ore. After coal extraction, the process of topographic recomposition of the area begins, which involves intense movement of heavy machinery aiming at filling the open pit by mining. Finally, the new soil profile in these areas presents, in general, two layers: a layer called overburden, composed of rock debris and eventually coal, and a layer of topsoil, composed of the mixture of horizons A, B, and C of the original soil. The main problems observed in the new profile of the constructed soil, which directly impact the revegetation of the mined areas, are the generation of acid mine drainage promoted by the presence of coal debris in the overburden layer and the compaction of the topsoil promoted by the machine traffic when the material used in the overburden cover is more clayey [3].

Acid mine drainage occurs when sulfite minerals, such as pyrite, are present in rock fragments used in topographic recomposition of the mined area [4]. Pyrite in contact with oxygen and water generates sulfuric acid [5], which drastically reduces pH [6], besides generating large concentrations of Fe, Mn, and Al in the solution [7], with negative implications in the revegetation of the degraded area [8]. Nunes [9] observed that construction methods that use low soil thickness originate from constructed soils with a large number of mining steriles and, consequently, with low pH values (around 2.4). Therefore, a greater thickness of topsoil over the overburden is critical during the construction of the new soil profile to minimize the occurrence of acid drainage. On the other hand, the improper handling and distribution of the topsoil can cause its compaction [10] and hinder the development of vegetation cover, the main starting point for the recovery of mined soils, since the accumulation of organic material results in positive changes in the physical–chemical properties of the new soil [11].

Revegetation is paramount in programs to recover degraded areas because the phytomass addition to the system provides a gradual increase in soil organic matter, which directly impacts microbiological activity and soil fauna diversity of these areas, promoting improvements in the ecological functions of the new ecosystem established after mining [12, 13]. In this sense, when evaluating the effect of revegetation on the microbiological attributes of soils impacted by mining, Longo et al. [14] found a significant increase in microbial biomass in open-pit mined areas after 3 years of legume revegetation. However, these were still below those found in forest soil (1344 and 1514 mg kg^{-1}, respectively). In India, when comparing constructed soils of different ages on forest species, Ahirwal et al. [15] also observed an increase in microbial biomass carbon over the years (5 years: 60 mg kg^{-1}; 7 years: 125 mg kg^{-1}; and 15 years: 270 mg kg^{-1}). Microbial biomass corresponds to about 80% of the living fraction of the soil, so it is considered an efficient indicator of the stage of degradation of soil, as it is directly related to the amount of organic matter added to the soil, in the form of live or dead plant residues, and participates strongly in soil formation processes, aggregation, cycling, and nutrient availability [16].

In relation to soil fauna, regardless of the revegetation used in the constructed soils, it has been observed population of mites and springtails predominance in these environments disturbed with different years of restoration [17–19]. On the other hand, in these soils strongly impacted by mining, even after a decade of revegetation, the populations of mites and springtails were still much lower than the populations observed in a natural soil without anthropic action [20]. The soil mesofauna represented by mites and springtails mainly plays an important role in the decomposition of plant residues, reducing the surface area of waste and facilitating the continuation of decomposition by microorganisms, especially bacteria [21]. As mites and springtails usually live in the soil pores closest to the surface, the physical changes that occur in the soil, such as compaction and consequently reduction of soil porosity, directly alter these populations [22]. On the other hand, pH changes also interfere in the diversity of these organisms because the presence or absence of some species may be related to the availability of specific ions in the soil solution. Therefore, soil management in order to improve fertility and decrease soil acidity can sometimes cause stress in these communities [23].

Therefore, the objective of this chapter was to analyze the impact of more than a decade of revegetation with different perennial grasses on the chemical, physical, and biological quality of constructed soil after coal mining in southern Brazil.

2. Materials and methods

The study was carried out in a coal mining area, under concession by Companhia Riograndense de Mineração (CRM), located in Candiota/RS with the following geographical coordinates: 31°33′56″S and 53°43′30″W (**Figure 1**). As described by Stumpf et al. [3], the soil was constructed in early 2003 and the experiment was installed in

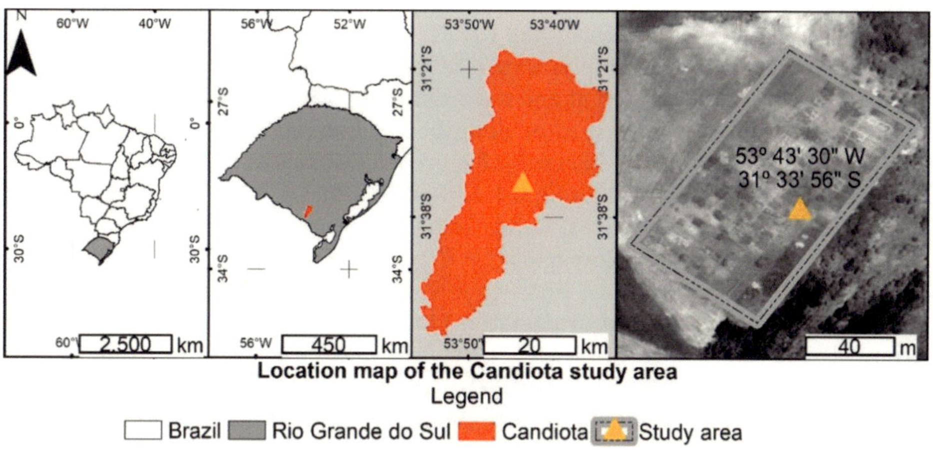

Figure 1.
Location map of Candiota study area.

November/December 2003 in plots of 20 m^2 (5 × 4 m) in a randomized block design with four replications. The soil layer replaced in the experimental area (Topsoil) comes from horizon B of the natural soil of the pre-mined area, a Rhodic Lixisol [24], as indicated by the clayey textural class, the dark red color (2.5 YR 3.5/6) and the low organic matter content (1.15%).

The perennial summer grasses, *Hemarthria altissima*; *Paspalum notatum* cv. Pensacola; *Cynodon dactylon* cv Tifton; and *Urochloa brizantha*, were implanted in November/December 2003. Before the implantation of plant species, the topsoil was scarified at a depth of 0.15 m, followed by a weight corresponding to 10.4 Mg ha^{-1} limestone with 100% PRNT and fertilization of 900 kg ha^{-1} of the formula Nitrogen, Phosphorus, Potassium 5-20-20 (NPK) based on results obtained by soil analysis. Annual fertilization in all plots was also performed annually by applying 250 kg ha^{-1} of the formula NPK 5-30-15 and 250 kg ha^{-1} of urea.

In March 2015 (11.5 years of revegetation), 16 soil samples were collected in the 0.00–0.10 m layer, with the aid of a cutting shovel, for the determination of chemical attributes: pH in water, calcium, magnesium, potassium, aluminum, and soil organic matter, and to determine the distribution of water-stable aggregates in different size classes.

Following the methodology of Tedesco et al. [25], the soil pH was determined in water at the ratio of 1:1 (soil:water); calcium (Ca^{+2}), magnesium (Mg^{+2}), and aluminum (Al^{+3}) exchangeable were extracted with KCl 1 mol L^{-1} and determined in the atomic absorption spectrophotometer (Ca^{+2} and Mg^{+2}) and by titration with NaOH (Al^{+3}). The available potassium content was estimated by the Mehlich^{-1} method and analyzed by flame photometry. The potential acidity was extracted with calcium acetate and determined by titration with NaOH. Based on the results of the analyses, base and aluminum saturation was calculated. The soil carbon content was determined by the Walkley Black combustion method, in the fine earth fraction.

To determine the distribution of water-stable aggregates in different size classes, the soil samples were placed on a wooden tray and air-dried at room temperature in the shade until the moisture reached the friability point, when the soil was gently broken into large clods along the natural planes of weakness to obtain the natural aggregate, passed in a sieve with a mesh size of 9.52 mm, and then air-dried for two weeks. After that, four sub-samples were taken with approximately 50 g, one used to determine the moisture content and the other three were submitted to wet sieving with vertical following the method described by Kemper and Rosenau [26] and adapted by Palmeira et al. The intervals of aggregates classes were: C1: 9.52–4.76 mm; C2: 4.76–2.0 mm; C3: 2.00–1.00 mm; C4: 1.00–0.25 mm; C5: 0.25–0.105 mm, and C6:<0.105 mm. From these classes, the aggregates were separated into macroaggregates (>0.25 mm) and microaggregates (<0.25 mm), according to Tisdall and Oades [27].

In March 2015 (11.5 years of revegetation) and November 2021 (18 years of revegetation), 16 soil samples were collected in the 0.00–0.10 m layer, with the aid of a cutting shovel, to evaluate the microbiological attributes: microbial biomass carbon and basal respiration. The soil samples were preserved at refrigeration temperature (4°C) and for analysis of the microbial biomass were weighed 32 g of moist soil, in duplicate, where one repetition was subjected to irradiation in a microwave oven and another not, according to the methodology proposed by Islam and Weil [28]. The samples were titrated Fe2SO4 0.25 molc L^{-1} solution. Following the methodology described by Anderson and Domsch [29], basal respiration was performed using 100 g of fresh soil, with known moisture, and 20 mL of NaOH was added in hermetically sealed vials. After 21 days the solution was removed for titration with HCl.

In May 2021 (17.6 years of revegetation), 32 samples were collected using steel cylinders (0.050 m high and 0.047 m in diameter) to determine soil fauna organisms. The total number of individuals of the soil fauna was counted in 169.4 cm^3 of soil and the constructed soil moisture at the time of collection fluctuated between 25.2 and 31.8%. For the determination of soil fauna organisms, the Tullgren Extractor Funnel method proposed by Bachelier [30] was used. The samples were carefully placed in sieves with 2 mm mesh at the top of each funnel and, at the base of the hoppers, collector cups containing 70% alcohol and four drops of glycerin were placed in order to avoid rapid evaporation of alcohol. The samples were identified in each funnel and remained under the luminosity of lamps of 25 watts for 7 days, so that with the action of light and heat, the organisms move down, and thus be captured by the collector cup with a capacity of 50 ml. The soil fauna was identified and quantified at the class level according to Gallo et al. [31] with the aid of Opton magnifying glass, model TNE-10TN, with magnification ranging from 0.8 to 5×. The relative frequency of each group of organisms found in relation to the total number of organisms counted was calculated.

3. Results and discussion

Figure 2 shows the positive effects of limestone incorporation up to the approximate 0.15 m depth, which occurred before the implantation of plant species (Nov/Dec 2003). That is, even after 11.5 years of revegetation, the soil pH values in the 0.00–0.10 m layer (**Figure 2a**) are very close to or higher than the reference value for perennial grasses (pH > 5.5), according to the Soil Chemistry and Fertility Commission in the state of Rio Grande do Sul and Santa Catarina [32]. Consequently, base saturation is still at medium levels (65–80%) (**Figure 2b**), with high levels of calcium (>4 $cmol_c kg^{-1}$) and magnesium (>1 $cmol_c kg^{-1}$) (**Figure 2c** and **d** respectively), while aluminum saturation (**Figure 2e**) is below the level considered critical to plant development (<20%).

Although the chemical condition in the surface layer of the constructed soil is still adequate for the grasses development, the acidification effect, promoted by rainwater infiltration and annual fertilization with urea, is promoting soil pH and base saturation reductions if we consider the notes of Stumpf et al. [33]. According to the authors, at 8.6 years of revegetation, the surface layer of the constructed soil presented a pH between 5.74 and 6.25 (currently ranging from 5.40 and 5.70 – **Figure 2a**), while the base saturation was higher than 80% in all treatments (currently ranging from 61 to 69.50% – **Figure 2b**). Aluminum saturation also increased from values below 1.50% (at 8.6 years of revegetation) to values ranging from 3.20 to 15.33% when the revegetation was completed 11.5 years (**Figure 2e**). These results show, in the medium term, a new corrective action should be considered in the mined area so that aluminum saturation does not become restrictive to the root development of plants.

The chemical quality of constructed soils is essential to ensure the full plant's development in the long term, mainly to enable their root expansion, which is responsible for improving the physical quality of these soils [3, 11], which are strongly impacted by topsoil compaction. Da Silva Barboza et al. [34] noted that the first machine traffic event for the clay topsoil placement under the overburden promoted

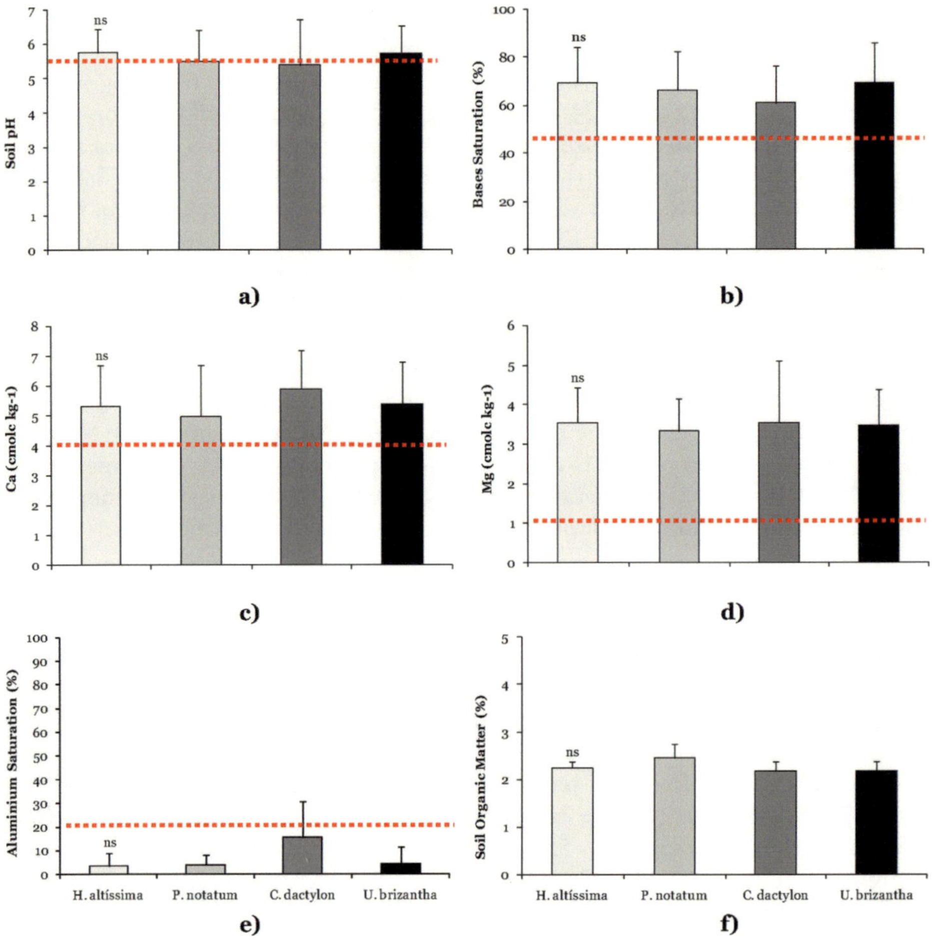

Figure 2.
Soil pH (a), base saturation (b), calcium (c) and magnesium contents (d), aluminum saturation (f), and organic matter content (e) in the 0.00–0.10 m layer of a constructed soil after coal mining and revegetated with perennial grasses for 11.5 years. Dashed red line indicates: (a) pH suitable for grass development; (b) base saturation considered low; (c) calcium and magnesium contents considered high (c, d respectively); (e) aluminum saturation considered limiting to root development; and (f) soil organic matter content considered low. Error bars mean standard deviation. Ns: Not significant to the Tuckey test ($p < 0.05$).

an increase in the bulk density of 23.5% in the 0.00–0.10 m layer in the minesoil newly formed. In addition, the soil particle's compression was evidenced after twelve machine traffic, with a significant increase in the percentage of soil macroaggregates from 22.56 (zero traffic event) to 36.58% (twelve traffic events). At the same time, it occurs a reduction in the percentage of microaggregates from 77.44 (zero traffic event) to 63.42% (twelve traffic events).

In our study, the descompaction effect of constructed soil through the root system grasses expansion can be evidenced by the similar proportion between the percentage of macro and microaggregates in the 0.00–0.10 m layer (**Figure 3**). That is, after 11.5 years of revegetation, the percentage of macroaggregates ranged between 52.60 and 58.86% (**Figure 3a**), while the percentage of microaggregates ranged between 41.13 and 47.41 (**Figure 3b**). This result is considered probably root effects, since at 8.6 years of revegetation this same layer of the constructed soil had a percentage of

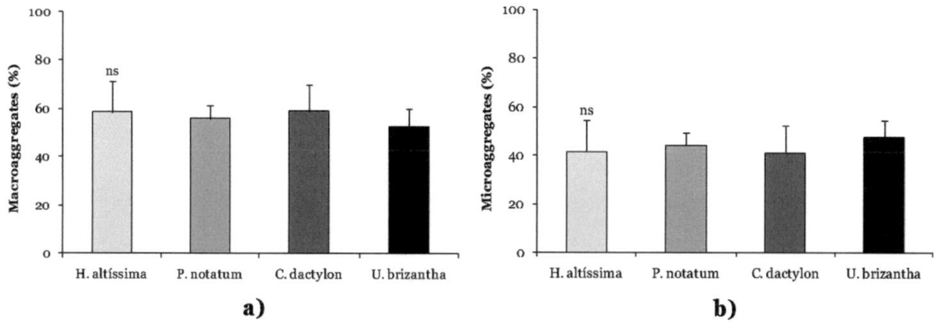

Figure 3.
Percentage of macroaggregates and microaggregates stable in water (a and b, respectively) in the 0.00–0.10 m layer of a constructed soil after coal mining and revegetated with perennial grasses for 11.5 years. Error bars mean standard deviation. Ns: Not significant to the Tuckey test ($p < 0.05$).

macroaggregates higher than 80%, while the percentage of microaggregates did not exceed 20%, according to Pinto et al. [35]. As the plants developed, the aggregates formed by compression were broken, increasing the presence of smaller aggregates, according to Stumpf et al. [3]. These results converge with Zhao et al. [11], which also observed improvement in the minesoil aggregation in the first 5–10 years of revegetation.

In addition to the improvements promoted by the grasses root system expansion, the phytomass deposition on the soil surface over the years also shows positive influence on soil biological attributes. Thus, **Figure 4a** shows that at 11.5 years of revegetation, microbial biomass was significantly higher in the soil under *Hemartria altissima* compared to *C. dactylon* cv. Tifton However, after 6.5 years of this evaluation, that is, after 18 years of revegetation, this difference promoted by grasses no longer exists. The microbial biomass indicates the carbon reserve potential of a soil [16], and its values may indicate the importance of vegetation cover in improving attributes related to soil biological quality.

Basal respiration, also known as microbial respiration, measures the amount of CO_2 released by microorganisms and is a parameter that, along with microbial biomass, is directly related to the amount of organic matter present in the soil. That is, the more organic material is added, the faster the "microbiological wheel" rotates, consuming more O_2, releasing nutrients and CO_2 from transformations, and producing more humus in the soil [16]. In this sense, at 11.5 years of revegetation, *H. altissima* and *U. brizantha* promoted the highest values of basal respiration in relation to others grasses. On the other hand, at 18 years of revegetation, only *U. brizantha* remained basal respiration superior to the other species (**Figure 4b**). This result may be a consequence of the high root exudation of this species, converging with the study by Stumpf et al. [3] who observed that *U. brizantha* had a root density of 13.29 Mg m^{-3} in the 0.00–0.10 m layer at 8.6 years of revegetation.

The evolution of basal respiration shows that there was a decrease of 10.7 to 49.8% of the values as the revegetation period progressed, regardless of the grass evaluated (**Figure 3b**). Possibly this may have occurred due to the fact that, after 18 years of revegetation, the accumulation of plant biomass on the soil surface caused microbial stress. That is, despite the high deposition of organic material in the soil, in general grasses have organic compounds of difficult degradation, requiring a range of more

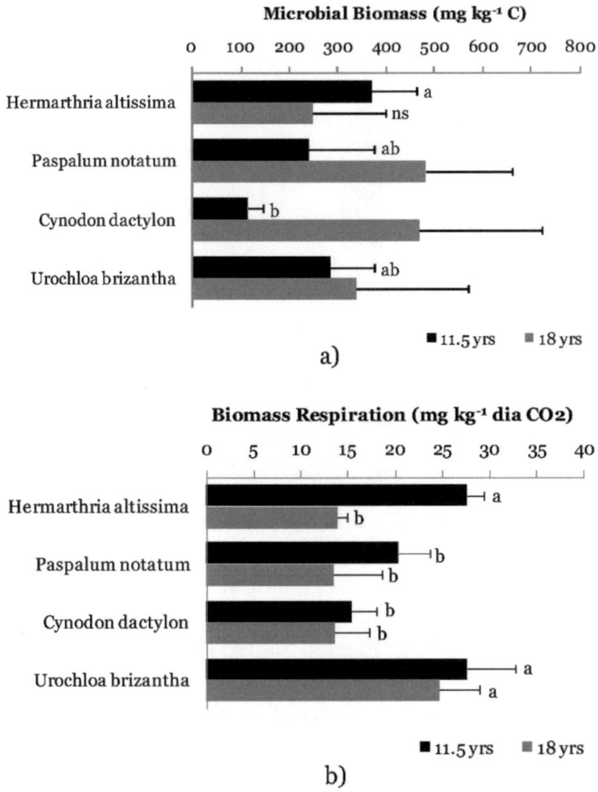

Figure 4.
Microbial biomass carbon (a) and basal respiration (b) of a constructed soil after 11.5 (black bars) and 18 years (gray bars) of revegetation with perennial grasses. Error bars mean standard deviation. Same letters in the black bars are not significantly different by the Tukey test ($p < 0.05$). Same letters in the gray bars are not significantly different by the Tukey test ($p < 0.05$).

specialized microorganisms for the decomposition of residues. Consequently, this also reflects on the low organic matter content of the constructed soil, which even after more than a decade of revegetation still has levels below 2.5% (**Figure 2f**).

Regarding the soil fauna of the constructed soil, at 17.6 years of revegetation, 1932 organisms were counted in 9 taxonomic groups in the four different perennial types of grass. The largest number of taxonomic groups (9) was observed in the soil under *C. dactylon* cv Tifton and under *U. brizantha*, followed by soil under *H. altissima* (8) *and P. notatum* cv Pensacola (7) (**Table 1**).

Mites were the most common soil fauna individuals among all taxonomic groups (RF% = 72.8), and when we analyzed the possible effect of each grass, we found that mites population in the constructed soil under *H. altissima* and *P. notatum* cv Pensacola was similar (281 and 289 individuals, respectively). However, in the constructed soil under *U. brizantha* there was a higher number of mites counted (478) compared to the other grasses, followed by *C. dactylon* cv Tifton (358) (**Table 1**). Possibly the largest number of mites in these grasses is due to the plant biomass quality deposited on the soil surface, contributing to a more favorable environment for these organisms. According to Urbanowski et al. [36], the increase in the mite population in the initial years of constructed soils restoration is due to the decomposition of organic matter; however, over the years, there may be a decrease in this population,

Soil fauna attributes	***Hemarthria altissima***	***Paspalum Notatum***	***Cynodon Dactylon***	***Urochloa Brizanta***	**RF (%)**
Mites	281	289	358	478	72.8
Springtails	86	69	148	110	21.4
Coleoptera	4	—	2	1	0.4
Dyptera	5	4	6	7	1.1
Dipluro	—	—	1	1	0.1
Enchytreid	1	4	1	4	0.5
Larva	3	3	11	3	1.0
Hymenoptera	6	18	5	1	1.5
Pupa	2	6	11	2	1.1

RF (%): Relative Frequence.

Table 1.
Total number of individuals of the edafica fauna of a constructed soil and revegetated with perennial grasses.

mainly affected by the type of organic material (litter) and not only by the amount of phytomass deposited on the soil surface.

Table 1 also observes that the second taxonomic group of the soil fauna most found were the springtails (RF% = 21.4) and the largest population was observed in the soil under *C. dactylon* cv Tifton (148), followed by the soil under *U. brizantha* (110). On the other hand, again, the grasses with the lowest number of individuals were *H. altissima* and *P. notatum* cv Pensacola (86 and 69, respectively). The springtails, along with the mites, are the organisms of greater abundance and diversity in the soil, considered the main indicators of soil fauna in constructed soils [17, 37]. Because they are the most representative groups of soil invertebrates, they control nutrient cycling through the predation of nematodes, protozoa, and fungi [38].

Among the other groups, which were numerically much lower than mites and springtails, the Hymenoptera group stands out, which obtained the third highest relative frequency in our study (RF% = 1.5) (**Table 1**). In descending order, the number of individuals in the different perennial grasses were: *P. notatum* cv Pensacola (18), *H. altissima* (6), *C. dactylon* cv Tifton (5), and *U. brizantha* (1). According to Rocha et al. [39], the richness of the Hymenoptera group tends to increase according to the complexity of the environments, and because it has a close relationship with vegetation, this group can be used as an indicator of degradation or alteration of the environment [40, 41]. Ants are fundamental soil organisms for soil "engineering processes" [42], they participate in the litter decomposition and incorporation in the soil, porosity maintenance and soil aggregates formation, community control, and microbial activities [43]. The predominance of mites, springtails, and hymenoptera groups observed in our study coincides with the study by Oliveira Filho et al. [19], who also observed areas in the process of recovery after coal mining, with different ages of restoration and revegetation, this same behavior.

4. Conclusion

This book chapter aimed to show an overview of the impact of more than a decade of revegetation with different perennial grasses on the chemical, physical, and biological quality of constructed soil after coal mining in southern Brazil. For the authors, it is clear that the liming is an important practice in the restoration of these strongly anthropized soils because this positively impacts the plant's development, facilitating the roots system expansion. Thus, the roots explore the constructed soil in search of nutrients and water, cracking the topsoil and improving its physical condition. On the other hand, biological attributes such as soil fauna and the microorganism's population are the attributes of the constructed soil that possibly takes longer to establish itself in these areas depending on the chemical and physical improvements of these areas, and it also depends on the phytomass quantity and quality added to the soil. Studies that monitor the evolution of the biological condition in these areas impacted by coal mining should be carried out in the long term always tied to the soil organic matter content for a better understanding of the actions of these organisms during the ecological recovery of these areas.

Author details

Lizete Stumpf[1*], Maria Bertaso De Garcia Fernandez[2], Pablo Miguel[1], Luiz Fernando Spinelli Pinto[1], Ryan Noremberg Schubert[1], Luís Carlos Iuñes de Oliveira Filho[3], Tania Hipolito Montiel[2], Lucas Da Silva Barbosa[2], Jeferson Diego Leidemer[2] and Thábata Barbosa Duarte[1]

1 Agronomy College, Federal University of Pelotas, Brazil

2 Soil and Water Management and Conservation Program, Federal University of Pelotas, Brazil

3 Soil Science Department, Santa Catarina State University, Brazil

*Address all correspondence to: zete.stumpf@gmail.com

References

[1] World Coal Association – WCA. 2021. Available in: https://www.worldcoal.org/

[2] Ministry of Mines and Energy – MME. Energy Research Company (Brazil). Rio de Janeiro: EPE; 2018

[3] Stumpf L, Pauletto EA, Pinto LFS. Soil aggregation and root growth of perennial grasses in a constructed clay minesoil. Soil and Tillage Research. 2016a;**161**:71-78

[4] Pinto LFS, Kämpf N. Contamination of constructed soils. In: Environment and Coal: Impacts of Exploitation and Use. Porto Alegre: State Foundation for Environmental Protection; 2002

[5] Finkelman RB, Gross PMK. The types of data needed for assessing the environmental and human health impacts of coal. International Journal of Coal Geology. 1999;**40**(1-2):91-101

[6] Daniels WL, Zipper CE. Creation and Management of Productive Mine Soils: Powell River Project. Virginia: Virginia State University; 2010

[7] Chen L, Tian Y, Stehouwer R, Kost D, Guo X, Bigham JM, et al. Surface coal mine land reclamation using a dry flue gas desulfurization product: Long-term biological response. Fuel. 2013;**105**:258-265

[8] Inda AV, Quinõnes ORG, Giassoni E, Bissani CA, Dick DP, Birth PC. Chemical attributes related to the process of sulfurization on soils after coal mining. Rural Science. 2010;**40**:1060-1067

[9] Nunes MCD. Physical Conditions of Soils Built in Coal Mining Area of Candiota/RS. Pelotas: Federal University of Pelotas; 2002

[10] Borůvka L, Kozák J, Mühlhanselová M, Donátová H, Nikodem A, Němeček K. Effect of covering with natural Topsoil in reclamation measure on brown-coal mining dumpsites. Journal of Geochemical Exploration. 2012;**113**:118-123

[11] Zhao Z, Shahrour I, Bai Z, Fan W, Feng L, Li H. Soils development in opencast coal mine spoils reclaimed for 1-13 years in the west-northern loess plateau of China. European Journal Soil Biology. 2013;**55**:40-46

[12] Courtney R, Feeney E, O'Grady A. An ecological assessment of rehabilitated bauxite residue. Ecological Engineering. 2014;**73**:373-379

[13] Orozco-Aceves M, Tibbett M, Standish RJ, Rica C. Correlation between soil development and native plant growth in forest restoration after surface mining. Ecological Engineering. 2017;**106**:209-218

[14] Longo RM, Ribeiro AÍ, Melo WJ. Recovery of degraded soils in cassiterite mineral exploration: Microbial biomass and dehydrogenase activity. Bragantia. 2011;**70**:132-138

[15] Ahirwal J, Kumar A, Pietrzykowski M, Maiti SK. Reclamation of coal mine spoil and its effect on Technosol quality and carbon sequestration: A case study from India. Environmental Science and Pollution Research. 2018;**25**:27992-28003

[16] Moreira FMS, Siqueira JO. Soil Microbiology and Biochemistry. Lavras: UFLA; 2006

[17] Barros YJ, Melo V, Sautter KD, Buschle B, de Oliveira EB, Azevedo JCR,

et al. Soil quality indicators of mining area and lead metallurgy. II – mesofauna and plants. Brazilian Journal of Soil Science. 2010;**34**:1413-1426

[18] Menta C, Conti FD, Pinto S, Bodini A. Soil biological quality index (QBS-ar): 15 years of application at global scale. Ecological Indicators. 2018;**85**:773-780

[19] Oliveira Filho LCI, Baretta D, Santos JCP. Influence of soil recovery processes after coal mining on edaficate mesofauna in Lauro Müller, Santa Catarina, Brazil. Biotemas. 2014;**27**:69-77

[20] Stumpf L, Pauletto EA, Pinto LFS, Geissler LO, Castilhos DD, de Souza DDL, et al. Biological and physical quality of a mined soil under revegetation with perennial grasses. Revista Brasileira de Ciências Agrárias. 2018;**13**:5498

[21] Brady NC, Weil RR. The Nature and Properties of Soils. 14th ed. Upper Saddle River, NJ: Pearson Education Inc.; 2008

[22] Baretta D, Santos JCP, Segat JC, Geremia EV, Oliveira Filho LCI, Alves MV. Edaphic fauna and soil quality. In: Fashion: Topics in Soil Science. Viçosa: Brazilian Society of Soil Science; 2011. pp. 119-170

[23] Silva PM, Carvalho F, Dirilgen T, Stone D, Creamer R, Bolger T, et al. Traits of collembolan an life-form indicate land use types and soil properties across an European transect. Applied Soil Ecology. 2015;**97**:69-77

[24] IUSS – Working Group WRB. World Reference Base for Soil Resources. International Soil Classification System or Naming Soils Creating Legends for Soils Maps. Rome: FAO; 2014

[25] Tedesco MJ, Gianello C, Bissani CA, Bohnen H, Volkweiss SJ. Analysis of Soil, Plants and Other Materials. Porto Alegre: Federal University of Rio Grande do Sul; 1995

[26] Kemper WD, Rosenau RC. Aggregate stability and size distribution. In: Klute A, editor. Methods of Soil Analysis. 2nd ed. Madison. Wisconsin USA: American Society of Agronomy. Soil Science Society of America; 1986

[27] Tisdall JM, Oades JM. Organic matter and waterstable aggregates in soil. European Journal of Soil Science. 1982;**33**:141-163

[28] Islam KR, Weil RR. Microwave irradiation of soil for routine measurement of microbial biomass carbon. Biology and Fertility of Soils. 1998;**27**:408-416

[29] Anderson JPE, Domsch KH. A physiological method for the quantitative measurement of microbial biomass in soils. Soil Biology and Biochemistry. 1978;**10**:215-221

[30] Bachelier G. La faune de ssols, sonécologie et son action. Paris: Initiations at Documents Techniques; 1978

[31] Gallo D, Nakano O, Silveira Neto S, Carvalho RPL, Batista GC, Berti Filho E, et al. Manual of Agricultural Entomology. 2nd ed. São Paulo: Publisher Agronomic Ceres; 1988

[32] Soil Chemistry and Fertility Commission – CQFS. Manual fertilization and cathem for the states of Rio Grande do Sul and Santa Catarina. Porto Alegre: Brazilian Society Soil Science; South Regional Center; 2004

[33] Stumpf L, Pauletto EA, Pinto LFS, Ambus JA, Garcia GF,

da Silva TS. Chemical characteristics of a constructed soil and its effects in the root development of perennial grasses. Agrarian – Brazilian Journal of Agrarian Sciences. 2016b;**11**:343-349

[34] da Silva Barboza F, Stumpf L, Pauletto EA, de Lima CLR, Pinto LFS, Jardim TM, et al. Impact of machine traffic events on the physical quality of a minesoil after topographic reconstruction. Soil and Tillage Research. 2021;**210**:104981

[35] Pinto LFS, Stumpf L, Miguel P, Junior LAD, Leidemer JD, da Silva Barbosa L, et al. Reclamation of soils degraded by surface coal mining. In: Mining Techniques-Past, Present and Future. London, UK: IntechOpen; 2020

[36] Urbanowski CK, Kamczyc J, Skorupski M. Geoderma does litter decomposition affect mite communities (Acari, Mesostigmata) A five-year litterbag experiment with 14 tree species in mixed forest stands growing on a postindustrial area. Geoderma. 2021;**391**:114963

[37] Menta C, Conti FD, Pinto S, Leoni A, Lozano-Fondón C. Monitoring soil restoration in an open-pit mine in northern Italy. Applied Soil Ecology. 2014;**83**:22-29

[38] Lavelle P, Bignell D, Lepage M, Wolters V, Roger P, Ineson P, et al. Soil function in a changing world: The role of invertebrate ecosystem engineers. European Journal of Soil Biology. 1997;**33**:159-193

[39] Rocha WDO, Dorval A, Peres Filho O, Vaez CDA, Ribeiro ES. Ants (Hymenoptera: Formicidae) bioindicators of environmental degradation in Poxoréu, Mato Grosso, Brazil. Forest and Environment. 2015;**22**:88-98

[40] Crepaldi RA, Portilho II, Silvestre R, Mercante F. Ants as bioindicators of soil quality in integrated crop-livestock system. Rural Science. 2014;**44**:781-787

[41] Oliveira MA. Diversity of Myrmecofauna and Forest Succession in the Amazon. Acre, Brazil. Viçosa: Federal University of Viçosa; 2009

[42] Lavelle P, Decaëns T, Aubert M, Barot S, Blouin M, Bureau F, et al. Soil invertebrates and ecosystem services. European Journal of Soil Biology. 2006;**42**:3-15

[43] Jouquet P, Dauber J, Lagerlöf J, Lavelle P, Lepage M. Soil invertebrates as ecosystem engineers: Intended and accidental effects on soil and feedback loops. Applied Soil Ecology. 2006;**32**:153-164

Chapter 3

Phyto-Sociological Attributes, Ecosystem Services and Conservation Dynamics of Three Protected Forests in Tropical Rainforest Ecosystem of Nigeria

Aladesanmi Daniel Agbelade

Abstract

This study aimed at determining ecosystem services and conservation dynamics of three protected forests in Nigeria. Using simple sampling technique, 24 plots with 25 m^2 were established in these protected forests to facilitate data collection. A total of 370 individual trees per hectare, disproportionately distributed between 53 different species in 25 families, were encountered in Omo biosphere reserve (BR) while 381 stems in 63 species in 24 families in strict nature reserve (SNR) and Okomu national park (NP) recorded 352 individual stems, 59 species, and 25 families. The three protected forests had high tree species diversity index (Shannon-Wiener diversity index of 3.19 for Omo BR, 3.90 for Akure SNR, and 3.45 for Okomu NP). The values for basal area (36.63, 72.39, and 32.47 m^2), volume (427.08, 929.05, and 366.71 m^3), above-ground biomass (153.20, 316.73, and 353.92 ton), below-ground biomass (30.64, 63.35, and 190.04 ton), and total carbon stock (70.78, 91.92, and 212.35 ton) for Omo BR, Akure SNR, and Okomu NP, respectively. This study serves as baseline information for management of protected forests in Nigeria and it shows the potential of *in-situ* conservation for the dynamism of the ecosystem services.

Keywords: biodiversity, conservation, *Celtis zenkeri*, ecosystem services, *in-situ*, national park, protected forest

1. Introduction

Nigeria as one of the developing countries of the world has high population density and high biodiversity forest associated with tropical rainforest ecosystem. Nigeria as tropical rainforest is known for large diverse forest ecosystems with lots of *in-situ* conservation methods such as forest reserves (FR), strict nature reserves (SNR), biosphere reserves (BR), national parks (NP), and enrichment plantation (EP), they contribute to biodiversity reservoir of the world. Forest ecosystems in Nigeria are very important for biodiversity conservation, ecosystem functions, watershed protection, mitigation of climate change, economic sustainability, habitats for wild animals, and nutrition and enhancement of rural livelihoods [1]. The tropical rainforest is the most diverse of the terrestrial ecosystems, containing more flora and fauna species than any other biome, which are important source of biodiversity, food, and carbon storage [2]. Tropical rainforest has the ability to accommodate 70% of animal and plant species, which serves as a reservoir of biodiversity [1, 3]. Carbon sequestration, production of oxygen and Ozone layer protection are part of many services that the forest delivers as ecosystem services for the goods of the environment [4, 5].

Over 60% of the population residing in rural areas depend on forest and other biomass resources for fuel-wood and timber and non-timber forest products for their energy needs and livelihood. Forests are repository of the biodiversity, gene pool resources and sequester carbon dioxide, and they also provide a lot of other environmental services [1]. They play a vital role in sustaining the life of people which are crucial for food, water, forest resources and livelihood security. This is ensured through better management practices and sustainable utilization of forestlands. Forest vegetation covers protect watersheds, conserve species diversity, serve as habitat for wildlife, and contribute to good quality environment [6]. More than half of the world's species diversity can be found in the tropics according to a study by [7]. However, this critically important role of tropical forests is being threatened by deforestation. The effects of deforestation on the environment are numerous. There is overwhelming evidence and consensus that the high rate of tropical deforestation is one of the major causes of climate change [8, 9]. Deforestation contributes to climate change by reducing forest vegetation cover and increasing surface temperature. An estimated 20 to 25% of global emissions stem from deforestation, predominantly in the tropics [8]. Strategies to drastically reduce the rate of deforestation and restoration of degraded forests must be urgently developed in order to save the tropical forests and humans from the catastrophic effects of climate change. The forest is regulator of the micro and macro environment as carbon stock and carbon sequestration [5]. Furthermore, forest vegetation covers protect watersheds and species diversity, serves as biodiversity reservoir and conservation, habitat to wildlife and thus contributes to effective ecosystem functioning and good quality ecosystem. In addition, healthy forest vegetation cover plays important role in climate change mitigation by sequestering carbon dioxide from the atmosphere.

Tropical forests are predominantly located in developing countries and are often subject to activities such as logging and conversion to agriculture [10]. Nigerian rainforest ecosystems occupy 95,372 km^2 (9.7%) of the country's land mass. It is the most densely populated part of Nigeria and source of the bulk of the country's timber needs [11]. The protected forest is a specific term to denote forests with some amount of legal and constitutional protection in certain countries, besides being a generic term to denote forests where the habitat and resident species are legally accorded protection and are protected from any further depletion. Protected areas provide habitat for the country's endangered, rare and endemic plant and animal species. There are seven strict nature reserves, one biosphere reserve, 160 constituted forest reserves and six national parks in Nigeria. There are no known studies on the Phytosociological characteristics, ecosystem services and conservation dynamics of these protected forests. The method adopted by the government to promote ecosystem sustainability makes the forest play essential role in carbon sequestration and function effectively as climate change regulator. Thus, an increase in carbon sequestration can be achieved through conservation and effective management of these protected forests. This study essentially provides information on structure, species diversity, biomass, and carbon stock as ecosystem services and function to the environment. This, therefore, represents an important contribution to ecosystem-wide carbon cycle and amount of carbon dioxide that can be sequestered annually from the atmosphere by these protected forests.

2. Methodology

2.1 Study area

The research was conducted in three protected forests (Omo biosphere reserve (BR), Akure strict nature reserve (SNR) and Okomu national park (NP)), in May 2021, located in the southern part of Nigeria. Strict nature reserve (SNR) is prominent among the methods for *in-situ* conservation of biodiversity in Nigeria in particular and the world in general [1]. SNRs are created to protect representative samples of natural ecosystems for preservation of biodiversity and ecological processes, scientific study, environmental monitoring, education and the maintenance of genetic resources in a dynamic and evolutionary state [12]. Biosphere reserve (BR) is areas of terrestrial and coastal ecosystems promoting solutions to reconcile the conservation of biodiversity with its sustainable use. They are internationally recognized, nominated by national governments and remain under the sovereign jurisdiction of the states where they are located. An example of biosphere reserve in Nigeria is the Omo biosphere reserve (BR). The Omo biosphere reserve is located in Ijebu area of Ogun State, Southwest, Nigeria. Established in 1977, roughly 6000 people live within the biosphere reserve boundaries, which cover a total area of 130,600 hectares. There are several national parks established by the Federal Government (FG) of Nigeria to serve as tourist attraction centres and conservation of plants and animals. These national parks consist of diverse species of fauna and flora that contribute to the ecosystem development and conservation status. Nigeria national park service is the body that oversees and responsible for preserving, enhancing, protecting and managing vegetation and wild animals in the national parks of Nigeria. Nigeria national park service works closely with the Nigerian tourism development corporation in the handling and management of the national parks. These national parks cover a total land area of approximately 20,156 km^2, which is about 3% of Nigeria's total land area. Okomu national park covers a land area of 200 hectares in Edo State, about 60 km Northwest of Benin City, Nigeria. The area is bounded by latitudes 6.08° and 6.30° N and longitudes 5.01° and 5.27° E. The climate of the region is characterized by a double maximal year-round rainfall pattern with a mean annual rainfall of about 2200 mm, which peaks between May and October and a mean monthly temperature of 27°C. Tropical hardwood tree species in the area include *Celtis zenkeri, Triplochiton scleroxylon, Pycnanthus angolensis, Alstonia congoensis, Khaya ivorensis* (African mahogany) and *Lovoa trichilioides* (African walnut).

2.2 Method of data collection

These three (3) protected forests were selected (Omo BR, Akure SNR and Okomu NP) for this study. In each protected forest, biodiversity and tree growth data were collected from two lines transects of 1000 m each in length laid approximately at the centre of the protected forest [3, 6]. The two line transects were separated by a distance of at least 1000 m. Four temporary sample plots of 25 m × 25 m were laid at alternate sides along each transect at every 250 m interval. Thus, there were 4 sample plots per transect, 8 per protected forest and twenty-four (24) for this study. Within each plot, all living trees with Dbh ≥ 10 cm were identified, their Db, Dbh measured with girth tape and Dm, Dt and total heights measured with Spiegel Relaskop [1, 6].

2.3 Data analysis for species diversity

Species relative density (RD) used to determine species relative distribution was computed using:

$$RD = \left(\frac{\mathrm{ni}}{\mathrm{N}}\right) \times 100 \tag{1}$$

where: RD (%) = species relative density; n_i = number of individuals of species i and N = total number of all individual trees of all species in the entire forest reserve.

Species relative dominance (RDo) was estimated using:

$$RD_0 = \left(\frac{\sum Ba_i \times 100}{\sum Ba_n}\right) \tag{2}$$

where: Ba_i = basal area of all trees belonging to a particular species *i* and Ba_n = basal area of all individual trees.

Importance value index (IVI) of each species was computed with the relationship:

$$\mathrm{IVI} = \left(\frac{\mathrm{RD} + \mathrm{RD_0}}{2}\right) \tag{3}$$

Species diversity index (H′) was computed using the Shannon-Wiener diversity index below:

$$\mathrm{H}' = -\sum_{\mathrm{i=1}}^{\mathrm{s}} Pi\ In(Pi) \tag{4}$$

where: H′ = Shannon-Wiener diversity index; S = total number of species in the protected forest; p_i = proportion of S made up of the ith species and ln = natural logarithm.

Simpson Concentration index

$$\mathrm{D} = \sum\left(\frac{\mathrm{ni}}{\mathrm{Ni}}\right)^2 = \mathrm{D} = \frac{1}{\sum_{\mathrm{i=1}}^{\mathrm{s}} \mathrm{P_i}^2} \tag{5}$$

In the Simpson index, p is the proportion (n/N) of individuals of one particular species found (n) divided by the total number of individuals found (N), Σ is still the sum of the calculations and (S) is the number of species.

Shannon's maximum diversity index was calculated using:

$$\mathrm{H_{Max}} = \mathrm{Ln}\ (\mathrm{S}) \tag{6}$$

Where: H_{max} = Shannon's maximum diversity index and S = total number of species in the protected forest.

Species evenness in each plot was determined using Shannon's equitability (E_H), which was obtained using:

$$E_H = \frac{H'}{H_{Max}} = \frac{\sum_{i=1}^{s} P_i \text{Ln } P_i}{\text{Ln } (S)} \tag{7}$$

2.4 Forest conservation structure

The following computations were computed for forest structure analyses. The basal area of each tree in the protected forest was calculated using

$$BA = \frac{\pi D^2}{4} \tag{8}$$

Where: BA = Basal area (m^2), D = Diameter at breast height (cm) and л = pie (3.142). The total basal area for the plot was obtained by adding basal area of all trees in the forest reserve. Volume of individual trees was estimated using

$$V = \pi h \frac{Db^2 + 4(Dm^2) + Dt^2}{24} \tag{9}$$

where: V = Tree volume (m^3), π = 3.142, h = tree height (m) measurement and Db, Dm and Dt = tree cross-sectional area at the base, at the middle and top of merchantable height, respectively. The total volume for the forest reserve was obtained by adding all individual trees volume computed.

2.5 Biomass and carbon stock

In determining the total carbon (TC) stocks, estimation of AGB and BGB were computed. Biomass expansion factor (BEF) of 1.74 was used to estimate tree above-ground biomass for tropical rainforest [13], multiple by volume over bark (m^3/ha) and wood density (kg/m^3).

$$\text{Above} - \text{Ground Biomass (AGB)} = \text{BEF} \times \text{VOB} \times \text{WD} \tag{10}$$

Where, BEF = Biomass expansion factor; VOB = Volume over bark (m^3) and WD = Wood density (kg^{-2}). Wood density for tree species was acquired from Global Wood Density Database. Arithmetic mean of (0.60 gcm^3) for a tropical African forest was used for species that were not found in the database following [14]. The carbon stock of the protected forests was determined by a fraction of 50% of biomass.

$$AGC = AGB \times 0.5 \tag{11}$$

Thus, above-ground carbon (AGC) was calculated as a conversion factor of 0.5 multiplied by AGB.

$$BGB = AGB \times 0.2 \tag{12}$$

where below-ground biomass was computed as 20% of AGB following MacDicken [15]; IPCC [16], using a synthesis of global data and a conservative ratio shoot-to-root biomass of 5:1 [17].

$$TCS = AGC + BGC \tag{13}$$

The estimation of carbon content in BGC is the same as that of AGC Eq. 12. Total carbon storage (TCS ton ha^{-1}) stock was calculated by summing up the carbon stock of AGC and BGC following [18].

3. Results

3.1 Phytosociological characteristics, diversity and biomass estimation

The phytosociological characteristics, diversity and biomass estimation of these protected forests were analysed in this study (**Table 1**). The table indicates that a total of 370 individual trees per hectare, disproportionately distributed between 53

Growth characteristics and biodiversity indices	Protected forests		
	Omo Biosphere reserve	Akure Strict nature reserve	Okomu National park
Density ha^{-1}	370b	381a	352c
Family	25a	24a	25a
Mean Diameter (cm)	26.08b	35.61a	26.31b
Maximum Diameter (cm)	168.50c	251.50a	190.20b
Mean height (m)	14.15b	17.41a	15.59b
Basal Area (m^2) ha^{-1}	36.63b	72.39a	32.47c
Volume (m^3) ha^{-1}	427.08b	929.05a	366.71c
Species richness	81b	119a	76c
Species diversity	3.19c	3.90a	3.45b
Tree species	53c	63a	59b
Simpson_1-D	0.954b	0.951c	0.969a
Evenness_e^H/S	0.574b	0.571c	0.707a
Menhinick	3.768c	3.905b	3.933a
Margalef	10.42a	10.91a	10.71a
Equitability_J	0.863b	0.864b	0.915a
Above-ground biomass (ton) ha^{-1}	153.20c	316.73b	353.92a
Below-ground biomass (ton) ha^{-1}	30.64c	63.35b	190.04a
Total carbon storage (ton) ha^{-1}	70.78c	91.92b	212.35a

Values followed by similar letters at not significantly different ($p \leq 0.05$).

Table 1.
Growth characteristics, biodiversity indices and ecosystem services.

different species in 25 families, were encountered in Omo biosphere reserve while 381 stems in 63 species in 24 families and Okomu national park recorded 352 individual stems, 59 species and 25 families. There were significant differences in all the growth variables investigated in this study across the three protected forests (**Table 1**). The three protected forests had high tree species diversity index (Shannon-Wiener diversity index of 3.19 for Omo BR, 3.90 for Akure SNR and 3.45 for Okomu NP). The Simpson concentration index and species evenness are assessed for these protected forests as 0.954, 0.951 and 0.969 while 0.574, 0.571 and 0.707 for Omo BR, Akure SNR and Okomu NP, respectively. There were significant differences in all the biodiversity indices investigated in this study across the three protected forests (**Table 1**), indicating that diversity is dissimilar between the forests. The biological diversity indices of the protected forests compared favorably with other protected natural forests, sacred groves and other natural forest formations. The values for basal area (36.63, 72.39, and 32.47 m^2), volume (427.08, 929.05, and 366.71 m^3), above-ground biomass (153.20, 316.73, and 353.92 ton), below-ground biomass (30.64, 63.35, and 190.04 ton) and total carbon stock (70.78, 91.92, and 212.35 ton) for Omo BR, Akure SNR and Okomu NP, respectively. There were significant differences in all the biomass and carbon storage investigated in this study across the three protected forests (**Table 1**), indicating the carbon sequestration potentials of the forests.

3.2 Diameter distribution and canopy structure of the protected forests

The diameter distribution of trees in these protected forests followed inverted J distribution pattern common with tropical forest ecosystems (**Figure 1**). The figure reveals that the highest number of trees (205 stems/ha) was in the diameter class of less than 15 cm in Okomu NP followed by (168 and 108 stems/ha) Akure SNR and Omo BR Osogbo, respectively. The vertical structure of the selected protected forests is shown in **Figure 2**. The structures of these protected forests were determined by the canopy distribution of the forests, which was calculated based on the height distributions of the tree species. The figure reveals that the highest numbers of trees (111 stems/ha Akure SNR and 99 stems/ha Okomu NP) were in the middle canopy

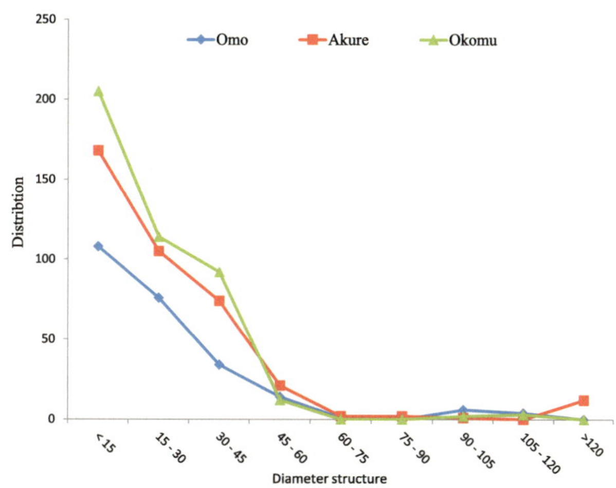

Figure 1.
Diameter distribution class for the protected forests.

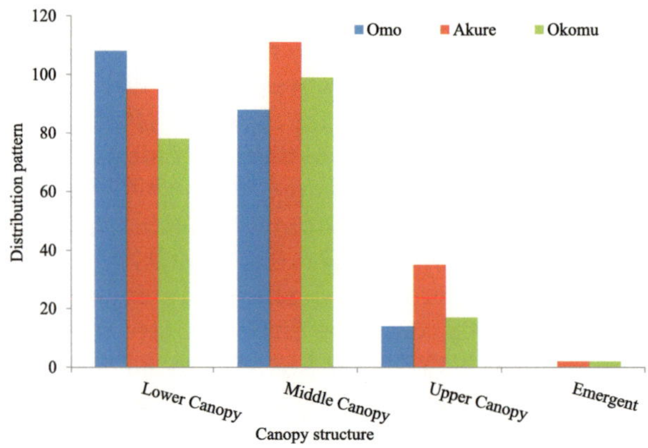

Figure 2.
Canopy structure based on height distribution for the protected forests.

structure of the forest followed by (108 stems/ha Omo BR) at the lower canopy structure of the forest.

4. Discussions

In these protected forests, individual tree density and species were recorded 370 (53) Omo BR, 381 (63) Akure SNR and 352 (59) Okomu NP. Similar tree compositions were reported in different researches across the tropical ecosystems, Onyekwelu *et al.* [6] reported 55, 73 and 78 for Osun Osogbo, Idanre hill and Ogun Onire sacred groves, respectively, in Nigeria, Baul *et al.* [19] reported 52 tree species in farm forests in Nepal, Chowdhury *et al.* [20] reported 55 tree species in village common forests of Khagrachari while Roy *et al.* [21] reported 62 species in the home gardens of Bangladesh [22]. The assessment of plant species revealed that all the encountered trees in the three protected forests were indigenous tropical hardwood species that are of economic value to humans and their environment. This is an indication that the three protected forests are repositories of many indigenous tropical hardwood tree species that are of high ecological, social and economic values. Forest habitats play an important role in the effective functioning of the forest ecosystem, and protected forests serve as *in-situ* conservation for rare plants being disturbed by anthropogenic activities [23]. This is crucial to the sustainable management and preservation of tree diversity resources. The tree species diversity and abundance of these protected forests as determined by the biodiversity indices indicated that these protected forests fulfilled the mandate of a biodiversity conservation strategy [24].

Biodiversity indices for these protected forests represent the diversity of floral compositions and their distribution in these forests. Tree species diversity and evenness indices found in this study are comparable to the study of Onyekwelu *et al.* [6], which reported that 3.19 (0.84), 3.25 (0.85) and 3.46 (0.86) for Osun Osogbo, Idanre hill and Ogun Onire sacred groves, respectively, in Nigeria. Lower Shannon-Wiener tree species diversity index of (1.80) was recorded in sacred grove of Igbo Olodumare, (1.23) was recorded for home gardens in Northern Bangladesh while (1.64) was recorded for protected forests of Bangladesh [25, 26], and (1.34) was recorded for collaborative forests in Nepal [27]. The Phytosociological attribute and floristic diversity of these selected protected forests were discovered to be comparable with other protected areas of tropical forest ecosystems of south-west, Nigeria [1]. The stand densities were also similar to those obtained for Garo hills, India [28], Borneo rainforest [29], Indonesian forest [30] and the Mexican tropical deciduous forest [31].

4.1 Biomass and Carbon stock estimation

Tropical forests are known to play an important role in regulating the global carbon cycle. The biomass of tropical forests plays a critical role in micro and macro absorption of carbon and carbon cycling of forest ecosystems [32]. However, tropical forest ecosystems particularly protected forests need to be adequately and regularly investigated for carbon stock accumulation. The total biomass of 183.83 kg ha^{-1}; 380.08 kg ha^{-1} and 543.96 kg ha^{-1} was recorded for Omo BR, Akure SNR and Okomu NP, respectively, which is lower than 164.82 ton ha^{-1} recorded for India forest and 156.73 ton $ha^{-1}$for Nigerian forest, respectively [4, 33]. The result of biomass in these protected forests is comparable with the studies of Wittmann *et al.* [32], which estimated 259.45 kg ha^{-1} for Southern Pantanal, Brazil. Agbelade and Adeagbo [5] estimated 617.85 kg ha^{-1} and 209.26 kg ha^{-1} for Akure SNR and Osun Osogbo sacred grove, respectively. The disparity in the values maybe a result of the different methods and equations adopted, this study uses BEF while other researchers used allometric equations. Carbon storage of forest biomass is an important attribute of a stable forest ecosystem and a key link in global carbon cycle. The total carbon stock estimated for these protected forests is 70.78 kg ha^{-1}, 91.92 kg ha^{-1} and 212.35 kg ha^{-1} for Omo BR, Akure SNR and Okomu NP, respectively, which is lower than 617.86 kg ha¯1 recorded for Akure strict nature reserve and 209.27 kg $ha^{-1}$for Osun Osogbo sacred grove [5]. Adekunle *et al.* [4]; IPCC [16] recorded 82.41 ton ha^{-1} for Indian forest and 78.29 kg ha^{-1} for Eda SNR, which can be compared with the result of this study. Tabue *et al.* [34] estimated 354.73 Mg ha^{-1} for Dja wildlife reserve in Cameroon. Munishi and Shear [35] reported over 300 Mg ha^{-1} carbon stock in Tanzanian Eastern forests. The above results indicated that protected forests would contribute significantly to carbon sequestration and climate change mitigation as long as the forest is adequately protected from deforestation and degradation. Thus, besides being a reservoir of biodiversity, protected forests also act as sink of atmospheric CO_2. The high biomass and carbon stock in this forest reserve is attributed to the effective conservation system that prevented the forest from degradation and deforestation as well as the federal government policy on National parks.

5. Conclusion

The use of protected forests is important for biodiversity and climate change mitigation. The scientific information provided by this research would further promote accurate estimation of the tree species diversity, stand volume and carbon stock in the Okomu protected forest. The results of the study indicated that there are many indigenous tropical tree species that are rare and in danger of extinction. There is strong evidence of active regeneration status, which indicates a good future for the Okomu protected forest. The study shows that the biodiversity-protected forests act as sink of atmospheric CO2 because of their high carbon stock and high biomass. The research revealed the abundance of relative dominant trees as well as their contributions to the preservation of the environment. This reference can be used to compare changes in carbon stocks over time. The current position of protected areas in terms of tree species abundance, evenness, carbon sequestration, productivity and structure shows the effectiveness of *in-situ* management. The forest has the potential to serve as a long-term carbon sink due to its good potential for carbon sequestration. To continue providing these ecosystem services and functions, the previous method that prevented the forest reserve from being degraded should be maintained and strengthened. According to the study, strict measures should be taken on identified protected areas so as to ensure their continuous impact on the environment.

Conflicts of interest

The manuscript was approved by the author. The author declares that there are no conflicts of interest or competing interests.

Author details

Aladesanmi Daniel Agbelade
Faculty of Agricultural Sciences, Department of Forest Resources and Wildlife Management, Ekiti State University, Ado Ekiti, Nigeria

*Address all correspondence to: aladesanmi.agbelade@eksu.edu.ng

References

[1] Adekunle VAJ, Adewole OO, Shadrach OA. Tree species diversity and structure of a Nigerian strict nature reserve. Tropical Ecology. 2013;**54**(3): 275-289

[2] Food and Agriculture Organization of the United Nations (FAO). Global forest resources assessment 2020, Country Report. Bangladesh: Food and Agriculture Organization of the United Nations; 2020

[3] Agbelade AD, Ojo BH. Species diversity, volume determination and structure of protected forests for *in-situ* biodiversity conservation. International Journal of Conservation Science. 2020;**11** (1):133-144

[4] Adekunle VAJ, Nair NK, Srivastava AK, Singh NK. Volume yield, tree species diversity and carbon hoard in protected areas of two developing countries. Forest Conservation Science. 2014;**11**(1):133-144

[5] Agbelade AD, Adeagbo DO. Relationship between aboveground biomass and tree species diversity in two protected forests. Journal of Forestry Research and Management. 2020;**17**(1): 13-25

[6] Onyekwelu JC, Lawal A, Mosandl R, Stimm B, Agbelade AD. Understory species diversity, regeneration and recruitment potential of sacred groves in south West Nigeria. Tropical Ecology. 2021;**62**(3):427-442

[7] May RM, Stump MPH. Species area relations in tropical forests. Science. 2000;**290**:2084-2086

[8] Archana K. Impact of deforestation on climate change. Journal of Environmental Science, Toxicology and Food Technology. 2013;**4**(2):24-28

[9] Lawrence D, Vandecar K. Effects of tropical deforestation on climate and agriculture. Nature Climate Change. 2014; **5**:27-36. DOI: 10.1038/nclimate2430

[10] Lewis SL, Phillips OL, Baker TR. Impacts of global change on the structure, dynamics and function of south American tropical forests. In: Laurance WF, Peres C, editors. Emerging Threats to Tropical Forests. Chicago: Chicago University Press; 2006. pp. 15-31

[11] Onyekwelu JC, Mosand R, Stimm B. Tree species diversity and soil status of primary and degraded tropical rainforest ecosystems in South-Western Nigeria. Journal of Tropical Forest Science. 2008; **20**(3):193-204

[12] Isichei AO. Omo Biosphere Reserve, Current Status, Utilization of Biological Resources and Sustainable Management (Nigeria). Paris: UNESCO; 1995

[13] Brown S, Lugo AE. Above ground biomass estimates for tropical moist forests of the Brazilia Amazo. Interciencia. 1992;**17**:8-18

[14] Chave J, Andalo C, Brown S, Cairns M, Chambers JC, Eamus D, et al. Tree allometry and improved estimation of carbon stock and balance in tropical forests. Oecologia. 2005;**145**:87-99

[15] Macdicken KG. A Guide of Monitoring Carbon Emission from Deforestation and Degradation in Developing Countries: An Examination of Issues Facing the Incorporation of REDD into Market-Based Climate Policies. Washington DC: Resource for Future; 1997. p. 84

[16] IPCC Inter-governmental Panel on Climate Change. In: Eggleston HS,

Buendia L, Miwa K, Ngara T, Tanabe K, editors. IPCC Guidelines for National Greenhouse gas Inventories. Prepared by the National Greenhouse Gas Inventories Programme. Japan: Institute for Global Environmental Strategies; 2006

[17] Meragiaw M, Woldu Z, Martinsen VBR. Floristic composition and structure of the Kibate forest along environmental gradients in Wonchi. Southwestern Ethiopia Journal of Forest Research. 2021;**32**:2669-2682

[18] Pearson TRH, Brown SL, Birdsey RA. Measurement Guidelines for the Sequestration of Forest Carbon. General Technical Report NRS-18. Vol. 18, No. 1. Delaware: United States Department of Agriculture Forest Service; 2007. p. 42. DOI: 10.1089/hum.2005.16.57

[19] Baul TK, Tiwari KR, Atique Ullah KM, McDonald MA. Exploring Agro-biodiversity on farm: A case from Middle-Hills of Nepal. Small-scale Forest. 2013;**12**

[20] Chowdhury MA, Islam KN, Hafiz N, Islam K. Diversity of trees in a community managed forest: The case of Komolchori VCF, Khagrachari, Bangladesh. Geological Ecology Landscapes. 2019;**3**:95-103

[21] Roy B, Rahman MH, Fardusi MJ. Status, diversity, and traditional uses of homestead gardens in northern Bangladesh: A means of sustainable biodiversity conservation. ISRN Biodiversity. 2013;**2013**:1-11

[22] Baul TK, Chakraborty A, Nandi R, Nath TK, Mohiuddin M. Phytosociological attributes and ecosystem services of homegardens of Maheshkhali island of Bangladesh. Trees, Forests and People. 2021;**5**:100092

[23] IIRS-NRSA (Indian Institute of Remote Sensing Dehradun). Biodiversity Characterization at Landscape Level in Northeast, Western Himalayas. Western Ghats Using Satellite Remote Sensing and Geographic Information System. 2002. p. 98

[24] UNEP-WCMC. State of the World's Protected Areas: An Annual Review of Global Conservation Progress. Cambridge, UK: UNEP-WCMC; 2008. Available From: http://www.unepwcmc.org/protected_areas/pdf/stateOfTheWorld%27sProtectedAreasLow.pdf

[25] Jaman MS, Hossain MF, Shariful IJ, Helal MG, Jamil M, Mizanur R. Quantification of carbon stock and tree diversity of Homegardens in Rangpur District Bangladesh. International Journal of Agricultural Forest. 2016;**6**:169-180

[26] Rahman MM, Mahmud MA, Al Shahidullah M, Nath TK, Jashimuddin M. The competitiveness of the phytosociological attributes of the protected areas in Bangladesh with that in the other tropical countries. Journal of Sustainable Forestry. 2016;**35**:431-450

[27] Mishra BK, Garkoti SC. Species diversity and regeneration status in Sabaiya Collaborative Forest, Nepal. In: Raju NJ, editor. Geostatistical and Geospatial Approaches for the Characterization of Natural Resources in the Environment: Challenges, Processes and Strategies. Vol. 2. 2016. pp. 427-433

[28] Kumar A, Marcot BG, Saxena A. Tree species diversity and distribution patterns in tropical forests of Garo Hills. Current Science. 2006;**91**:1370-1381

[29] Small A, Martin TG, Kitching RL, Wong KM. Contribution of tree species to the biodiversity of a 1 ha Old World rainforest in Brunei. Borneo.

Biodiversity and Conservation. 2004;**13**: 2067-2088

[30] Kessler M, Keber PJA, Gradstein SR, Bach K, Schmull M, Pitopand R. Tree diversity in primary forest and different land use systems in Central Sulawesi, Indonesia. Biodiversity and Conservation. 2005;**14**:547-560

[31] Duran E, Meave JA, Lott DJ, Segura G. Structure and tree diversity patterns at landscape level in a Mexican tropical deciduous forest. Boletin de la Sociedad Botanica Mexico. 2006;**79**:43-60

[32] Wittmann F, Zorzi BT, Tizianel FAT, Urquiza MVS, Faria RR, Sousa NM, et al. Tree Species Composition, Structure and Aboveground Wood Biomass of a Riparian Forest of the Lower Miranda River, Southern Pantanal. Brazi; 2008. p. 4

[33] Chandrashekara UM, Sankar S. Ecology and management of sacred groves IKeralan, India. Forest Ecology and Management. 1998;**112**:165-177

[34] Tabue MRB, Zapfack L, Noiha NV, Nyeck B, Meyan-Ya DRG, Ngoma LR, et al. Plant diversity and carbon storage assessment in an African protected forest: A case of the eastern part of Dja wildlife Reserve in Cameroon. Journal of Plant Science. 2016;**4**:95-101

[35] Munishi PKT, Shear TH. Carbon storage in Afromontane rain forests of the eastern arc of Tazania: Their net contribution to atmospheric carbon. Journal of Tropical Forest Science. 2004: 78-93

Chapter 4

Integrated Conservation Approaches for Rescuing, Regeneration and Adaptive Management of Critically Endangered Asteraceae Florae in Africa: A Case of *Bothriocline auriculata* Species in Uganda

Semwanga Mohammed, Nakiguli Fatumah, Nasejje Shadia, Kigozi Abas, Ashraf Nkumba and Nakimwanyi Shamirah

Abstract

Among the 62 Bothriocline plant species, *Bothriocline auriculata* is the only endemic species in Uganda. Although this species is capable of thriving in diverse agroecosystems including mountainous areas, bamboo thickets, montane and tropical rainforests, it is only sited along Mt. Elgon slopes, the species' native ecosystem. Unfortunately, for the last two decades, the species' native ecosystem is undergoing very rapid deterioration as the increasing human populations have no option for survival but to clear protective forests and vegetation for arable farming, grazing and settlement. Despite the proven ecological importance of *Bothriocline auriculata* as a fodder plant, nutrient recycling and biodiversity in contributing to ecosystem balance, the species is declining at unprecedented rates. Consequently, the *Bothriocline auriculata* is now on the blink of extinction and is classified as critically endangered. This project aimed to rescue and conserve this species to ensure its full recovery, restoration and conservation under protected ecosystems. The specimens were rescued from the native ecosystem and multiplied into 150 juvenile seedlings. The seedlings exhibited a faster growth rate under well-nourished and moist soil conditions and vice versa. The data confirm the species' ability to thrive in protected ecosystems with favorable weather and soil conditions for ex-situ conservation.

Keywords: biodiversity, conservation, endangered species, endemic species, Africa

1. Introduction

1.1 Background and context

Bothriocline is a genus of angiosperms belonging to the Aster family called Asteraceae [1, 2]. The *Bothriocline*s are hairy herbs with purple flowers whose species range from annual, bi-annual and perennial growth cycles. There are 62 known *Bothrine species*, most of which are native to tropical Africa and a few are native to isolated islands of the Indian Ocean [2]. Among the 62 species [1], *Bothriocline auriculata* (M. Taylor) hereafter referred to as *BAT* is the only endemic species in Uganda [3]. The *BAT* (**Figure 1**) thrives well in the montane forest, mountainous areas and bamboo thickets. The only known locations of *BAT* in Uganda are within Mount Elgon areas between 2745 and 3150 meters above sea level [4]. **Figure 1** presents photos of the *Bothriocline auriculata* (BAT) species.

However, the BAT natural habitat (**Figure 1**) is undergoing an unprecedented rate of ecosystem deterioration and destruction over the last two decades. The rapid ecosystem loss is attributed to exponentially increasing anthropogenic activities from exponentially increasing local populations, who have no livelihood option but to clear the remaining forest and protective vegetation cover to open up more land for arable farming and human settlement [5]. There is also increased frequency and intensity of other anthropogenic activities mainly deforestation for fuelwood and timber as well as overgrazing of livestock along the Mount Elgon slopes, which further destroys the species and its native ecosystems. For instance, the increasing livestock numbers, – which surpass the carrying capacity of the traditional rangelands, force farmers to drive their livestock upwards on the Mount Elgon slopes in search of dense vegetation [5, 6]. Farmers have shifted cultivation upwards because of the virgin fertile lands along the mount slopes [6].

Figure 1.
Bothrine species in native Mount Elgon ecosystems, Eastern Uganda.

The aforementioned anthropogenic activities coupled with climate change-induced disasters like prolonged droughts, more frequent and intense flooding, landslides, and erratic precipitation are the leading threats to *BAT* species survival [5, 7]. The more severe and frequent flooding episodes within the species' natural habitat, – the Mount Elgon biosphere reserve, have not only damaged the micro-climate but also resulted in severe landslides that significantly destroy the species' protective vegetation [5, 6]. In summary, the aforementioned environmental challenges are responsible for the observed and continuing decline in the quality of native habitats for the *BAT* taxon, which is now categorized as critically endangered by the International Union for Conservation of Nature (IUCN): "Red List of Threatened Species" [4, 7].

1.2 Objectives of the action

The main project goal was to assess the potential of using integrated-conservation approaches in Rescuing, Regeneration, and Protection of critically-endangered plant species, using *Bothriocline auriculata* as a case study. In this project, the target Asteraceae florae species was the *Bothriocline auriculata* M. Taylor (BAT). The BAT species were regenerated into mass seedlings. The clean (disease-free) juvenile species seedlings were propagated back into native ecosystems. Sone of the species seedlings was simultaneously introduced into the new Mabira tropical rainforest, to ensure full recovery under a protected natural environment.

i. In light of the above, the specific objectives were to;

ii. Evaluate BAT species conservation status, spatial distribution, and richness;

iii. Identify and rank ecological threats to the survival of the BAT species;

iv. Rescue the BAT species from its native ecosystem and regenerate them into mass seedlings;

Propagate and assess the growth performance of the newly regenerated BAT species under the different ecosystems and conservation management approaches.

1.3 Justification of the interventions

The *BAT* plant species are ecologically important in providing soil cover against erosion and nutrient recycling and are also palatable fodder crops for livestock. The BAT plants are also among the herbal medicine used by the local population [8]. Under this project, the plant specimens from existing populations of the critically-endangered BAT species, which were on threatened with extension or were on the brink of extinction from their protective native ecosystem were surveyed and rescued from their harsh native ecosystems.

The rescued *BAT* species were regenerated and multiplied into mass seedlings, using germination and tissue culture techniques respectively. In order to support full species recovery and further in-situ conservation under the natural environment, the regenerated *BAT* seedlings were propagated and re-introduced back into their native ecosystems along Mount Elgon slopes. Similarly, replicate BAT species seedlings were introduced into a new forest ecosystem within Mabira tropical rainforest sites which have more favorable ecological conditions (e.g., soil fertility, moisture and weather).

The new sites are also protected from human activities and encroachment thereby providing suitable ground for mass regeneration and full recovery of the BAT species under natural environments.

Finally, the juvenile *BAT* species seedlings were shared with farmers as well as the national institutions mainly the National Environment Management Authority (NEMA), and the National Forestry Resources Research Institute (NaFORRI); as the key line stakeholders for inclusion in both the present and future conservation programs such as agroforestry, herbariums, gene banks, and re-afforestation programs.

2. Materials and methods

2.1 Study sites

In this project, field activities were conducted in 2 ecological sites; 1) Mount Elgon slopes, and 2) Mabira forest reserve in Eastern and central Uganda; respectively (**Figure 2**). The Mt. Elgon and Mabira forest areas are selected based on the availability of the surviving *BAT* species specimens, and relative climate suitability to regenerate the BAT species; respectively.

2.1.1 Description of site 1: Mount Elgon with montane

Site 1 (the species' native ecosystem): is located from mid to high slopes (2745–3150 meters) of Mount Elgon with montane forests and dense vegetation in Bududa districts, Eastern Uganda. The latitude and longitudinal spatial coverage of the study site stretch from 0°59'N, 34°17′ E to 1°04'N, 34°25′E (**Figure 2**). The area receives bimodal rainfall patterns with a mean annual rainfall of 1900 mm, where the long and short rainy seasons range from September to November, and March to April 2021–2022. respectively. The mean maximum and minimum daily temperatures are 23° and 15°C; respectively.

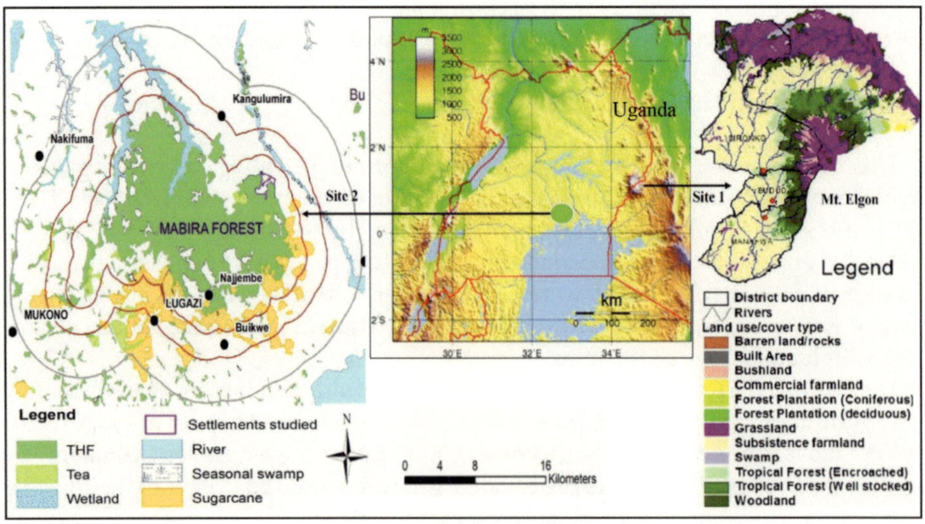

Figure 2.
Location of Mt. Elgon slopes (Site 1) and Mabira Forest (Site 2), Uganda.

Vegetation along the slopes of Mt. Elgon is associated with large mountain massifs and is altitudinally controlled zonal belts, which are classified into four zones namely; moorland - above 3500 m, bamboo and low canopy montane forest (from 2400 to 3000 m); and mixed montane forest -up to 2500 m elevation [9].

2.1.2 Description of site 2: tropical rainforest

Site Two (Tropical rainforest) is located in the Mabira forest reserve, the largest rainforest in Uganda covering over 30,000 hectares, in the Buikwe district of Central Uganda (**Figure 2**). The forest is located at latitude and longitudes of 0°23′54″N and 33°0′59″ E; respectively with several watersheds draining from Lake Victoria. The Mabila forest is elevated between 1000 and 1340 m A.S.L., and nearly 5% of the forest is made of four gently sloping hills: Dangala, Namusa, Ntunda, and Wakobe, which rise up to 1340 m [8]. The forest has an equatorial climate type with a bimodal rainfall pattern with two distinct short and long rainy seasons, from March to May, and September to November; respectively. The area receives a mean annual rainfall of 1300 mm, which is generally distributed throughout the year. The mean annual temperature within the Mabira forest ranges from 21 to 25°C, with minimum and maximum mean annual temperature ranges of 16–17°C, and 28–29°C; respectively [10].

Like the climate along Mt. Elgon slopes, the general climate of Mabira forest zones displays comparably small inter-seasonal variations in wind, rainfall, humidity, and temperatures throughout the year. This is because the forest is in close proximity to Lake Victoria and at an elevation of up to 1340 m, which moderates the area's micro-climatic conditions (**Figure 2**), including the warming effect despite being within the equatorial region [10].

2.2 Methodology and activities undertaken

2.2.1 Objective 1: Evaluating the current conservation status, spatial distribution, and richness of the Bothriocline auriculata species

The conservation status, distribution and species richness of the *Bothriocline auriculata* (BAT) species were studied in its native ecosystem on the slopes of Mt. Elgon in the Bududa district (Site 1). The study site, site 2 (**Figure 2**) was subdivided into three big zones, separated by elevation and vegetation differences. In each zone, 10 quadrants of 250x300ft size (covering 1.7acres) were randomly distributed across each zone, from which the BAT specimens were collected and studied.

The spatial distribution and species richness of the *Bothriocline auriculata* species in each zone were recorded through a detailed ecogeographical field survey [11], in which the BAT species' geographical range and ecological status were defined and recorded. During the surveys, spatial locations of the BAT species were geographically marked with a new Global Positioning System, GPS datum WGS84 using Garmin 12XL receiver. Study sites were mapped based on the species' spatial occurrence, the frequency of habitation, mature and young individuals, and population size using ArcGIS 9.2 software. These parameters were used to estimate BAT conservation status, vegetation structure, spatial distribution and species richness using the IUCN criteria [7].

Conversely, soil samples from each zone were collected using a soil auger, packed in plastic bags and transported to Makerere University for analysis of biophysical and chemical characteristics, so as to determine the conditions needed for the growth and regeneration of BAT. Similarly, both weather and climate data from each study site/zone were collected from the nearest meteorological station. The vegetation structure, soil and climatic data were used to spot the best habitat preferences for BAT species to enhance in-situ conservation within the study sites.

2.2.2 Objective 2: identification and ranking of different threats to BAT conservation and their causes

Different threats to BAT conservation in its natural ecosystems and their causes, Mt. Elgon areas (Site 1) were identified in each zone and classified under three major threat categories, namely human-use threat, livestock threat and habitat destruction threat.

a. Human-use threat: any sign of BAT collection for purposes of human socio-economic use: medicinal trade, medicinal plant trade, fuel, agriculture use (as fodder, mulch, compositing etc.) were recorded. Additional assessments of the conservation threats and their causes were made through oral interviews and informal meetings with community stakeholders namely farmers and herbalists, within and around the habitat zones.

b. Livestock/wild threat: vulnerability to animal destruction was measured by the grazing level in each zone (overgrazing), which were determined spatially by dung presence, frequency and density methods of Omar *et al.* [12].

c. Habitat destruction: the level of anthropogenic habitat destruction for arable farming, infrastructural development (roads, urbanization), deforestation for fuelwood, human settlement, and other land use/change activities, were assessed using high-resolution satellite imagery of LANDSAT and google earth. Likewise, natural causes of habitat destruction such as floods, prolonged dry spells, and climate change (extreme shifts in temperature and precipitation) were identified using LANDSAT imagery from 1990 to 2017, and data processing in GIS.

In each study zone, the identified threats and their causes were arranged and ranked as; very high, high, medium, low and very low when the threat in question is capable of destroying at least 80%, 70–80%, 50–69%, 30–49%, and below 30% of in-situ BAT species, respectively [13]. The threats are ranked as Very High: when a very large area of conservation target species and site is likely to be destroyed and High: when many locations of the target conservation site and the number is likely to be destroyed. Medium thread occurs when it's localized in its scope with a moderate degradation level and affects limited locations of the target conservation zone. Low and very low threats occur when the threat only impairs small and insignificant portions of the target conservation area and species; respectively [13].

2.2.3 Objective 3: undertake the BAT species regeneration, and mass seedling propagation and test the in-situ rescue performance

Mass regeneration of BAT species was done by rapid in-vitro micropropagation, ex-vitro plantlet growth, and in-situ re-establishment of the species in the wild. The BAT plant materials: vegetative portions and mature fruits, were collected from the Mt. Elgon region (Site 1) and transported to Makerere University Microbiology

Laboratory where the species were regenerated into mass seedlings using micropropagation. The plant part(s) were cut open or seeds extracted from pods for fruits and rinsed in distilled water three times. The parts/seeds were washed in 2%(v/V) Tween-20 detergent solution for 8 minutes, and later sterilized with 1%NaClO solution containing 2-drops of Tween-20 for 15 minutes. The plant part tissues/seeds were re-washed with sterile distilled water under aseptic conditions.

The extracted plant parts and seeds were regenerated into mass seedlings using tissue culture and in-vitro seed germination techniques; respectively [14], – where they were cultured on growth media and incubated under aseptic conditions. The growth media were prepared by mixing equal volumes of 3%Sucrose and 0.8%Agar into and diluting to 50% concentration while keeping the pH at 5.8. About 20 ml of the growth media will each be dispersed into sterilized 100 ml culture jars, where the plant tissues/seeds were inoculated. Non-absorbent cotton wrapped in cheesecloth was plugged into the culture jars, and the jars were autoclaved at 1.06kgcm^{-2} and 121°C for 15 minutes [14]. The inoculated tissues/seed samples were kept under aseptic conditions, at a temperature of 25 ± 2°C and 16/8 light–dark photoperiod supplied by white fluorescent lights under tissue culture rooms, until full tissue regeneration/seed germination. At least 20 seeds and 50 plant tissues were regenerated during 1st micropropagation cycle and were exponentially increased for five successive seedling generations, so that mass BAT seedlings were rescued during the five micropropagation phases.

The newly germinated BAT seedlings, after 3–4 weeks from inoculation in the Tissue Culture Laboratory, were removed from the laboratory to artificially induced shoot and root formation. The regenerated shoot tips from the seedlings were cut using a sterile blade, and cultured in Murashige and Skoog (MS) medium supplemented with varying concentrations of BA/KA: 0.5, 1.0, 1.5 and 2 mg/l for 5–6 weeks [15]. The best resulting micro-shoots were incubated on a half-strength MS medium containing varying concentrations of indole-3-butyric acid (IBA)/naphthalene-1-acetic acid (NAA): 0.5, 1.0, 1.5 and 2 mgl^{-1} to induce root formation and growth [15]. After the mass micropropagation process, regeneration success was assessed as the percentage ratio of fully developed seedlings to the total number of seedlings cultured.

Seedling hardening: the most vigorous BAT plantlets were transplanted into PE-plastic pots (dimensions: 10 × 15 × 7cm), filled with soil growth media artificially sterilized by autoclaving and mixed peat-moss, perlite and soil in ratios of 1:1:1(v/v). During hardening, the potted plantlets were irrigated with distilled water for about 10 days under culture room conditions before being moved to greenhouse and field weather conditions for further in-situ re-establishment.

2.2.4 Objective 4: propagating and assessing the growth performance of the regenerated BAT species under the different ecosystems and conservation management approaches

The growth performance of the regenerated and potted plantlets was assessed in terms of the quantity and quality of their morphological traits, re-established/grown under the completely randomized design (CRD) experiments. The three main in-situ conservation practices for BAT recovery management namely: active management, habitat restoration and habitat preservation. The conservation practices were classified based on their management intensity for the endangered species as high, medium, and low intensity for the active, habitat preservation and restoration management practices; respectively [4, 7].

2.2.4.1 Active management: Greenhouse production at Makerere University

About 100 BAT plantlets were transferred into a specially designed greenhouse, constructed at Makerere University, where they were monitored to full maturity. The plantlets, while still in pots, were regularly watered during the first 3–4 weeks (twice a day in the morning and evening), so as to get acclimatized to the harsh greenhouse weather conditions. Afterwards, the plantlets were transplanted from their conditioning pots, and re-planted about 5–10 cm in depth into previously fumigated soils, so as to grow under normal greenhouse soils and weather to maturity. Active management assumes that the species in question is critically endangered, and requires delicate management beyond its natural ecosystems to ensure rescue and full recovery [13].

2.2.4.2 Habitat preservation: BAT production in Mabira forest (Site 2)

At least a 1-acre area on one of the 4-gently sloping hills in Mabira forest, study site 2 (**Figure 1**) was randomly selected and isolated for BAT species growth trails. The plantlets were established in the preserved habitat of site 2, which are relatively identical soil and weather conditions of the BAT native ecosystem in site 1 using the same procedures in active management. The species under habitat preservation management were to grow under natural environment conditions: soil, temperature, humidity and rainfall to support species recovery. The habitat preservation management assumes that, if it is successful in supporting full species re-establishment, any other unoccupied ecological habitat with relatively similar environmental conditions will facilitate full species recovery by natural dispersal mechanism [11].

2.2.4.3 Habitat restoration: BAT production in Mt. Elgon slopes (Site 1)

About 1 acre of the experimental land in site 2 will also be isolated and BAT plantlets were planted in its native ecosystem using the same procedures described above in habitat preservation practices. The plantlets were monitored in their native environment until their full recovery and maturity. The habitat restoration management assumes that the endangered species is capable of full establishment once the native habitat is restored [11].

2.3 Data collection and analyses

Under each conservation management practice experiment, the 100 BAT plantlets were sowed at a standard spacing of 15 × 20cm, in a completely randomized design (CRD) experiment and were replicated three times [16]. Soil samples were collected and taken to Makerere University Soil Science Laboratory, for analysis of biophysical and chemical parameters such as pH, humus, electric conductivity, and nutrient (N, P, K and micronutrients), using the calorimetry method [17]. The daily ambient weather conditions, including the area temperature, rainfall and humidity were collected using digital thermometers, rain gauges and hydrometers; respectively.

Likewise, data on the morphological traits of the regenerated BAT plants such as leaf number and size, plant height, number of branches per plant and number of mature plants, as well as reproductive aspects of the mature plants (number of flowering stems, flowers, fruits and seeds per plant) were collected. The morphological and reproductive data were used for a detailed study of the species' recovery, regeneration and adaptation under the different in-situ conservation management practices and ecosystems.

The species' performance under the different in-situ conservation management practices was assessed by establishing the correlation relationships between the species' morphological and reproductive traits, and the management aspects or environment data: management practice, soil and weather parameters. Besides correlation relationships, Analysis of Variance (ANOVA) between the BAT species performance, morphological and reproductive traits under different conservation management against the soil and weather parameters were analyzed using Genstat statistical package [18]. The statistical significance for the data analyses was set at a 95% ($p = 0.05$) confidence interval. The most feasible management practice(s) for BAT species rescue, recovery and establishment was recommended not only for adaptive management and conservation of BAT species but also for conservation of other critically endangered plant species in Uganda.

2.4 Justification of the methodology and approach taken

Bothriocline auriculata (BAT) species is endemic in Uganda and is exclusively sited along Elgon slopes (Site 1), where field surveys were conducted. But because the species is near extinction, we envisage collecting a few species samples which were regenerated into mass seedlings to support full species recovery. The critically-endangered species have not been studied until now, and hence no scientific data: mainly on the species reproductive modes, germination and seed regeneration capacity, species plant physiology, growth performance and adaptive capacity to abiotic stress etc., exist to guide the species conservation actions. Faced with this uncertainty, natural germination and tissue culture methods were employed to regenerate the species samples into mass seedlings [15]. The tissue culture method was used as a backup against natural germination which takes an extended time and its success is not 100% guaranteed.

While on the contrarily, the tissue culture method is capable of exponentially multiplying a few plant tissues into mass seedlings under aseptic conditions in a short time and produces disease-free seedlings [15]. However, artificial crossings between the plantlets produced by tissue culture and a few produced by the germination process were made to address the challenge of genetically identical seedlings produced via tissue culture and simultaneously increase of genetic diversity of the species' offspring over successive generations during conservation.

The BAT taxon is expected to have varying adaptive capacities to abiotic stress and growth rates when subjected to varying weather and soil conditions during its in-situ conservation, as well as under different ecosystems and conservation management practices. Therefore, the BAT species' growth performance was monitored from sowing to maturity, under CRD field experiments [16]. The CRD trials were set up in the species' native habitat and Mabira tropical rainforest, as the 2 sites coincidently have nearly identical weather patterns, vegetation types, soil conditions, and hydrological regimes, – which are likely to favor BAT species' successful growth. For instance, both sites receive bimodal rainfall patterns with a mean annual rainfall of 1300 mm, which is evenly distributed within the two distinct long and short wet seasons from September to December, and March to May, respectively. The mean annual temperatures of both sites range from 21 to 25°C, with both the minimum and maximum mean annual temperature ranges of 16–17°C, and 28–29°C, respectively [19].

Data on BAT species morphological traits: leaf number/size, vigor, seedling rate, plant number/height, branches per plant as well as reproductive traits; flowering stems, flowers, fruits, and seeds per plant were recorded on a bi-weekly basis for growth performance analysis. Soil samples were collected and analyzed for pH, EC, and nutrient composition using the calorimetry method [16], and the weather variables: temperature, rainfall and humidity etc., were collected from site weather stations. The soil and weather parameters will depict local soil health and weather conditions that are suitable for the successful in-situ conservation of the BAT species in the study sites.

Using GenStat software [18], Spearman's correlation was employed to assess the relationship between species growth performance, soil parameters, and weather variables under different conservation management practices and the environment. Analysis of Variance (ANOVA) was employed to determine statistical significances ($p = 0.05$), and model relationships between different conservation practices (as the independent variables), and soil and weather parameters, – as the dependent variables. The ANOVA test will identify the most responsive conservation management practice(s) that support the BAT species rescue, recovery and regeneration in the natural environment.

3. Results and discussion

3.1 Description of the suitability of the ecosystems

Study site: although study site 1 is located along the slopes of Mount Elgon at higher altitudes from 2745 to 3150 m while study site 2 is within Mabira Forest is located at a significantly lower altitude from 1000 to 1340 meters Above Sea Level, both sites coincidently have nearly identical weather patterns, vegetation types, soil and hydrological regimes. Thus the environmental conditions of the 2 sites are most likely to favor the successful regeneration and conservation of the target species: *Bothriocline auriculata* (BAT) species, and hence their choice.

For instance, both sites receive bimodal rainfall patterns with a mean annual rainfall of 1900 mm and 1300 mm for sites 1 and 2 respectively, where the long and short rainy seasons range from September to December, and March to May, respectively [19]. The mean annual temperatures for the two sites also range from 21 to 25°C, with minimum and maximum mean annual temperature ranges of 16–17°C, and 28–29°C, respectively [19]. Like climate along Mt. Elgon slopes (Site 1), MWE [10] independently observed that the general climate of the Mabira forest area (site 2) displays comparably small inter-seasonal variations in wind, rainfall, humidity, and temperatures throughout the year. This is because the forest is in close proximity to Lake Victoria and at an elevation of up to 1340 m, which moderates the area's microclimatic conditions including the warming effect despite being within the equatorial region [10]. It is against this background, that the two sites were selected for the study and conservation of the BAT species.

3.2 Conservation status of the *Bothriocline auriculata* species in terms of the spatial distribution, species richness and abundance of the individual plants

Like many endangered species, BAT species is susceptible to many environmental threats within its natural ecosystems. Thus, the ecological threats to the species' survival within its native ecosystem (site 1), were identified and classified under three major categories namely human-use threat, livestock threat and habitat destruction threat, using the frequency and density method [12].

Conservation status, distribution and species richness of *Bothriocline auriculata* species were studied in its native ecosystem on the slopes of Mt. Elgon (Site 1) following an eco-geographical field survey [11, 20]. The survey method was successfully used by Choudhury and Khan [20] used the survey method to assess the population structure, species richness and conservation status of endemic and critically endangered plant species of *Aquilaria malaccensis, Gleditsia assamica*, and *Gymnocladus assamicus,* in India.

In this study, the type of host vegetation and flora at the landscape level are very important ecosystem components in the study of the conservation status of the inhabiting species, including species diversity, spatial distribution and ecological patterns in spatial variability. *Bothriocline auriculata* species host native ecosystems along the slopes of Mount Elgon differ considerably from the rest of the mountainous zones across Uganda due to the unique climatic conditions and physiography. The type of vegetation and flora diversity also vary with the elevation along the slopes of Mount Elgon. There are seven main classes of vegetation namely; mixed montane forest, bamboo, shrubs and thickets, low canopy montane forest, high canopy montane forest, and low, mid and high moorland ecosystems [3].

Table 1 presents the current *Bothriocline auriculata* (BAT) species conservation status in terms of relative abundance and spatial distribution of the individual plants within the host native ecosystems and vegetation types along the slopes of Mount Elgon, Uganda.

The number of *Bothriocline auriculata* plants was more abundant in the midland altitude and ecosystems having bamboo, shrubs and tickets followed by the high mixed montane forest vegetation (**Table 1**). The *Bothriocline auriculata* exhibited a very low species richness and limited ecological range along the slopes of Mount Elgon because the remaining plant populations were on the verge of extinction from their native ecosystems. The results are consistent with the findings by the International Union for Conservation of Nature [3, 7] who reported that the *Bothriocline auriculata* species are critically endangered and are on the verge of existing. The IUCN [7] also reported that *Bothriocline auriculata* species was also endemic

Zones ID	Vegetation type and ecosystems	Elevation (m)	Number of plants (n)
1	Mixed montane Forest	500–2000	43 ± 1^a
2	Bamboo, shrubs and thickets	2001–2500	52 ± 1^b
3	Low canopy montane forest	2501–3000	19 ± 1^c
4	High canopy montane forest	3001–3500	6 ± 1^d
5	Moorland – Low	3501–3800	1 ± 1^e
6	Moorland – Middle	3801–4000	0 ± 1^e
7	Moorland – High	Above 4000	0 ± 1^e

Values are arithmetic means with standard deviation (±SD) computed for values taken from every sampling area or zone along the line transect. Comparisons were made between the sampling zones, and n = number of plants collected per zone. Means in the same row bearing different superscript alphabetic letters are significantly different at a 5% (p = 0.05) confidence level.

Table 1.
Spatial distribution of the Bothriocline auriculata plants in the species' native ecosystems.

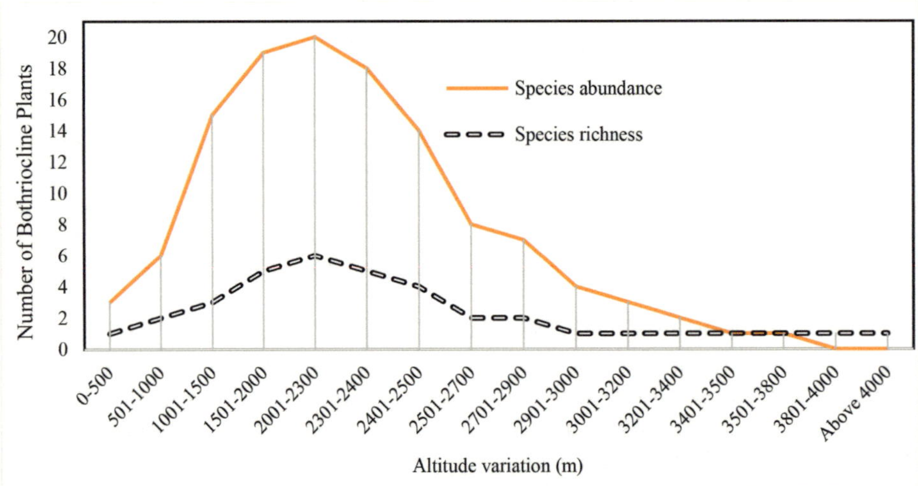

Figure 3.
Distribution of plant species in different altitudes of Mt. Elgon.

in Uganda whose ecological range was restricted to a few sites along the slopes of Mount Elgon between 2745 and 3150 meters Above Sea Level (A.S.L).

Figure 3 shows variations in the spatial distribution of *Bothriocline* plants with the altitude along the Mount Elgon slopes. The *Bothriocline* species richness and abundance increase with altitude along the gradient transect, and are the highest mid-altitude zones between 1500 and 2500 m above sea level. The mid-altitudes are composed of the mixed montane forest, bamboo, shrubs and thickets, as well as low canopy montane forest with a more favorable protective vegetative cover and microclimate as opposed to the highland altitudes (**Figure 3**).

On the other hand, the *Bothriocline* species richness and abundance begin to decline at higher altitudes from 2501 m and above. The vegetation type at higher altitudes is dominated by low and high canopy montane forest and moorland coupled with mild to cool temperatures; that is not favorable for the protection and proliferation of the *Bothriocline* species (**Figure 3**).

Overall, the richness of *Bothriocline auriculata* species is low with the narrow ecological range within the species' native ecosystems along the slopes of Mount Elgon (**Figure 3**). The species' spatial distribution and plant populations are restricted to mid-altitudes having mixed montane forest, bamboo, shrubs and thickets. Previous studies also confirm that the *Bothriocline auriculata* species thrive well under tropical ecosystems with dense vegetative cover, and the species' plant populations are abundant in the montane forest, mountainous areas and bamboo thickets [3, 7].

The low species richness and narrow ecological range suggest the high vulnerability of the *Bothriocline auriculata* species which could be primarily driven by environmental and human-induced threats to the species' survival, existing population and protective ecosystem. The possible environmental or natural factors could be soil biotic and abiotic factors (fertility, pH, microbiota, moisture etc.), extreme weather episodes like heat waves or high temperatures, erratic rainfall and humidity.

Similarly, the likely human-induced threats to the *Bothriocline* species' survival and population could include among others, anthropogenic climate change episodes such as dry spells, flooding coupled with mismanagement of the fragile ecosystems namely, over-harvest/grazing, arable farming and deforestation for timber and fuel wood; that often destroy the species protective ecosystems.

In this context, future studies should focus on profiling the environmental and human-induced threats that compromise the survival of the *Bothriocline auriculata* species, as well as species richness, spatial distribution and ecological range across its native protective ecosystems. If so, the data will inform future species conservation options and adaptive management practices within the same ecosystems or find better alternative host environments.

3.3 Rescue the critically-endangered *Bothriocline auriculata* species from its native ecosystem

Intensive eco-geographical surveys and scouting were performed along the slopes of Mount Elgon, the species' native ecosystem as well as on nearby farmlands, bushes and thickets to find, locate and rescue any remaining species specimens. Despite several repetitive surveys, 114 species specimens were found and positively identified within its native ecosystem along the slopes of Mount Elgon.

Regrettably, the species plants did not have seeds and seedlings in the wild so seed germination or artificial propagation of seedlings could be carried out for conservation purposes. Failure to produce seeds suggests that the *Bothriocline auriculata* species does not attain a complete growth cycle where the species enter flowering phases to produce seeds after attaining full maturity but rather remains in the vegetative phase. Besides being a survival mechanism in a fragile ecosystem, unfavorable weather mainly dry spells and humidity put additional stress on the plants, suppressing hormones for flowering and seed production thereby remaining in vegetative phases ([7] and IUCN, 2020).

Failure to enter into the flowering phase and complete the growth cycle could partially explain the low species richness and limited ecological, as well as the fast declining *Bothriocline auriculata* plant population within its endemic and native ecosystems; where the species is classified by the IUCN as critically endangered ([7] and 2020). In response, future studies should identify both human-induced and ecological factors that affect the growth performance of *Bothriocline auriculata,* and propel the species not to enter into flowering phases but rather remain in vegetative phases. Ecological biotic and abiotic factors limit the species' dispersal and geographical range.

Due to the small populations of the *Bothriocline auriculata* species and the absence of the seeds, only a few vegetative portions of the species plants were collected from the species. For this study, 15 vegetative portions (leaves and branches) of the specimens were collected, and the donor parent plants were left living in the wild. As such, the 15 specimen portions were rescued and regenerated into mass seedlings using the lab tissue culture and micropropagation protocols at Makerere University.

3.4 Perform regeneration of the rescued *Bothriocline auriculata* species specimens into many seedlings and propagation of the species seedlings for in-situ conservation

The extracted BAT plant parts and seeds were regenerated into mass seedlings using tissue culture and in-vitro seed germination techniques respectively [14]. But because the species is endemic and critically endangered, a few plants were got whose seedling and multiplication rates by the natural process of germination, – to raise the required number of seedlings for in-situ conservation, is not 100% guaranteed.

Thus, artificial micropropagation of the collected seedlings or plant tissues by tissue culture [15], were included as a backup process just in case the seeds generated by germination are not enough to raise the required number of seedlings over successive generations. But because the tissue culture method despite multiplying seedlings in mass numbers at a faster rate and producing disease-free plantlets, the method is challenged by the production of genetically identical seedlings [15]. To increase genetic diversity, artificial crossings between the tissue cultured plantlets with those produced from the natural germination process were made. Afterwards, the progeny/F1 generation of the crossed seedlings was back-crossed with germinated parents [21], to increase genetic diversity and encourage additional crossovers during recovery and regeneration in the natural environment.

The *Bothriocline auriculata* species specimens were collected and rescued from their fragile ecosystems along the slopes of Mount Elgon for multiplication into many seedlings for propagation and conservation. The collected specimens were put into specialized aerated specimen sampling bags. The bags containing the species specimens were in a cool box and taken to the Makerere University Tissue Culture laboratory for processing. In the lab, the specimens were cleaned with tap water, disinfected with ethanol and cut into replicate 25 smaller vegetative portions of length 2–3 cm. The specimens were regenerated and multiplied into many seedlings using plant tissue culture and micropropagation protocols. The vegetative species specimens were successfully regenerated and multiplied into at least 150 juvenile seedlings under aseptic conditions, artificial lighting and growth hormones. The seedlings produced are identical to the parent donor *Bothriocline auriculata* plants. **Figure 4** shows some of the pictures for species tissue cultured *Bothriocline auriculata* seedlings.

Figure 4.
Photos of the Bothriocline auriculata *seedlings tissue cultured in situ.*

Over three tissue cultures, seedling propagation cycles were performed during the generation of BAT specimens. A total of 150 seedlings were propagated and regenerated during tissue culture propagation cycles.

3.5 Assessing the growth performance of the *Bothriocline auriculata* species, and its response to abiotic stress under different conservation management schemes and environments

Availability of high-quality seedlings that are disease-free, fast-growing, biotic and abiotic stress-tolerant and are genetically identical to the parent species is one of the primary challenges facing the conservation biologists and forest nursery operators in Uganda. As such, the lack of a reliable supply of quality seedlings for the threatened and most critically endangered species such as *Bothriocline auriculata* species makes them get excluded from the local species restoration, conservation and adaptive management programs such as agroforestry, forest nurseries, parks, plant genetic resources, gene banks and forest ecosystems.

In response, quality seedlings of the critically-endangered *Bothriocline auriculata* species were produced in an aseptic (disease-free environment in the lab) using micro-propagation and tissue culture protocols at Makerere University (**Figure 4**). After tissue culture and micropropagation, the juvenile *Bothriocline auriculata* species seedlings were transplanted into PE-plastic pots for hardening. During hardening, the potted seedlings were irrigated with distilled water for about 5 weeks under culture room conditions in a screen house. The growth performance parameters were root collar diameter and shoot height.

The growth performance of the seedlings was studied in the screen house under artificial conditions namely, lighting, watering and growth media. **Table 2** shows a summary of the growth performance of the seedlings in response to the abiotic conditions of the growth media.

The growth performance parameters, namely shoot height, root collar diameter and number of leaves per seedlings increase with increasing duration as well as bulk density of the soil media and soil moisture level (**Table 2**).

The results suggest a normal seedling growth rate of the *Bothriocline auriculata* species under favorable growth media soil biophysical conditions; as represented by soil moisture and artificial lighting) in the screen house. Out planting success of plant

Time/ Treatment parameters	Root collar diameter (mm)	Shoot height (cm)	Bulk density of soil media (g × cm – 3)	Moisture content (%)	Mean number of leaves/plants
Week 1	2.12 ± 0.01	10.4 ± 0.3	0–5	0–5	1
Week 2	3.23 ± 0.02	21.3 ± 0.4	6–10	6–10	2
Week 3	5.75 ± 0.02	25.2 ± 0.2	11–15	11–15	3
Week 4	7.53 ± 0.03	43.8 ± 0.3	16–20	16–20	4
Week 5	6.85 ± 0.04	45.6 ± 0.2	21–25	21–25	4
SD	*0.48*	*15.12*	—	—	
CV	*0.07*	*0.52*	—	—	

Table 2.
Growth rate of the Bothriocline auriculata *species seedlings.*

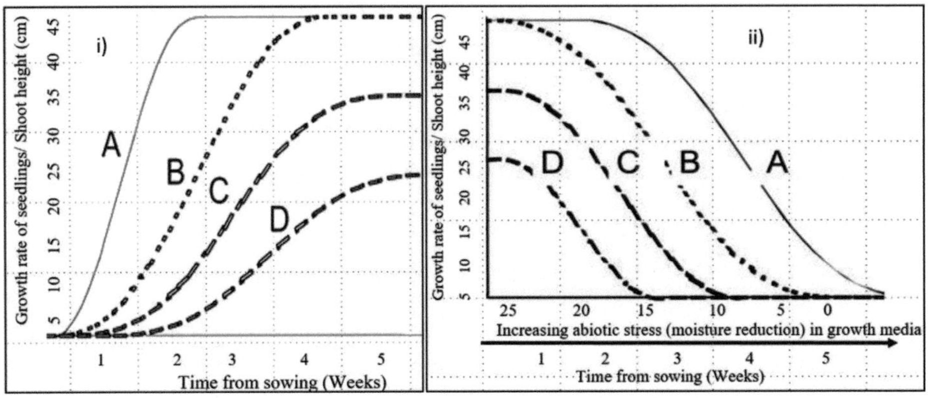

Figure 5.
Variations in the growth rate of the seedlings (shoot height) for the Bothriocline auriculata species seedlings under screen house conditions with: i) normal soil moisture and ii) moisture deficient.

species depends on several species' interaction between light, soil moisture other ecological conditions [13]. It has also been argued that the phenotypes of different plant species are characterized by morphological and physiological attributes, and are hence more important in predicting species-specific survival and growth during ex-situ conservation [15].

Conversely, the *Bothriocline auriculata* species seedlings were also subjected to stressful conditions to ascertain their growth response under soil moisture deficient conditions. The growth response of the species seedlings under both normal soil moisture and soil moisture deficient conditions were analyzed. The results describing the growth of the seedlings in terms of shoot length and root collar diameter are given, by a comparative approach (**Figure 5**).

Under optimum light and soil moisture conditions in a screen house, the growth rate (shoot height) of the *Bothriocline auriculata* species seedlings exponentially increases with time from the 1st week and attains a peak growth rate after the 5th week. The results suggest that the *Bothriocline auriculata* species complete the juvenile seedling growth cycle just after 1 month from sowing in appropriate soil media. On the contrary, the growth rate of the *Bothriocline auriculata* species seedlings declines rapidly with increasing deficiency of the soil moisture in the growth media. The data/results further confirm the high susceptibility of the *Bothriocline auriculata* species seedlings to abiotic stress conditions in the soil ecosystems. The results are consistent with the findings of IUCN [7] and [17]; who also reported positive correlational relationships between growth rate, soil moisture, temperature and other abiotic ecological conditions.

Nonetheless, the growth performance of the *Bothriocline auriculata* species was only assessed at the screen house level. Performance at the greenhouse and field level in the different ecosystems was NOT done due to a gap in funding. To this end, only 150 seedlings were regenerated enough for only replicate trials in the screen house. The data collected for the species growth performance was inconclusive because it did NOT include greenhouse conditions and under natural ambient weather and other environmental conditions in the fields. Therefore, future studies and efforts should focus on assessing the growth performance of the *Bothriocline auriculata* species seedlings beyond the artificial screen house conditions but rather under the natural ecological conditions; where the species will undergo ex-situ conservation in protected ecosystems and/or adaptive management in the agro-ecosystems.

4. Conclusions and recommendations

The *Bothriocline auriculata* species is endemic along the slopes of Mount Elgon The species has a limited ecological range despite its native ecosystem being at stake because of massive deforestation, arable farming and climate change. As a result, the *Bothriocline auriculata* species are among the critically endangered species and are currently threatened with extinction.

In this project, a few species specimens were rescued from their fragile native ecosystem and multiplied into massive juvenile quality (fast-growing and disease-free) seedlings that are genetically identical to the parent *Bothriocline auriculata* species. The species seedling exhibited a faster growth rate when subjected to artificial lighting and soil media under aseptic conditions in a screen house.

The most important lesson we learnt was that, in searching for any critically endangered plant species, the involvement of local farming communities and leadership is very important and should never be underestimated. Future efforts should be directed toward building capacity for a robust and sustainable supply of quality seedlings for the *Bothriocline auriculata* species in Uganda and beyond. If so, the seedlings should be supplied to local actors (farmers, conservationists and forestry bodies) for it to be streamlined, and included in the ongoing species conservation and restoration programs. Future programs should also be directed toward building the capacity of the local actors in conservation and adaptive management of the *Bothriocline auriculata* species to ensure full species recovery and restoration.

Author details

Semwanga Mohammed[1*], Nakiguli Fatumah[2,3], Nasejje Shadia[3], Kigozi Abas[4], Ashraf Nkumba[1] and Nakimwanyi Shamirah[2]

1 Agriculture Environment and Ecosystems (AGRENES), Kampala, Uganda

2 Agriculture Environment and Livelihood (AGRILIV), Kampala, Uganda

3 Makerere University, Kampala, Uganda

4 National Livestock Resources Research Institute (NaLIRRI), Kampala, Uganda

*Address all correspondence to: ssemwangaali@gmail.com

References

[1] Flann C. Global Compositae Checklist [WWW Document]. http//compositae. Landcare research.co.NZ/. 2009. Available from: https://en.wikipedia.org/wiki/Bothriocline. [Accessed: November 2, 2018]

[2] TICA. The Plant List: A Working List of All Plant Species [WWW Document]. actahort.org. 2012. Available from: http://www.theplantlist.org/. [Accessed: November 2, 18]

[3] Prinsloo S, Plumptre AJ, Kityo R, Behangana M. Nationally Threatened Species for Uganda - National Red Lists: National Red List for Uganda for the following Taxa: Mammals, Birds, Reptiles, Amphibians, Butterflies, Dragonflies and Vascular Plants, Wildlife Conservation Society. 2016

[4] IUCN. IUCN SSC East African Plants Red List Authority- 2013. Bothriocline Auriculata. 2013. DOI: 10.2305/IUCN.UK.2013-2.RLTS.T47347541A47347544.en

[5] Mugagga F, Kakembo V, Buyinza M. Land use changes on the slopes of Mount Elgon and the implications for the occurrence of landslides. Elsevier - Catena. 2012;**90**:39-46

[6] Knapen A, Kitutu M, Poesen J. Landslides in a densely populated county at the foot slopes of Mount Elgon (Uganda): Characteristics and causal factors. Elsevier-Geomorphology. 2006;**73**:149-165

[7] IUCN, 2018. The IUCN Red List of Threatened Species [WWW Document]. Available from: http://www.iucnredlist.org/details/47347541/0. [Accessed: November 2, 2018]

[8] Howard PC, Davenport TRB, Kigenyi FW, Viskanic P, Baltzer MC, Dickinson CJ, et al. Protected area planning in the tropics: uganda's national system of forest nature reserves. Conservation Biology. 2000;**14**:858-875. DOI: 10.1046/j.1523-1739.2000.99180.x

[9] Scott P. Assessment of natural resource use by communities from Mt. Elgon National Park. UNDP/Technical Report No. 15, 1994; 1994. DOI: 10.1103/PhysRevLett.13.585

[10] MWE. Mabira Central Forest Reserves - Ministry of Water and Environment, Ministry of Water and Environment (MWE). Kampala, Uganda; 2017

[11] Schemske DW, Husband BC, Ruckelshaus MH, Goodwillie C, Parker IM, Bishop JG. Evaluating approaches to the conservation of rare and endangered plants. Ecology. 1994;**75**:584-606. DOI: 10.2307/1941718

[12] Omar K, Khafagi O, Elkholy M. Eco-Geographical Analysis on Mountain Plants: A Case Study of Nepeta Septemcrenata in South Sinai, Egypt. L. LAMBERT Acad. Publ.; 2012

[13] Groom M, Meffe G, Carroll C. Principles of Conservation Biology. Sinauer Associates; 2006

[14] George E, Hall M. Culture, G.D.K.-P. propagation by tissue, 2008, U. In: Plant Tissue Culture Procedure-Background. Springer; 2008

[15] Trigiano R, Gray D. Plant Tissue Culture, Development, and Biotechnology. 2011

[16] Petersen R. 1994. Agricultural Field Experiments: Design and Analysis

[17] Rowell D. Soil Science: Methods & Applications. 2014

[18] Payne R. Genstat 5 Release 3 Reference Manual. Oxford University Press; 1993

[19] UBOS. Uganda Statistical Abstract 2017. Kampala: Uganda Bureau of Statistics; 2017 UNDP: Climate change Profiles. http://country-profiles.geog.ox.ac.uk

[20] Baharul C, Latif Khan M. Conservation and management of endangered Plant species: A Case Study of Northern India. Bioremediation, Biodiversity and Bioavailability. 2010;**4**(1)

[21] Ellstrand NC, Elam DR. Population genetic consequences of small population size: Implications for plant conservation. Annual Review of Ecology and Systematics. 1993;**24**(1):217-242

Chapter 5

Analysis of Anthropogenic Impediments to African Forest Ecosystems Conservation: Case of Gambari Forest Ecosystem, Ibadan, Nigeria

Tolulope Ayodeji Olatoye, Oluwayemi IbukunOluwa Olatoye, Sonwabo Perez Mazinyo, Gbadebo Abidemi Odularu and Akinwunmi Sunday Odeyemi

Abstract

Gambari Forest Reserve (GFR) is located in Oyo State, in the south-western region of Nigeria, in the Mamu locality (Gambari Forest), co-ordinate 3.7 and 3.9E" and latitude 7°26 1 N and longitude 3°5 1 E. i.e. 17 km South-East of Ibadan, along the Ibadan/Ijebu-Ode road. The major taxa studies for this research include the forest tree species forest ecosystem in Gambari Forest Reserve, such as: *Leucaena leucocephala, Leucaena glauca, Gliricidia sepium,Tectona grandis, Gmelina arborea, Swietenia macrophylla, Acacia spp., Albizia spp., Cassia siamea, and Pithecellobium saman*. 200 key respondents participated in this study, which were drawn from the seven main communities namely Ibusogboro, Oloowa, Daley North and south, Onipe, Mamu, Olubi and Onipanu respectively. The results revealed that there are significant anthropogenic interventions taking place in the study area. It is therefore imperative to conserve and safeguard GFR ecosystem resources, as ensuring that ecosystem services and biodiversity function at optimum levels. This study therefore recommends continued research to be undertaken, in addition to consistent monitoring and conserving our fragile forest resources, with the aim of achieving optimum functioning and service delivery.

Keywords: conservation, deforestation, ecology, forest ecosystem, forest degradation, environment, Gambari forest reserve (GFR), non-wood forest products (NWFPs)

1. Introduction

It is expedient to elucidate that on the average, forest ecosystems all over the world contain not less than an estimated 80% of the earth's biodiversity. According to Olatoye, [1], tropical forests are particularly rich in species. Furthermore, forests cover not less than 10% of the total terrestrial surface, and approximately 50% of the total area covered by the world's forests serve as habitat to considerably more than 60% of all terrestrial and freshwater biodiversity [2–4]. From the foregoing, Benayas, [5] opined that about 10 million people are employed in forest management and conservation all over the world. In the same vein, over 1.6 billion people- including more than 2000 indigenous cultures depend on forest Land Use Land Cover (LULC) as a means of livelihood [6], thereby providing a wide range of environmental services [3, 4], including biodiversity conservation, water supply, carbon sequestration, flood control [1], and protection against soil erosion and desertification [5].

According to Green, [7]; FAO, [8], forests offer wide- ranging services for the marketing of wood and non-wood forest products (NWFPs), examples include timber, fuelwood, nuts, fruits, as well as medicinal plants. As elucidated by Seto, [9], forest products accounted for more than US$300 billion, or 4% of the total value of international trade in commodities in 2004. Additionally, many people also place spiritual, religious and cultural values on forests as well as NTFPs, while others utilize them for leisure and recreational purposes [3, 4]. Despite the decline in the global forest area over centuries, the rate of deforestation has accelerated to alarming proportions. Gardner, [6] reports that the change in the African forest cover over the last three decades has largely been dependent on forest-clearing for agriculture/pastures, as well as for firewood collection. Further, population pressures in rural African communities were recognized as the main driving force accountable for the forest land- use and land- cover changes, while most of the deforested areas are unsuitable for long-term farming or grazing. It is on this premise that FAO, [8] stated that the very few of the remaining African tropical forests have little potential for sustainable agriculture.

One of the major enemies of forests is its use for domestic fuel consumption. Demand for fuel destroys near villages and towns in many countries [7]. Fuel wood gathered from the forest is the most important source of domestic energy in the rural areas of many developing countries. The collection and consumption of fuel wood are complexly linked to the management of environmental and natural resources. There exists a two-way relationship between fuel wood collection and deforestation. On one hand, the excessive demand for fuel wood causes forest degradation, thereby exacerbating fuelwood collection above its sustainable conservation. The degradation of forests, on the other hand, concomitantly results in fuel wood scarcity, thereby leading to global fuel wood crisis [2]. From the foregoing, there are a several consequences of forest degradation, such as the loss of biodiversity, watershed deterioration, the release of excessive atmospheric carbon dioxide (CO_2) as well as soil erosion.

The de-reservation of forest areas for community expansion purposes, commercial agriculture as well as infrastructural development have assumed a startling dimensions in many African rainforest regions from the end of the 20th century. According to Nahuelhual, [10] de-reservation, lack of articulate policies on forest conservation/management, the preponderance of illegal selective logging as well as unsustainable harvesting of NWFPs, inadequate funding and under-staffing of government forest management parastatals, excessive bureaucracy, lack of harmonized coordination as well as inadequate reliable inventory for planning and forest regeneration activities, are major challenges to the sustainability of African forest conservation. Furthermore, it should be noted that the extinction of flora and fauna species, as well as the conversion of fragile forest ecosystems to other land-uses have culminated in fragmentation, serial extinction and genetic erosion of rare and endemic timber species.

2. Statement of problem

Forests today are destructively exploited in Africa for timber, fuelwood, fibers, ornamentals and pharmaceutical products without genuine effort to replenish them [11]. From the foregoing, the destruction of essential forest resources such as vegetation cover, fertile soils wildlife, as well as environmental pollution emanating from oil spillage are alarmingly evident in several African countries, this is coupled with the high rate of population increase which is responsible for agricultural and industrial growth, resulting in the continuous expansion of land. This has gone a long way in the degrading of the environment causing different types of problems [1, 3, 4].

Forests exhibits a fragile ecosystem in Africa in particular, and all over the world in general, and in Ref. to the study area, this requires constant monitoring and update information for just in time interventions to promote sustainability [6]. This is because, forests not only sustain high biodiversity and productivity but captures sediments and nutrients, stabilize and provides food and livelihoods for the human populations that surround them. In Africa, loss of forest resources is caused by both anthropogenic and/or naturally occurring factors. It was observed that forest resources at the study area are threatened by numerous anthropogenic activities leading to degradation of the most heavily threatened natural systems. However, the present status in terms of extent, nature, at species richness and local people's perception are not known despite the increasing population growth and urbanization in the area. Further, Olatoye, [1] opined that forests are greatly threatened and are depleting fast in the study area due to human factors. These have been found to jeopardize these natural habitats and triggering climate change [9].

It is also imperative to state that forest loss has continually reduced the GFR ecosystems biodiversity, which the surrounding rural communities depend upon. Hence, it is urgent need to know the status of the resource to ensure the sustainability of the GFR ecosystem. Further, there are various environmental problems requiring urgent attention in the study area, and these include degradation of forests, expansion of rural settlements around GFR conservation areas, degradation and erosion of fertile soils on account of unsustainable forest LULC practices. Other environmental problems include the preponderance of invasive/alien flora species, as well as the illegal dumping of refuse in the study area [12].

3. Aim of the study

This study investigated the anthropogenic impediments to the conservation of Gambari Forest Ecosystem, Ibadan, Nigeria.

3.1 The study area

Gambari Forest Reserve (GFR) is located in Oyo State, in the south-western region of Nigeria, in the Mamu locality (Gambari Forest), Co-ordinate 3.7 and 3.9E" and latitude 7°26 1 N and longitude 3°5 1 E. i.e. 17 km South-East of Ibadan, along the Ibadan/Ijebu-Ode road. GFR falls within the low land semi- deciduous forest belt of Nigeria and covers a total land area of 17,984 hectares, between River Ona on the West and the main road from Ibadan to Ijebu-Ode on the East. The effective productive area of the plantation is therefore 306.9 acres, no allowance being made for subsidiary roads and compartment boundaries. The Forest Reserve is owned by Oyo/ Ogun State governments and forms part of what was the Mamu Government Forest Reserve. Formerly Ibadan District Council Forest Reserve, GFR is divided into five series namely Onigambari, Busogboro, Onipe, Olonde and Mamu. It was originally 12,535.6 ha of which 1036 ha was de-reserved by the Oyo State Government for the Cocoa Research Institute of Nigeria. Later in 1986, another 1000 hectares was given to Safa Splints Nigeria Limited for its industrial plantation. Presently, the working area in the reserve is 10,429.6 ha (**Figure 1**).

3.2 Justification for the study

This research seeks to bring to the awareness of government and stakeholders on the uncontrolled threat to the ecological system in the study area, and hence the need to asses, monitor and prevent the threatened and endangered endemic forest resources from extinction. Therefore, this study will provide information to guide

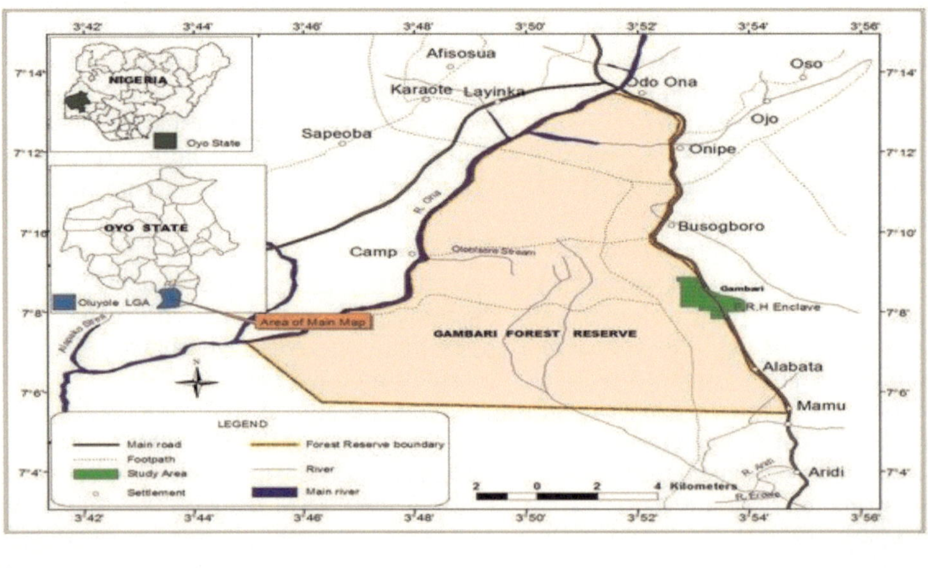

Figure 1.
Map of study area within Oyo state and Nigeria.

improvements to the management and restoration of forest vegetation in GFR to commercial activities and local economies [13]. This research also seeks to examine related causes and other basic driving forces on account of forest loss as well as proffer solutions towards optimal conservation and sustenance of biodiversity in the study area. Finally, this study calls for the need to provide vivid information on threats encountered at the study area, such as threats due to population pressure, climate change, nutrient loading, pollution and over-exploitation.

4. Materials and methods

The methodology adopted in the course of this research is the mixed method, i.e. qualitative and quantitative methods. In this study, a total of 200 key respondents participated in it, which were drawn from the seven main communities namely Ibusogboro, Oloowa, Daley North and south, Onipe, Mamu, Olubi and Onipanu respectively. These included government officials, civil servants (related to vegetation conservation), headsmen, local leaders, traditional healers, farmers, traders, artisans, grass root dwellers, fishermen, hunters, lumbers, community members and the general public residing in the study area. Out of the 200 copies of the questionnaire distributed, 194 were returned, giving a response rate of 97%. This offered rich information about the impacts of forest loss at the study area. Copies of the questionnaire were distributed covering areas listed in **Table 1** below.

4.1 Features of Forest loss changes in study area

This section analyses the respondents reaction to the features of forest loss in GFR,

4.1.1 Causes of Forest loss in GFR

This section analyzed respondents' opinion to the causes of forest loss in GFR. The analysis of the responses is tabulated in **Table 2**.

Table 2 depicts the total number of causes of forest loss which were received from the GFR respondents, chief of which is illegal logging (190/97.9%), and this assertion is consistent with existing literature by Mantyka-Pringle [14]. This is seconded by deforestation (169/87.1%). Also, climate change, which is caused by human activities in the study area, (e.g. through carbon emissions, deforestation, urbanization, population increase, etc.) accounts for 128 (65.9%).

4.1.2 The anthropogenic interventions in the study area

In response to the survey form, the researcher sought to know the anthropogenic factors that occur in GFR. The study reveals the viewpoints of the respondents, as presented in **Table 3**.

Table 3 above reveals the depicts the total number of anthropogenic interventions that take place in GFR, which reveals that most respondents (i.e. 188 or 96.9%) were of the opinion that illegal logging was a major anthropogenic action in the study area, and is evident in locations such as Mamu, Ibusogboro, and environs. This assertion is further buttressed by Adedeji [15], who stressed that over 5240 hectares of forest vegetation was lost to other land uses from year 1984 to 2014 (average of over 140 hectares/year loss). Additionally, land development and illegal woodcutting accounts

Variable	Frequency	Percentage
Sex		
Males	101	52.1
Females	93	47.9
Age		
21-30	39	20.1
31-40	51	26.3
41-50	47	24.2
51 and above	57	29.4
Marital status		
Single	76	39.2
Married	83	42.8
Divorced	12	6.2
Widow/Widower	23	11.9
Educational level		
No formal education	38	19.6
Primary education	29	14.9
Secondary education	73	37.6
Tertiary education	14	7.2
Household size		
≤ 3	11	5.7
4-6	65	33.5
7-9	54	27.8
10-12	46	23.7
Above 12	18	9.3
Income level		
≤$100	54	27.8
$101-$300	61	31.4
$301-$500	33	17
$501-$700	28	14.4
Above $700	18	9.3

Table 1.
Gender respondents characterization.

for 162 (57.9%) and 188 (96.9%) responses respectively, while conversion of ecosystem for agriculture accounts for 167 (86.1%). Also, judging from the responses, it is evident that illegal waste disposal at vegetation land cover areas is still an environmental challenge in the study area, and this poses negative environmental effects, and this include inhibition in the population and activities of soil microbes as well as enzyme activities required for soil fertility, [16], reduction in the percentage of soil organic matter, decreased soil basal respiration [17], etc.

(a) Deforestation 169 (87.1%)	(d) Government Policy 106 (54.6%)
(b) Illegal Logging 190 (97.9%)	(e) Climate change 128 (65.9%)
(c) Crop Cultivation 120 (61.9%)	
Total	254 (100%)

Table 2.
Causes of Forest loss in the study area.

Illegal woodcutting/selective logging 188 (96.9%)	Informal settlements 66 (34.02%)
Conversion of forestland for crop cultivation 167 (86.1%)	Bush fires 74 (38.1%)
Land development 162 (57.9%)	Illegal waste disposal 71 (36.6%)
Overgrazing 104 (53.6%)	Illegal sand mining 128 (66%)
TOTAL	**194 (100%)**

Table 3.
Anthropogenic interventions in GFR.

4.1.3 The reasons for anthropogenic influences in GFR

The analysis of question 17 of the questionnaire is centred on the reasons for the anthropogenic influences of the inhabitants in GFR, which centred on poverty, weak implementation of government policy and ignorance. The result is presented as pie chart in **Figure 2** below.

Figure 2 illustrates the reasons for anthropogenic factors by the respondents. The highest frequency related to the aforementioned is poverty, which is 117 (60.3%). Poverty is still a major challenge in the study area, and this viewpoint is in conformity with Adeoye [18], who opined that about over 33% of surveyed respondents in the study area earn below N25, 000 monthly. Also, 41 (21.1%) respondents underpinned the weak implementation of government policies, while 36 (18.6%) claimed it could be due to ignorance.

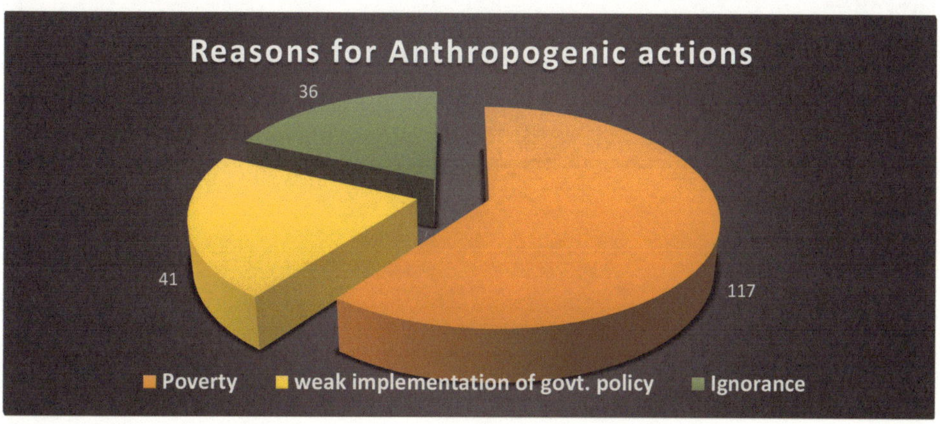

Figure 2.
Pie chart depicting reasons for anthropogenic actions in the study area.

4.1.4 The perception of respondents to forest reserves

This section epitomizes the perception of Gambari respondents to forest reserves. This is purposed to decipher the respondents' level of understanding, care as well as the determination of how significant they uphold conservation of FRs. From the foregoing, the results are pictorially presented as pie chart in **Figure 3** below.

At this juncture, it is expedient to state that the importance of a phenomenon is a function of the value placed on it. In laying credence to this assertion, **Figure 3** above diagrammatically illustrates the perception of respondents regarding their impression of forest reserves (FRs) in the study area. The result reveals that 156 (80.4%) respondents are of the opinion that FRs are not urban spaces for development. This group of respondents are those that place great value on FRs on account of its numerous ecological benefits to man, wildlife and the environment. On the contrary, 38 (19.6%) respondents are of the opinion that FRs are wastelands or virgin territories awaiting clearing/vegetation removal for other land uses such as infrastructural development, agriculture and other land uses. This situation is consistent with existing literature such as Akinluyi [19]; Raheem [20] and Adekola [21]. In addition, the respondents also responded to question 23 on whether it is necessary to protect vegetation resources, the results states that 225 (90.4%) responded in the affirmative, while 23 (9.2%) respondents stated otherwise. From the foregoing, it is expedient to orientate community dwellers, and more especially, decision makers on the need to consider the conservation of ecosystems as sacrosanct.

4.1.5 Proposed strategies towards the conservation of timber species in GFR

This study also sought to proffer solutions to illegal logging as well as the degradation in the study area and control strategies were proposed in this regard. The result is tabulated and graphically illustrated in **Figure 4** below.

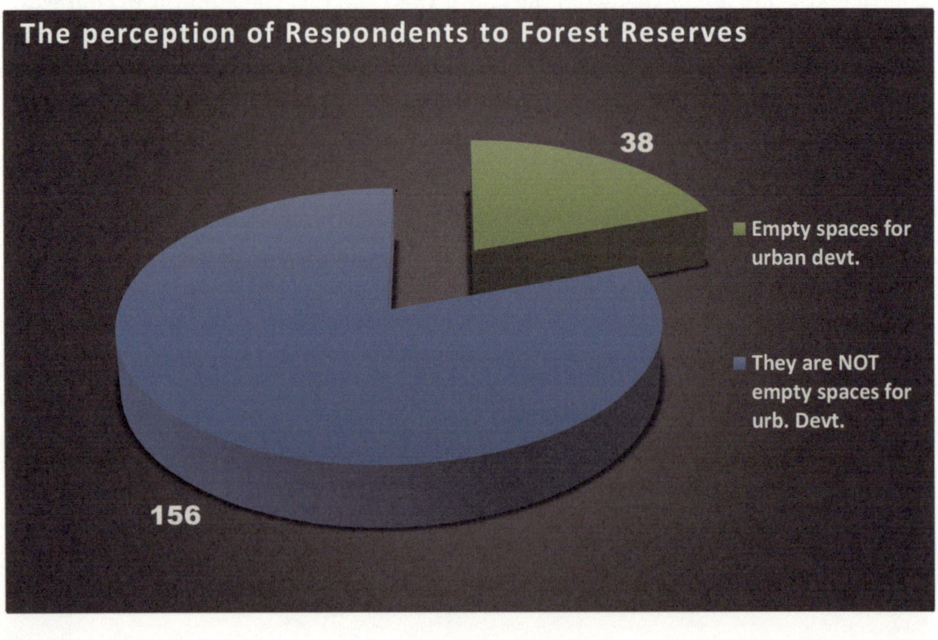

Figure 3.
Pie chart depicting the perception of respondents to the conservation of Gambari Forest reserve.

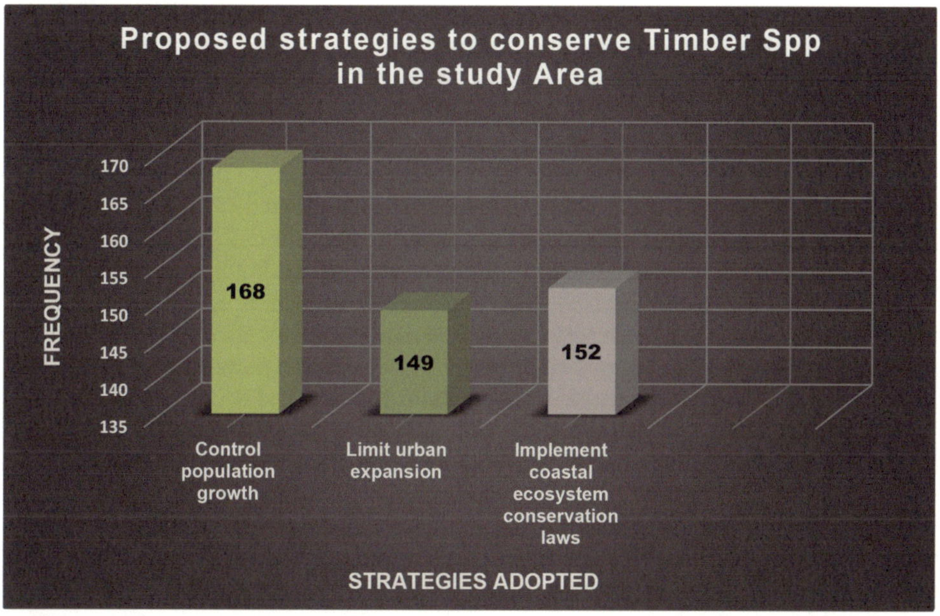

Figure 4.
Bar chart depicting proposed strategies towards the conservation of timber species in Gambari Forest reserve.

Figure 4 presents a graphic illustration of the suggested strategies proposed by the respondents for the conservation of timber species in GFR. The greatest frequency of respondents (i.e. 168 or 86.5%) supported population growth control. This assertion is consistent with existing literature such as Potatov, [22]; Olatoye, [3, 4]; Olatoye, [1], who stated that there is a high correlation between population growth and ecosystem/biodiversity loss. They further opined that controlling human population growth levels will gradually reduce stress on ecosystems. Further, 149 (58.9%) respondents were of the opinion that limits should be placed on urban expansion, for the purpose of ensuring sustainability of ecosystem resources, and this confirms the position of ecosystem scholars such Sadorsky [23]; Deng [24] On a global scale, Sadorsky [23]; posited that global urban expansion will continue unabated, even as the world's human urban population is estimated to rise by more than 3 billion people by year 2050, and this will be due to high urban fertility rates and re-categorization of rural territories into urban centers [23, 25].

5. Conclusion

The impact of anthropogenic actions on forest ecosystems is alarming, as it ultimately results in exploitation of virgin territories for the purpose of meeting higher demands on wood utilization, agriculture, housing, transportation and other basic human infrastructure to cater for the teeming population. In the same vein, population has an adverse effect on the preservation of biological diversity, and the consequences of population increase is decline in ecosystem service delivery as well as flora and fauna habitat loss. For example, over 5240 hectares of forest vegetation was lost to other land uses from year 1984 to 2014 (average of over 140 hectares/year loss) in the study area due to poverty, unemployment, government negligence and inadequate security at the reserve. Another observation of important significance is the attitude of fuel wood collectors to issues of environmental degradation. Despite government policies and awareness on the need for forest conservation, protection and sustainable forest management practices, some of them still regard fuel wood as a free gift of nature. This in consequence makes it difficult for them to believe in the adverse effects of their action. The low literacy level in the study area and environs and the general belief in the traditional practice of tree felling and bush clearing have also contributed to environmental degradation in the study area. While healthy ecosystems are highly resilient to impacts of environmental change on account of their capability to sustain ecosystem services on one hand, ecosystems are more vulnerable due to their poor state and they cannot provide the required delivery of service. It is on this premise, therefore that this study calls for prompt and proactive attention as regards conservation of the highly vulnerable GFR ecosystem as population increase is fast encroaching on the study area, a situation which continues unabated. Further, rapid growth in population results in multifaceted landscape changes, which further culminates in the altering of forest ecosystems structure and functioning, no wonder they have been imperiled by severe stresses from human interventions due to the uncontrolled and largely mismanaged exploitations of forest resources which has culminated in resource loss, degradation, degeneration, reduced productivity levels, and socio-economic opportunity costs. From the foregoing, the environmental pollutants that are released due to high levels of deforestation, energy consumption and environmental degradation in the study area pose negative implications on human health and well-being. Also, it is clearly stated in the referenced literature that deforestation in the study area significantly alters soil microbial and hydrological activities, and consequentially raising budgets devoted to land surface radiation. It is therefore imperative at this juncture to advocate for the control of deforestation in its totality, and in addition, it is germane for town planners to ensure that sustainable urban development policies are implemented in state laws. Finally, it is important to state that the future of our societies will depict our decisions, values and interests, and ultimately, our actions on the ecosystem, and these will consequentially determine the fate of the human, flora and fauna species. Therefore, the exploitation of forest resources have to embrace the sustainability of ecological systems in scientific practice and as its ultimate goal.

6. Recommendations

In conclusion, the African forestry sector can be developed in the following ways:

- Firstly, there is need for government to restore public awareness campaign in the area of deforestation and environmental resource management.

- Increasing allocation to fund forestry research.

- There is need for sustainable exploitation and management of forest resources. More forests need to be established through afforestation and reforestation, while the existing ones should be protected. If there is need for exploitation, it should be done in a sustainable manner.

- Recycling and reduction of wood wastes and establishment of gene banks to prevent forest species extinction, introduction of domestication programmes and workable legislation in the forestry sector.
- Enforcement of a properly conducted Environmental Impact Assessment (EIA) as a pre-condition for approving projects in various sectors of the African economy.
- Updating of forest legislation among member countries.

Focusing on the unique products and services required locally and globally and strengthening local institutions can be important ways of addressing forest resource depletion in the African continent. Such efforts should build on successful experience with locally based sustainable resource management integrating agriculture, animal husbandry and forestry, taking advantage of local/indigenous knowledge. The growing demand for environmental services- especially biodiversity and carbon sequestration provides particular opportunities in Africa.

On account of population pressure around forest land uses in Africa, it is therefore imperative to conserve and safeguard forest resources across the world, as well as ensuring that ecosystem services and biodiversity function at optimum levels. It is on this premise that this research therefore makes a clarion call for continued research to be undertaken, in addition to consistent monitoring and conserving our fragile forest resources, with the aim of achieving optimum functioning and service delivery.

A. Appendix- Questionnaire

A.1 Analysis of anthropogenic impediments to the conservation of gambari forest ecosystem, Ibadan, Nigeria

A.1.1 Section A: demographic characteristics

1. Gender.

Male	Female

2. **Age** ... ...
3. **What is your highest level of education?**

(a) No education	(f) Matric
(b) Primary	(g) University degree
(c) Middle school	(h) Undergraduate Student
	(i) National Diploma

A.1.2 Section B: conservation of forest resources

4. Do you know about the conservation of forest resources and the environment?

(a) Yes	(b) No

5. Do you care about the conservation of forest resources and the environment?

(a) Yes	(b) No

6. Which benefits do you derive forest resources in the study area? (Tick as many choices).

(a) Food	d) Economic
(b) Raw Material	(e) Others (Please specify)
(c) Medicinal purposes	

7. In your perception, which of these have changed regarding the conservation of forest resources in your area?

the overall quality	species abundance	Ecosystem diversity	Do not Know

8. With regards to your answer in number 8, what is the degree of change? It has.

Increased	No change	Decreased	Do not know

9. On a scale of 1 to 10, how do you rate the management and conservation of the coastal vegetation resources in your location? (1 = extremely poor, 10 = Excellent), please provide your score in the box below:

A.1.3 Section C: challenges encountered in the conservation of gambari forest ecosystem

11. Are there environmental challenges in your location?

a. Yes	(b) No

12. What are the causes of forest loss in your area? Tick as many as possible.

a. Deforestation	(e) Government Policy

(b) Urban expansion	(f) Climate Change
(c) Crop Cultivation	
(d) Others, specify	

13. On a scale of 1 to 10, How has the environmental challenges in your location affected you? (1 = Extremely Adverse Impact; 10 = Excellently Positive Impact), please provide your score in the box below:

14. Select the anthropogenic activities that take place in your location:

(a)Illegal woodcutting/selective logging	(f)Informal settlements
(b) Conversion of ecosystem land use for crop cultivation	(g) Bush fires
(c) Land development	(h) Illegal waste disposal
(d) Overgrazing	(i) Others (specify)
(e) Illegal sand mining	

15. In your own perception, what are the reasons for these anthropogenic influences? (Tick as many as applicable).

(a) agriculture	(d) Weak implementation of conservation policies
(b) urban expansion	(e) Ignorance
c) Poverty	(f) Others (specify)

16. As an individual, have you been able to manage or adapt to the environmental problems in your location?

a. Yes	(b) No

A.1.4 Section D: ecosystem goods and services

17. What are the major ecosystem services provided by the study area? (Tick as many as possible)

a. **Provisioning:**

Timber	**Fuelwood**	**Pines**	**Genetic Resources**
Medicinal/Cosmetic Plants	Livestock	Fiber Crops	Tree Plantations
Food			

b. **Cultural**

Recreational	**Tourism/Ecotourism**	**Landscape beauty**	**Education**
Scientific research	Traditional knowledge	Cultural heritage	Religious

c. **Regulating**

Erosion control	**Hydrological regulation**	**Climate regulation**	**Soil purification**
Water purification	Waste treatment	Flood buffering	Pest prevention
Air Quality	Habitat maintenance	Carbon sequestration	Environmental stabilization
Nursery			

18. Which ecosystem goods and services do you know that are no longer available, or diminished in this area and state why?

A.1.5 Section E: urban expansion and conservation of gambari forest ecosystem

19. In your opinion, what are the reasons for urban expansion: Tick as many as applicable?

a. Population growth	c. Need for development
b. Agriculture (Food security)	d. Others

20. What are the impacts of urban expansion? (Tick as many as appropriate).

Pollution	**Environmental degradation**
Ecosystem biodiversity loss	Infrastructural development
	Others (Specify)

21. Do you consider Gambari Forest Reserve as empty space for urban development?

(a) Yes	(b) No

22. On a scale of 1 to 10, what impact does urban expansion have on the conservation of coastal vegetation? (1 = adversely poor impact; 10 = excellently positive impact), please provide your score in the box below:

23. Do you think that forest resources should be conserved in the study area the so that they do not cause long term depletion or affect the diversity of the ecosystem?

(a) Yes	(b) No

24. In your opinion, what strategies have been adopted to contain urban expansion and conserve forest resources in the study area? (Tick as many as appropriate).

(a) Control population growth	(c) Implement forest conservation laws
(b) Limit urban expansion	

Thank you.

Author details

Tolulope Ayodeji Olatoye[1*], Oluwayemi IbukunOluwa Olatoye[2], Sonwabo Perez Mazinyo[1], Gbadebo Abidemi Odularu[3] and Akinwunmi Sunday Odeyemi[1]

1 Faculty of Science and Agriculture, Department of Geography, University of Fort Hare, Alice, South Africa

2 Library and Information Science, Walter Sisulu University, Mthatha, South Africa

3 American Heritage University of Southern California, San Bernadino Ontario, CA, United States of America

*Address all correspondence to: olatoyetolu@gmail.com

References

[1] Olatoye TA. Impact of Urban Expansion on the Conservation of Coastal Vegetation in Buffalo City Metropolitan Municipality; a Doctoral Thesis Submitted to the Department of Geography and Environmental Science, Faculty of Science and Agriculture. Alice Campus, South Africa: University of fort Hare; 2020. 330pp

[2] Hosonuma N, Herold M, De Sy V, De Fries RS, Brockhaus M, Verchot L, et al. An assessment of deforestation and forest degradation drivers in developing countries. Environmental Research Letters. 2012;**7**(4):044009

[3] Olatoye TA, Kalumba AM & Mazinyo SP 2019a. Exploratory review of urban expansion, coastal vegetation environments (CVEs) and sustainability paradox. Being a paper published in the Journal of Human Ecology, India. REF NO: JHE-041-19/3152. **67**(1-3): 21-30

[4] Olatoye TA, Mazinyo SP, Odeyemi AS. Forest Systems Services Provisioning In Africa: Case of Gambari Forest Reserve, Ibadan, Nigeria. Hindawi Limited, Adam House, Third Floor, 1 Fitzroy Square, London W1T 5HF, United Kingdom: Published in the International Journal of Forestry Research; 2019b

[5] Benayas JMR, Bullock JM. Restoration of biodiversity and ecosystem services on agricultural land. Ecosystems. 2012; **15**(6):883-899

[6] Gardner T. Monitoring Forest Biodiversity: Improving Conservation through Ecologically-Responsible Management. Routledge; 2010

[7] Green JM, Larrosa C, Burgess ND, Balmford A, Johnston A, Mbilinyi BP, et al. Deforestation in an African biodiversity hotspot: Extent, variation and the effectiveness of protected areas. Biological Conservation. 2013;**164**:62-72

[8] Food and Agriculture Organization of the United Nations. 2017. USA: The Future of Food and Agriculture: Trends and Challenges

[9] Seto KC, Güneralp B, Hutyra LR. Global forecasts of urban expansion to 2030 and direct impacts on biodiversity and carbon pools. Proceedings of the National Academy of Sciences. 2012; **109**(40):16083-16088

[10] Nahuelhual L, Carmona A, Lara A, Echeverría C, González ME. Land-cover change to forest plantations: Proximate causes and implications for the landscape in south-Central Chile. Landscape and Urban Planning. 2012;**107**(1):12-20

[11] Bradshaw CJ. Little left to lose: Deforestation and forest degradation in Australia since European colonization. Journal of Plant Ecology. 2012;**5**(1):109-120

[12] Olatoye TA. Analysis of land cover change and fuelwood production in gambari forest reserve, Ibadan, Nigeria: A Masters dissertation submitted to the Department of Geography, Faculty of the Social Sciences, University of Ibadan: Nigeria; 2013. 76pp

[13] Mcleod E, Chmura GL, Bouillon S, Salm R, Björk M, Duarte CM, et al. A blueprint for blue carbon: Toward an improved understanding of the role of vegetated coastal habitats in sequestering CO_2. Frontiers in Ecology and the Environment. 2011;**9**(10):552-560

[14] Mantyka-Pringle CS, Visconti P, Di Marco M, Martin TG, Rondinini C, Rhodes JR. Climate change modifies risk

of global biodiversity loss due to land-cover change. Biological Conservation. 2015;**187**:103-111

[15] Adedeji OH, Tope-Ajayi OO, Abegunde OL. Assessing and predicting changes in the status of Gambari forest reserve, Nigeria using remote sensing and GIS techniques. Journal of Geographic information system. 2015; 7(03):301

[16] Thavamani P, Malik S, Beer M, Megharaj M, Naidu R. Microbial activity and diversity in long-term mixed contaminated soils with respect to polyaromatic hydrocarbons and heavy metals. Journal of Environmental Management. 2012;**99**:10-17

[17] Wangai PW, Burkhard B, Müller F. A review of studies on ecosystem services in Africa. International Journal of Sustainable Built Environment. 2016;**5** (2):225-245

[18] Gbiri IA, Adeoye NO. Analysis of pattern and extent of deforestation in Akure Forest Reserve, Ondo State, Nigeria. Journal of Environmental Geography. Nigeria: Academic Publishers; 2019;**12**(1-2):1-11

[19] Akinluyi ML, Adedokun A. Urbanization, Environment and Homelessness in the Developing world: The Sustainable Housing Development. Mediterranean Journal of Social Sciences. 2014;**5**(2):261

[20] Raheem ID, Tiwari AK, Balsalobre-Lorente D. The role of ICT and financial development in CO_2 emissions and economic growth. Environmental Science and Pollution Research. 2020; **27**(2):1912-1922

[21] Adekola J, Fischbacher-Smith M, Fischbacher-Smith D, Adekola O. Health risks from environmental degradation in the Niger Delta, Nigeria. Environment and Planning C: Politics and Space. 2017; **35**(2):334-354

[22] Potapov PV, Turubanova SA, Hansen MC, Adusei B, Broich M, Altstatt A, et al. Quantifying forest cover loss in Democratic Republic of the Congo, 2000–2010, with Landsat ETM+ data. Remote Sensing of Environment. 2012; **122**:106-116

[23] Sadorsky P. The effect of urbanization on CO_2 emissions in emerging economies. Energy Economics. 2014;**41**:147-153

[24] Deng X, Shi Q, Zhang Q, Shi C, Yin F. Impacts of land use and land cover changes on surface energy and water balance in the Heihe River Basin of China, 2000–2010. Physics and Chemistry of the Earth, China: Parts A/B/C. 2015;**79**:2-10

[25] Jiang L, Deng X, Seto KC. The impact of urban expansion on agricultural land use intensity in China. Land Use Policy. 2013;**35**:33-39

Chapter 6

Contextualizing the Factors Affecting Species Diversity and Composition in the African Savanna

Kondwani Kapinga, Jules Christian Zekeng, Lackson Chama, Nalukui Matakala, Stanford Siachoono, Obote Shakacite, Concilia Monde and Stephen Syampungani

Abstract

Recently, sustainable forest management has been the top priority for many international forest conservation organizations, governing authorities, and interest groups. Forest conversion to farmland for fuel wood removal, charcoal production, and woodland grazing is the principal mechanism of forest degradation, habitat change, and loss of biodiversity. Despite the increasing acknowledgment of conservation values of savanna, our understanding of the factors affecting species diversity and composition for the African savanna remains limited. This chapter provides a systematic review of the factors affecting species diversity and composition in an African savanna. However, in order to reduce this inadequacy, a careful examination of the existing literature was conducted. After a thorough review, it was revealed that species diversity and composition in savanna are significantly shaped by grazing, fire, and resource availability, that is, rainfall and soil nutrients, as well as anthropogenic activities. Understanding the diversity and composition of tree species is vital since they provide resources and habitats for several other species. Botanical assessments, such as floristic composition, species diversity, and structural analysis studies, are significant for providing accurate information on species richness, which is valuable for sustainable forest management and helps to understand forest ecology and ecosystem functions.

Keywords: floristic composition, drivers, species richness, species diversity, savanna ecosystem

1. Introduction

Africa houses the largest area of savanna, covering about 65% of the continent [1]. However, the natural savanna ecosystems in Africa are undergoing a serious transformation, which is due to the conversion of land parcels into wide-ranging livestock production foraging arenas, slash and burn for the production of crops, and creation of municipal cities [2–4]. Despite these changes having enormous impacts on the capability of the ecological system to exchange and store carbon [4, 5] as well as on species diversity and composition, the factors affecting the species diversity and composition are still poorly understood. Furthermore, poor grazing management, linked with livestock overstocking, results in extreme treading, which has a detrimental impact on the vegetation [6–8]. This also leads to an increase in the bulk density of the soil and reduces its water permeation [8, 9]. Hence, it results in alterations in the composition of the plant species, reducing the rates of photosynthetic activities in the leaf and reducing the storage of the carbon stock in the ecosystem [7, 10].

On the contrary, the aboveground biomass may be stimulated by light grazing via increased cultivation of grasses [8, 11, 12]. Indeed, increased germination of young shoots from the cultivators leads to a better and improved generation of fresh energetically photosynthesizing green vegetation. This ultimately, when combined with reduced accrual of dead biomass, leads to an improvement in the penetration of light into the canopy and intensifies the uptake of carbon dioxide and storage of carbon by the grasslands [10, 12].

Moreover, degradation of habitats, excessive utilization, harmful non-native species, contamination, and alteration in the climate have affected and still impact the biosphere's ecological systems [13, 14]. Sixty percent (60%) of the globe's ecosystems are estimated by Refs. [15, 16] to be utilized in an unsustainable way. Seventy-five percent (75%) of the stocks of fish resources are being used excessively, which may lead to their depletion. Moreover, about 13 million hectares of tropical forest ecosystems are cleared annually [15, 16], significantly affecting the biotic diversity and composition, especially in African savannas.

The rate at which the loss of biological diversity proceeds if continues, we may face a mass extinction [13]. The decline in biological diversity represents a planet's irreversible loss. It poses a serious menace to the life support system of humanity, and some authors describe it as the amenities that are delivered by natural systems representing the entirety from the food we consume to the air we breathe (e.g., [17–19]). This chapter, therefore, contextualizes the factors affecting species diversity and composition in an African savanna. Understanding these factors will provide policymakers with a strong basis for planning and managing the forest ecosystems in an African savanna.

1.1 Concepts of species diversity and composition and ecosystem functioning

The term "biodiversity" was first used in its long version (biological diversity) by Ref, [20] and is most commonly used to describe the number of species. It came into effect in the mid-1980s, prefigured by a convention in 1986, followed by the book called biodiversity [21]. These proceedings, in most cases, are construed as the inauguration of the story of biodiversity.

Faith [22] defines biodiversity as a variation in living things, from genetically modified characters to species and, ultimately, ecological systems. The term "biodiversity" energized some essential concepts formulated over the preceding decade. Moreover, the concept of the variation in living things itself has current value since it avails the chance for humanity to benefit. The International Union for the Conservation of Nature [23], summed up these initial concepts about diversity as affording both "insurance" as well as "investment" benefits. The convention on biological diversity (CBD) and intergovernmental platform on biodiversity and ecosystem services (IPBES) echoed the focus on biodiversity. The IPBES conceptual framework describes nature's contribution to humanity [24], which encompassed maintaining future generations' choices availed by biotic diversity as variety. This significance of diversity in living things supplements the recognized worthiness of singular species and solidifies the concept that biotic diversity may denote a singular assemblage of species (or other units) and the quantity of disparity as a property of that assemblage.

The other prominent concept with regard to species diversity, composition, and ecosystem functioning is that modern energy and water have a direct influence on the richness of plant species. This is due to the fact that they control the primary productivity of an ecosystem, which helps in the formation of the food chain and drives other indirect effects (i.e., primary productivity hypothesis) and also by the degree to which an organism can tolerate one or more environmental factors that control the distribution of species (i.e., physiological tolerance hypothesis) [25–28].

Therefore, the theory of the connection of water and energy dynamics postulates that there is an impact of the environment on the distribution of abiotic resources (e.g., water, temperature, and ultraviolet-B (UV-B)) among various species and that this is what principally controls regional biological diversity patterns [29]. Its basic view is that water and energy control the extensive species diversity patterns [30, 31].

Global climate variations could also impact the allocation of species, composition of communities, and structure of an ecosystem [32–34]. Global climate change is not anymore a debatable matter, though various scientists may disagree on the precise forecasts from several models [35], and coming up with viable solutions to the connections between vegetation and climate dynamics is an exceptionally precarious issue of current research [36, 37]. Ref. [38] predicts a rise in the average temperature of between 0.3 and 4.5°C by the end of the twenty-first century, with the Arctic having experienced speedy warming over the current decades.

1.2 The distribution of the African savannas

Worldwide, savanna is the second largest biome, covering about 33 million km^2 or nearly 20% of the earth's land surface [39–41]. Africa inhabits the largest savanna, which occupies about half of the continent with approximately 15.1 million km^2 [42]. Besides, Ref. [1] note Africa to encompass undoubtedly the largest area of savanna, covering 65% of the African content. Tropical savannas are found in the move between the deserts and the tropical rainforests, where there is limited rainfall to support the forest.

Savanna ecosystems predominantly are comprised of a combination of open grasslands, closed coppices with closed thickets consisting of trees and shrubs, which have broad leaves and woodlands that have dispersed trees [43]. This according to Ref. [44] can have implications on human livelihood globally in an even to any alteration to the ecosystem.

The western central rainforest is surrounded by tropical savannas. To the north and south, it is surrounded by deserts. For African savannas, systematic numerical classifications based on climate and physiognomy have been used. The bioclimatic categorization separates four physiognomies of savanna (**Table 1**) [46] cited in [45]).

Despite it being challenging to describe the confines of the African savanna accurately, Okigbo (1985) estimates that it occupies over 12 million km^2 and covers approximately 60% of tropical Africa. It encompasses all or parts of closely all the 45 countries of tropical Africa (**Figure 1**).

Bioclimatic zone	Equivalent ecological region		Annual rainfall average (mm)	Growing season length (days)
	West Africa	Eastern and southern Africa		
Arid savanna	Southern Sahelian	Acacia woodland	300–600	60–90
Subarid savanna	Sudanian	Southern miombo woodland	600–900	90–140
Subhumid savanna	Northern Guinean	Northern miombo woodland	900–1200	140–190
Humid savanna	Southern Guinean	Derived savanna	1200–1500	190–230

Note: adapted from Ref. [45].

Table 1.
Bioclimatic zones of the African savanna.

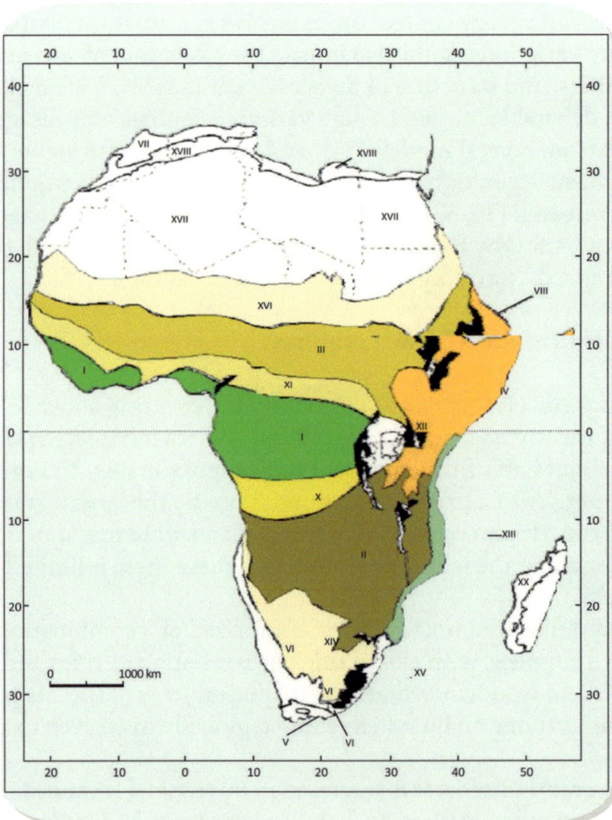

Figure 1.
The distribution of African savannas (in color) around the Congo Basin with tropical moist forests, in dark green. (adapted from White, 1983; Geldenhuys and Golding, 2008). The symbols indicate the following vegetation units (phytochoria): I = Guineo-Congolian regional centre of endemism (RCE); II = Zambezian RCE; III = Sudanian RCE; IV = Somalia-Masai RCE; X = Guinea-Congolia/Zambezian regional transition zone (RTZ); XIO = Guinea-Congolia/Sudanian RTZ; XII = Lake Victoria regional mosaic (RM); XIII = Zanzivar-Inhambane RM; XIV = Kalahari-highveld transtion zone; XVI = Sahel RTZ (adapted from Geldenhuys & Golding, 2008).

1.2.1 Arid savanna zone

The arid savanna zone includes a large portion of central Mali, northern Burkina Faso, southern Niger, northeast Nigeria, Chad, Sudan, and Ethiopia. It also includes a large portion of northern Senegal from Dakar to south of the Senegal river. This region reaches southern Ethiopia, central Tanzania, and Somalia via a thin strip that passes into Kenya. Along with eastern Zambia and southwest Angola, it is also common in southern Mozambique, Zimbabwe, and eastern and northern Botswana. The predominant plant species in this region include *Acacia* spp., including *Acacia* Senegal (gum Arabic), *Acacia vaddiana*, *Leptadenia pyrotechnica*, *Salvadora* spp., *Grewia* spp., and *Acacia seyal* in low places susceptible to flooding, as well as Sahelian grasses, such *Aristida* and *Chloris* spp.

This zone is characterized by low rainfall (300–600 mm/year), a brief rainy season (2–3 months), and frequent droughts. The leguminous Faidherbia albida trees, which shed their leaves in the wet season and manure the soil to provide feed for cattle in the dry season, have historically been preserved by the Serer people of Senegal.

Additionally, significant in eastern Africa are *Commiphora* species. Nearly every tree in the Sahel seems to have a purpose, whether it is for fruit (*Balanites aegyptiiaca*, *Phoenix dactylifera*), fodder (Acacia spp.), or both (Okigbo, 1986). During the protracted dry season, when grazing is scarce, a large number of trees and bushes are severely pruned for food and subsistence.

Dune sands are common in this area, especially in West Africa, and it has historically been subject to wind erosion to a large extent. However, other soils with marginally higher productivity are also present. Although there is little or inconsistent rainfall, this region is crucial for agriculture. A sizeable portion of it is cultivated in West Africa, where pearl millet is mostly grown along with big herds of domestic cattle. As a result, the long rainy season in the northern half of the zone fits the early mature types, known as Souna in Senegal (Bilquez, 1975).

1.2.2 Subarid savanna zone

From Dakar, the subarid savanna zone extends through the majority of northern Nigeria, central Chad, southern Burkina Faso, central Mali, and southern Burkina Faso. In Sudan, it broadens to encompass a sizable portion of the central rain lands as well as a few blue Nile irrigation projects. The region includes a significant portion of the Rift Valley in the south of Addis Ababa in Ethiopia, the Karamoja district in northeastern Uganda, a small portion of the Rift Valley in Kenya to the east of Nairobi, and a portion of central Tanzania. This region also includes the majority of western Zimbabwe, southern Zambia, and southern Angola.

Although alfisols are the most common upland soils in this region, there is significant regional diversity, including dune sands in some regions of West Africa, vertisols around lake Chad and in Ethiopia, Somalia, and Sudan, and oxisols and entisols in southern Africa. F. albida and Hyphaene thebiaca are indicator species, and mixed combretaceous and acacia tree savanna make up the majority of the vegetation in this zone. These species are protected and used as browse, together with Parkia spp. Mango trees are grown in areas with shallow groundwater. Cenchrus ciliaris, Cenchrus biflorus, Eragrostis tremula, and Pennisetum pedicillatum are examples of grasses (Okigbo, 1986).

With significant populations concentrated throughout most of both western and eastern Africa, this region is crucial for agriculture. An essential component of Africa's primary belt for producing cereals is the subarid savanna region.

1.2.3 Subhumid savanna zone

Another significant agricultural region in the African savanna is the subhumid zone, which has the greatest potential for the production of annual crops, particularly cereals.

It stretches from southern Senegal, Gambia, and Guineas through southern Mali and Burkina Faso, northern Cote d'Ivore, Ghana, Togo, Benin, central Nigeria, Chad, and the Central African Republic, before entering southern Sudan, the majority of Uganda, and western Kenya.

There are also significant portions of Tanzania, Mozambique, northern Zambia, southern Zaire, and Angola.

The soils in a large portion of this zone are classified as alfisols, despite the fact that the soil associations in this zone are complex and there are numerous diverse catenary sequences, similar to the subarid zone. However, there are sizable areas of ultisols in Guinea and Uganda and oxisols in Angola, the Central African Republic, and Zaire. Isoberlinia spp., Burkea Africana, and Afzekia Africana are prevalent in West Africa, while in the miombo woodlands, Brachystegia spp. and Julbernardia spp. define the natural vegetation in this zone. Parkia clappertoniana and Butyrospermum spp., sometimes known as the shea butter but, which is a popular source of fat and oil, are protected species. Milicia (previously Chlorophora) excels and Entandrophragma spp. are examples of timber trees. Andropogon gayanus, Hyparrhenia spp., and Pennisetum spp. are examples of tall grasses (Okigbo, 1986).

1.2.4 Humid savanna zone

The humid savanna zone, which begins in West Africa, includes Guinea Bissau, most of Guinea, a portion of southern Mali, Cote d'Ivoire, Ghana, Togo, Benin, a sizable portion of central Nigeria, Cameroon, the majority of the Central African Republic, a small portion of southwest Sudan, the majority of Uganda, the Kenyan highlands, the majority of Rwanda, Burundi, a portion of western and southeastern Tanzania. Although the soils of this region are diverse, a significant portion of the West African coast, as well as Burundi, Rwanda, Uganda, and some areas of Zaire, are classified as utisols. Other large areas are classified as alfsols in much of Benin, Cote d'Ivoire, Ghana, Nigeria, Tanzania, and Togo, and oxisols in Cameroon, Central African Republic, and Zaire.

The majority of the zone's natural climax vegetation would be light forest, with some open woodland in the drier regions and on the less fertile soils, but due to human activity, the majority of the forest has been transformed into derived savanna. This is a transition zone between savanna and forest, and between a unimodal rainfall distribution in roughly 190–200 days in the savanna part and a bimodal distribution in two rainy seasons, each lasting 2–4 months and adding up to 210–230 days in the area derived from the forest.

Some areas of the zone, especially those with light or shallow soils, can occasionally make it difficult to plant annual crops because, even if it is sometimes possible to grow two crops per year, there may not be enough rain in season to result in an ideal crop. The humid savanna is well adapted to a variety of crops, both annual and perennial because it is a transition zone between unimodal and bimodal rainfall distribution, as well as between savanna and forest.

2. Floristic composition in the African savanna

2.1 Factors driving the structure of the African savanna vegetation

The vegetation structure is coined as a three-dimensional distribution of plant biomass. The structure typically denotes perpendicular distribution despite having a flat configuration. More significantly, the species composition and fluctuations in temporal scales can be used to characterize the vegetation structure. According to Ref. [47] factors, including geologic substrate, terrain, human activity, fire, and large herbivores, particularly elephants, all influence the woody vegetative structure of savanna ecosystems.

Plant species' adaptations to the environment include both biotic and abiotic elements, resulting in the structure of the vegetation [48]. The vegetation structure studies mainly concentrate on density, canopy cover, and standing crops of various species [49]. Resource managers use the measurements derived from these factors to understand better how vegetation responds to different management approaches [49].

Several researchers (e.g., [50–52]) have investigated how woody canopy cover varies throughout African savannas in response to environmental conditions. Given that tropical savannas span roughly one-eighth of the land surface [53] and play a significant role in the global carbon cycle [54], it is critical to know how these differences in crown widths and tree densities are formed.

Vegetation structure and the composition of plant species play a trivial role in animal habitat suitability. This is due to the fact that the sensitivity of diverse animal species depends on the quantity of vegetation and bare patches in the landscape, making the condition of spatial heterogeneity a valuable indicator of the suitability of animal habitat [55]. Grass plains, tree savannas, woods, and thickets can all be distinguished by their vegetative structure. Besides, according to Ref. [56], structural vegetation differences exist within a particular habitat, like between short-grass and tall-grass plains.

Variations in the structure of savanna vegetation can be undergone seasonally due to the deciduous nature of woody plants. This may happen when the tall grasses are burnt so as to transform the area to a short-grass area or due to the impact of animals which may defoliate or trample on the vegetation [57–61]. Savanna trees are reported to have both positive and negative competition on the individual tree scale and, therefore, influence the growing grasses beneath their canopies in relation to grasses in inter-canopy areas. Hence, plant species composition in sub-canopy areas can be impacted by large trees, thereby modifying the vegetation structure of the nearby environment [62, 63].

However, this has a positive impact: rich soil nutrients and grass leave in sub-canopy areas, an increase in the availability of soil water due to the hydraulic lift, a reduction in evapotranspiration, and an increase in the productivity of grass [62]. Furthermore, animals are supplied with shade, shelter from the elements, and huge woody plants to browse. Large woody plants also provide higher sub-habitats, a higher yield of highly palatable grass species, and many more rewards [63–65].

Holdo [44] denotes key determinant variables of vegetation structure, such as fire, browsing, and grazers. These were pointed out by Ref. [44] to determine the vegetation structure across rainfall and fertility gradient. Elephants were identified as keystone browsing species. Although the current elephant population does not control the woody vegetation on its own, [44] postulate that the population density has an impact on the tree-size class distribution of woody vegetation because an increase in population density causes a shift in tree size distribution from mature to small height classes, as well as a shift in woodland to grasslands. Refs. [44, 66, 67] identified fire to have a detrimental impact on the type and structure of woody vegetation. It was also noted that grazers affect the woody vegetation by regulating the quantity of fuel suitable for fire and enhancing competition between grasses and woody plants, especially in the regeneration of nutrients and water. Besides, browsers were identified by Ref. [43] to have the greatest influence on woody species regeneration. Using *Euclea divinorum* seedlings, it was discovered that fire reduces seedling survival by 50% and browsers reduce it by 70%, with elephants having no effect.

2.2 Diversity and richness in an African savanna

African savanna ecosystems have experienced severe variations in their vegetative composition, diversity, and species richness due to the influence of human land use and changes in climate conditions. According to Ref. [68], there is a challenge in comprehending the factors affecting the variations in spatial patterns of composition, diversity, and structure of woody species.

2.3 Factors influencing species diversity and composition

Savanna ecosystems play a trivial role in the entirety of the biosphere's activities. Trees and grasses are two divergent life forms characterizing savanna ecosystems. The structure and purpose of savanna are said by Ref. [69] to be influenced by the availability of resources, such as rainfall and nutrients. This is also stated to be affected by natural and anthropogenic disturbances, such as fire and herbivory.

The supply of water is detected by rainfall through the quantity that is then consequently accessible by plants and is subjected to the drainage aspects and storage, which comprises topography, vegetation cover and losses, soil texture, and compaction as a result of evaporation and evapotranspiration. There is a high variation of rainfall in the spatial and temporal scales in the savanna ecosystem. This is exacerbated by aridity, with many places experiencing regular droughts, which according to Ref. [70], can be a chief influencer of the variations in the composition of the vegetation. Recruitment and growth of trees are preferred to grasses in years of high rainfall compared to the times experiencing droughts where growth and recruitment of trees are limited [69].

Species diversity and composition are also affected by soil nutrients. Mainly, there are limited nutrients in the soil since soils in most tropical savannas are obtained from old weathered acid crystalline igneous rock leading to leaching sandy soils with low fertility and CEC. Specifically, low nitrogen and phosphorous availability limit numerous savanna ecological systems [71]. In addition, the water in the soil also hinders the plants from accessing the nutrients because mineralization of nutrients, transport, and root uptake are all factors of soil water content.

Fire has conventionally been utilized as an instrument for managing and conserving savanna ecosystems. This is mainly because there is an exposure of the woody meristems within the flame zone (< 5 m) to the fire damage compared to the grass meristems. Besides, according to Ref. [53], grass meristems can recuperate more efficiently in the short term. Hence, the frequency of fires overpowers the recruitment of mature woody plants. It is noted that there can be interactive impacts on the savanna ecosystem structure by fire and grazing where the high grass biomass buildup is permitted by low grazing, which can impact the biomass and population of the tree by fueling intense fires. Ref. [53] denotes that woody plants are kept within the flame zone by heavy browsing; hence a combination of tree-grass is strongly affected by a strong grazer-browser fire relationship.

Fire has since time immemorial been a significant feature of the ecology in African savanna ecosystems. Fire, according to Refs. [39, 72, 73], has been used both as a selective and regulatory agent, as well as a destructive force. For instance, 25 to 50% of an area has been reported to be subjected to burning on an annual basis. Furthermore, the entire African zone has been estimated to undergo burning for about 2.5 years due to anthropogenic activities [74, 75].

Fire is noted to be vital for a lot of farmers in Africa and it is said to be a cheap hunting tool for clearing vegetation that is not wanted, maintenance of grasslands, and removal of dry vegetation and crop residues to promote the productivity of agricultural produces and permit better visibility [76, 77]. Most fires occur at the beginning of the dry spell when herbaceous biomass has dried out. Refs. [73, 78] reports fire to be known for shaping the savanna ecosystem. As a result, all savanna vegetation communities virtually display fire dependence or tolerance [79, 80].

Plant species respond differently to fire. Some are totally resilient, while the aboveground biomass of other plants may be damaged but can shoot from below-ground structures after the fire, and still, other plant species rely on the seed to recover [78]. The frequency of fire consumes the accumulated production of grass and litter in most tropical savannas, and this favors the predominant grassland vegetation to develop and be maintained by decreasing the natural regeneration of trees and shrubs [81].

Fire can also inspire growth in the flowering of plants and, consequently, the production of seeds among shrubs and herbaceous species [82]. Fire is depended upon for the germination of seeds of a variety of plant species in that it provides one or more physical cues, such as light and temperature as well as chemical cues, including smoke, gases, and nutrients [83–85].

Fire suppression on a temporal basis is regarded as being induced by human disturbance [86]. This is because some woody and herbaceous species, which are generally resilient in burnt areas, often become feeble in protected areas. Lack of fire occurrence may disturb their productiveness either by sustaining a dense layer of standing dead material that hinders young suckers or by keeping alive buds on old axes and thereby wearying the plant [87].

Savanna structure and composition are also stated by Ref. [88] to be impacted by the physical effects of defoliation and selective feeding. The viability of some woody plant populations may be compromised due to the excessive pressure emanating from heavy browsing, such as the elephant. This may result in community fluctuations coupled with species diversity and structural diversity loss. Besides, herbivory plays a trivial role in the cycling of nutrients, dispersal of seeds, and creation of microsites and space, thereby improving and increasing the recruitment of shrubs [88].

Plant species diversity in African savanna is determined by some factors, including competitive exclusion, disturbance processes, and environmental heterogeneity [89–91]. The diversity in species is decreased by competitive exclusion since it is considered as strong competitors first overpower lesser competitors and later cause them to become extinct locally. Moreover, environmental heterogeneity can stimulate species diversity by promoting spatial niche apportioning and heterogeneity in plant community composition across environmental slopes [89]. Disturbances can diminish the diversity of plant species by eradicating disturbance-sensitive species. This can also be done by increasing the diversity of species by opening up growing space and resources for utilization by the colonizing species, maintaining species richness by retarding or averting competitive exclusion, and modifying spatial heterogeneity in the composition of the plant community [91].

The competition was backed by Ref. [92] to influence the savanna ecosystem's vegetation types and structure. Their study exhibited that areas receiving less rainfall are dominated by grasslands, whereas those receiving rainfall of more than 650 mm annually are mostly composed of woodlands except for areas where moisture is transformed by a landscape factor, such as topography.

Sharam et al. & Pellew [43, 67] additionally view vegetative structure in the savanna ecosystem as being affected by demographic factors. Ref. [67] in his study of African savannas, postulates factors, such as elephants, giraffes, and fire, directly impact the type of structure of the woodland. He further states tree height affects woody vegetation structure. He sees trees surpassing 6 m as ecologically mature because they escape the influence of fire due to the fact that the fire threshold is 3 m and also browsing by giraffes, whose threshold is approximately 5.75 m. According to Ref. [67], fires influence the regeneration of trees, whereas elephants lead to mortality in trees of all size classes of regeneration (< 1 m), recruitment (1–5.75 m), and mature phases (> 5.75 m).

3. Species diversity, composition, and ecosystem functioning

Loss in biotic diversity is occurring so that we may experience mass extinction if the trend continues, which may affect the functioning of the ecosystem and ultimately affect the diversity of species and composition [13]. The decline of biological diversity not only represent irrevocable damage to the globe but also poses a menace to what offers support to the lives of human beings owing to the fact that services that nature affords signify everything from the food we eat to the air we breathe [17–19].

4. Impact of anthropogenic activities on the species diversity, composition, and ecosystem functioning

For more than two million years, the structure of the African savanna ecosystems has been affected by anthropogenic activities [93]. Human beings impact the structure of savanna ecosystems directly by cultivating and cutting woody species. They also indirectly impact the structure because of their ability to alter fire and herbivore numbers and distribution through hunting and livestock management [79].

4.1 Anthropogenic activities and species diversity

Human activities have time in memorial influenced the diversity of species in African savannas. Refs. [94, 95] reported that the annual burning of savannas tends to increase species richness. On the other hand, Ref. [96] reported a negative correlation between fire frequencies and woody plant species richness where it was stated that high frequencies of fire reduce woody plant species devoid of producing substantial upsurges in species richness of grass and forb. Fires have been said to alter and restrain tree crown cover and diversity by sporadically killing the tree crown canopy. Fires also restrict the sapling growth into the crown cover canopy and eradicate the regeneration of the tree species, which are intolerant to fire but tolerant to shade [96].

The dominance of the woody plant growing on savannas with finer textured soils may be decreased due to fires which may ultimately reduce the species' richness. The occurrence of fires may harm the woody plants and reduce their cover of the woody plants. It may also reduce the frequencies of woody species, thereby declining the species diversity [97–99]. On the other hand, low fire frequencies have been reported to maximize the understory woody species richness because high fire frequencies damage the seedlings and saplings, which are species sensitive to fire. This inhibits seedlings and saplings' growth into tree crown cover, harming overstory trees over time and limiting the species mix of overstory trees to extremely fire-resistant oaks [96]. The savanna ecosystem houses huge heterogeneous communities because the plant composition communities differ across fine-scale environmental slopes connected to the variability in tree canopy cover throughout space [53, 100].

4.2 Anthropogenic activities and species composition and ecosystem functioning

Ceballos et al. & Steffen et al. [101, 102] denoted that anthropogenic activities have rapidly exacerbated the rate of loss in global biological diversity in the past. Loss of habitat has been the principal driver of this decline globally and is reported to be accountable for almost 67% of the terrestrial land surface, having passed a suggested "safe limit" of the extinction of local species [103].

These declines in biological diversity adversely affect the local functioning and services of an ecosystem and pose a major threat to mankind [12]. On the other hand, biodiversity in flora and fauna has a positive relationship with the productivity of plants as well as soil health [104, 105], thereby enhancing the sequestration ability of atmospheric carbon [104]. Besides, mitigation of alarming carbon dioxide levels may be decelerated due to the loss of diversity which may undermine the progress on limiting climate change. On the other hand, increased crop production and resilience to distress are connected to higher species diversity in agricultural lands [106]. Conservation of biodiversity is thus key to safeguarding the availability of food in light of rising demand and fluctuating environmental conditions.

The structure and functioning of the savanna ecosystem are also affected by the selective removal of trees in several ways. Mostly, the selective removal of trees forms gaps in the crown canopy may lead to increased diversity and abundance because the competition for water and nutrients, as well as increased availability of light, may be reduced [79]. Loss in biological diversity has also been reported by Ref. [18] to reduce the efficiency by which ecological communities capture biologically essential resources, produce biomass, decompose and recycle biologically essential nutrients. Advances have also been made in under-studied areas, such as soils and exhibiting, for instance, that cycling of nitrogen and diversity of plants can be reduced due to the reduced soil biodiversity [107–109].

4.3 Land use change and species diversity and composition

In the past 20 years, the emergence of change in land science has sought to understand the dynamics of anthropogenic activities on the earth through the changes in land use and land cover [110]. Conversion of land to agriculture is reported by Refs. [111, 112] to be a key driver of universal biological loss, with implications for the functioning of an ecosystem, thereby disrupting species diversity and composition. According to Ref. [113], the human population is much higher across the savannas, implying that the more anthropogenic activities, the more pressure on savanna degradation, cover loss, a continued downward trend in species diversity, and composition. Ref. [114] adds that globally land-use change is the principal influence of biological diversity loss.

Taubert et al. [115] note that agricultural expansion and its impact on the loss and disintegration of native habitats causes a reduction in the habitat areas. This also leads to increased predation, a decline in population, extinctions in species, and alterations in species compositions [112, 116]. Nevertheless, these effects change according to the species characteristics and the spatial structure of the habitat. The effects also change according to the surrounding human-modified landscape (matrix) [117]. However, land-use change in African savannas poses a threat to the tenacity of biotic diversity through wildlife grazing loss and dispersal area to agriculture, as well as enlarged disturbance of wildlife around human inhabitation [118].

5. Implications for management of the African woodland savannas

Understanding the diversity and composition of tree species is vital since they provide resources and habitats for a number of other species. Assessments in botanical studies, such as floristic composition, species diversity, and structural analysis studies, are significant for providing accurate information on the species richness, which is valuable for sustainable forest management and helps to understand forest ecology and functioning of the ecosystem. Understanding the long-term response of biological diversity as a result of land use and cover change is vital if species extinctions and decline in biotic diversity were to be minimized. This can be enhanced by implementation of sound and timely conservation and restoration efforts by international forest conservation organizations, governing authorities, interest groups, etc. The knowledge provided in this chapter is expected to be beneficial for planning purposes and management intended for the conservation of biotic diversity and the sustenance of local livelihoods.

6. Conclusion

The evidence presented in this chapter shows how anthropogenic activities, such as forest conversion to woodland for fuel wood removal, charcoal production, woodland grazing, herbivory, habitat degradation, overexploitation, invasive alien species, pollution and climate change, exploitation through illegal and legal logging, and set forest fires, have impacted on the African savanna's species diversity and composition. The chapter also noted that the availability of resources, that is, rainfall and soil nutrients, is a significant factor in species diversity and composition. The chapter also highlights the influence of selective feeding and the physical effects of defoliation, competitive exclusion, and environmental heterogeneity as other significant factors affecting species diversity and composition in an African savanna. This has clearly degraded most parcels of African savanna and has disturbed the species' habitat, thereby significantly exacerbating its decline.

Author details

Kondwani Kapinga[1*], Jules Christian Zekeng[2], Lackson Chama[3], Nalukui Matakala[2], Stanford Siachoono[3], Obote Shakacite[3], Concilia Monde[4] and Stephen Syampungani[2,5]

1 Dag Hammarskjöld, Institute for Peace and Conflict Studies (DHIPS), Copperbelt University, Kitwe, Zambia

2 ORTARChI Chair Environment and Development, Copperbelt University, Kitwe, Zambia

3 Department of Plant and Environmental Sciences, Copperbelt University, Kitwe, Zambia

4 Department of Zoology and Aquatic Sciences, Copperbelt University, Zambia

5 Department of Plant and Soil Sciences, Plant Sciences Complex, University of Pretoria, South Africa

*Address all correspondence to: kondwani.kapinga@cbu.ac.zm; mwanzakonny@gmail.com

References

[1] Huntley BJ, Walker BH. Ecology of Tropical Savannas. Ecological Studies. Berlin: Springer; 1982. p. 42

[2] Bombelli A, Henry M, Castaldi S, Adu-Bredu S, Arneth A, De Grandcourt A, et al. An outlook on the sub-Saharan Africa carbon balance. Biogeosciences. 2009;**6**:2193-2205

[3] Ciais P, Piao S-L, Cadule P, Friedlingstein P, Chédin A. Variability and recent trends in the African terrestrial carbon balance. Biogeosciences. 2009;**6**:1935-1948

[4] DeFries RS, Bounoua L, Collatz GJ. Human modification of landscape and surface climate in the next fifty years. Global Change Biology. 2002;**8**:438-458

[5] Ciais P, Bombelli A, Williams M, Piao SL, Chave J, Ryan CM, et al. The carbon balance of Africa: Synthesis of recent research studies. Philosophical Transactions of the Royal Society A: Mathematical, Physical and Engineering Sciences. 2011;**369**:2038-2057

[6] Abdel-Magid AH, Trlica MJ, Hart RH. Soil and vegetation responses to simulated trampling. Journal of Range Management. 1978;**40**:303-306

[7] Hodgkinson KC, Ludlow MM, Mott JJ, Baruch Z. Comparative responses of the savanna grasses *Cenchrus ciliaris* and *Themeda triandra* to defoliation. Oecologia. 1989;**79**:45-52

[8] Patton BD, Dong X, Nyren PE, Nyren A. Effects of grazing intensity, precipitation, and temperature on forage production. Rangeland Ecology and Management. 2007;**60**:656-665

[9] Abdel-Magid AH, Schuman GE, Hart RH. Soil bulk density and water infiltration as affected by grazing systems. Journal of Range Management. 1978;**40**:307-309

[10] Leriche H, Le Roux X, Desnoyers E, Benest D, Simioni G, Abbadie L. Grass response to clipping in an African savanna: Testing the grazing optimization hypothesis. Ecological Applications. 2003;**13**:1346-1354

[11] Silva JF, Raventos J. Effects of end of dry season shoot removal on the growth of three savanna grasses with different phonologies. Biotropica. 1999;**31**:430-438

[12] Turner CL, Seastedt TR, Dyer MI. Maximization of aboveground grassland production: The role of defoliation frequency, intensity, and history. Ecological Applications. 1993;**3**:175-186

[13] Barnosky AD, Matzke N, Tomiya S, Wogan GOU, Swartz B, Quental TB, et al. Has the Earth's sixth mass extinction already arrived? Nature. 2011;**471**(7336):51-57

[14] Pereira HM, Navarro LM, Martins IS. Global biodiversity change: The bad, the Good, and the unknown. Annual Review of Environment and Resources. 2012;**37**:25-50

[15] MA. Ecosystems and Human Well-Being: Synthesis. Washington, DC: Island Press; 2005

[16] UN FAO. Payments for Ecosystem Services and Food Security. Rome: UN FAO, Office of Knowledge Exchange, Research and Extension; 2011

[17] Díaz S, Fargione J, Chapin FS, Tilman D. Biodiversity loss threatens human well-being. PLoS Biology. 2006;**4**(8):1300-1305

[18] Cardinale BJ, Duffy JE, Gonzalez A, Hooper DU, Perrings C, Venail P, et al. Biodiversity loss and its impact on humanity. Nature. 2012;**486**:59-67

[19] Hooper DU, Adair EC, Cardinale BJ, Byrnes JEK, Hungate BA, Matulich KL, et al. A global synthesis reveals biodiversity loss as a major driver of ecosystem change. Nature. 2012;**486**(7401):105-108

[20] Lovejoy TE. The global 2000 report to the president. In: Barney GO, editor. The Technical Report. Vol. 2. New York: Penguin; 1980. pp. 327-332

[21] Wilson EO. The current state of biological diversity. In: Wilson EO, Peter FM, editors. Biodiversity. Washington, DC: Natl. Acad. Press; 1988. pp. 3-18

[22] Faith DP. Biodiversity: The Stanford Encyclopedia of Philosophy. In: Zalta EN, editor. (Spring 2021 Edition). The Metaphysics Research Lab, Philosophy Department, Stanford University, Stanford, CA. 2021

[23] IUCN. World Conservation Strategy: Living resource conservation for sustainable development. Gland, Switzerland: UNESCO, International Union for Conservation of Nature and Natural Resources (IUCN). 1980

[24] Díaz S, Pascual U, Stenseke M, Martín-López B, Watson RT, Molnár Z, et al. Assessing Nature's contributions to people. Science. 2018;**359**(6373):270-272. DOI: 10.1126/science.aap8826

[25] Currie DJ. Energy and large-scale patterns of animal and plant–species richness. American Naturalist. 1991;**137**:27-49

[26] Latham RE, Ricklefs RE. Global patterns of tree species richness in moist forests: Energy-diversity theory does not account for variation in species richness. Oikos. 1993;**67**:325-333

[27] Qian H, Ricklefs RE. Large-scale processes and the Asian bias in species diversity of temperate plants. Nature. 2000;**407**(6801):180-182

[28] Buckley LB, Davies TJ, Ackerly DD, Kraft NJ, Harrison SP, Anacker BL, et al. Phylogeny, niche conservatism and the latitudinal diversity gradient in mammals. Proceedings of the Royal Society. 2010;**277**:2131-2138

[29] Xu X, Zhao H, Zhang X, Hanninen H, Korpelainen H, Li C. Different growth sensitivity to enhanced UV-B radiation between male and female populus cathayana. Tree Physiology. 2010;**30**:1489-1498

[30] Levsen N, Tiffin P, Olson M. Pleistocene speciation in the genus populus (salicaceae). Systematic Biology. 2012;**61**:401-412

[31] Clarke A, Gaston KJ. Climate, energy and diversity. Proceedings of the Royal Society. 2006;**273**:2257-2266

[32] Qin H, Dong G, Zhang Y, Zhang F, Wang M. Patterns of species and phylogenetic diversity of *Pinus tabuliformis* forests in the eastern loess plateau, China. Forest Ecology and Management. 2017;**394**:42-51

[33] Thuiller W, Lavergne S, Roquet C, Boulangeat I, Lafourcade B, Araujo MB. Consequences of climate change on the tree of life in europe. Nature. 2011;**470**:531-534

[34] Pio DV, Engler R, Linder HP, Monadjem A, Cotterill FPD, Taylor PJ, et al. Climate change effects on animal and plant phylogenetic diversity in southern Africa. Global Change Biology. 2014;**20**:1538-1549

[35] Hultine KR, Burtch KG, Ehleringer JR. Gender specific patterns of carbon uptake and water use in a dominant riparian tree species exposed to a warming climate. Global Change Biology. 2013;**19**:3390-3405

[36] González-Orozco CE, Pollock LJ, Thornhill AH, Mishler BD, Knerr N, Laffan SW, et al. Phylogenetic approaches reveal biodiversity threats under climate change. Nature Climate Change. 2016;**6**:1110-1114

[37] Bellard C, Bertelsmeier C, Leadley P, Thuiller W, Courchamp F. Impacts of climate change on the future of biodiversity. Ecological Letters. 2012;**15**:365-377

[38] Stocker TF, Qin D, Plattner GK, Tignor M, Allen SK, Boschung J, et al. IPCC: Climate change 2013: The physical science basis. Contribution of working group I to the fifth assessment report of the intergovernmental panel on climate change. Computational Geometry. 2013;**18**:95-123

[39] Scholes RJ, Walker BH. An African Savanna: Synthesis of the Nylsvley Study. Cambridge, United Kingdom: Cambridge University Press; 1993. p. 306

[40] Ramankutty N, Foley JA. Estimating historical changes in global land cover: Croplands from 1700 to 1992. Global Biogeochemical Cycles. 1999;**13**:997-1027

[41] Beerling DJ, Osborne CP. The origin of the savanna biome. Global Change Biology. 2006;**12**:2023-2031

[42] Grace J, Jose JS, Meir P, Miranda HS, Montes RA. Productivity and carbon fluxes of tropical savannas. Journal of Biogeography. 2006;**33**:387-400

[43] Sharam G, Sinclair AR, Turkington R. Establishment of broad-leaved thickets in Serengeti, Tanzania: The influence of fire, browsers, grass competition, and elephants 1. Biotropica. 2006;**38**(5):599-605

[44] Holdo MR, Holt DR, Fryxell MJ. Grazers, browsers and fire influence the extent and spatial tree cover in the Serengeti. Ecological Applications. 2009;**19**(1):95-109

[45] Ker A. Farming systems of the African savanna: A Continent in Crisis. Canada: IDRC, Ottawa; 1995

[46] Philips J. Agriculture and Ecology in Africa: A Study of Actual and Potential Development South of the Sahara. London, UK: Faber and Faber; 1959. p. 424

[47] Wessels KJ, Mathieu R, Erasmus B, Asner G, Smit I, van Aardt J, et al. Impact of communal land use and conservation on woody vegetation structure in the Lowveld savannas of South Africa. Forest Ecology and Management. 2011;**262**:19-29

[48] Sala O. The Effect of Herbivory on Vegetation Structure. Plant Form and Vegetation Structure. Hague, Netherlands: SBP Academic Publishers; 1988

[49] Schulz TT, Leininger WC. Differences in riparian vegetation structure between grazed areas and exclosures. Rangeland Ecology & Management/ Journal of Range Management Archives. 1990;**43**(4):295-299

[50] Good SP, Caylor KK. Climatological determinants of woody cover in Africa. Proceedings of the National Academy of Sciences. 2011;**108**(12):4902-4907

[51] Sankaran M. Determinants of woody cover in African savannas. Nature. 2005;**438**(8):846-849

[52] Staver AC, Archibald S, Levin SA. The global extent and determinants of savanna and forest as alternative biome states. Science. 2011;**334**(6053):230-232

[53] Scholes RJ, Archer SR. Tree-grass interactions in savannas. Annual Review of Ecology, Evolution, and Systematics. 1997;**28**:517-544

[54] Poulter B, Frank D, Ciais P, Myneni RB, Andela N, Bi J, et al. Contribution of semi-arid ecosystems to interannual variability of the global carbon cycle. Nature. 2014;**509**(7502):600-603

[55] Turner MG. Landscape ecology: The effect of pattern on process. Annual Review of Ecological Systems. 1989;**20**:171-197. DOI: 10.1146/annurev.es.20.110189.001131

[56] Venter FJ, Scholes RJ, Eckhardt HC. The abiotic template and its associated vegetation pattern', In: J. Du Toit, H. Biggs & K.H. Rogers (eds.), The Kruger Experience: Ecology and Management of Savanna Heterogeneity. Island Press, London. 2003; 83-129.

[57] Bond WJ. What limits trees in C4 grasslands and savannas? Annual Review of Ecology, Evolution and Systematics. 2008;**39**(1):641-659. DOI: 10.1146/annurev.ecolsys.39.110707.173411

[58] Gordijn PJ, Rice E, Ward D. The effects of fire on woody plant encroachment are exacerbated by succession of trees of decreased palatability: Perspectives in plant ecology. Evolution and Systematics. 2012;**14**(6):411-422. DOI: 10.1016/j.ppees.2012.09.005

[59] Holdo RM, Holt RD, Coughenour MB, Ritchie ME. Plant productivity and soil nitrogen as a function of grazing, migration and fire in an African savanna. Journal of Ecology. 2007;**95**:115-128

[60] Janecke BB, Smit GN. Phenology of woody plants in riverine thicket and its impact on browse availability to game species. African Journal of Range & Forage Science. 2011;**28**(3):139-148. DOI: 10.2989/10220119.2011.642075

[61] Joubert SCJ. 'Animal behaviour', In: J. Du, P. Bothma & J.T. Du Toit (eds.), Game ranch management. Van Schaik Publishers, Pretoria. 2016; 385-392.

[62] Riginos C, Grace JB, Augustine DJ, Young TP. Local versus landscape-scale effects of savanna trees on grasses. Journal of Ecology. 2009;**97**(6):1337-1345. DOI: 10.1111/j.1365-2745.2009.01563.x

[63] Treydte AC, Heitkönig IMA, Prins HHT, Ludwig F. Trees improve grass quality for herbivores in African savannas. Perspectives in Plant Ecology, Evolution and Systematics. 2007;**8**(4):197-205. DOI: 10.1016/j.ppees.2007.03.001

[64] Smit GN, Swart JS. The influence of leguminous and non-leguminous woody plants on the herbaceous layer and soil under varying competition regimes in mixed bushveld. African Journal of Range and Forage Science. 1994;**11**(1):27-33. DOI: 10.1080/10220119.1994.9638350

[65] Treydte AC, Riginos C, Jeltsch F. Enhanced use of beneath-canopy vegetation by grazing ungulates in African savannahs. Journal of Arid Environments. 2010;**74**(12):1597-1603. DOI: 10.1016/j.jaridenv.2010.07.003

[66] Norton-Griffith M. The influence of grazing browsing and fire on the vegetation dynamics of the Serengeti. In: Sinclair ARES, Griffith N, editors. Serengeti: Dynamics of an Ecosystem. Chicago: University of Chicago Press; 1979

[67] Pellew RA. The impacts of elephant, giraffe and fire upon the Acacia tortilis woodlands of the Serengeti. African Journal of Ecology. 1983;**21**(1):41-74

[68] Lomolino MV. Elevation gradients of species-density: Historical and prospective views. Global Ecology and Biogeography. 2011;**10**(1):3-13

[69] Sankaran M, Ratnam J, Hanan NP. Tree grass coexistence in savannas revisited - insights from an examination of assumptions and mechanisms invoked in existing models. Ecological Letters. 2004;**7**:480-490

[70] Ellis JE, Swift DM. Stability of African pastoral ecosystems: Alternative paradigms and implications for development. Journal of Range Management. 1988;**41**:450-459

[71] House JI, Hall DO. Productivity of tropical savannas and grasslands. In: Roy J, Saugier B, Mooney HA, editors. Terrestrial Global Productivity. San Diego: Academic Press; 2001:363-400

[72] Goldammer JG. Fire in the tropical biota: Ecosystem process and global changes. In: JG, editor. Papers from the Third Symposium on Fire Ecology Held at Freiburg University. Goldammer ed. Berlin, Germany: Springer-Verlag; 1990 497 pp

[73] Van Langevelde F, Van De Vijver CA, Kumar L, Van De Koppel J, De Ridder N, Van Andel J, et al. Effects of fire and herbivory on the stability of savanna ecosystems. Ecology. 2003;**84**(2):337-350

[74] Delmas RA, Loudjani P, Podaire A, Menaut JC. Biomass burning in Africa: An assessment of annually burned biomass. In: Levine SJ, editor. Global Biomass Burning. Atmospheric, Climatic and Biospheric Implications. Cambridge, Massachusetts, London, England: The MIT Press; 1991. pp. 126-132

[75] Menaut JC, Lepage M, Abbadie L. Savannas, woodlands and dry forests in Africa. In: Bullock SH, Mooney HA, Medina EE, editors. Seasonally Dry Tropical Forests. Cambridge, England: Cambridge University Press; 1995. pp. 64-92

[76] Ehrlich D, Lambin EF, Malingreau JP. Biomass burning and broad-scale land cover changes in western Africa. Remote Sensing of Environment. 1997;**61**:201-209

[77] Laris P, Wardell DA. Good, bad or 'necessary evil'? Reinterpreting the colonial burning experiments in the savanna landscapes of West Africa. Geographical Journal. 2006;**172**:271-290

[78] Bond WJ, van Wilgen BW. Fire and Plants. 1st ed. United Kingdom, London: Chapman & Hall; 1996. p. 263

[79] Frost P, Menaut JC, Walker B, Medina E, Solbrig OT, Swift M. Responses of savannas to stress and disturbance. A proposal for a collaborative programme of research. In: IUBS-UNESCO-MAB, Biology International (special issue), Paris; 1986;**10**:82

[80] Swaine MD. Characteristics of dry forest in West Africa and the influence of fire. Journal of Vegetation Science. 1992;**3**:365-374

[81] Hoffmann WA. Fire and population dynamics of woody plants in a Neotropical savanna: Matrix model projections. Ecology. 1999;**80**:1354-1369

[82] Baskin CC, Baskin JM. A geographical perspective on germination ecology: Tropical and subtropical zones. In: Seeds: Ecology, Biogeography and Evolution of Dormancy and Germination. London, United Kingdom: Academic Press; 1998. pp. 239-329

[83] Dixon KW, Roche S, Pate JS. The promotive effect of smoke derived from burnt native vegetation on seed-germination of western-Australian plants. Oecologia. 1995;**101**:185-192

[84] Brown NAC, van Staden J. Smoke as a germination cue: A review. Plant Growth Regulation. 1997;**22**:115-124

[85] van Staden J, Brown NAC, Jäger AK, Johnson TA. Smoke as a germination cue. Plant Species Biology. 2000;**15**:167-178

[86] Menaut JC, Abbadie L, Loudjani P, Podaire A. Biomass burning in West Africa savannahs. In: Levine SJ, editor. Global Biomass Burning. Atmospheric, Climatic, and Biospheric Implications. Cambridge, Massachusetts, London, England: The MIT Press; 1991. pp. 133-142

[87] Menaut JC, Cesar J. The structure and dynamics of a west African savanna. In: Huntley B, Walker BH, editors. Ecology of Tropical Savannas. Berlin: Springer-Verlag; 1982. pp. 80-100

[88] Mworia JK, Wambua JK, Omari JK, Kinyamario JI. Patterns of seed dispersal and establishment of the invader *Prosopis juliflora* in the upper floodplain of Tana River, Kenya. African Journal of Range and Forage Science. 2011;**28**(1):35-41

[89] Whittaker RH. Communities and Ecosystems. 2nd ed. New York: McMillan Publishing Company; 1975

[90] Connell JH. Diversity in tropical rain forests and coral reefs. Science. 1978;**199**:1302-1310

[91] Huston MA. Biological Diversity: The Coexistence of Species on Changing Landscapes. Cambridge: Cambridge University Press; 1994

[92] Reed DN, Anderson TM, Demplewolf J, Metzger K, Serneel S. The spatial distribution of vegetation types in the Serengeti ecosystem: The influence of rainfall and topographic relief on vegetation patch characteristics. Journal of Biogeography. 2008;**36**:770-782

[93] Harris DR. Commentary: Human occupation and exploitation of savanna environments. In: Harris DR, editor. Human Ecology in Savanna Environments. London: Academic Press; 1980. pp. 31-39

[94] Lewis CE, Harshbarger TJ. Shrub and herbaceous vegetation after 20 years of prescribed burning in the South Carolina coastal plain. Journal of Range Management. 1976;**29**:13-18

[95] Walker J, Peet RK. Composition and species diversity of pine-wiregrass savannas of the green swamp. North Carolina. Vegetatio. 1983;**55**:163-179

[96] Peterson DW, Reich PB. Prescribed fire in oak savanna: Fire frequency effects on stand structure and dynamics. Ecological Applications. 2001;**11**:914-927

[97] Axelrod AN, Irving FD. Some effects of prescribed fire at Cedar Creek natural history area. Journal of the Minnesota Academy of Science. 1978;**44**:9-11

[98] Waldrop TA, White DL, Jones SM. Fire regimes for pine-grassland communities in the southeastern United States. Forest Ecology and Management. 1992;**47**:195-210

[99] Peterson DW. Fire Effects on Oak Savanna and Woodland Vegetation in Minnesota [Dissertation]. Saint Paul, Minnesota: University of Minnesota; 1998

[100] Vetaas OR. Micro-site effects of trees and shrubs in dry savannas. Journal of Vegetation Science. 1992;**3**:337-344

[101] Ceballos G, Ehrlich PR, Barnosky AD, et al. Accelerated modern human-induced species losses: Entering the sixth mass extinction. Science Advances. 2015;**1**:e1400253

[102] Steffen W, Broadgate W, Deutsch L, Gaffney O, Ludwig C. The Trajectory of the Anthropocene: The Great Acceleration. The Anthropocene Review. 2015;**2**:81-98

[103] Newbold T, Lawrence NH, Andrew PA, Sara C, Adriana DP, Samantha LLH, et al. Has land use pushed terrestrial biodiversity beyond the planetary boundary?A global assessment. Science. 2016;**353**:288-291

[104] Lal R. Soil carbon sequestration to mitigate climate change. Geoderma. 2004;**123**:1-22

[105] Maestre FT, Quero JL, Gotelli NJ, Escudero A, Ochoa V, Delgado-Baquerizo M, et al. Plant species richness and ecosystem multifunctionality in global drylands. Science. 2012;**335**:214-218

[106] Di Falco S. On the value of agricultural biodiversity. Annual Review of Resource Economics. 2012;**4**:207-223

[107] Bardgett RD, Van Der Putten WH. Belowground biodiversity and ecosystem functioning. Nature. 2014;**515**(7528):505-511

[108] Wagg C, Bender SF, Widmer F, van der Heijden MG. Soil biodiversity and soil community composition determine ecosystem multifunctionality. Proceedings of the National Academy of Sciences of the United States of America. 2014;**111**(14):5266-5270

[109] De Vries FT, Thébault E, Liiri M, Birkhofer K, Tsiafouli MA, Bjørnlund L, et al. Soil food web properties explain ecosystem services across European land use systems. Proceedings of the National Academy of Sciences of the United States of America. 2013;**110**(35):14296-14301

[110] Brown RM, Siler CD, Oliveros CH, Esselstyn JA, Diesmos AC, Hosner PA, et al. Evolutionary processes of diversification in a model island archipelago. Annual Review of Ecology, Evolution, and Systematics. 2013;**44**:411-435

[111] Haddad NM, Brudvig LA, Clobert J, Davies KF, Gonzalez A, Holt RD, et al. Habitat fragmentation and its lasting impact on Earth's ecosystems. Science Advances. 2015;**1**:1-9

[112] Pfeifer M, Lefebvre V, Peres CA, Banks-Leite C, Wearn OR, Marsh CJ, et al. Creation of forest edges has a global impact on forest vertebrates. Nature. 2017;**551**:187-191. DOI: 10.1038/natur e24457

[113] Campbell Grant EH, Lowe WH, Fagan WF. Living in the branches: Population dynamics and ecological processes in dendritic networks. Ecology Letters. 2007;**10**(2):165-175

[114] Hemant GT, Emily SW, Mariana C, Catherine LP, Casey MR. Agricultural expansion in African savannas: Effects on diversity and composition of trees and mammals. Biodiversity and Conservation. 2021;**30**:3279-3297

[115] Taubert F, Fischer R, Groeneveld J, Lehmann S, Muller MS, Rodig E, et al. Global patterns of tropical forest fragmentation. Nature. 2018;**554**(7693):519-522

[116] Fahrig L. Why do several small patches hold more species than few large patches? Global Ecology and Biogeography. 2020;**29**(4):615-628

[117] Ewers R, Didham R. Confounding factors in the detection of species responses to habitat fragmentation. Biological reviews of the Cambridge Philosophical Society. 2006;**81**:117-142. DOI: 10.1017/s1464 79310 50069 49

[118] Aboud AA, Kisoyan PK, Said MY, Notenbaert A, de Leeuw J, Gitau JW, et al. Payment for Wildlife Conservation in the Masai Mara Ecosystem. Entebbe: Natural Resource Management and Biodiversity Conservation in the Drylands of Eastern and Central Africa. ASARECA; 2012

Chapter 7

Vegetation and Avifauna Distribution in the Serengeti National Park

Ally K. Nkwabi and Pius Yoram Kavana

Abstract

In order to examine the bird species changes within different vegetation structures, the variations were compared between *Commiphora*-dominated vegetations with those of *Vachellia tortilis* and *Vachellia robusta*-dominated vegetations, and also compared the birds of grassland with those of *Vachellia drepanolobium* and *Vachellia seyal*-dominated vegetations. This study was conducted between February 2010 and April 2012. A total of 40 plots of 100 m × 100 m were established. Nonparametric Mann-Whitney U-test was used to examine differences in bird species between vegetations. Species richness estimates were obtained using the Species Diversity and Richness. A total of 171 bird species representing 103 genera, 12 orders, and 54 families were recorded. We found differences in bird species distribution whereby *V. tortilis* has higher bird species richness (102 species), abundance, and diversity when compared with *Commiphora* with 66 species and *V. robusta* with 59 species. These results suggest that variations in bird species abundance, diversity, and distribution could be attributed to differences in the structural diversity of vegetation. Therefore it is important to maintain different types of vegetation by keeping the frequency of fire to a minimum and prescribed fire should be employed and encouraged to control wildfire and so maintain a diversity of vegetation and birds community.

Keywords: vegetation changes, abundance, richness, conservation, protected area

1. Introduction

1.1 Vegetation change and birds' distribution

One of the most important concerns in conservation is the cause of change in species abundance and diversity in various vegetation types over time. Conservation relies in part on protecting species in legally fixed areas, but protection is jeopardized if species leave those areas and move into other areas where they are more vulnerable. For example, there is evidence of range shifts in Tanzanian birds [1].

There has been a recent expansion of geographical range in savanna birds such as the White-bellied Go-away-bird *Criniferoides leucogaster* and Taita Fiscal *Lanius dorsalis* (A. Nkwabi pers. obs.). These are dryland species previously confined to the drier north-east areas of the East African central plateau but have now spread westwards into the much moister parts of the Serengeti National Park by crossing the birds' barrier of the eastern Rift Valley wall and the associated crater highlands [2, 3].

Another species that has shown changes in distribution is the Black-throated Barbet (*Tricholaema melanocephala*) whose center of distribution was the savanna surrounding the Wembere grasslands, south of the Maswa Game Reserve [4]. This species has spread north into Serengeti National Park in 2004 and up to date, such change in birds distribution resulted due to a change in density of its main vegetation such as *Vachellia tortilis* and *Vachellia robusta* woodland, which was low in the 1970s but had increased by the early 2000s (Prof. A. R. Sinclair pers. com.).

1.2 What causes a change in bird species' geographical range?

There are several possible causes for bird changes in geographical distribution and range. Natural vegetation succession, wildlife grazing, and human activities are all possible causes. Change in vegetation in the Serengeti National Park has been well documented [5–8]. Park vegetations have experienced major changes, alternating between open grassland and dense woodland. However, how these vegetation structure changes influence animal distribution, particularly birds is not well documented. Such influences are important for the conservation and management efforts of birds.

The structure of African vegetation is predominantly determined by rainfall, fire, nutrients, and grazing of herbivores [9, 10]. Bird distributions may be related to these differences in vegetation structure due to gradients of latitude, elevation, and isolation [11, 12]. Vegetation structure is known to have a major influence on the abundance and diversity of birds [11, 13–15] and this applies to the Serengeti National Park [16, 17]. Bird species diversity generally increases with increased foliage height, diversity, or increased woody vegetation [18]; vegetation species composition (or floristic) may also strongly affect bird communities [19]. Furthermore, individual bird species often demonstrate strong preferences for certain vegetation types, thus permitting vegetation parameters to describe bird habitats [20].

The diversity of bird species is a function of the diverse and complex vegetations of the Serengeti National Park in which the bird species respond to differences in the structural components of vegetation and availability of food [16, 21, 22]. Ecological studies have described the Serengeti National Park over the past 40 years and much is known about the impact of natural and human disturbances on large mammals [23]. The resource requirements of some species of birds in the Serengeti National Park have been documented [16, 22, 24, 25]. The impact of natural and human disturbances through small-scale agriculture and human habitation on birds has been reported by Estes [21], Sinclair et al. [26], and Nkwabi et al. [27] who examined the influence of disturbances, such as burning and grazing, on bird distribution within the Serengeti National Park.

Studying birds in the Serengeti National Park will open a room for birds-tourism activities and conserve vegetation dynamics. This chapter examines how differences in vegetation structure might influence the richness, abundance, and diversity of birds in the Serengeti National Park. It is firstly predicted that *V. tortilis* and *V. robusta* increase the abundance and diversity of bird species than *Commiphora* spp. Secondly, predicted that *Vachellia drepanolobium* and *Vachellia seyal* increase the abundance and diversity of birds than grassland areas. Thirdly, predicted that bird abundance and diversity decline with changing grass height from tall to short grass.

2. Materials and methods

2.1 Study area

The study was conducted in the Serengeti National Park (14,763 km^2) situated between 33°50′ and 35°20′E and 1°28′ and 3°17′S (**Figure 1**). The park occupies a vast upland area varying in elevation from 1162 m at sea level at the Speke Gulf to 1860 m at sea level in the northeast. The park is bordered by the Maswa Game Reserve to the southwest, Ngorongoro Conservation Area to the southeast, Loliondo Game Controlled Area to the northeast, Maasai-Mara National Reserve in the north, and Ikorongo-Grumeti Game Reserves in the west (**Figure 2**).

2.2 Temperature and rainfall

The temperature of the Serengeti National Park shows a relatively constant monthly mean and maximum temperature of 27–28°C taken daily in the morning (9:00 am) and afternoon (3:00 pm) at the Serengeti National Park weather station and Serengeti Research Station [28]. The minimum temperature varies from 16°C in the hot months of October-March to 13°C during May-August. Rain typically falls in a bimodal pattern, with the long rains during March-May and the short rains in November-December [28]. However, the rains can fuse into one long period, particularly in the north, or the short rains can fail entirely, especially in the southeast of the park [28].

2.3 General sampling design

A total of 40 plots of 100 m × 100 m were established and locations were marked using the global positioning system (GPS) manufacturer model Garmin 12 XL.

2.4 Vegetation categorization

Vegetations were categorized on the basis of physiognomic features and dominant plant species when present.

1. Grassland (*mbuga*): This was devoid of trees (**Figure 3**) except for an occasional *Vachellia spp.* or *Balanites aegyptiaca* and one or two regenerating *V. seyal*, *V. drepanolobium*, or *Vachellia hockii*. The grass layer was dominated by *Themeda triandra* and *Pennisetum mezianum*. Open grasslands were found on poorly drained soils on the Ndabaka flood plain in the west and at Kogatende in the northern part of the Serengeti National Park.

2. *V. robusta* (**Figure 4**) and *V. tortilis* (**Figure 5**) dominated vegetations: This vegetation is comprised of *V. robusta* and *V. tortilis* that grew up to 14 m in height and is characterized by a wide canopy, umbrella-shaped crown, which could reach 21 m in diameter [29, 30]. This vegetation was found on various sites from ridge tops, and gently sloping valley sides to alluvial benches beside major streams, but it sometimes reached valley bottoms around Banagi areas in the central Serengeti National Park. The soil layer in this vegetation is covered by grass species that form a mixture of *T. triandra*, *P. mezianum*, and *Digitaria macroblephara*.

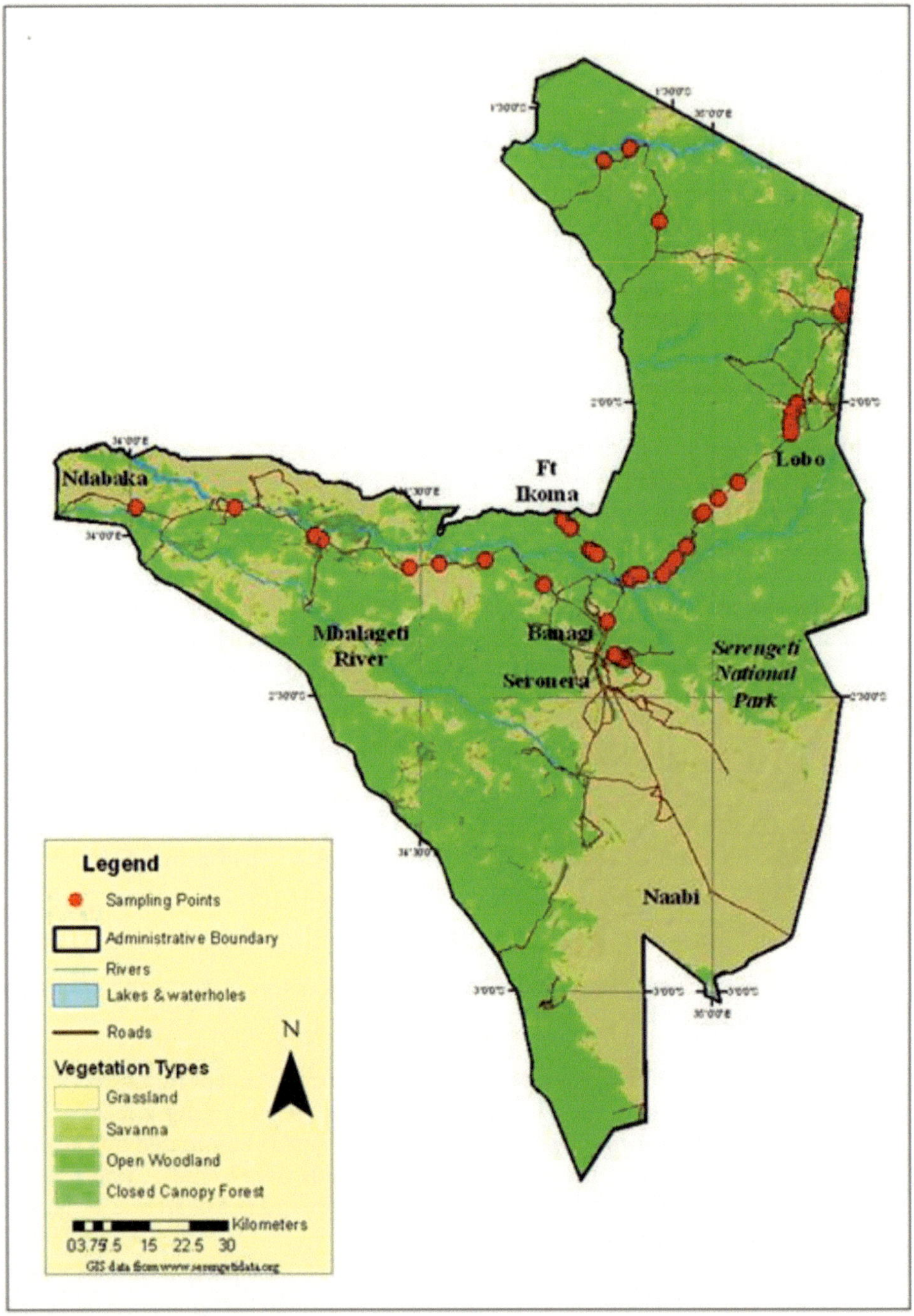

Figure 1.
Map of Serengeti National Park showing vegetation types. Red dots show birds sampling plots (Map produced by the Serengeti GIS Office, 2022).

3. *Commiphora* spp.: This was found mainly on ridge tops in shallow stony soils along the road to Fort Ikoma village, in the north to Lobo, and in the eastern woodlands [29, 31, 32]. This vegetation is comprised of *Commiphora schimperi*,

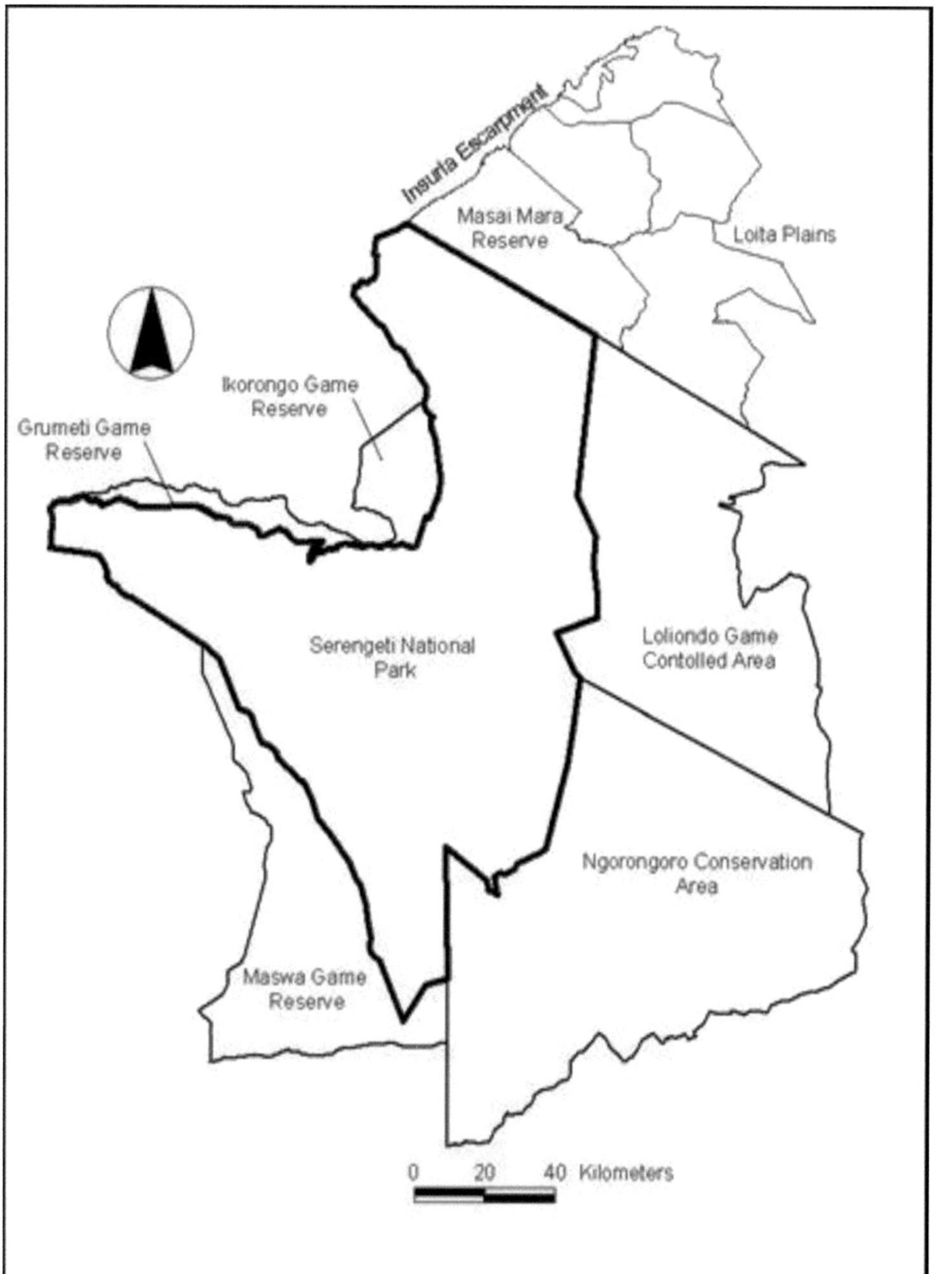

Figure 2.
Map of Serengeti National Park (solid line) and adjacent protected areas (Source: Serengeti GIS Office, 2022).

Commiphora trothae, and *Commiphora africana* mixed with *Senegalia senegal* (**Figure 6**). Trees of this vegetation grew to 5 m high and were characterized by a small canopy that may reach 6.5 m in diameter [29, 30]. Some *V. hockii*, a tree of similar height, sometimes occurs with *S. senegal*. The grass layer within the stand is dominated by *P. mezianum* and *Sporobolus pyramidalis* [29].

Figure 3.
One of the grassland areas in Serengeti National Park.

Figure 4.
A pure stand of Vachellia robusta at Seronera area in Serengeti National Park.

Figure 5.
Vachellia tortilis observed near Serengeti Wildlife Research Centre (SWRC) in Serengeti National Park.

4. *V. drepanolobium* and *V. seyal*-dominated habitats: This vegetation occurs on poorly drained soils in valley bottoms, and on foot slopes at the base of rocky hills [29]. The two tree species were found in different regions of the park; *V. drepanolobium* occurs widely in the north and west of the park, whereas *V. seyal* was found largely on the Ndabaka flood plain of the far western Serengeti National Park. Both are characterized by a small canopy that may reach 6.5 m in diameter and grow to 6 m in height. The grass layer is dominated by *T. triandra*, *D. macroblephara*, and *P. mezianum* [29].

2.5 Birds assessment

A total of 10 plots in each of the 4 habitat categories mentioned above were established. Grassland in the park was divided into western at Ndabaka and northern

Figure 6.
Stand of Commiphora spp and Senegalia Senegal observed near Banagi hill, central Serengeti National Park.

at Kogatende, each having five plots and these were paired with their equivalent *V. drepanolobium* and *V. seyal*-dominated habitats, namely five plots of *V. seyal* in the west and five plots of *V. drepanolobium* in the north. Five plots in each of *V. tortilis* and *V. robusta*-dominated vegetations were established, and these were compared with five plots in *Commiphora* spp.-dominated vegetation.

The locations of the sampling plots in the woodland were situated at least 500–800 m apart, marked using the global positioning system (GPS). Plots were surveyed on a monthly basis between February 2010 and February 2012, with 20 min per plot between 6:30 am and 10:30 am. The time of survey for each plot was rotated so that all plots were surveyed equally during the morning from 6:30 am to 10:30 am to avoid bias for the time of day. For example, if a plot was surveyed early on one visit then on the next visit was surveyed last so that effects of time of day were distributed equally over all plots. Birds were counted using the total count method by walking slowly across the plot and back again [33]. Birds were identified by both sight and call, and numbers were recorded. Bird naming was adopted from those published by Zimmerman et al. [4] and Gill and Wright [34].

2.6 Grass height measurement and grass cover estimates

The grass layer included both grasses and forbs, but as the grass was usually the dominant group, the term "grass" was used in the following analysis and the data from grasses and forbs were pooled. Measurements of grass height were obtained in all habitats where birds were counted. At each of the plots, grass height and grass cover were estimated. A small hole was made at the center of the plastic plate of 37 cm diameter and inserted in a 1 m ruler, which was placed vertically in the grass, and detailed explanations of the method on how to prepare and use a plastic plate were well explained by McNaughton [35]. The plastic plate rested on the surface of the grass, and its position on the meter ruler was recorded as the standardized grass height [35]. Three height measurements were made and the average was used for each plot, only three points were selected in each plot due to the large area coverage in the Serengeti National Park and surrounding areas. Grass cover was estimated by direct observation and recorded as a percentage cover. Measurements were taken during dry and wet seasons from February 2010 to April 2012.

2.7 Data analysis

Birds were classified according to feeding guild (insectivores, granivores, omnivores, and frugivores) and foraging guild (ground vs. tree canopy feeding) [36, 37]. Bird species that spent time on ground foraging were grouped as ground feeding and those that spent time on trees searching for food were termed as tree feeding [36, 37]. All ground, aerial, water, and tree canopy feeding birds were recorded, but the analysis concentrated on ground- and tree-feeding birds because the study concentrated on habitat structure in the terrestrial environment. All bird species that fed on insects, seeds, and fruits, and foraged either on the ground or in the grass layer were classified as ground-feeding birds; and all species that foraged in bushes, shrubs, and tree canopy were classified as tree-feeding birds [38, 39]. Ground-feeding birds were also sorted based on the grass height they were associated with.

2.8 Species richness, diversity, and abundance of the ground- and tree-feeding birds

Data were tested for normality by using the Shapiro-Wilk (W) test and the Kolmogorov-Smirnov (KS) test. The data were found not to be normally distributed even after transformation. Therefore a nonparametric Mann-Whitney U-test was used to examine differences in independent samples [40]. Species richness estimates were obtained using the Species Diversity and Richness (SDR) computer program [41]. The first-order Jackknife estimator was chosen for species richness because it has been shown to perform well on bird communities distribution [42, 43].

Abundance was measured as the number of individuals found per unit time. Comparisons between pairs of habitats for an abundance of birds at similar grass heights were tested for normality using Kolmogorov-Smirnov (KS) test. In this case, the data were found to be normally distributed. Therefore, differences in the abundance of ground bird species at similar grass heights were compared using the parametric paired sample t-test from the computer software Statistix-10 [44].

The diversity of bird species was determined by the Shannon-Wiener diversity (H′) index denoted as:

$$H' = -\sum_{i=1}^{k} p_i \ln p_i, \qquad (1)$$

where k is the total number of species and p_i is the proportion of individuals found in the ith species. The Shannon-Wiener diversity (H′) value obtained from each habitat type was tested for differences by using the randomization test [41, 45] to observe whether bird diversity of ground- or tree-feeding bird species differ in the *Commiphora* spp. habitat when compared with those in *V. tortilis* and *V. robusta* and when grassland compared with *V. drepanolobium* and *V. seyal* vegetations.

3. Results

3.1 Birds in *Commiphora* spp. compared with those in *Vachellia*-dominated vegetations

A total of 171 species in 103 genera, 12 orders, and 54 families of birds were recorded. The results show that *V. tortilis*-dominated vegetation supported the highest species richness of ground-feeding birds, followed by the *Commiphora spp.* vegetations. In contrast, *V. robusta* remained with lower bird species richness (**Table 1**). In terms of abundance, *Commiphora*-dominated vegetation supported a higher abundance of birds followed by *V. tortilis* and then *V. robusta*-dominated vegetations (**Table 1**). There was no significant difference in the abundance of bird species between *Commiphora* and *V. tortilis* ($U = 8896.5, n_1 = 34, n_2 = 29, P = 0.246$). However, the difference in abundance between *Commiphora*- and *V. robusta*-dominated vegetations was significant ($U = 8085.5, n_1 = 34, n_2 = 30, P = 0.017$, **Table 1**). The Shannon-Wiener index shows that *V. tortilis* and *Commiphora* accommodated higher bird species diversity values than *V. robusta*-dominated habitat and the randomization test showed that these values of diversity were significantly different ($P = 0.0001$, **Table 1**).

The result from tree-feeding bird species shows that *V. tortilis*-dominated vegetation has the highest bird species richness with 102 species, followed by *Commiphora spp.* with 66 species and *V. robusta*-dominated vegetation with 59 species (**Table 2**). Bird abundance of tree-feeding species in *V. tortilis*-dominated vegetation showed the highest value of 25 individual birds. The *V. robusta*-dominated vegetation remained with 20 individual birds when compared with *Commiphora* spp. which had 19 individual birds, the difference of individual birds between the two vegetations was not significant ($U = 5151, n_1 = 28, n_2 = 24, P = 0.905$, **Table 2**). The abundance of birds in *V. tortilis*-dominated vegetation was significantly higher than that in *Commiphora* spp. ($U = 4123, n_1 = 28, n_2 = 22, P = 0.0102$). The randomization test results showed that the index of diversity for birds was higher in *V. tortilis* than that for *V. robusta*- and *Commiphora spp.*-dominated vegetations ($P = 0.0001$, **Table 2**).

Index	*V. tortilis*	*V. robusta*	*Commiphora spp.*	*V. tortilis* vs. *Commiphora spp.*	*V. robusta* vs. *Commiphora spp.*
Richness	62	45	56		
Abundance	20	15	29	$U = 8896.5$, $P = 0.246$	$U = 8085.5$, $P = 0.017$
Shannon-Wiener (H′)	3.117	2.651	3.097	$P = 0.65$	$P = 0.0001$

Table 1.
Richness, abundance, and diversity of ground-feeding birds in V. tortilis, Commiphora spp., and V. robusta of the Serengeti National Park.

Index	*V. tortilis*	*V. robusta*	*Commiphora spp.*	*V. tortilis* vs. *Commiphora spp.*	*V. robusta* vs. *Commiphora spp.*
Richness	102	59	66		
Abundance	25	20	19	$U = 4123$, $P = 0.0102$	$U = 5151$, $P = 0.905$
Shannon-Wiener (H′)	3.444	2.926	3.167	$P = 0.0001$	$P = 0.0001$

Table 2.
Richness, abundance, and diversity of tree-feeding birds in V. tortilis, Commiphora spp., and V. robusta of the Serengeti National Park.

3.2 Grassland compared with *V. drepanolobium* and *V. seyal*

V. drepanolobium and *V. seyal*-dominated vegetations supported higher ground-feeding bird species richness but lower overall abundance compared with that of grassland vegetation (**Table 3**). The difference in abundance between *V. seyal* and the western grasslands was significantly different ($U = 7684.0$, $n_1 = 48$, $n_2 = 40$, $P = 0.003$), but not between *V. drepanolobium*-dominated vegetation and the northern grasslands ($U = 8791.5$, $n_1 = 44$, $n_2 = 42$, $P = 0.186$, **Table 3**). However, the results from the randomization test showed that there was a significant difference in diversity between both the *V. drepanolobium* and *V. seyal*-dominated vegetations and their adjacent grassland plots (**Table 3**).

3.3 Tree-feeding bird species in *V. drepanolobium* and *Acacia seyal*

Concerning tree-feeding bird species both *V. drepanolobium* and *V. seyal*-dominated vegetations accommodated higher bird species richness compared with that in grassland habitats (**Table 4**). However, the abundance of birds showed a different pattern: abundance of tree-feeding birds was higher in grassland than in *V. drepanolobium*-dominated vegetations, *V. seyal*-dominated habitats supported a higher abundance of birds when compared with that of grassland vegetation (**Table 4**).

Index	*V. drepanolobium*	Northern grassland	*V. seyal*	Western grassland	*V. drepanolobium* vs. Northern grassland	*V. seyal* vs. Western grassland
Richness	73	53	112	70		
Abundance	13	19	33	43	$U = 8791.5$, $P = 0.186$	$U = 7684$, $P = 0.003$
Shannon–Wiener (H′)	2.730	2.878	2.910	2.759	$P = 0.004$	$P = 0.0002$

Table 3.
Abundance of birds, species richness, and diversity of ground foraging birds in V. drepanolobium, V. seyal, and grasslands of western and northern Serengeti National Park.

Index	*V. drepanolobium*	Northern grassland	*V. seyal*	Western grassland	*V. drepanolobium* vs. Northern grassland	*V. seyal* vs. Western grassland
Richness	38	12	69	27		
Abundance	6	7	22	20	$U = 4329$, $P = 0.0314$	$U = 3628$, $P = 0.0002$
Shannon–Wiener (H′)	2.778	1.571	3.045	1.73	$P = 0.0001$	$P = 0.0001$

Table 4.
Estimation of richness, abundance, and diversity of tree-feeding birds in V. drepanolobium, V. seyal, and grassland in the western and northern Serengeti National Park.

Grass height (cm)	Abundance of birds						
	Commiphora spp	*V. robusta*	*V. tortilis*	*V. drepanolobium*	Northern grassland	*V. seyal*	Western grassland
0–10	147 (21.3)	16 (4.3)	36.4 (11.9)	48.6 (17.7)	21.8 (4.6)	252.6 (17.9)	202.6 (61.9)
10–25	137.2 (15.0)	47.6 (17.5)	73.8 (15.5)	53.8 (12.2)	54.2 (5.5)	188.6 (94.3)	300.2 (97.7)
25–50	93 (23.8)	50 (7.7)	45.2 (16.0)	47.6 (3.3)	62.8 (12.4)	168.8 (44.7)	171 (93.4)
50–75	19.4 (2.8)	49.6 (12.4)	0	16.6 (6.6)	98.4 (20.6)	12.8 (5.3)	28.6 (3.5)
75–100	0	2.4 (2.4)	0	9.2 (2.4)	12 (5.8)	3.8 (3.8)	63.2 (9.6)
>100	0	0	0	0	0	6.2 (4.1)	0

Table 5.
Abundance of ground feeding birds in different habitat types relative to grass height in the Serengeti National Park (n = 6, standard error in brackets).

3.4 The effect of grass height on the abundance of ground-feeding bird species

The abundance of ground-feeding birds in *Commiphora spp.*-dominated vegetations was higher in short grass height and declined consistently as the grass grew taller (from 75 to >100 cm). The abundance of birds in short grass (≤25 cm) in *Commiphora* spp. was significantly higher than that in *V. tortilis* and *V. robusta*-dominated vegetations. In contrast, the abundance of birds in tall grass in *Commiphora spp.*-dominated vegetations was lower than those in *V. tortilis* and *V. robusta*-dominated habitats (**Table 5**). The highest abundance of birds in *V. tortilis* and *V. robusta*-dominated habitats were at the intermediate level of grass heights (10–25 and 25–50 cm levels, **Table 5**). However, the abundance of birds did not decrease linearly with grass height; it reached a peak at the intermediate level of grass heights (10–25 cm, **Figure** 7).

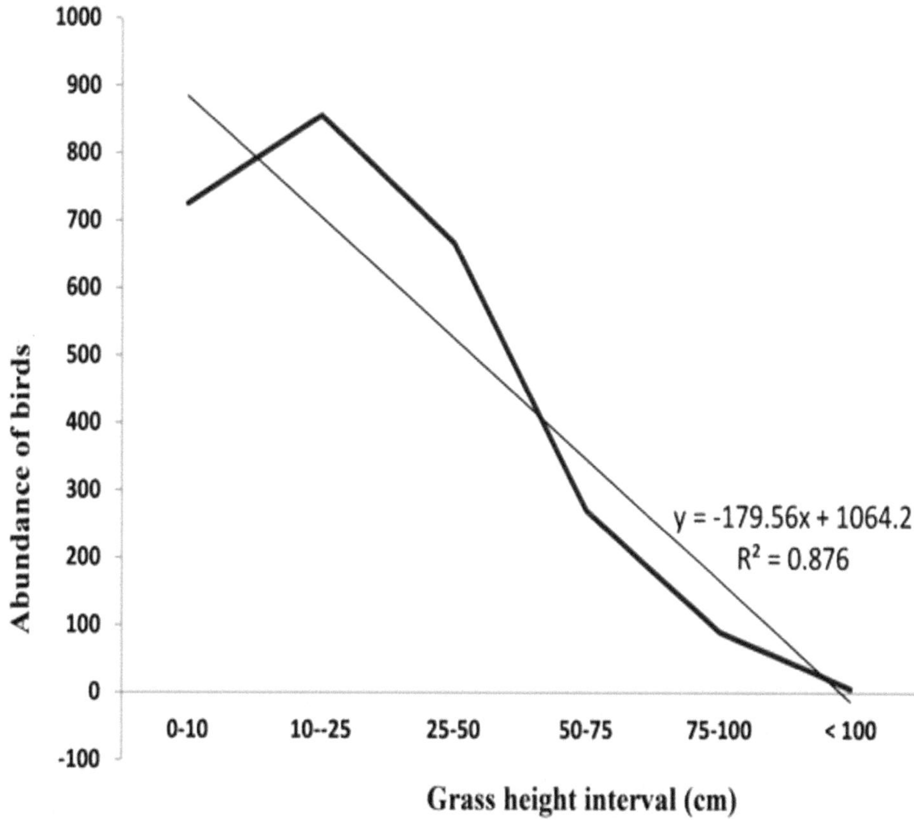

Figure 7.
Linear estimate of birds' abundance in different grass height interval in the Serengeti National Park habitats.

4. Discussion

4.1 Birds' distribution in *Commiphora spp.* compared with *V. tortilis-* and *V. robusta*-dominated vegetations

The higher species richness, diversity, and abundance of tree-feeding birds in *V. tortilis*-dominated vegetation were due to greater tree density and underlying grass cover; which attracted birds for feeding and perching. The results are consistent with the first prediction which suggested that the greater structural complexity of the *Vachellia* trees provided more resources for birds to exploit, and thus increases species diversity [46, 47]. However, the results for *V. robusta*-dominated vegetation, which showed lower richness, abundance, and diversity than those for *Commiphora*-dominated vegetation, are not consistent with this prediction. This suggests that it is not just canopy size that matters but also other factors such as the abundance of food from insects and fruits. The tannin content of leaves in *V. tortilis*-dominated vegetation is much lower than that in *V. robusta*-dominated vegetation making the former more palatable and attractive to insects and hence attracting birds [48, 49].

The higher abundance and diversity of ground-feeding birds in *Commiphora*-dominated vegetation was due to the low canopy cover of *Commiphora* spp. which provided bare ground for birds to forage. It was observed that *Commiphora*-dominated vegetation changed by succession into *Vachellia* vegetation in the same area over the 1970–2000 period [8], change in geology and soil properties and a change in the herb layer species communities probably affected ground-feeding birds. The herb layer does differ between these vegetation type [50–52] in most habitats, vegetation determine the physical structure of the environment, and therefore, have a considerable influence on the distributions and interactions of bird species. Renwald [53], however, pointed out that the ecological distinction between plant communities does not uniformly correlate with differences in animal communities. Such apparent inconsistency is explained by the observation that wildlife species including birds most often select vegetation on the basis of structure rather than plant species composition [15, 54].

4.2 Birds in grassland habitats compared with *V. drepanolobium* and *V. seyal*

Secondly, it is predicted that higher diversity and abundance of birds in *V. drepanolobium*- and *V. seyal*-dominated vegetations than in grassland. The higher abundance of birds in *A. seyal*-dominated vegetation than those in the western grassland of the Serengeti National Park was consistent with this prediction; the differences observed were due to the trees providing greater food resources, trees providing hiding areas from predators, and provision of nesting sites compared with the grassland. However, the opposite result occurred in the northern Serengeti National Park with grassland having a higher abundance than that in *V. drepanolobium*-dominated vegetation, which was inconsistent with the hypothesis. Some of this effect was due to tree-feeding birds moving into grasslands to exploit the available resources, for example, Rufous-tailed weaver *Histurgops ruficauda* and buffalo weavers were frequently observed to forage in grasslands next to *V. drepanolobium* but not *V. seyal*-dominated vegetations. The reason why the mentioned bird species avoid foraging very close to *V. seyal*-dominated vegetation needs more investigation. Mwangomo et al. [22] observed that Rufous-tailed weaver preferred to forage in grassland and avoided feeding in most *Vachellia*-dominated vegetation. This pattern of birds that nest and roost in *V. seyal* moving into grassland to feed was not observed suggesting more research to determine the reason why did not forage close to *V. seyal*-dominated vegetation. So some but not all of the results were consistent with the second hypothesis which predicted that changing habitat structure from grassland to *V. drepanolobium* or *V. seyal*-dominated vegetations may influence increases in the abundance of the birds' fauna. Thus, individual tree species provide particular resources for birds not found in grassland.

4.3 The effect of grass height on the abundance of ground-feeding bird species

The lack of difference in abundance with grass height between this vegetation was due to different groups of birds compensating for each other. For example, widowbird and whydahs were abundant in tall grass but larks, lapwing, coursers, storks, and pipits were abundant in short grass. The higher abundance of ground-feeding birds observed in shorter grass levels (0–25 cm) was contrary to the third hypothesis which predicted that bird abundance and diversity would decline with changing grass height from tall to short grass due to lower grass cover and structure. Grass structure and cover may not be the only factors determining the types of bird species using the grassland. Some species prefer tall grass while others prefer short grass and densities may not be related to the biomass and structure.

Lower grass height in the Serengeti National Park was created by herbivore grazing and anthropogenic fires, which changed the vegetation cover by exposing ground and creating short grass habitats for specialized birds such as the Red-capped Lark *Calandrella cinerea*, Crowned Lapwing *Vanellus coronatus*, African Pipit *Anthus cinnamomeus*, and Fischer's Sparrow-lark *Eremopterix leucopareia*. Although grass height was lower, habitat heterogeneity was greater than in tall grass. In addition, short grass increases the visibility of food sources for birds. Thus, the short grass community was abundant or even more than that in grass height taller than 75 cm community, one may replace the other, a replacement can be of species that prefer to forage in tall grass replaced with those forage in short grass as has been observed on the Serengeti National Park plains [55].

However, there was a higher abundance of birds on short grass areas in *Commiphora*-dominated vegetation than in *V. tortilis* and *V. robusta*-dominated vegetations. Grass height could not explain this observation, which suggests that it is the vegetation structure that is affecting ground bird abundance rather than just the height of the herb layer.

This study found that the abundance of birds did not decrease with grass height; it reached a peak at an intermediate level of grass heights interval of 10–25 cm, contrary to the initial prediction. This was especially clear in the western part of the Serengeti National Park but was also evident in the northern part of the park where bird diversity increased as grass height became shorter (≤25) due to fire and grazing by ungulates. These findings are consistent with the generally accepted pattern that ecological succession following intermediate disturbance is characterized by increases in species richness, equitability, and similarity due to an increase in patchiness of habitat [56–58]. It is generally accepted that animal species richness increases with increasing habitat complexity, given that more complex habitats offer a greater variety for potential exploitation of resources. For example, Brown [59] reported that bird species diversity increased with an increase in grass height in America. However, the present results conform more to the intermediate disturbance hypothesis developed by Connell [60] which suggests that the highest diversity of living things is maintained at intermediate scales of disturbance. Nkwabi et al. [61] reported that both bird species diversity and composition change with grass height. There have been similar results in southern Africa by Jansen et al. [62] who described that both density and species richness of francolins (*Francolinus* spp.) change with changes in grass height. Several studies have revealed that the structure of the vegetation, its complexity, and vertical arrangement are primary defining factors in bird community nesting and foraging [46, 60, 63–68].

5. Conclusion

The results show that *V. tortilis*-dominated habitat has greater species richness, abundance, and diversity compared with *Commiphora*-dominated vegetation or *V. robusta* suggesting that the species of the tree itself, was an important factor determining the distribution of birds. Grasslands in the western and northern part of the Serengeti National Park accommodated a higher abundance of birds due to tree-feeding birds moving into grasslands to exploit the available resources. This pattern of birds nesting and roosting in *Vachellia*-dominated vegetation and moving into grassland to feed was not observed in the grassland and *V. seyal*-dominated vegetation. In addition, *V. seyal*-dominated vegetation provided different food resources from those of *V. drepanolobium*-dominated vegetation so it was the difference in tree species that explained the differences in the abundance of birds. This study found that bird diversity did not decrease linearly with grass height, it reached a peak at intermediate grass heights. In the western and northern parts of the park, bird diversity increased as grass height became shorter due to fire and grazing by ungulates.

Acknowledgements

This work was funded by a Canadian Natural Sciences and Engineering Research Council grant to A.R.E. Sinclair. We thank the Serengeti Wildlife Research Centre, Serengeti National Park, University of Dar es Salaam, Tanzania Wildlife Research Institute, Tanzania National Parks, and Tanzania Commission for Science and Technology for their help and permission to conduct this research. We are grateful to Stephen Makacha, John Mchetto, Joseph Masoy, and Dr. Bukombe John for their assistance with fieldwork.

Conflict of interest

The authors declare no conflict of interest.

Author details

Ally K. Nkwabi* and Pius Yoram Kavana
Western Wildlife Research Centre, Tanzania Wildlife Research Institute, Kigoma, Tanzania

*Address all correspondence to: nkwabikiy@yahoo.com

References

[1] Beale CM, Baker NE, Brewer MJ, Lennon JJ. Protected area networks and savannah bird biodiversity in the face of climate change and land degradation. Ecology Letters. 2013;**16**(8):1061-1068

[2] Stevenson T, Fanshawe J. Field Guide to the Birds of East Africa: Kenya, Tanzania, Uganda, Rwanda, Burundi. London: Bloomsbury Publishing; 2020

[3] Stevenson T, Fanshawe J. A Field Guide to the Birds of East Africa. London: T and A. D. Poyser; 2004

[4] Zimmerman DA, Turner DA, Pearson DJ. Birds of Kenya and Northern Tanzania. Helm Field Guides. Johannesburg: Russel Friedman Books; 1999

[5] Metzger K, Coughenour M, Reich R, Boone R. Effects of seasonal grazing on plant species diversity and vegetation structure in a semi-arid ecosystem. Journal of Arid Environments. 2005;**61**(1):147-160

[6] Holdo RM, Sinclair ARE, Dobson AP, Metzger KL, Bolker BM, Ritchie ME, et al. A disease-mediated trophic cascade in the Serengeti and its implications for ecosystem. Public Library of Science Biology. 2009;**7**(9):e1000210. DOI: 10.1371/journal.pbio.1000210

[7] Homewood K, Lambin EF, Coast E, Kariuki A, Kikula I, Kivelia J, et al. Long-term changes in Serengeti-Mara wildebeest and land cover: Pastoralism, population, or policies? Proceedings of the National Academy of Sciences. 2001;**98**(22):12544-12549

[8] Sinclair ARE, Mduma SRA, Hopcraft JGC, Galvin K, Sharam GJ. Historical and future changes to the Serengeti ecosystem. In: Sinclair ARE, Packer C, Mduma SRA, Fryxell JM, editors. Serengeti III: Human Impacts on Ecosystem Dynamics. Chicago: University of Chicago Press; 2008. pp. 7-46

[9] Roques K, O'connor T, Watkinson A. Dynamics of shrub encroachment in an African savanna: Relative influences of fire, herbivory, rainfall and density dependence. Journal of Applied Ecology. 2001;**38**(2):268-280

[10] Scholes RJ, Walker BH. An African Savanna: Synthesis of the Nylsvley Study. Cambridge: Cambridge University Press; 1993

[11] Benson CW, Irwin MPS. The Brachystegia avifauna. Ostrich. 1966;**37**(S1):297-321

[12] Milewski A, Campbell B. Bird diversity in relation to vegetation types in the Moremi Wildlife Reserve. Transactions of the Royal Society of South Africa. 1976;**42**(2):173-184

[13] Rotenberry JT, Wiens JA. Habitat structure, patchiness, and avian communities in North American steppe vegetation: A multivariate analysis. Ecology. 1980;**61**(5):1228-1250

[14] Bamford AJ, Monadjem A, Hardy IC. An effect of vegetation structure on carcass exploitation by vultures in an African savanna. Ostrich. 2009;**80**(3):135-137

[15] Cody ML. Habitat selection in birds: The roles of vegetation structure, competitors, and productivity. Bioscience. 1981;**31**(2):107-113

[16] Gottschalk TK, Ekschmitt K, Bairlein F. Relationships between

vegetation and bird community composition in grasslands of the Serengeti. African Journal of Ecology. 2007;**45**(4):557-565

[17] Folse LJ. Avifauna-Resource Relationships on the Serengeti Plains. Michigan: Texas A&M University; 1978

[18] Pomeroy DE, Dranzoa C. Methods of studying the distribution, diversity and abundance of birds in East Africa-some quantitative approaches. African Journal of Ecology. 1997;**35**(2):110-123

[19] Oindo BO, De By RA, Skidmore AK. Environmental factors influencing bird species diversity in Kenya. African Journal of Ecology. 2001;**39**(3):295-302

[20] Oindo BO, Skidmore AK, de Salvo P. Mapping habitat and biological diversity in the Maasai Mara ecosystem. International Journal of Remote Sensing. 2003;**24**(5):1053-1069

[21] Estes AB. Avian Diversity in Serengeti National Park: How do Community Differ Between Interior Savanna and Grassland and Exterior Agricultural Landscape? Madison: University of Wisconsin, Madison; 2000

[22] Mwangomo EA, Hardesty LH, Sinclair ARE, Mduma SRA, Metzger KL. Habitat selection, diet and interspecific associations of the Rufous-tailed weaver and Fischer's lovebird. African Journal of Ecology. 2007;**46**(3):267-275

[23] Packer C, Hilborn R, Mosser A, Kissui B, Borner M, Hopcraft G, et al. Ecological change, group territoriality, and population dynamics in Serengeti lions. Science. 2005;**307**(5708):390-393

[24] Folse LJ. An analysis of avifauna-resource relationships on the Serengeti plains. Ecological Monographs. 1982;**52**(2):112-127

[25] Sinclair ARE. Factors affecting the food supply and breeding season of resident birds and movements of Palaearctic migrants in a tropical African savannah. Ibis. 1978;**120**(4):480-497

[26] Sinclair ARE, Mduma SRA, Arcese P. Protected areas as biodiversity benchmarks for human impact: Agriculture and the Serengeti avifauna. Proceedings of the Royal Society of London. Series B: Biological Sciences. 2002;**269**(1508):2401-2405

[27] Nkwabi AK, Sinclair ARE, Metzger KL, Mduma SAR. Disturbance, species loss and compensation: Wildfire and grazing effects on the avian community and its food supply in the Serengeti Ecosystem, Tanzania. Australian Ecology. 2011;**36**(4):403-412

[28] Norton-Griffiths M, Herlocker D, Pennycuick L. The patterns of rainfall in the Serengeti ecosystem, Tanzania. East African Wildlife Journal. 1975;**13**(3-4):347-374

[29] Herlocker DJ. Structure, Composition, and Environment of Some Woodland Vegetation Types of the Serengeti National Park, Tanzania. Michigan: Texas A&M University; 1976

[30] Dublin HT. Vegetation dynamics in the Serengeti-Mara ecosystem: The role of elephants, fire, and other factors. In: Sinclair ARE, Arcese P, editors. Serengeti II: Dynamics, Management, and Conservation of an Ecosystem. Chicago: University of Chicago Press; 1995. pp. 71-90

[31] Dublin HT. Dynamics of the Serengeti-Mara woodlands: An historical perspective. Forest and Conservation History. 1991;**35**(4):169-178

[32] Dublin HT, Sinclair ARE, McGlade J. Elephants and fire as causes of multiple

stable states in the Serengeti-Mara woodlands. The Journal of Animal Ecology. 1990;**59**(3):1147-1164

[33] Bibby C, Jones M, Marsden S. Bird Surveys. Cambridge, UK: Birdlife International; 2000. p. 137

[34] Gill F, Wright M. Birds of the world: Recommended English names. London: Christopher Helm; 2006

[35] McNaughton S. Grazing lawns: Animals in herds, plant form, and coevolution. American Naturalist. 1984;**124**(6):863-886

[36] Greenberg R, Bichier P, Sterling J. Bird populations in rustic and planted shade coffee plantations of eastern Chiapas, Mexico. Biotropica. 1997;**29**(4):501-514

[37] Faria D, Laps RR, Baumgarten J, Cetra M. Bat and bird assemblages from forests and shade cacao plantations in two contrasting landscapes in the Atlantic Forest of southern Bahia, Brazil. Biodiversity and Conservation. 2006;**15**(2):587-612

[38] Fry CH, Keith S. The Birds of Africa Volume VII. Sparrows to Buntings. London: Christopher Helm; 2004

[39] Fry CH, Keith S, Urban EK. Birds of Africa Volume VI. Picathartes to Oxpeckers. San Diego, CA: Academic Press; 2000

[40] Motulsky HJ. GraphPad InStat version 3.00. San Diego, CA: GraphPad Software, Inc.; 2009

[41] Seaby RMH, Henderson PA. Species Diversity and Richness Version 4.1.2. Lymington: Pisces Conservation Ltd; 2007

[42] Walther BA, Martin J-L. Species richness estimation of bird communities: How to control for sampling effort? Ibis. 2001;**143**(4):413-419

[43] Hammer Ø, Harper DAT, Ryan PD. PAST: Paleontological statistics software package for education and data analysis. Palaeontologia Electronica. 2001;**4**(1):1-9

[44] Statistix-10. Analytical Software. Tallahassee, Florida: Statistix-10; 2013

[45] Solow AR. A simple test for change in community structure. Journal of Animal Ecology. 1993;**62**(1):191-193

[46] Tews J, Brose U, Grimm V, Tielbörger K, Wichmann M, Schwager M, et al. Animal species diversity driven by habitat heterogeneity/diversity: The importance of keystone structures. Journal of Biogeography. 2004;**31**(1):79-92

[47] Kristan WB. The role of habitat selection behavior in population dynamics: Source–sink systems and ecological traps. Oikos. 2003;**103**(3):457-468

[48] Feeny P. Seasonal changes in oak leaf tannins and nutrients as a cause of spring feeding by wintr moth caterpillars. Ecology. 1970;**51**(4):565-581

[49] Mduma SRA, Sinclair ARE, Turkington R. The role of rainfall and predators in determining synchrony in reproduction of savanna trees in Serengeti National Park, Tanzania. Journal of Ecology. 2007;**95**(1):184-196

[50] Anderson MT, Metzger KL, McNaughton SJ. Multi-scale analysis of plant species richness in Serengeti grasslands. Journal of Biogeography. 2007;**34**(2):313-323

[51] Anderson MT, Dempewolf J, Metzger KL, Reed DN, Serneels S. Generation and maintenance of

heterogeneity in the Serengeti ecosystem. In: Sinclair ARE, Packer C, Mduma SRA, Fryxell JM, editors. Serengeti III: Human Impacts on Ecosystem Dynamics. Chicago: University of Chicago Press; 2008. pp. 135-182

[52] Metzger KL. The Serengeti Ecosystem: Species Richness Patterns, Grazing, and Land-Use, Natural Resource Ecology Laboratory. Fort Collins, CO: Colorado State University; 2002

[53] Renwald JD. The effect of fire on woody plant selection by nesting nongame birds. Journal of Range Management. 1978;**31**(6):467-468

[54] Rice J, Anderson BW, Ohmart RD. Comparison of the importance of different habitat attributes to avian community organization. The Journal of Wildlife Management. 1984;**48**(3):895-911

[55] Nkwabi AK, Sinclair ARE, Metzger KL, Mduma SRA. Disturbance, species loss and compensation: Wildfire and grazing effects on the avian community and its food supply in the Serengeti ecosystem, Tanzania. Austral Ecology. 2011;**36**(4):403-412

[56] Bazzaz F. Plant species diversity in old-field successional ecosystems in southern Illinois. Ecology. 1975;**56**:485-488

[57] Ferreira SM, Van Aarde RJ. Changes in community characteristics of small mammals in rehabilitating coastal dune forests in northern KwaZulu/Natal. African Journal of Ecology. 1996;**34**(2):113-130

[58] Brown VK, Southwood TRE. Trophic diversity, niche breadth and generation times of exopterygote insects in a secondary succession. Oecologia. 1983;**56**(2-3):220-225

[59] Brown K. Effects of Tree Species, Number of Trees, Basal Area and Understory Vegetation on the Abundance and Diversity of Avian Species. Unpublished Bird Project Paper. Biological Station, University of Michigan; 2008

[60] Connell JH. Diversity in tropical rain forests and coral reefs. Science. 1978;**199**(4335):1302-1310

[61] Nkwabi AK, Sinclair ARE, Metzger KL, Mduma SRA. The effect of natural disturbances on the avian community of the Serengeti woodlands. In: Sinclair ARE, Metzger K, Mduma SRA, Fryxell J, editors. Serengeti IV: Sustaining Biodiversity in a Coupled Human-Natural System. Chicago: University of Chicago; 2015. pp. 395-418

[62] Jansen R, Little R, Crowe T. Implications of grazing and burning of grasslands on the sustainable use of francolins (Francolinus spp.) and on overall bird conservation in the highlands of Mpumalanga province, South Africa. Biodiversity and Conservation. 1999;**8**(5):587-602

[63] MacArthur RH, MacArthur JW. On bird species diversity. Ecology. 1961;**42**(3):594-598

[64] Willson MF. Avian community organization and habitat structure. Ecology. 1974;**55**:1017-1029

[65] Willson MF, De Santo TL, Sabag C, Armesto JJ. Avian communities of fragmented south-temperate rainforests in Chile. Conservation Biology. 1994;**8**(2):508-520

[66] Loiselle BA, Blake JG. Annual variation in birds and plants of a tropical second-growth woodland. Condor. 1994;**96**:368-380

[67] McCoy ED, Mushinsky HR. Effects of fragmentation on the richness of vertebrates in the Florida scrub habitat. Ecology. 1994;**75**:446-457

[68] Whelan CJ. Foliage structure influences foraging of insectivorous forest birds: An experimental study. Ecology. 2001;**82**(1):219-231

Chapter 8

Impact on Forest and Vegetation Due to Human Interventions

Ramesh Prasad Bhatt

Abstract

Forest and vegetation play an important role in balancing ecosystem patterns, providing food security, and blessing the environment for living beings, so the status of global forests and biodiversity, their impact and change overtime with climatic effects and challenges is important. This study's methods include a review of global forest cover and status; distribution, and assessment; biodiversity, forest carbon assessment; causes of forest loss; and the impacts and implications of CO_2 emissions. Forests encompass 31% of the world's forests, are home to 2 million to 1 trillion species, and provide habitat for 80% of amphibian species, 75% of bird species, 68% of mammalian species, and so on. Deforestation is the major cause of forest loss, with a decrease of 4.7 million ha. From 2010 to 2020, only in the Asia Pacific region and from 2000 to 2010, 13 million ha of world forests were lost. All flora, fauna, and microbes are slowly degrading and disappearing due to human activities such as deforestation, intensive use, inappropriate forest management, agriculture, encroachment of forest land, slash burn practices, forest fires, urbanization, overharvesting, environmental deterioration, etc. Because the globe has emitted over 1.5 trillion tonnes of CO_2 since 1751, the persistence of biodiversity in human-modified habitats is crucial for conservation and the provision of ecosystem services.

Keywords: forest and vegetation, forest cover, biodiversity, impacts, interventions

1. Introduction

Humans have changed forests, browbeaten species, fragmented wildlife's, altered habitats, imported exotic pests and rivals, and domesticated favored species. All have had an impact on genetic diversity (both within and across species) through their effect on evolutionary developments such as destruction, assortment, implication, gene flow, and transformation. Several literatures illustrate the main impacts of human intervention causing deforestation, fragmentation, grazing, forest clearance for the development, intensive use, desertification, inappropriate forest management, agriculture, encroachment of forest land, slash burn practices, forest fire, urbanization, overharvesting, environmental deterioration, and so on.

Deforestation is one of the most significant drivers of rising greenhouse gas emissions because forests also remove CO_2 from the atmosphere. Limiting global warming to 1.5°C is out of reach without immediate and significant reductions in emissions across all sectors [1]. Southeast Asia has lost regional forest cover at a rate of 1% per year over the last 15 years [2]. Deforestation has a greater impact on tropical forests than ever before, accounting for 60% of forest loss in Latin America and Southeast Asia [3]. According to the research, logging has a disproportionate impact on deforestation processes in Southeast Asia, whereas deforestation in arid and populated regions of East Africa and South Asia appears to be driven mostly by demand for fuelwood [4].

Several studies explain that potential drivers of deforestation are international trade in Brazil [5], deforestation in Amazonia due to the comparative advantages of agriculture in South America [6], agricultural products, leading to agricultural land expansion and in turn promoting deforestation [7], weak governance in developing countries with forests often leads to higher rates of deforestation [8], and about 80% of current global deforestation is supposedly due to agricultural production [9]. Thus, agriculture, grazing, the use of firewood and charcoal, and forest fires are the primary causes of deforestation. The main reasons for deforestation are poverty and rapid population growth [10].

Fragmentation occurs due to the continuous development of cities and related infrastructure. Ledig et al. [11–13], forest fragmentation is a broad issue that affects global forest biodiversity, ecosystem function, and ecosystem services [14], forest fragmentation can affect the forest ecosystem's long-term health and vitality, leading to species extinction [15], increases in agriculture, logging, and urban growth during the past decades caused unprecedented losses of tropical forest [16]. Europe had the most fragmentation caused by humans, while South America had the least. Humans have fragmented or eliminated about half of the temperate broadleaf and mixed forest biomes, as well as roughly one-quarter of the tropical rainforest biome [17].

Grazing is responsible for the loss of forest and vegetation in many parts of the world where conventional forest management practices are used. Forest grazing is also common in Bhutan and the Himalayan coniferous forests [18]. Forest vegetation depletion is particularly severe in northern Ethiopia's highlands [19, 20]; practically all the available area is under cultivation or used for pasture and reported severe deforestation due to forest clearances [21]. The Swiss Alps have a long history of forest grazing [22]. Wood pastures are a distinguishing feature of the traditional European rural landscape [23]. Other significant factors of forest loss and degradation in the Siwaliks and midhills include overgrazing, which is a major driver in Nepal's Siwaliks and high highlands [24].

Agricultural growth is responsible for approximately 80% of worldwide deforestation, with infrastructure improvements such as roads and dams, as well as mining and urbanization, accounting for the remaining sources of deforestation [25]. Forest removal, as well as accompanying grazing and mining activities, has increased erosion and landslides in the Dolomites, the Maritime Alps, and the south-central Italian Alps [26]. According to the mining businesses, individual miners remove significant sections of forest in Ecuador, Peru, and Venezuela [27, 28]. Forests and woods cover 22% of Africa's total land area. Firewood is the most important forest product, as well as the primary source of energy for the majority of African families. East Africa's annual rate of deforestation increased from 0.7% between 1981 and 1990 to 1% between 1990 and 2000 [29, 30].

Drylands represent around 38% of the Earth's land area, including much of North and southern Africa, western North America, Australia, the Middle East, and Central Asia. Approximately 2.7 billion people live in drylands. Because of scant and irregular rainfall as well as inadequate soil fertility, drylands are especially vulnerable to land degradation. Plowing, grazing, or deforestation, as well as poor land management and agricultural growth, all contribute to this. As a result, India, Pakistan, Zimbabwe, and Mexico have been identified as being particularly vulnerable to degradation [31].

Forest fires contribute to global greenhouse gas emissions and have the potential to harm human health. Fires are a natural aspect of the dynamics in boreal forests, while they are mostly man-made in the humid tropics. Due to a lack of trustworthy data, global trends in fire-related forest loss remain ambiguous [32]. The Forest Resources Assessment (FRA) 2020 [33] has reported a regional total of "tree cover area burned." This was calculated by crossing a 500-m resolution burned area map [34] and a global 30-m tree cover map from the year 2000 by Hansen et al. [35].

Inadequate forest management is one of the factors contributing to the detrimental human effect on urban and suburban forests [36]. Another issue associated with the decline of forest vegetation and its consequences for human health and biodiversity is environmental degradation. Environmental degradation is due to industrial and urban emissions as well as the presence of pollutants in the atmosphere (sulfur dioxide, ozone, nitrogen oxides, and particulate matter PM_{10} and $PM_{2.5}$) and soil contaminants (heavy metals and acid deposition) [11]. Together with SO_2, NH_3 and NOx contribute to soil acidification, habitat alteration, and biodiversity loss. Ground-level O_3 harms forests by slowing their growth [37].

Protecting, restoring, and encouraging the protection and sustainable use of terrestrial and other ecosystems are all required for the survival of various sorts of life on land. Thus, Goal 15 focuses on sustainable forest management, minimizing and reversing land and natural habitat degradation, combating desertification effectively, and ending biodiversity loss. All of these projects seek to ensure that the benefits of land-based ecosystems, such as sustainable livelihoods, are available to future generations.

2. Material and methods

2.1 Review of literature

The important literature concerning the loss of forest and biodiversity was reviewed from different dimensions. Forest assessment biodiversity related information is made by the global forest assessment reports, books, research articles, and proceedings. Climate change is a significant threat to forests, as is the link between forests and climate change, which was assessed by UNEP, the World Bank, and other published works. Other man-made implications were assessed through assessment studies, research, and publications, including difficulties in preserving ecological integrity. The assessment finds gaps in the enabling conditions that necessitate additional research and action.

2.2 Use of ArcGIS, global forest maps and database

Although global deforestation rates average 13 million hectares per year, around 30 percent of the world's land surface is currently forested [38]. The UN Convention on Biological Diversity (CBD) has set a target of 10% protected area coverage. The MODIS05 VCF dataset identified forest zones that did not exist in the original GFM. Plantations, shrub lands, and agricultural areas are examples of non-natural regions that could be included on the GLC 2000 map.

2.3 Assessment of geographical distribution

The world's forests are critical for biodiversity conservation as well as climate mitigation. The use of remotely sensed data to create new forest status and forest change spatial layers has revolutionized forest monitoring around the world. The review of simulated biodiversity values uses remote sensing data on tree cover to create worldwide maps of the importance of forest biodiversity.

Many vulnerable species rely on intact forest landscapes, including "primary" forests. For more than 40 years, remote sensing has been recognized as an essential tool for understanding land cover and land use. The phrase "urban/suburban forest" is used to refer to a type of urban or suburban forest in Europe as well as one of the following categories: location; forest type (woodland), the documented, or at least indicated, problem (pressure and threat to nature), and the quality of the information source.

2.4 Assessment of forest cover and biodiversity

The USGS Land Cover Institute provided tree cover data for 2010–2017 [39]. The FAO utilizes a 10% MCC criteria to evaluate if an area has been deforested [40].

According to Hansen et al., a number of 25% can be used to calculate global deforestation [41]. As Whittaker outlined, three commonly used criteria for measuring species-level biodiversity, including measurements such as species diversity, endemism, and genetic variety, were considered to assess the consequences [42]. Because of this heterogeneity, comparably sized areas of tree cover mapped via remote sensing can vary dramatically in biological value. Remote sensing represents an important tool for looking at ecosystem diversity, forest cover, and various structural aspects of individual ecosystems. Many different forms of remote sensing sources are reviewed and assessed to provide a means to make assessments across several different spatial scales and changes in ecosystem patterns over time.

3. Results and discussion

3.1 Global Forest assessment

Forests are the most important ecosystems on the planet because they include a diverse range of plant species and are home to a diverse range of animal species, including microorganism. Forests cover over 31% of the total surface area on the planet or 4.06 billion hectares (40 million square kilometers). With around 1015 million hectares of forest cover, Europe is the second-smallest continent by area. With 842 million hectares of forest cover, South America comes in forest area in second category. With 593 ha of forest, Asia, the world's largest continent, has the second-smallest forest area. More than half of the world's forest cover is accounted for by five countries: Russia, Brazil, Canada, the United States, and China. When it comes to our world's forests, deforestation and degradation of forest significant issues: by 2030, nearly 47% of the world's forests will be deforested or degraded. More than half of the world's forest cover is accounted for by five countries: Russia, Brazil, Canada, the United States, and China. However, due to their huge size, forests cover just a small percentage of the land areas in most of these countries. Suriname, Guyana, the Federated States of Micronesia, and Gabon have the largest proportion of forest cover, with forest covering at least 90% of their land areas. More than two-thirds of the world's total forest acreage is shared by 10 countries [43].

Russia, the world's largest country, has by far the most forest cover. With 815 million hectares, the country possesses more than one-fifth of the world's forest acreage (20.1 percent), making it the most wooded country on the planet. Russia's forest cover accounts for around 45% of its entire land area and 5.5% of the global land area. Only Canada, the United States, China, and Brazil have a larger overall land area than Russia. It also accounts for around 81% of Europe's total forest area and is the sole reason that the continent has the largest forest area among the seven continents. Russia has four categories of forest: recreational, reserve, field, and waterproof. The Russian forestry industry provides approximately $200 billion each year. The 10 countries with the most forest cover are shown in **Figure 1**.

Forests are important resources for climate regulation. Every square kilometer of land in Amazonia emits 20 billion tons of water into the sky every day. That is 3 billion tons greater than the amount of water that pours out of the Amazon, the world's most plentiful river [44]. The most serious dangers to forests around the world are deforestation and forest degradation. Deforestation occurs when forests are converted to non-forest uses such as agriculture and road development. Forest ecosystems are said to be degraded when they lose their ability to provide vital goods and services to humans and the environment. More than half of the world's tropical forests have been destroyed since the 1960s, with more than 1 hectare destroyed or severely degraded every second. Cattle, insects, illnesses, forest fires, and other human-related activities affect an estimated 3.7 million hectares of Europe's woods [45].

3.1.1 Deforestation

Tropical deforestation is caused by a complex interplay of natural forces (social, ecological, economic, environmental, and biophysical). The particular mix of drivers differs by region of the world, country, and locality. Population growth, density, and

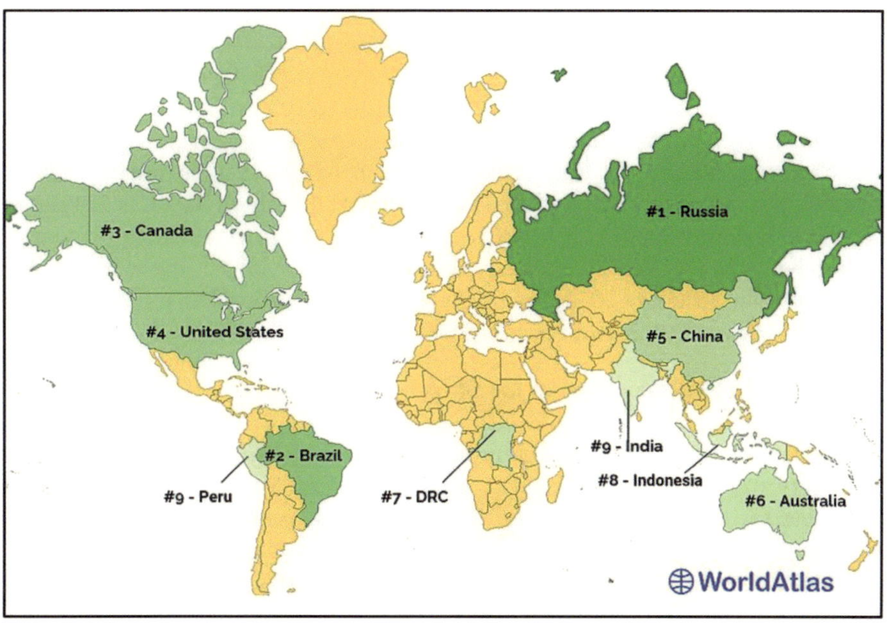

Figure 1.
Forest cover in the world (%).

spatial dispersion are rarely the primary causes of deforestation. Taxation, subsidies, corruption, property rights, and other institutional elements are usually linked to economic forces. Cultural and sociopolitical variables, such as a lack of public support for forest protection and sustainable use, are also important [46]. Many areas continue to experience high rates of forest loss and degradation. Tools that can offer an integrated assessment of human impacts on forest biodiversity are needed. The modeling toolkit method proposed by Sturvetant et al. could be useful in these situations [47].

Deforestation and forest degradation are both harmful to forest health, but there is a distinction to be made. Brazil has around 497 million hectares of forest, accounting for about 12.2% of the world's total forest area. Around 92% of Brazil's forest is categorized as primary, which means it is carbon-dense and diversified. Deforestation in the Brazilian Amazon has hit the highest annual level in a decade. However, Brazil has set a goal of slowing the pace of deforestation to 3900 sq. km annually by 2020.

In the Amazon, forest degradation is a widespread occurrence that often affects a significantly larger area than clear-cut deforestation. Each year between 2007 and 2016, an average of 11,000 km^2 of forest was degraded. In the same time period, this is twice the annual average for deforested lands. While deforestation progressed at a reasonably consistent rate during the study period, degradation fluctuated greatly over time, particularly from 2009 to 2016. The total degraded area per year fluctuated from a low of 2700 km^2 in 2014 and a high of 23,700 km^2 in 2016 [48].

The Asia-Pacific Region's total forest area in 1990 was 733.4 million ha, 726.3 million ha in 2000, 737.8 million ha in 2005, and 740.4 million ha in 2010, accounting for approximately 18.3% of the global forest area. Deforestation in the Asia-Pacific Region has decreased from an annual loss of more than 0.7 million hectares of forest from 1990 to 2000 to an annual increase of 0.5 million hectares from 2005 to 2010. There has been a considerable decrease in the net annual gain in forest area since 2005, from about 2.3 million hectares between 2000 and 2005 to approximately 0.5 million hectares between 2005 and 2010, **Figure 2**, [49].

According to FRA 2020, the rate of net forest loss decreased from 7.8 million ha per year in the 1990–2000 decade to 5.2 million ha per year in the 2000–2010 decade and 4.7 million ha per year in the 2010–2020 decade. In 2015, a statistical profile of the world's forest assessment revealed 3999 individuals and 234 nations and territories have a total forest area of 2 million hectares, with an annual change rate of 0.13%. Since 1990, the ration has dropped by 31.6%, from 4128 million hectares in 1990 to 3999 million hectares in 2015 [50]. According to the FAO, forests span approximately 3.9 billion hectares (or 9.6 billion acres), or approximately 30% of the world's land surface. Between 2000 and 2010, the FAO estimates that around 13 million hectares of forest were converted to other uses or lost due to natural causes.

3.1.2 Biodiversity in human-modified landscapes

The persistence of biodiversity in human-modified environments is critical for conservation and the maintenance of ecosystem services. Studies of biodiversity in settings where humans live, work, and extract resources could aid in the development of sound policies. However, research should cover relevant areas, and study topic biases should not result in gaps in the evidence base. Biodiversity is an important resource on the earth, but the world is in a declining stage of biodiversity. All flora, fauna, and microbes are slowly degrading and disappearing due to human activities. Global terrestrial forests account for 75% of terrestrial gross primary output and

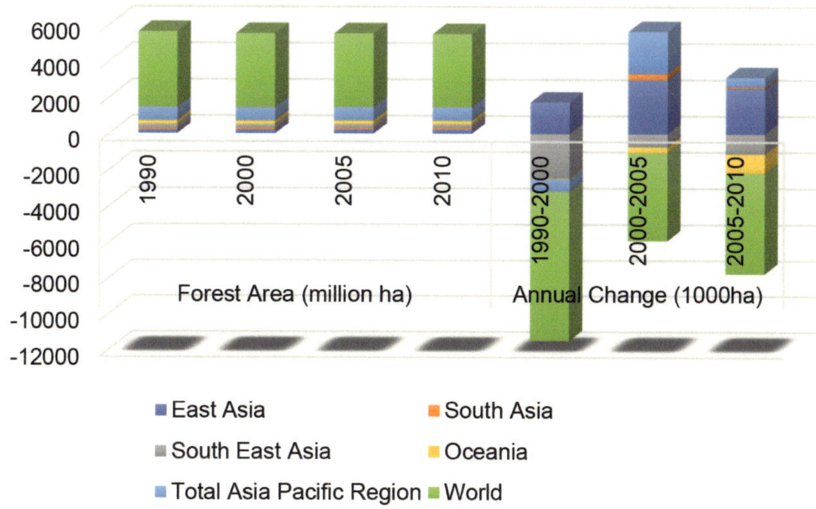

Figure 2.
Forest area change.

80% of Earth's total plant biomass, encompassing 4.03 billion hectares or 30% of the planet's total land area [51].

3.1.3 Assessment of carbon and human alteration

The long-term or permanent transfer of forest to other land uses is referred to as deforestation. Deforestation and forest degradation constitute roughly one-fifth of total greenhouse gas emissions globally. Deforestation can have far-reaching consequences for society and the environment, from the local to the global. Forest users and managers are concerned about deforestation because it threatens their livelihoods [52]. Forest managers can help to reduce deforestation by improving knowledge of the importance of forests in landscapes. Deforestation can also harm the production, biodiversity, and health of neighboring forests. Forests are both a source of carbon emissions and a carbon sink due to forest fires, the carbon imbalance of old trees, and other non-cleaning forest processes. Forest investment and management of existing forests to address environmental issues are potential carbon-reduction strategies [53–55].

Forests play an important role in the Earth's carbon cycle, storing and releasing this vital element in a dynamic cycle of growth, decay, disturbance, and rejuvenation. Forests have helped to mitigate climate change by absorbing roughly one-quarter of the carbon released by human activities such as fossil fuel combustion. The carbon balance of the Earth is the ratio of CO_2 emissions to CO_2 uptake by oceans and terrestrial systems. Photosynthesis absorbs carbon from the atmosphere and deposits it in forests. Carbon sequestration refers to the process of carbon absorption and deposition. Since 1970, the net carbon balance has risen from 280 parts per million to more than 390 parts per million [56].

Terrestrial ecosystems are important players in the global carbon cycle. Annually, an estimated 125 Gt of carbon is exchanged between vegetation, soils, and the atmosphere. Forests account for over 80% of this exchange; research suggests that deforestation in the 1980s may have accounted for a quarter of all human carbon emissions.

Carbon is stored in both live and dead biomass, including standing timber, branches, foliage, and roots, as well as litter and woody debris. Any activity that changes the amount of biomass in vegetation and soil has the ability to either absorb carbon from the atmosphere or release carbon into it. CO_2 emissions were totaled. Globally, there are approximately 3870 million acres of forest, with over 95% of it being natural. While forest areas in rich countries have stabilized, deforestation in underdeveloped countries has continued. The 2001 edition of The State of the World's Forests highlights two recent causes of forest destruction [57].

The CO_2 levels in the atmosphere have risen from 400 parts per million (ppm) for the first time in 55 years of measurements to over 410.79 ppm in the latest CO_2 reading [58]. Human activities have emitted almost 400 petagrams of carbon (C) into the atmosphere. Human activities, such as the combustion of fossil fuels and land usage, contribute to the atmospheric CO_2 content. Plants and soils retain about 2000 PgC, with forests and forest soils containing 60% of this amount. Changes in human activities could aid in the preservation of forest carbon stores and promote more CO_2 uptake and storage [59].

3.2 Forest biodiversity

The range of living organisms that occupy forests, as well as the ecological responsibilities they play in an ecosystem, is referred to as forest biological diversity. It includes not just trees, but also the numerous plants, animals, microorganisms, and species that live within them. Forest biological diversity can be considered at several levels, including ecosystem, landscape, species, population, and genetic. According to the state of the world's forests 2020, the majority of the Earth's terrestrial biodiversity is found in forests. Forests provide habitat for 80% of amphibian species, 75% of bird species, and 68% of mammalian species. A number of fish and shellfish species use mangroves as breeding grounds and nurseries. They help to collect sediments that would otherwise harm seagrass meadows and coral reefs.

It listed 2.12 million species in the world in 2020. **Figure 3** shows that the number of described species in the world is 105 million insects, over 11,000 birds, over 11,000 reptiles, and over 6000 mammals [60].

The overall variability of life on Earth is characterized as global biodiversity. The current number of species on Earth is estimated to be between 2 million and 1 trillion [61]. Biodiversity has increased and decreased over time for (supposedly) abiotic reasons such as climate change. Biodiversity loss involves both the global extinction of many species and the local decline or loss of species in a specific environment. The latter phenomena can be either temporary or permanent, depending on whether the environmental deterioration that causes the loss is reversible via ecological restoration or ecological resilience or is effectively permanent [62, 63].

3.3 Impacts and changes due to human intervention

Ecological succession is the relatively predictable shift in forest types over time, typically decades. Environmental factors such as soil type, water regimes, vegetation history, climate, and invasive species all have an impact on succession. All of these characteristics are influenced by humans, yet the relationship between them and humans might be ambiguous.

Forest lands are increasingly under development pressure, which may result in parcelization and fragmentation. When the forest canopy is dissected for houses,

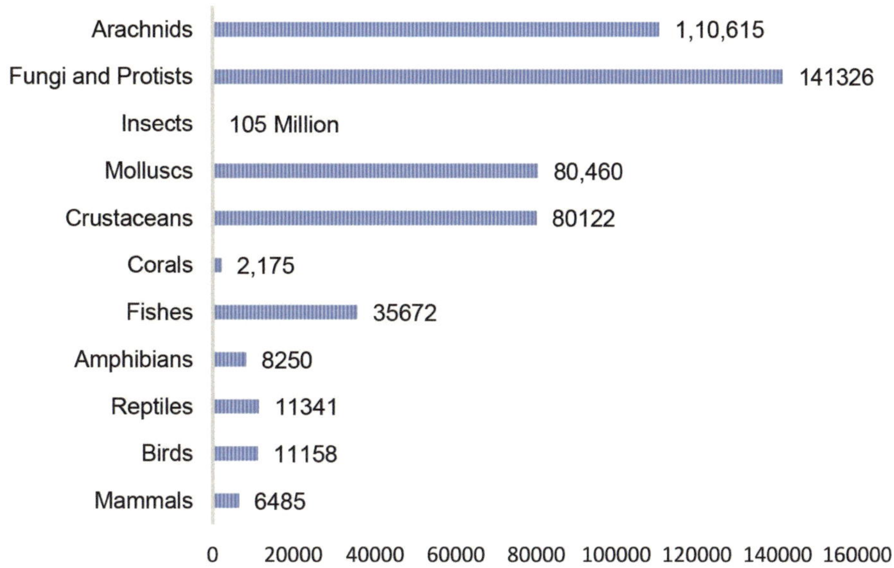

Figure 3.
Numbers of described species in the world.

lawns, roadways, and other infrastructure, this is referred to as fragmentation. The annual net loss of forest area decreased from 7.8 million hectares in the 1990s to 4.7 million hectares from 2010 to 2020 [64]. The presence of more humans in the landscape raises the chance of exotic invasive species spreading. Invasive characteristics in native species can sometimes be promoted. Human impacts on forests have altered key biological traits, allowing species such as deer, Pennsylvania sedge, and ironwood to become invasive at times.

These native species have, in turn, weakened ecological dynamics even further. Although the effects of climate change have been clearly documented, the effects on forests have been more difficult to determine. Predictions of future forest effects are much less trustworthy. Changes in carbon dioxide levels, land use, natural cycles, and other factors all have an impact on climate change. Temperature, precipitation, and extreme events are all showing the consequences.

The best way for forest owners to prepare their woods for change and work with change is to actively manage to lessen environmental stress. Forest management, including timber harvesting, has been shown to increase commodities and services. Management leads to a more resilient and healthier forest.

The increase and contraction of forest cover are erratic. Deserts, farms, and urban areas flow all over the world, and although some countries are rapidly removing trees from their ecosystems, others are increasing their forest cover. Since 1990, the world's forested land area has shrunk by 2 million square miles (3.1 million square kilometers), with the majority of the losses occurring in South America and Sub-Saharan Africa. Human activities have put a significant burden on the Amazon Rainforest, one of the world's most important carbon sinks, in recent decades. Brazil's expanding road network has been critical to economic success, but the landscape has frequently suffered as the country's GDP per capita rises [65].

Deforestation disrupts ecosystems that are essential to both animals and humans. Every year, we take down more than 15 billion trees. Humans have transformed 420 million hectares of wooded land into different uses since 1990. Over one billion acres of forest have been removed to make space for strip mining, cattle grazing, and industrial sprawl. Animal feces from factory farms pollute the air, water, and land, hastening climate change. More greenhouse gas emissions from industrial agriculture remain in the atmosphere when forests are cut down. Forests operate as a "carbon sink," collecting CO_2 and converting it into the oxygen we breathe [66].

During Australia's "Black Summer" season, which began on January 1, 2019, more than 24 million hectares (59 million acres) were burned. Fires raged through forests in Victoria, Queensland, and New South Wales for 8 months. More than 510,000 hectares were burned in one incident (1.26 million acres). The total area burned during the Black Summer is believed to be 24 million hectares (59 million acres), nearly the size of the whole United Kingdom [67].

The main causes of deforestation are forest fire, livestock grazing, commercial agriculture, growing animal feed, excessive use of palm oil, illegal logging, mining extraction, paper production, urbanization, and desertification of land. People who live near woods bear the brunt of deforestation's consequences. Forests are home to millions of wild animal and plant species. When humans destroy trees for short-term economic gain, we endanger our species' long-term survival. Thus, each nation must first protect the natural forests and biodiversity, cope with development in an environmentally friendly manner, mitigate long-term impacts onsite, promote plantation, protect natural habitats, and control environmental pollution.

4. Conclusion

Forest lands are increasingly being pressured for development, which may result in parcelization and fragmentation. Fragmentation occurs when the forest canopy is cut up for houses, lawns, roadways, and other infrastructure. The increased presence of people increases the likelihood of exotic invasive species spreading. Since 1990, the world's forested acreage has shrunk by 2 million square miles (3.1 million square kilometers). Forests act as "carbon sinks," absorbing CO_2 and turning it into the oxygen we breathe. More than one billion acres of forest have been cleared to make way for strip mining, cattle grazing, and industrial sprawl.

One of the major contributors to increased greenhouse gas emissions is deforestation. Deforestation is responsible for 60% of forest loss in Latin America and Southeast Asia. Poverty and rapid population expansion are the primary causes of deforestation. Agricultural output is said to be responsible for over 80% of current world deforestation. Forest fragmentation can have a long-term impact on the health and vitality of the forest ecosystem. Human-caused fragmentation was greatest in Europe, whereas it was least in South America. Forest removal, as well as accompanying grazing and mining activities, has exacerbated erosion and landslides in the Dolomites, the Maritime Alps, and the south-central Italian Alps.

Drylands cover around 38% of the Earth's land surface. Much of northern and southern Africa, western North America, Australia, the Middle East, and Central Asia are among them. Land degradation has been noted as a specific threat in India, Pakistan, Zimbabwe, and Mexico. Inadequate forest management is one of the elements contributing to the negative human impact on urban and suburban forests. Pollutants in the atmosphere, as well as emissions from industry and cities (sulfur dioxide, ozone, nitrogen oxides, and particulate matter PM_{10} and $PM_{2.5}$), all contribute to environmental degradation.

Biodiversity is a valuable resource on the planet, but the globe is experiencing a decline in biodiversity. Biodiversity research in areas where humans live, work, and extract resources may contribute to the formulation of sensible policy. However, research should cover important topics, and study topic biases should not result in evidence gaps. Extinction and speciation have an impact on global biodiversity. Mammal species, for example, have a mean life span of 1 million years. Biodiversity has increased and decreased over time for (supposedly) abiotic reasons.

The primary causes of global deforestation are logging, shifting agriculture, agricultural expansion, and urbanization. To reverse deforestation and biodiversity loss, we must alter our food systems. Agribusinesses must follow through on their commitments to deforestation-free commodity chains. People who live near woodlands bear the brunt of the repercussions of deforestation. Millions of wild animals and plant species live in forests. When humans damage trees for short-term economic benefit, we threaten the long-term existence of our species. Each country must first safeguard its natural forests and wildlife.

Author details

Ramesh Prasad Bhatt
Atlantic International University, Honolulu, Hawaii, USA

*Address all correspondence to: drrameshbhatta@gmail.com

References

[1] IPCC. The evidence is clear: the time for action is now. We can halve emissions by 2030 [Internet]. 2022. Available from: https://www.ipcc.ch/2022. https://www.ipcc.ch/2022/04/04/ipcc-ar6-wgiii-pressrelease

[2] Miettinen J, Shi C, Liew SC. Deforestation rates in insular Southeast Asia between 2000 and 2010. Global Change Biology. 2010;**17**:2261-2270

[3] CDP. The Money Tree, the Role of Corporate Action in the Fight Againest Deforestration. London, UK: CDP Worldwide; 2019

[4] Rudel TK, Flesher K, Bates D, Baptista S, Holmgren P. Tropical Deforestation Literature: Geographical and Historical Patterns. New Brunswick, New Jersey, United States: Department of Human Ecology, Rutgers University; 2000

[5] Almeida, Faria WR, Alexandre. Relationship between openness to trade and deforestation: Empirical evidence from the Brazilian Amazon. In: Ecological Economics. 2016;**121**:85-97. DOI: 10.1016/j.ecolecon

[6] Schmitz C, Kreidenweis U, Lotze-Campen H, Popp A, Krause M, Dietrich JP, et al. Agricultural trade and tropical deforestation: Interactions and related policy options. Regional Environmental Change. 2015;**15**:1757-1772

[7] Angelsen A, Kaimowitz D. Rethinking the causes of deforestation: Lessons from economic models. The World Bank Research Observer. 1999;**14**(1):73-98

[8] Barbier EB, Damania R, Léonard D. Corruption, trade and resource conversion. Journal of Environmental Economics and Management. 2005;**50**(2):276-299

[9] FAO. Global Forest Resources Assessment 2015; How Have the world's Forests Changed? Rome, Italy: FAO; 2015. p. 352

[10] Bhatt RP. Consequences of Climate Change Impacts and Implications on Ecosystem and Biodiversity; Impacts of Developmental Projects and Mitigation Strategy in Nepal. Dr. John P. Tiefenbacher. Climate Issues in Asia and Africa - Examining Climate, its Flux, the Consequences. and Society's Responses. s.l.: London: IntecOpen; DOI: 10.5772/intechopen.96455. 2021

[11] Ledig FT. Human impacts on genetic diversity in Forest ecosystems. Oikos. 1992;**63**(1):87-108

[12] Hunter IR. What do people want from urban forestry?—The European experience. Urban Ecosystems. 2001;**5**:277-284. DOI: 10.1023/a:1025691812497

[13] Młynarski W, Kaliszewski A. The current state of forest management in cities and associated problems in the Mazowieckie Province. Forest Research Papers. 2013;**74**:315-321

[14] Kettle CJ, Koh LP. Global Forest Fragmentation. ETH Zurich, Zurich, Switzerland: Department of Environmental System Science; 2014

[15] Baltic. Global forest fragmentation: Baltic 21 Series No 1/2000, 2006. Graphics from the year

[16] Lewis SL, Edwards DP, Galbraith D. Increasing human dominance of tropical. Sciecne. 2015;**349**(6250):827-832

[17] Wade TG et al. Distribution and causes of global Forest fragmentation. Conservation Ecology. 2003;**7**(2):7

[18] Walter R, Gratzer G, Wangdi K. Cattle grazing in the conifer forests of Bhutan. Mountain Research and Development. 2002;**22**(4):368-374

[19] Jan N, Poesen J, Moeyersons J, Deckers J, Haile M, Lang A. Human impact on the environment in the Ethiopian and Eritrean highlands—A state of the art. Earth-Science Reviews. 2004;**64**(3-4):273-320

[20] Berhanu G, Pender J, Girmay T. Community natural resource management: The case of woodlots in northern Ethiopia. Environ. Prod. Technol. Div. Discuss. Pap. 2000;**60**:1-33

[21] FAO. Global Forest Resources Assessment 2010-Country Report Ethiopia: Food and Agriculture Organisation (FAO). Rome, Italy: FAO; 2010

[22] Mekasha A, Gerard B, Tesfaye K, Nigatu L. Inter-connection between land use/land cover change and herders'/farmers' livestock feed resource management strategies: A case study from three Ethiopian eco-environments. Agriculture, Ecosystems and Environment. 2014;**188**:150-162

[23] Wiezik M, Lepeška T, Gallay I, Modranský J, Olah B, Wieziková A. Wood pastures in Central Slovakia–collapse of a traditional land use form. Acta Scientiarum Polonorum. Formatio Circumiectus. 2018;**17**(4):109-119

[24] WWF. Drivers of Deforatration and Forest Degradration. Kathmandu, Nepal: WWF Nepal, Hariyo Ban Program; 2013

[25] FAO. What Is Deforestation? Definition, Causes, Consequences, Solutio [Internet]. 2020. Available from: https://youmatter.world. https://youmatter.world/en/definition/definitions-what-is-definition-deforestation-causes-effects/.

[26] Walsh K, Giguet Covex C. A History of Human Exploitation of Alpine Regions. DellaSala DA, Goldstein MI, editors. Encyclopedia of the World's Biomes. 1st edition. Elsevier; 2020. p. 555-573

[27] Mine Watch. Mining and oil exploration. San Jose, Costa Rica: World Commission on Forests and Sustainable Development. 1997

[28] Miranda M, Blanco-Uribe AQ, Hernandez L. All that glitters is not gold. Balancing conservation and development in Venezuela's frontier forests. Washington, D.C: World Resources Institute. 1998:53. ISBN 1-56973-252-3

[29] FAO. Global Forest Resources Assessment 2000: Main Report. Rome, Italy: FAO; 2001

[30] FAO. Global Forest Resource Assessment. Rome, Italy: FAO; 1993

[31] Carbon Brief. Explainer: 'Desertification' and the role of climate change [Internet]. 2019. Available from: https://www.carbonbrief.org. https://www.carbonbrief.org/explainer-desertification-and-the-role-of-climate-change.

[32] Tyukavina A, Potapov P, Hansen MC, Pickens AH, Stehman SV, Turubanova S, et al. Global trends of Forest loss due to fire from 2001 to 2019. Front. Remote Sens. 2022;**3**:825190

[33] FAO. Global Forest Resources Assessment 2020: Main Report. Rome: Food and Agriculture Organization of the United Nations; 2020

[34] Giglio L, Boschetti L, Roy DP, Humber ML, Justice CO. The collection 6 MODIS burned area mapping algorithm and product. Remote Sensing of Environment. 2018;**217**:72-85

[35] Hansen MC, Potapov PV, Moore R, Hancher M, Turubanova SA, Tyukavina A, et al. High-resolution global maps of 21st-century Forest cover change. Science. 2013;**342**:850-853. DOI: 10.1126/science.1244693

[36] Pedrotti F. Plant and vegetation mapping. In: Types of Vegetation Maps. Berlin/Heidelberg, Germany: Springer; 2013. pp. 103-181

[37] (EEA), European Environment Agent. Air Quality in Europe—2017 Report. Luxembourg: European Union; 2017 EEA Report No 13/2017

[38] Achard F, Eva HD, Stibig HJ, et al. Determination of deforestation rates of the World's humid tropical forests. Science. 2002;**297**:999-1002

[39] USGS. Land Cover Institute 2017. Tree Cover for 2010. Available from: https://landcover.usgs.gov/glc/TreeCoverDescriptionAndDownloads.php

[40] FAO. On Definitions of Forest and Forest Change. Rome: FAO: FRA Working Paper 33; 2000

[41] Hansen MC, Stehman SV, Potapov PV. Qantification of global forest cover loss Proc. Natl. Acad. Sci. U.S.A. 2010;**107**:8650-8655. DOI: 10.1073/pnas.0912668107

[42] Whittaker RH. Evolution and measurement of species diversity. Taxon. 1972;**21**:213-251

[43] Where Are The World's Forests? [Internet]. 2021. Available from: https://www.worldatlas.com. World Atlas, 08 June 2021. https://www.worldatlas.com/articles/where-are-the-world-s-forests.html.

[44] The Amazonian Effect: how the rainforest sustains life in South America [Internet]. Available from: https://www.regnskog.no/en. Rainforest Foundation Norway. https://www.regnskog.no/en/long-reads-about-life-in-the-rainforest/the-amazonian-effect-how-the-rainforest-sustains-life-in-south-america#:~:text=A%20calculation.

[45] IUCN. Deforestation and Forest Degradation. Feb. 2021. Available online: https://www.iucn.org/resources/issues-briefs/deforestation-and-forest-degradation

[46] Shvidenko A, Jorgensen SE, Fath BD. In: Jorgensen SE, Fath BD, editors. Encyclopedia of Ecology. Oxford: Academic Press; 2008. pp. 853-859

[47] Newton AC et al. Toward integrated analysis of human impacts on forest biodiversity: Lessons from Latin America. Ecology and Society. 2009;**2**:14

[48] SDG 15, 2030. SDG 15, 2030. Life on Land-Ted Talks: You Tube [Internet]. 2017. Available from: Youtube.com. https://www.ted.com/talks/tasso_azeve....

[49] FAO. Global forest resources assessment. Rome: FAO; 2010

[50] FAO. Global Forest Resources Assessment 2015. How are the worlds forest changing? Second edition. Rome, Italy: Food and Agriculture Organization of the United Nations; 2016:163

[51] Yude P, Richard BA, Philips L, Robert JB. The structure, distribution,

and biomass of the World's forests. Annual Review of Ecology, Evolution, and Systematics. 2013;**44**:593-562

[52] FAO. Reducing Deforestation. s.l.: Villar MR - FAO, Forestry Department, 2018. Available online: https://www.fao.org/sustainable-forest-management/toolbox/modules/reducing-deforestation.

[53] Van der Werf G, Morton D, DeFries R. CO2 emissions from Forest loss. Nature Geoscience. 2009;**2**:737-738. DOI: 10.1038/ngeo671

[54] What's The Leading Cause of Wildfires In The U.S.? Humans. The Two Way [internat]. 2017. Available from: https://www.npr.org [Accessed: December 20, 2020] https://www.npr.org/sections/thetwo-way/2017/02/27/517100594/whats-the-leading-cause-of-wildfires-in-the-u-s-humans.

[55] Sarwar S, Shahzad U, Chang D, Tang B. Economic and non economic sector reforms in carbon mitigation: Empirical evidence from Chinese provinces. Structural Change and Economic Dynamics. 2019;**49**:146-154. DOI: 10.1016/j.strueco.2019.01.003

[56] Forest Carbon [Internet]. 2022. Available from: https://www.nrcan.gc.ca, Canada.ca, 14 04 2022. https://www.nrcan.gc.ca/climate-change-adapting-impacts-and-reducing-emissions/climate-change-impacts-forests/forest-carbon/13085.

[57] FAO. State of the World's Forests. Italy, Rome: Food and Agriculture Organization (FAO); 2001

[58] Society. Climate Milestone: Earth's CO2 Level Passes 400 ppm [Internet]. 2019. Available from: https://www.nationalgeographic.org/article. Mauna Loa Observatory, March 25, 2019. https://www.nationalgeographic.org/article/climate-milestone-earths-co2-level-passes-400-ppm/#:~:text=Today%2C%20greenhouse%20gasses%20in%20the,and%20the%20Earth%20was%20warmer..

[59] IPCC. In: Contribution of Working Groups I, II and III to the Fourth Assessment Report of the Intergovernmental Panel on Climate Change, editor. Climate Change 2007: Synthesis Report. Geneva, Switzerland: IPCC; 2007

[60] Roser, Hannah Ritchie and Max. Biodiversity and Wildlife [Internet]. 2021. Available from: https://ourworldindata.org. Our World in Data, 2021. https://ourworldindata.org/biodiversity-and-wildlife.

[61] Locey KJ, Lennon JT. Scaling laws predict global microbial diversity. Proceedings of the National Academy of Sciences. 2016;**113**(21):5970-5975

[62] Cowie RH, Bouchet P, Fontaine B, s.l. The sixth mass extinction: Fact, fiction or speculation? Biological Reviews of the Cambridge Philosophical Society. 2022;**97**(2):640-663

[63] Ripple WJ, Wolf C, Newsome TM, Galetti M, Alamgir M, Crist E, et al. World Scientists' warning to humanity: A second notice. Bio Science. 2017;**67**(12):1026-1028

[64] UNEP, FAO. The State of the World's Forests 2020: Forests, Biodiversity and People. Rome: FAO; 2020

[65] Forum, World Economic. This is how humans have impacted the world's forests [Internet]. https://www.weforum.org. August 30, 2018. https://www.weforum.

org/agenda/2018/08/the-human-impact-on-the-world-s-forests/.

[66] Nunez, Christina. Climate 101: deforestation [Internet]. https://www.nationalgeographic.com. National Geographic, 2022. https://www.nationalgeographic.com/environment/article/deforestration.

[67] As Australia faces new fire reality, forest restoration tactics reevaluated [Internet]. 2022. Available from: https://news.mongabay.com Mongabay [Accessed: March 15, 2022] https://news.mongabay.com/2022/02//as-australia-faces-new-fire-reality-forest-restoration.

Chapter 9

Land Use Impacts on Diversity and Abundance of Insect Species

Akinbi Olarewaju John, Akinbowale Akinlolu Sylvester, Ajayi Olalekan Kehinde and Agbeje Abiodun Michael

Abstract

Land use is a major constraint to the population of insect species. Insects have provided essential and irreplaceable services ranging from pollination to decomposition of large organic matters. However, these roles provided have been jettisoned as a result of human anthropogenic activities. In recent times, the conversion of existing natural forest ecosystem to other land use types has become a menace that requires urgent attention due to its effects on the population of plant and animal species. Many factors such as land-use changes, deforestation, pollution, intensive agriculture, among others have been reported to contribute to the decline in the population of insect species. As a result of changes, insects are threatened and vulnerable to extinction. Loss of key stone species also affects the function, structure and population of other species in the ecosystem. This is because of the level of interdependency between these insect species and other components of natural forest ecosystem. Such losses lead to the decline in ecosystem goods and services which human depends on for sustenance. Apart from decline in all levels of biodiversity, land use also causes climate change and environmental pollution which in turns affects the population of insect species. However, protection of area of high biodiversity hotspots should be encouraged by resources managers. Also natural forest that has been disturbed should be left alone in order to recuperate and get back to its original state.

Keywords: land use, insect, biodiversity, natural forest, climate change

1. Introduction

Insects play a vital role in both natural and man-made settings. They perform critical and functional roles in ensuring the delivery of numerous ecosystem services that are critical for human life in areas such as agriculture, tourism, natural resource utilization etc. [1, 2]. Scientific investigations have shown that insect populations have been declining globally for decades [3]. Environmental scientists as well as the press, non-governmental organizations, and policymakers, have been alerted to the eminence reports of impoverished insect faunas from Europe and around the world [4]. West Africa countries are afro-tropical with a diverse flora and fauna, as do other tropical countries throughout the world. The tropics, which are home to over 70% of world biodiversity, are also a treasure trove of insect diversity, which is thought to be comparable to the region's plant richness [5].

2. Land use and insects dynamics

Land use is the conversion of natural environment to a built environment and this is majorly caused by human activities [6]. Natural rainforests are one of the most species-rich and functionally important terrestrial ecosystems [7]. Tropical forests are the most bio-diverse terrestrial habitat, with 50% of the world's species [8, 9], but about 68,000 km2 of tropical forest is lost annually [6], an amount that could be increasing by 3% (2000 km^2) each year [10], remains 11 million km^2, virtually half (5 million km2) is considered to be either deforested area [11] or degraded forest that has to regenerate after human use (e.g., cropped land abandonment and clear-felling: [12, 13] or natural disturbances (e.g., fires and windstorm) [14]. Insects are essential components in most natural and transformed landscapes. They play crucial functional roles that ensure the delivery of various ecosystem services necessary for several aspects of human livelihoods, such as agriculture, tourism, and natural resource use [2, 15, 16]. They also control populations of other organisms and provide a significant food source for other taxa [17]. However, they are also disease vectors to many other organisms, including humans, and they can alter the rates and directions of energy and matter fluxes in an ecosystem [18]. Recent reports imply devastating declines in the decline of all levels of biodiversity with potentially dire implications on ecosystem functioning [19, 20].

The tropical forest with high species composition, and high level of biodiversity is currently faced with a lot of challenges when it comes to its conservation [21]. Natural forests have been reported to contain a high diversity of insect and trees species diversity than other land-use types and this has been attributed to the high level of relationship between the various components of the forest ecosystem [22]. According to [23], conversion of forest to other land use types is one of greatest threats to insects, causing population declines and shifts in community composition. The tropical forests have been subjected to extensive disturbance by logging and clearance for agriculture; especially palm oil, cocoa, pasture, and rubber plantation [24, 25]. OT et al. [26] reported that tree species, which is a significant component of forest ecosystem plays vital roles in forest biodiversity conservation. Conversion of this forest to other land use type will affect the population and result to loss of key stone species that are very important to the forest ecosystem balance. Land-use change, particularly the conversion of natural forest to agriculture to sustain the growing global population, has negative effect on biodiversity conservation and contributes to climate change.

3. Contributions of insects to human nutrition

The decline in insect biodiversity concurrently presents an immediate threat to food security [17] and permanently affects humans' health and wealth. All terrestrial insects provide resources for higher trophic levels, especially for many vertebrates [18]. Then, the vertebrates and other beneficiaries in the trophic groups are subject to humans. It is expected that by 2050 the projected world population will be 9.8 billion and 11.2 billion by 2100 [27]. In particular, pollinating insects have experienced significant declines for several decades in many parts around the world [28]. This is a serious concern because pollination represents a critical ecosystem service [29], and declines in pollinators have been directly linked with reductions in the plants with which they interact [30]. Though, the pollination services of honeybees are just as valuable as their production of honey and wax which are useful humans. Meanwhile, plants serve several purposes in human existence on Earth. This is of particular importance because most insects interact with plants, and an estimated 85% of angiosperms, including most tropical tree species, are directly pollinated by insects [31]. Most insect species are associated with living plants [32] and are highly specialized in food resources and micro-habitats [33].

3.1 Major edible insects

Insect are rich protein source. Practically 100 analyzed edible insects at egg, larva, pupa or adult stages and the raw protein content is generally 20–70% [34, 35]. Globally, Most commonly consumed insect orders are beetles (Coleoptera), caterpillars (Lepidoptera), Bees, wasps, ants (Hymenoptera), grasshoppers, locusts and crickets (Orthoptera), cicadas, leafhoppers, planthoppers, scale insects, true bugs (Hemiptera), termites (Isoptera), dragonflies (Odonata), flies (Diptera). Consumption of these insects differs to each region of the world depends on stages of insects. Lepidoptera are consumed completely as caterpillars and Hymenoptera are consumed regularly in their larval or pupal stages [36]. Both adults and larvae of the Coleoptera order are eaten, while the Orthoptera, Homoptera, Isoptera and Hemiptera orders are mostly eaten in the mature stage [37]. Besides serving as sources of food, insects provide humans with a variety of other valuable products. Honey and silk are the most commonly known insect products. Bees produce about 1.2 million tonnes of commercial honey per year [38, 39], while silkworms produce more than 90 000 tonnes of silk [40].

4. Impact of land use on insects population

The conversion of Natural forest ecosystem has led to decline in the population of insect species. Many factors such as land-use changes, deforestation, pollution, urbanization, intensive agriculture, among others have been reported to contribute to this decline. The activities of humans in natural ecosystems have lead to habitat fragmentation, isolated from each other by prevailing conditions of hostile lands created by human activities [41]. This fragmentation leads to smaller habitat areas and decreased biodiversity [42]. As a result of changes, species are threatened and vulnerable to extinction. Loss of key stone insect species has also affected the population of other species and other components of the forest ecosystem, and it's because of the level of inter-dependency between these insect species and the natural forest ecosystem. Such losses lead to the reduction in ecosystem goods and services which human depend on for their survival (**Table 1**).

4.1 Urbanization

Urbanization is increasing in most developed and developing nations of the world, which leads to habitat fragmentation and converting the significant habitats of insect species into smaller areas and converting the forest into agricultural areas

	Order	Common name	NF	RP	OP
1	Araneae	Spider	14	2	40
2	Coleoptera	Beetle	24	14	12
3	Dermaptera	Earwig	2	0	2
4	Diptera	cranyfly	2	0	0
5	Hemiptera	Bug	18	10	10
6	Hymenoptera	Ant	114	14	43
7	Lepidoptera	Butterfly	48	6	14
8	Mantodae	Ants/Fly	2	0	0
9	Odonta	Dragonfly	56	4	6
10	Orthoptera	Criket/Grassh	14	36	45
12	Scolopendromorphra	Centipede	4	0	0
13	Spirostreptida	Millipede	20	34	62
14	polydesmida	Millipede	4	6	8
			322	**126**	**242**

Source; Adeduntan et al., 2021.
NF; Natural forest, RP; Rubber plantation, OP; Oil palm plantation.

Table 1.
Abundance of insect composition in each ecosystem in Okomu Edo state Nigeria.

and communities [43]. In tropical West Africa, a considerable decline in some insect populations was observed due to Urbanization [43, 44]. Globally, vegetation has been converted into an urbanization setting to bridge the craving for urban population increase, which alters the habitat of insect species and leads to insect decline except in bees keeping [43, 45] in the most fragment of the forest habitats today. Then, insect crossing on the roads during construction, collusion of insects with vehicles, and death of soil-borne insects during road constructions [46] are not major factors that decrease insect species in the forest habitat.

5. Insects and organic matter decomposition

Apart from bacteria and fungi, invertebrate (insects) herbivores as well as decomposers breaks down and feeds on living, dead or decaying plant or animal matters, making organic nutrients (recycling) available to the ecosystem [47]. As a result to renew forests by reduce the old and vulnerable trees, also provide new habitat and food for wildlife. However, food resource (detritus) that supports trophic food chains in almost every territory of the heterotrophic habitat, among them is countless species of insects [48]. Meanwhile, there are three major processes through which decomposition occurs [49, 50], (a) fragmentation of litter into smaller sizes, (b) leaching of soluble compounds into soil, and (c) catabolism by decomposer organisms. Insects are mostly engaged more in the fragmentation of litter into smaller sizes. Several decomposers have special relationship with plant species and are specialized to breakdown the litter of these materials [51]. Given that the decomposer food web consisting of fauna and microbial communities also varies in the underneath of different forest floors and affects the rates various litter fractions are mineralized in an ecosystem [52]. Most of these orders (larvae and adult) are engaged, for example Diplopoda, Isopoda, Collembola, Diptera, Coleoptera, Acari among few others.

6. Insects as pollinators

As pollinators, insects play an important role in ecosystem services. Majority of flowering plant species depend essentially on animal vectors for pollination [31]. Entomophilous (pollination by insects) serve as an important life-support mechanism that contributes to biodiversity and ecosystem services [53]. In addition to constant visitation of flowering plants by honeybees (Apis melifera), which has been considered as the most common single species of pollinator for crops in the world [54], various insects such as stingless and solitary bees, bumblebees, and a variety of beetles, flies, butterflies, and moths are significant pollinators [55–57]. Entomophilous plant species have regularly evolved ways to increase their attractiveness to insects. Many flowering plant species have developed structures or exudates that ensure insect pollinators will return to transfer pollen on a regular basis [58]. However, many types of land use (for example, agriculture and urbanization) drastically modify land cover, resulting in habitat loss for various species [59] and studies [53, 60, 61] have shown that habitat loss reduces the population sizes, composition, and species diversity of insect pollinators. The loss of habitat for native bees has been found to be much greater than that of other insect groups, as measured by species–area relationships [62].

7. Effects of pollution on insect abundance

Industrial pollution has long been known to reduce insect potential [31, 63]. Carbon monoxide from cars and engine exhaust also has its effect, primarily due to high levels of NOx [53, 64]. Though the most critical forms of industrial pollution are heavy metals in soils and waterways [54, 65], air pollution [55, 66], aquatic pollution [56, 67], and light pollution [57, 68]. Although it is somewhat reduced now in many developed nations [58, 69], it is yet a global threat to insect protection. This has drastically affected the population of the insect community, as most insects are sensitive to smell while most have disappeared due to their habitat loss.

8. Effects of climate change on insect population

The climate is crucial in determining various characteristics and distributions of managed and natural systems, including hydrology and water resources, cryptology, marine and freshwater ecosystems, terrestrial ecosystems, forestry, and agriculture [59, 70]. As a result of increased temperatures, climate extremes, increased CO2 and other greenhouse gases (GHGs), and altered precipitation patterns, global food production is under severe threat [60, 71]. Global warming is a severe problem facing the world today. It has reached record-breaking levels, as evidenced by exceptional rates of increase in atmospheric temperature and sea level [61, 72]. The World Meteorological Organization (WMO) reported that the world is now about one degree warmer than before widespread industrialization. Insects are poikilothermic organisms; the temperature of their body depends on the temperature of the environment. Thus, the temperature is probably the most important environmental factor affecting insect behavior, distribution, development, and reproduction [62, 73]. Then, it is most likely that the main drivers of climate change (increased atmospheric CO2, increased temperature, and decreased soil moisture) could significantly affect the population dynamics of insects [63, 74]. Climate change creates new ecological niches that allow insects to react and shift from one region to another [64, 75]. The Warm and dry conditions associated with climate change have affected most forest insect species population and activity in recent years [65, 76]. However, [65, 76] reported that temperature or drought on insects community is one of the essential climatic drivers of habitat vulnerability to forest insects [66, 77]. As a result, insect populations in tropical zones are predicted to experience a decrease in growth rate due to climate changes of current temperature level for insect development and growth [67, 78].

9. Conclusion

Study showed that Land use stand a significant on diversity and abundance of insect. The gradual disappearance of natural forest habitat has resulted loss of keystone species whose function in an ecosystem cannot be overestimated. Apart from decline in all levels of biodiversity, land use also induces climate change and pollution which in turns affects the population of insect species. Since they are the major contributor to biodiversity in most habitats, except in the sea, they accordingly play a variety of tremendously ecological functions of an ecosystem. Protection of an area in high biodiversity hotspots should be encouraged by resources managers. Also natural forest that has been disturbed should be left alone in order to regenerate and get back to its original state.

Conflict of interest

The authors declare no conflict of interest.

Author details

Akinbi Olarewaju John[1*], Akinbowale Akinlolu Sylvester[2], Ajayi Olalekan Kehinde[1] and Agbeje Abiodun Michael[1]

1 Forestry Research Institute of Nigeria, Ibadan, Nigeria

2 Department of Forestry and Wood Technology, Federal University of Technology Akure, Nigeria

*Address all correspondence to: akinbilanre3@gmail.com

References

[1] Samways MJ. Insect Conservation Biology. New York, USA: Champman & Hall; 1994. p. 358

[2] Tscharntke TC, Klein T, Kruess AM, Steffan A, Dewenter I. Landscape perspectives on agricultural intensification and biodiversity: ecosystem service management. Ecology Letters. 2005;**8**:857-874

[3] Wagner D, Grames DL, Forister EM, Berenbaum ML, Stopak MR. Insect decline in the Anthropocene: Death by a thousand cuts. Proceedings of the National Academy of Sciences. 2021;**118**(2):e2023989118

[4] Wagner DL. Insect declines in the Anthropocene. Annual Review of Entomology. 2020;**65**:457-480

[5] Bradshaw BBW, Sohdi CJA. Tropical turmoil: A biodiversity tragedy inprogress. Frontiers in Ecology and the Environment. 2009;7:79-87

[6] FAO J. Global forest land-use change 1990-2005. Rome: Food and Agriculture Organization of the United Nations and European Commission Joint Research Centre; 2012

[7] Myers KJ, Mittelmeier N, Mittelmeier RA, da Fonseca Gab CG. Biodiversity hotspots for conservation priorities. Nature. 2000:853-858

[8] Dirzo RP. Global state of biodiversity and loss. Annual Review of Environment and Resources. 2003;**28**:137-167

[9] WSJ. Tropical forests in a changing environment. l. Trends in Ecology & Evolution. 2005;**20**:553-560

[10] Hansen MC, Potapov PV, Moore R, et al. High-resolution global maps of 21st-century forest cover change. Sciences (New York). 2013;**342**:850-853

[11] SITTO. ITTO Guidelines for the Restoration, Management and Rehabilitation of Degraded and Secondary Tropical Forests. ITTO Policy Development Series No.13. Yokohama, Japan: ITTO; 2002. p. 84. ISBN: 4-902045-01-X

[12] Wright M-LH. The future of tropical forest species. Bio Tropica. 2006;**38**:287-301

[13] Lewis GD, Edwards DP. Increasing human dominance of tropical forests. Science. 2015;**3349**:827-832

[14] Chazdon RL, Peres CA, Dent D, et al. The potential for species conservation in tropical secondary forests. Conservation Biology. 2009;**23**:1406-1417

[15] Ramesh PM, Hussain KJ, Selvanayagam M, Satpathy KK. Patterns of diversity, abundance, and habitat association of butterflies communities in heterogeneous landscapes of Department of Atomic Energy (DAE) Campus at Alpakkam, South India. International Journal of Biodiversity and Conservation. 2010;**2**:75-85

[16] Choi MJ. Species Richness and Abundance among Macro Moths: Comparing Taxonomic, Temporal and Spatial Patterns in Oregon and South Korea. Entomological Society of Korea and Wiley Publishing Asia Pty Ltd; 2013

[17] TG WLB, Bieg C, Harwatt H, Pudasaini R. Food System Impacts on Biodiversity Loss. Three levers for food

system transformation in support of nature. London: Chatham House; 2021

[18] Timothy DS. Insect Ecology: An Ecosystem Approach. 4th ed. Academic Press; 2016

[19] CA GDH, Sorg M, Jongejans E, Siepel H, Hofland N, Schwan H, et al. More than 75 percent decline over 27 years in total flying insect biomass in protected areas. PLoS ONE. 2017;**12**(10):e0185809

[20] Lister A, Garcia BC. Climate-driven declines in arthropod abundance restructure a rainforest food web. PNAS. 2018;**115**:241-242

[21] Stanley DA, Stout JC. Quantifying the impacts of bioenergy crops on pollinating insect abundance and diversity: A field-scale evaluation reveals taxon-specific responses. Journal of Applied Ecology. 2013;**50**:335-344. DOI: 10.1111/1365-2664.12060

[22] Adeduntan SA, Akinbi OJ, Osabiya OS, Olusola JA. Diversity and seasonal dynamics of arthropods in Okomu forest reserve. Asian Journal of Biology. 2021;**13**:23-36. DOI: 10.9734/ajob/2021/v13i330187

[23] Sanchez-Bayo KAG, Wyckhuys F. Worldwide decline of the entomofauna: A review of its drivers. Biological Conservation. 2019;**232**:8-27

[24] Gibbs HK, Ruesch AS, Achard F, Clayton MK, Holmgren P, Ramankutty N, et al. Tropical forests were the primary sources of new agricultural land in the 1980s and 1990s. Proceedings of the National Academy of Sciences of the United States of America. 2010;**107**:16732-16737

[25] Gibson L, Lee TM, Koh LP, et al. Primary forests are irreplaceable for sustaining tropical biodiversity. Nature. 2011;**478**:378-381

[26] OT AFO, Akinbowale AS, Olugbadieye OG. Diversity and volume assessment of tree species in the tropical forest at Obanla. Asian Journal of Research in Agriculture and Forestry. 2020;**5**(4):11-19

[27] FAO. Composition database for biodiversity version. Bio Food Comp. 2012;**2**:408-412

[28] Carvalheiro LG et al. Species richness declines and biotic homogenization have slowed down for NW-European pollinators and plants. Ecology Letters. 2013;**16**(7):870-878

[29] Garibaldi LA et al. Wild pollinators enhance the fruit set of crops regardless of honey bees' abundance. Science. 2013;**339**(6127):1608-1611

[30] Potts SG, Biesmeijer JC, et al. Global pollinator declines trends, impacts, and drivers. Trends in Ecology & Evolution. 2010;**25**:345-353

[31] Ollerton S, Winfree J, Tarrant R. How many flowering plants are pollinated by animals? Oikos. 2011;**120**:321-326

[32] Price PW. Resource-driven terrestrial interaction webs. Ecological Research. 2002;**17**:241-247

[33] Novotny Y, Basset V. Rare species in communities of tropical insect herbivores: Pondering the mystery of singletons. Oikos. 2000;**89**:564-572

[34] Xiaoming Z, Ying C, Hong F, Zhiyong C. Review of the nutritive value of edible insects: Forest insects as food: Humans bite back. In: Proceedings of a Workshop on Asia-Pacific Resources and their Potential for Development Chiang Mai, Thailand. Rome: FAO; 2010. pp. 85-92

[35] Seni A. Edible insects: Future prospects for dietary regimen. International Journal of Current

Microbiology and Applied Sciences. 2017;**6**(8):1302-1314

[36] Huis V. Edible insects. Future prospects for food and feed security. Rome: Food and Agriculture Organization of the United Nations; 2013. pp. 1-201

[37] Cerritos R. Insects as food: An ecological, social and economic approach. Biology (Basel). 2009;**4**(27): 1-10. DOI: 10.1079/PAVSNNR20094027

[38] Vijay V, Pimm SL, Jenkins CN, Smith SJ. The impacts of oil palm on recent deforestation and biodiversity loss. PLoS ONE. 2016;**11**(7):e0159668. DOI: 10.1371/journal.pone.0159668

[39] FAO. Biodiversity and nutrition, a common path. Rome: FAO; 2009

[40] Yong-woo L. Silk reeling and testing manual. Agricultural Services Bulletin. FAO. 1999;**136**:1010-1365

[41] Weisner E. An Inordinate Fondness for Beetles: A Study of Insect Species Diversity and Abundance in Mazumbai Forest Reserve Versus Nearby Agricultural Areas. Independent Study Project (ISP) Collection. 2018. p. 2872. Available from: https://digitalcollections.sit.edu/isp_collection/2872

[42] Sherry T. Mechanisms and Consequences of Forest Fragmentation. In: Conservation Biology, Tulane University, New Orleans, LA. Class Lecture. Spring; 2016

[43] Guenat DM, Kunin S, Dougill WE. Effects of urbanization and management practices on pollinators in tropical Africa. Journal of Applied Ecology. 2019;**56**:214-224

[44] Cardoso HM, Barton P, Birkhofer PS, Chichorro K, Deacon F, Fartmann C, et al. Scientists' warning to humanity on insect extinctions. Biological Conservation. 2020;**242**:108426

[45] Akinbi OJ, Adeduntan SA, Toyinbo EO, Alamu OT. Assessment of Gmelina, Danta solid wood and plywood hive types for beekeeping. Journal of Applied Sciences and Environmental Management. 2021;**25**(4):625-629. DOI: 10.4314/jasem.v25i4.22

[46] Seibert CJ. Mortality of vertebrates and invertebrates on an Athens County. The Ohio Journal of Science. 1991;**91**(4):163-166

[47] Bignell DE, Eggleton P. Termites in ecosystems. In: Abe T, Bignell DE, Higashi M, editors. Termites: Evolution, Sociality, Symbioses, Ecology. Dordrecht: Kluwer Academic Publishers; 2000. pp. 363-387

[48] Louzada JNC, Nichols ES. Detritivorous insects. Insect Bioecology and Nutrition for Integrated Pest Management. 2012:397-415. DOI: 10.1201/b11713-20. Available from: https://www.researchgate.net/publication/300814104

[49] Hättenschwiler S, Jørgensen HB. Carbon quality rather than stoichiometry controls litter decomposition in a tropical rain forest. Journal of Ecology. 2010;**98**(4):754-763

[50] UC BRG, Martins KG, Brandle M, Schadler M, Marques R. Lack of home-field advantage in the decomposition of leaf litter in the Atlantic Rainforest of Brazil. Applied Soil Ecology. 2011;**49**:5-10

[51] LV, Austin AT. Tree species identity alters forest litter decomposition through long-term plant and soil interactions in Patagonia, Argentina. Journal of Ecology. 2008;**96**(4):727-736

[52] Laganière J, Angers D, Par D. Carbon accumulation in agricultural soils after afforestation: A meta-analysis. Global Change Biology. 2010;**16**(1):439-453

[53] Vanbergen AJ, Initiative TI. Threats to an ecosystem service: pressures on pollinators. Frontiers in Ecology and the Environment. 2013;**11**(5):251-259

[54] Garibaldi LA, Steffan-Dewenter I, Winfree R, Aizen MA, Bommarco R, Cunningham SA, et al. Wild pollinators enhance fruit set of crops regardless of honey bee abundance. Science. 2013;**339**(6127):1608-1611

[55] Chai SK, Wong SY. Five pollination guilds of aroids (Araceae) at Mulu National Park (Sarawak, Malaysian Borneo). Webbia. 2019;**74**(2):353-371

[56] Ojija F, Arnold SE, Treydte AC. Impacts of alien invasive Parthenium hysterophorus on flower visitation by insects to co-flowering plants. Arthropod-Plant Interactions. 2019;**5**:719-734

[57] Roubik DW, Sakai S, Gattesco F. In: Basset Y, Kitching R, Miller S, Novotny V, editors. Canopy flowers and certainty: loose niches revisited. Cambridge, UK: Cambridge University Press; 2003. pp. 360-368

[58] Simpson MG. Plant systematics. Academic Press; 2019

[59] Fischer J, Lindenmayer DB. Landscape modification and habitat fragmentation: A synthesis. Global Ecology and Biogeography. 2007;**16**(3): 265-280

[60] Steffan-Dewenter I, Potts SG, Packer L. Pollinator diversity and crop pollination services are at risk. Trends in Ecology & Evolution. 2005;**20**(12):651-652

[61] Taki H, Kevan PG. Does habitat loss affect the communities of plants and insects equally in plant–pollinator interactions? Preliminary findings. Biodiversity and Conservation. 2007;**11**:3147-3161

[62] Steffan-Dewenter I, Klein AM, Gaebele V, Alfert T, Tscharntke T. Bee diversity and plant-pollinator interactions in fragmented landscapes. Specialization and generalization in plant-pollinator interactions. 2006:387-410

[63] W. JB, "Presidential address: Insects and plants in a changing atmosphere.," Journal of Ecology, vol. 89, no. 4, pp. 507-518, 2001

[64] Campbell VD. Plant defences mediate interactions between herbivory and the direct foliar uptake of atmospheric reactive nitrogen. Nature Communications. 2018;**9**:4743

[65] Sun ZG, Liu Y. Effects of heavy metal pollution on insects. Acta Entomologica Sinica. 2007;**50**:178-185

[66] FP H. "Interactions of insects, trees and air pollutants." Tree Physiology. 1987;**3**:93-102

[67] Arimoto IR. Ecological integrity of upper Warri River, Niger Delta using aquatic insects as bioindicators. Col. Indic. 2009;**9**:455-461

[68] Owens LS. The impact of artificial light at night on nocturnal insects: a review and synthesis. Ecology and Evolution. 2018;**8**:58

[69] Majewski CP. Pesticides in the atmosphere; distribution, trends, and governing factors. Reston, VA: U.S. Geol. Surv; 1995. pp. 94-506

[70] M. M. climate change risks in N. Y. C. Rosenzweig, C, Major DC, Demong K,

Stanton C, Horton R, Stults. Managing climate change risks in New York City's water system: Assessment and adaptation planning. Mitigation and Adaptation Strategies for Global Change. 2007;**12**:1391-1409

[71] Shrestha S. Effects of climate change in agricultural insect pest. Acta Scientific Agriculture. 2019;**3**:74-80

[72] Field RC, Barros CB, Dokken VR, Mach DJ, Mastrandrea KJ, Bilir MD, et al. IPCC Summary for policymakers. In Climate Change: Impacts, Adaptation, and Vulnerability, Part A: Global and Sectoral Aspects; Contribution of Working Group II to the Fifth Assessment Report of the Intergovernmental Panel on Climate Change. New York, NY, USA: Cambridge Univ. Press Cambridge, UK; 2014, 2014. pp. 1-32

[73] Kocmánková A, Trnka E, Juroch M, Dubrovský J, Semerádová M, Možný D, et al. Impact of climate change on the occurrence and activity of harmful organisms. Plant Protection Science. 2010;**45**:48-52

[74] Fand M, Kamble BB, Kumar AL. Will climate change pose serious threat to crop pest management. A Critical Reviews in International Journal of Scientific Research. 2012;**2**:1-14

[75] FAO. Climate-Related Transboundary Pests and Diseases. FAO; 2020

[76] Ramsfield TD, Bentz BJ, Faccoli M, Jactel H, Brockerhoff EG. Forest health in a changing world: Effects of globalization and climate change on forest insect and pathogen impacts. Forestry. 2016;**89**:245-252

[77] Jactel JK, Petit H, Desprez-Loustau J, Delzon ML, Piou S, Battisti D. Drought effects on damage by forest insects and pathogens: A meta-analysis. Global Change Biology. 2012;**18**:267-276

[78] Deutsch RL, Tewksbury CA, Tigchelaar JJ, Battisti M, Merrill DS, Huey SC, et al. Increase in crop losses to insect pests in a warming climate. Science. 2018;**361**:916-919

Chapter 10

Nexus between Savannah Woodland Degradation and Climate Change in Northern Ghana

Raymond Aabeyir, Kenneth Peprah and Gervase Kuuwaabong

Abstract

The Savannah woodland and forest ecosystems are considered as fragile ecosystems in Ghana. They are located in pro-poor areas of the country. They serve as livelihood support systems for the poor in those areas. In the midst of climate change, the same woods are expected to provide mitigation support against climate change. These woodland and forest ecosystems are in a state of dilemma: providing adaptation support to humans and at the same time providing mitigation support against climate change in the midst of climatic and seasonal challenges: low rainfall, excessive heat, harmattan and annual bushfires. The sustainability of these ecosystems depends on the net effect resulting from the pressures of adaptation, efforts of mitigation, resilience of the ecosystems and other natural support systems. This chapter explores the relationship among woodland, adaptation and mitigation activities. In this relationship, the human face has played a central role, thus influencing the direction of the net effects of the pressures on woodland ecosystems. Adaptation is over-emphasised, misunderstood and decoupled from mitigation resulting in maladaptation. This has contributed to the worsening impacts of climate change. Climate change adaptation needs to be re-emphasised to ensure mitigation is considered in every adaptation measure.

Keywords: adaptation, Giddens Paradox, mitigation, maladaptation, climate change

1. Introduction

The meaning people ascribe to climate change in terms of their understanding of the phenomenon, perception of the risks involved, value judgments, and emotional reactions, is closely related to how they adapt to and mitigate the impacts of climate change [1]. For instance, climate change is influenced by many factors, and it also influences many aspects of life and nature, which make it a complex issue to understand and deal with. Global climate change and its impacts are major environmental issues the world is currently battling with [2]. It is considered the most serious global environmental issue of today and the most difficult issue to manage by both scientists and policymakers [2, 3]. It affects every aspect of human society and worst of all, human well-being. It requires both individual and global efforts towards a sustainable solution (idem). The arguments and counter arguments about its existence, causes, and impacts are converging and there is a high level of global consensus about its existence, impacts, and the need to address it [4]. It is agreed that the solution lies in the ability to keep global temperature rise within a 2°C limit although there are disagreements on the strategies to reduce global temperature rise to below 2°C. This requires strong and sustained reductions in emissions of carbon dioxide (CO_2) and other greenhouse gases that would limit climate change [5].

The global community is relying so much on the mitigating capacity of woodland and forest ecosystems through the sequestration of atmospheric CO_2, and reduction in emissions from anthropogenic activities while encouraging the rural communities to adopt other ways of livelihoods. Current adaptation strategies are resulting in negative feedback on climate change and the woodland and forest ecosystems are under pressure to support alternative livelihoods. This suggests that the relationship between climate change, woodland/forest ecosystems, and adaptation is still not clearly understood. It is therefore important to understand climate change from a system perspective of how it influences woodland/forest ecosystems and anthropogenic activities and how they interact to influence climate change.

Though climate change is not new in human history, the current changes in the global climate change have serious implications for human life, other living organisms, and ecosystems in general. Scientific observations and model simulations have indicated that the climate of the Earth is now changing at unprecedented rates which are attributed to human activities such as burning fossil fuels clearing woodlands/forests for farmlands and cities, and grazing activities [5]. These activities release a huge amount of greenhouse gases (GHGs) (e.g. carbon dioxide, methane, halo-carbons, and nitrous oxide) into the atmosphere with dire consequences for the climate system of the Earth. Lots of efforts in various forms have been expended with the aim of reducing GHGs and improving the climate system of the Earth thereby making the Earth a conducive place for humans. These efforts are mainly in four areas:

a. The Science of climate change which is geared towards understanding climate and its associated changes. This has been debated over the past three decades and an acceptable consensus has been reached.

b. The Impact of climate change which is aimed at understanding the current and future effects of climate change on man and the environment.

c. Adaptation to climate change, which is aimed at understanding the coping strategies of man to lessen the effects of climate change on man, his livelihoods, and the environment.

d. Mitigation of climate change geared towards understanding the efforts of a man towards slowing down climate change and its negative effects on man and his environment.

In all these efforts, it is how humans adapt and mitigate the current impacts of climate change that will inform the intensity and extent of future climate change.

Although the scientific effort has yielded a number of good results in terms of scientific and political agreement on the reality of anthropogenic climate change and the need to tackle it, the adaptation and mitigation efforts are not yielding the needed results. One reason could be the inadequate understanding of the relationship between the forces of anthropogenic climate change, woodland/forest ecosystems, and adaptation strategies in rural areas. Also, there is no consensus among the rural, national and global communities on the mitigation strategies for tackling climate change. The global and national communities placed Reducing Emission from Deforestation and forest Degradation (REDD) and later Reducing Emission from Deforestation and forest Degradation coupled with forest conservation, sustainable forest management and enhancement of forest carbon stock (REDD+) top on the mitigation agenda and imposed that on the rural communities, which they see as a threat to their means of adaptation to same climate change. These strategies are currently resisted at the local levels due to the economic issues relating to adaptation. The caveat to overcome the challenges encountered from the implementation of REDD and REDD+ such as Payments for Ecosystem Services (PES) where beneficiaries or users of an ecosystem services make payments to the providers of those services [6], carbon markets cannot still overcome the strong land tenure and adaptation barriers mounted by the rural communities. The effects of resistance against these interventions are contributing significantly to the negative feedback effects of climate change adaptation on climate change and its impacts across the world.

In Ghana, the relationship between adaptation, woodland and forest ecosystems, and mitigation at the rural community level is still not clearly understood. This is leading to maladaptation due to over-dependence on woodland and forest ecosystems for adaptation, with very little effort in woodland and forest-based mitigation activities.

This chapter throws more light on the relationship between adaptation, woodland and forest ecosystems, and mitigation, with a focus on charcoal production and logging in the Savannah areas of Ghana. These adaptation and mitigation activities have a direct influence on the state of the woodland and forest ecosystems in the area.

2. Climate change and human ineptitude in the rural communities of Ghana

Scientists have explained that the impacts of climate change will continue to be devastating for both natural and human systems. In northern Ghana, the evidence of climate change is clear. The trend of minimum and maximum monthly temperatures of some parts of Northern Ghana (Wa, Navrongo, and Tamale) for a 45-year period (1961–2005) revealed increasing trends of minimum and maximum temperature values. The trend is of concern because current minimum temperature values are approaching past maximum temperature values. The trend lines for the monthly minimum and maxim for the period all indicate positive gradients, signifying a general increase in the average temperature values for three locations (**Figures 1** and **2**). In the case of the Tamale station, there is a dip in the trend line suggesting that there was a slide reduction in the minimum temperatures within the period and it then started increasing gradually again (**Figure 1**).

However, the trend of the annual rainfall amounts for the 45 years showed dwindling annual rainfall amounts (**Figure 3**). The trend lines for the three locations showed negative gradients, suggesting a continuous reduction in the average annual rainfall amounts for the 45 years.

These trends in the minimum, maximum, and annual rainfall amounts point to the fact that the efforts of natural systems alone would not be able to reverse the trends but there is the need for concerted human effort to complement the effort of the natural systems. Human intervention must emphasise the change in consumption

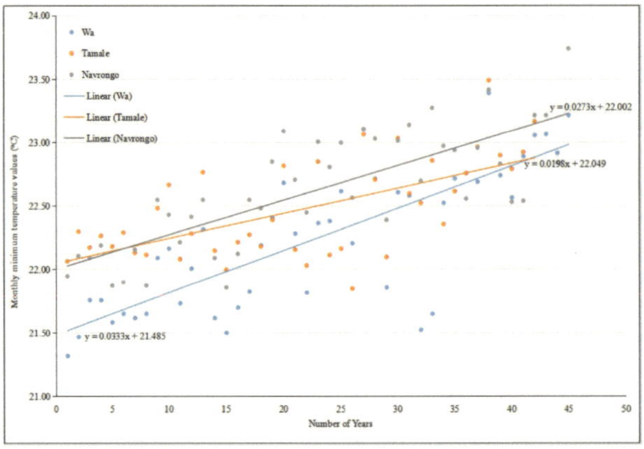

Figure 1.
Trend of minimum monthly temperature values from 1961 to 2005 of selected weather stations in Northern Ghana. (Data source: Ghana Meteorological Agency).

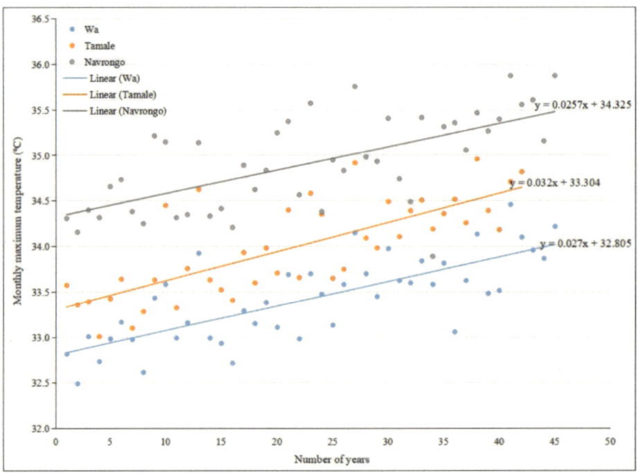

Figure 2.
Trend of maximum monthly temperature values from 1961 to 2005 of selected weather stations in Northern Ghana. (Data source: Ghana Meteorological Agency).

and sustainability attitudes towards natural resources, namely forest resources, and oil and gas products that influence the climate system.

The advocacy for change to mitigate the current and future impacts of climate change is well advertised through REDD, REDD+, AFLU (Agriculture, Forestry, Land Use), etc. The expectation is that there should have been a visible decline in the impacts of climate change, with increasing mitigation activities at the local community level but that is not the case. This leaves many people in the Giddens Paradox, wondering why a threat of such magnitude is ignored by some sections of society. Giddens [7] explained that if there is a lack of tangible and immediate dangers from a threatening phenomenon such as climate change, most people will do nothing to respond immediately. Consequently, by the time the dangers of the phenomenon become visible and dangerous to everyone, it would be too late to act. This suggests that if this is the current perception about climate change in the rural Ghanaian communities, then the worst of the impacts of climate change is yet to happen in the country because some people are yet to experience the tangible and immediate dangers of

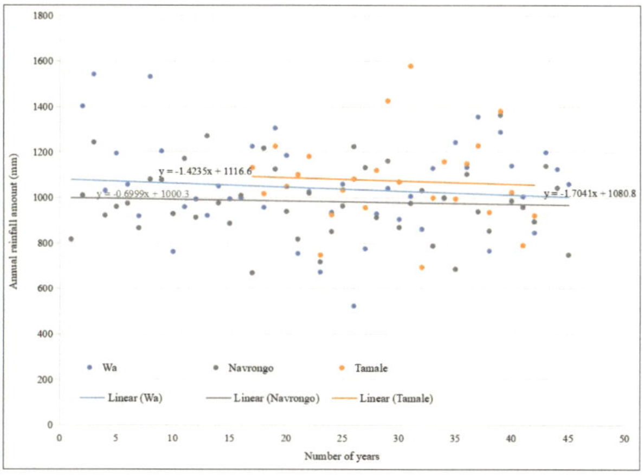

Figure 3.
Trend of annual rainfall amounts (mm) from 1961 to 2005 for Wa, Navrongo, and Tamale weather stations. (Data source: Ghana Meteorological Agency).

it and until then life is still business as usual [4]. However, the Giddens Paradox does not sufficiently explain the ineptness of some people towards climate change although they experience the impacts of climate change and acknowledge that it is real, and everyone is at risk (**Figure 4**). Some people feel that the risk of poverty today is more detrimental to their survival than the risk of climate change today and tomorrow; no matter what or who is the cause of the change. In the case of the rural poor, if the rains fail and government policies do not support them, their mode of adaptation can be the cause of climate change, but they will continue to ensure their daily survival rather than climate change mitigation. For instance, in the case of charcoal production and timber logging, the rural poor will not allow the last tree to sequester CO_2 while they suffer in poverty on daily basis (**Figure 4**). They prefer to fight poverty through charcoal production and sale, and unsustainable logging of timber at the expense of climate change mitigation strategies such as REDD+, sustainable harvesting of wood, and protection of the woodland ecosystems. Such people feel that the effects of poverty are experienced now while the effects of climate change mitigation are in anticipation and for them, the choice between mitigation and adaptation is obvious [8–10]. Charcoal production has become a major climate change adaptation strategy for many people in the charcoal production value chain. This is because it generates quick income for those involved with low startup capital requirements for the producers, many of whom rely on natural woodland for the raw materials for the production of charcoal. The production of charcoal in Ghana is on the rise.

3. Climate change adaptation in the rural areas

Adaptation is a necessity in climate change but how do those who lack or have insufficient capacity to adapt, do so sustainably? Without the appropriate capacity to adapt, the intended adaptation would be maladaptation i.e. more negative feedback effects on the climate system than the benefits to cope with the impacts of climate change. For the poorest of the poor, adaptation is meaningless without the capacity to adapt. When there is no capacity support from state institutions, non-governmental

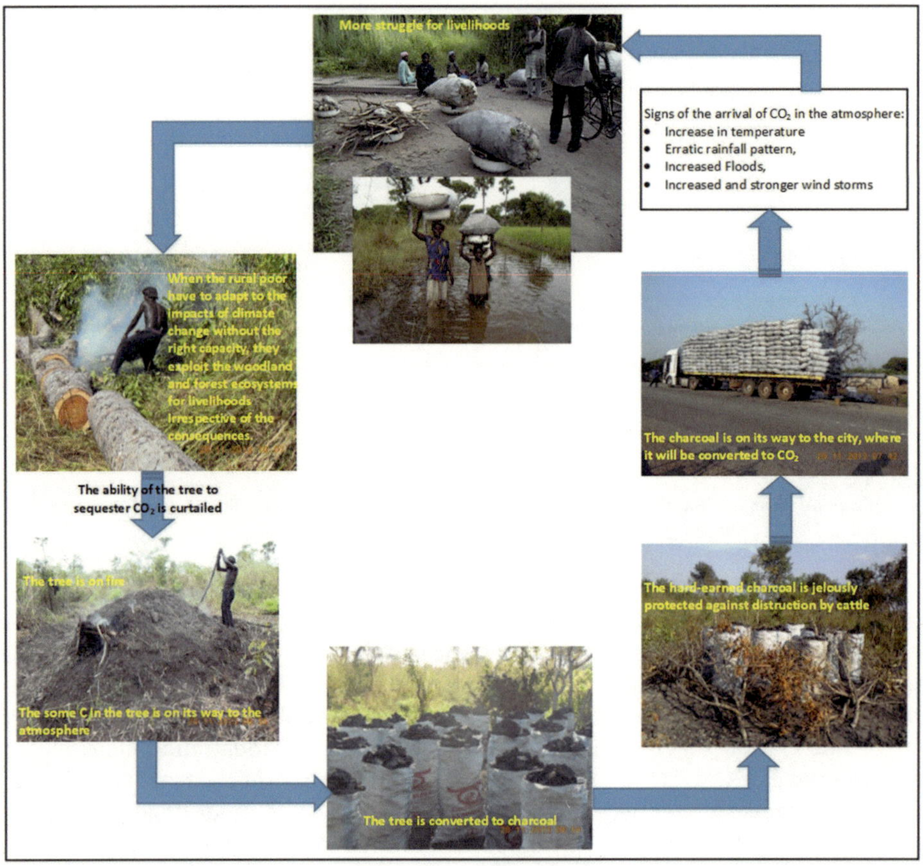

Figure 4.
The struggle against rural poverty at the expense of climate change mitigation. (Source of pictures: field data, 2014).

organisations (NGOs), and philanthropists, the capacity of the poor to adapt to climate change impacts would largely be dependent on woodland and forest ecosystems (**Figures 4** and **5**). However, over-emphasis on these systems would place undue stress on the capacity of these ecosystems.

The issue with the initial climate change discourse is that the need for the rural poor to adapt was over-trumpeted without providing such people the capacity to adapt sustainably. Thus, everything about climate change to the rural poor is an adaptation, adaptation, and adaptation. Adaptation is then misunderstood and decoupled from mitigation activities resulting in the worsening climate change impacts. The past misconception of adaptation by the rural poor later conflicted with the view of scientists about climate change as illustrated in **Figure 5**. The crust of the conflict is when it was realised that they (the scientists) and the rural poor are competing for the same woodland and forest ecosystems in addressing both adaptation and mitigation issues. The rural poor in Ghana, through no fault of theirs, have opened Pandora's box on the woodland and forest ecosystems for various adaptation activities thereby loosening existing access arrangements for these ecosystems (see [11]).

The woodland ecosystems require human support to either maintain or enhance their capacity to support the woodland-based adaptation activities of humans.

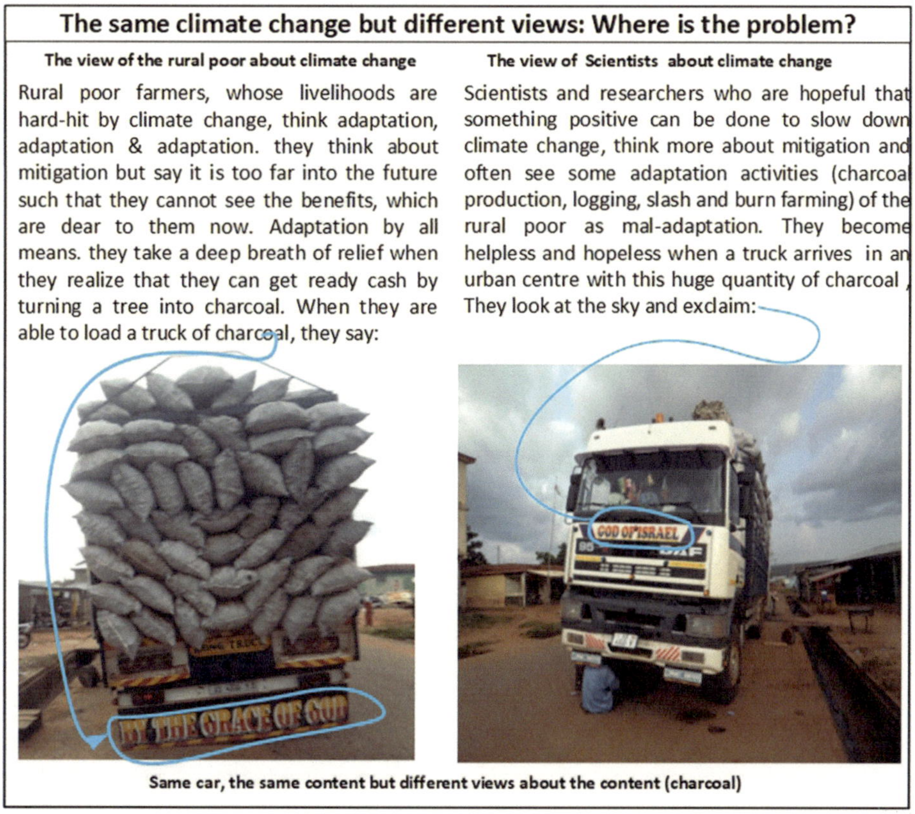

Figure 5.
Conflicting views of climate change about adaptation and mitigation. (Source of pictures: Field data, 2014).

Unfortunately, they are not receiving the needed support, they are rather exploited because of their fundamental role in rural livelihoods in the country [12]. In principle, it is clear that scientists are losing hope in the woodland and forest ecosystems mitigation strategies as a panacea to the increasing impacts of climate change because the dependence on the woodland and forest ecosystems' livelihoods is overwhelming.

4. Woodlands and forest ecosystems, adaptation and mitigation

Forest is an ecosystem that is dominated by trees (perennial woody plants) that are taller than 5 m at maturity and occupy a land area larger than 0.5 ha, with a tree crown cover of more than 10% [13, 14]. The forest ecosystem as defined could be used for production, protection, conservation, or multiple uses. Woodland is an ecosystem with a tree crown cover of 5–10% of trees able to reach a height of 5 m at maturity, a crown cover of more than 10% of trees not able to reach a height of 5 m at maturity, or shrub and bush cover of more than 10% [15]. These ecosystems in Ghana dominate the northern part of the country and have played important ecological and socio-economic roles in the development of the country. They have served as a buffer zone for both livelihood adaptation and mitigation activities (see **Figure 6**) [12].

The part of the country often described as Northern Ghana lies between latitudes 8°0′N and 11°5′N and longitudes 3°0′W and 0°32′E (**Figure 6**). It covers about 41%

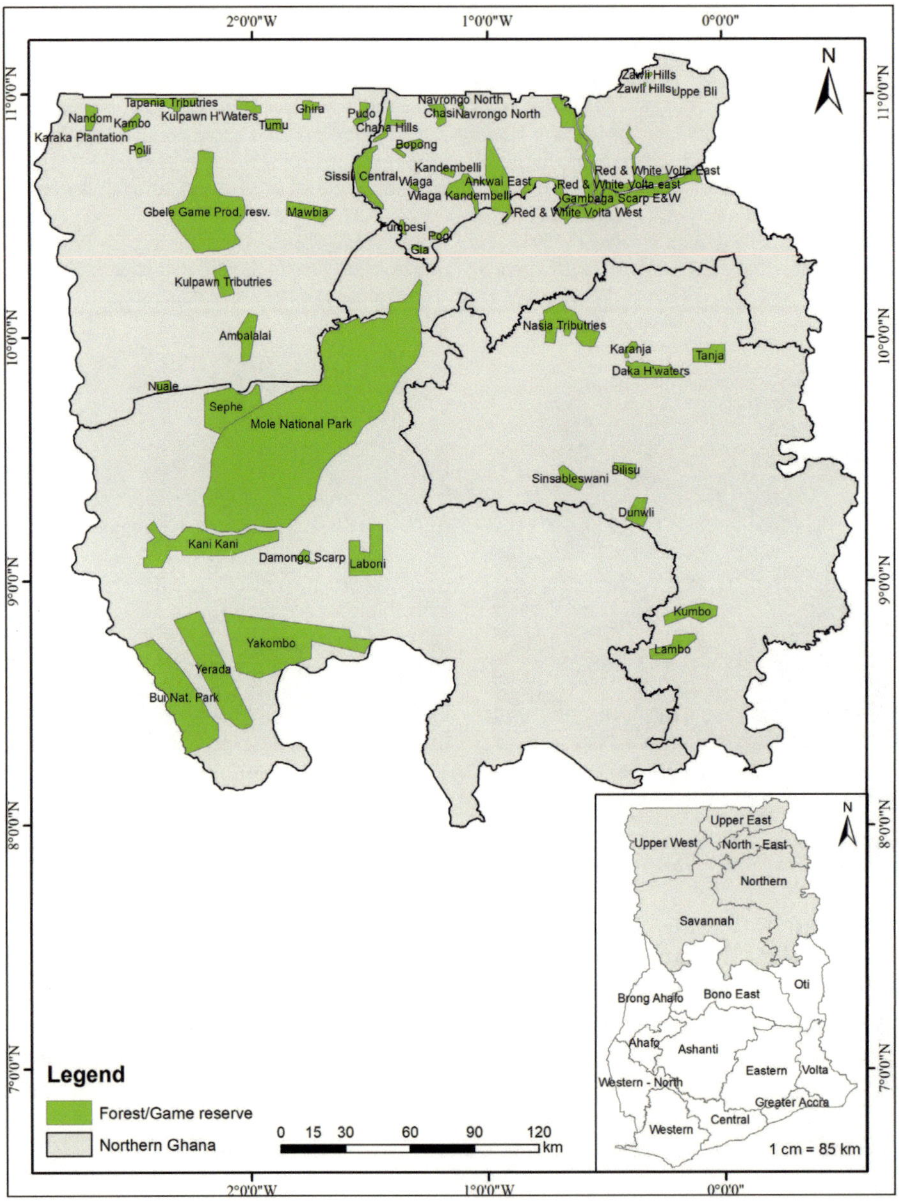

Figure 6.
Northern Ghana in the national context (source: authors' construct, 2022).

of the total area of Ghana with 52 patches of forests/game reserves. Poverty is high in rural areas with high dependence on woodland and forest ecosystems. While these ecosystems, through natural regeneration and human effort, mitigate the effects of climate change; humans through the same ecosystems adapt to climate change by deriving alternative livelihoods such as charcoal production, firewood harvesting, timber logging, and mineral exploitation [12, 16].

DOI: http://dx.doi.org/10.5772/intechopen.107265

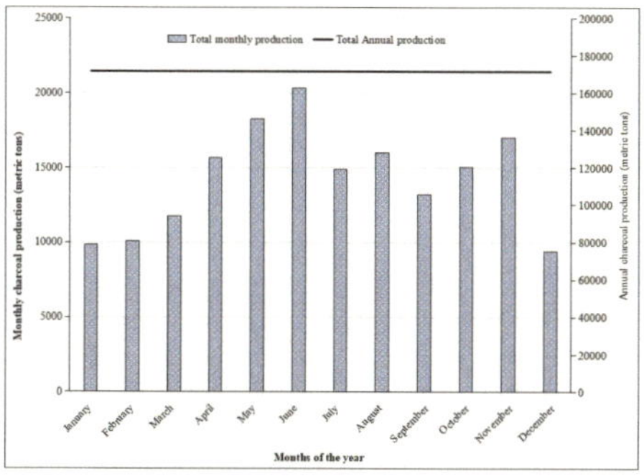

Figure 7.
Monthly trend of charcoal production compared to the annual production in Northern Ghana for 2016 (Sou rce of data: [21]).

Woodland and forest ecosystems in the Savannah areas are fundamental to the livelihoods of those who reside in rural woodland areas in Ghana [12]. Woodlands provide economic, social, cultural, and spiritual services, which are critical in the era of increasing climate change impacts [17]. For instance, charcoal production and logging of trees for timber and firewood in the country have become major climate change adaption strategies in the rural areas rather than conventional livelihood activities. Most charcoal producers in the 1970s and 1980s were perceived as the Sissala people [18, 19] but today, many other tribes are engaged in charcoal production as a supplement to the failing mainstream food crop farming (see [19, 20]). Charcoal production has been on the increase in northern Ghana, question the woodland and forest-based climate change mitigation strategies and efforts because the rate at which the trees are harvested for charcoal production is certainly greater than the rate of regeneration of shrubs. For instance, the trend of the monthly charcoal production in Northern Ghana for 2016 is presented in **Figure** 7, with an annual production of 171,624 metric tons. Other woodland related activities namely logging of rosewood for timber became lamplight livelihood activity in the early 2000s and escalated around 2017.

Adaptation and mitigation activities largely target the woodland and the forest ecosystems (**Figure** 8). Unfortunately, the intensity, extent, and rate of adaptation are

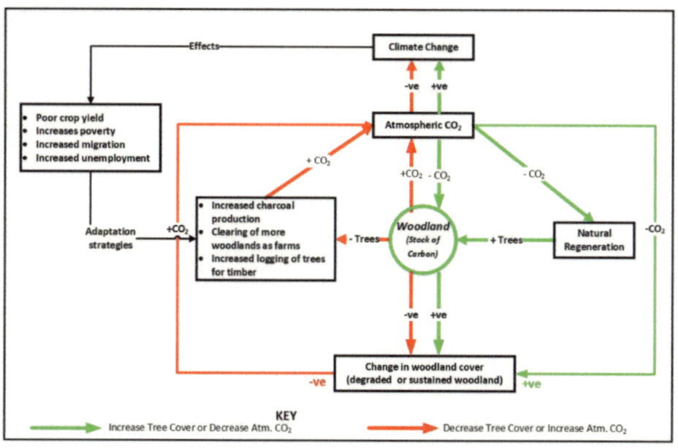

Figure 8.
Relationship between climate change and woodland-based adaptation strategies in Ghana. Source: Adapted from [22].

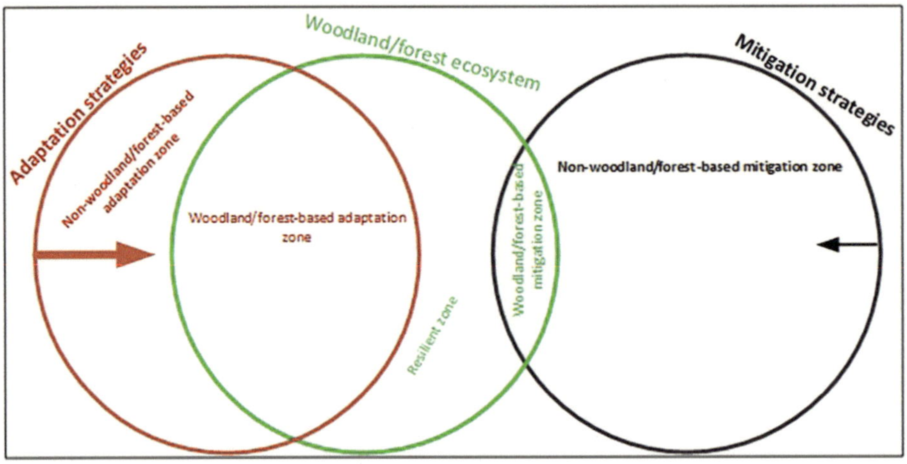

Figure 9.
Relationship among climate change adaptation strategies, mitigation strategies, and woodland and forest ecosystems at the rural Ghanaian community level.

not same as that of mitigation, thus creating a net imbalance between the effects of adaptation and mitigation activities.

Ideally, a net effect in mitigation is desirable in the relationship between adaptation, mitigation strategies, and woodland and forest ecosystems taking into account the strenghs, weaknesses, opportunities and threats of the adaptation and mitigation activities (**Table 1**). Many poor people are seeking woodland and forest-based adaptation strategies with high intensity compared to those working to improve in the capacity of the woodland and forest ecosystems at the community level (see **Figure 9**). This indicates that many more portions of the woodland and forest ecosystems in the Savannah areas are exploited for adaptation strategies than those being developed for mitigation strategies.

Thus, these ecosystems are under-stressed to support adaptation strategies namely charcoal production, logging rosewood, hunting for wildlife, and recently minerals exploitation [16]. For instance, it is indicated that a lot of the charcoal produced in the country comes from woodland areas of the country. Also, the rosewood that is exported to other countries is harvested in the same woodland and forest ecosystems.

The effects of climate change have been overwhelming and the element of the human face has greatly influenced adaption and mitigation activities disproportionately resulting in maladaptation.

The need to adapt to climate change impacts was given media and policy hype which created anxiety among the vulnerable and affected population. In the climate change–adaptation and mitigation nexus, adaptation is heightened to the detriment of mitigation (**Table 1**). The negative impacts on livelihoods were blown out of proportion creating fear and anxiety in people, especially the poor and the vulnerable. It is difficult to have effective adaptation to anxiety because people under duress in emergencies can make dangerous, unsafe choices, which may have tragic or fatal consequences [23]. The current climate change impacts are partly fatal consequences of the maladaptation decisions that were taken in anxiety in the past. Adaptation without mitigation is adaptation without resilience and sustainability. Maladaptation is therefore the bane of mitigation and further over-emphasis on adaptation without mitigation is a fight in support of worsening impacts of climate change. A win-win situation

Climate Change Response	Strengths	Weaknesses	Opportunities	Threats
Adaptation	• Ability to diversify sources of livelihood. • Availability of policies to construct dams and dugouts for irrigation and rearing. • Vast grassland and water bodies for the rearing of animals. • Provision of climate insensitive jobs in urban areas. • Presence of tourist attractions.	• Inability to be cautious of the feedback effects on climate change. • Inability to maintain existing irrigation schemes • High cost of irrigation inputs. • High population growth rate in an urban area. • Inadequate capacity to adapt to available alternative livelihoods • Lack of awareness of carbon trade	• Government policies provide opportunities for diversification of livelihoods, e.g. planting and rearing for food and jobs • Increase donor support for capacity enhancement on adaptation strategies • Development of adaptive crop varieties • Funding support for research in capacity development	• Endemic poverty in rural areas • Dwindling rainfall amounts • Increasing temperatures, increasing population, • Increased frequency of pests and diseases
Mitigation	• Establishment of agroforestry systems in Northern Ghana. • Ability of degraded woodlands to regenerate naturally. • Formulation of policies to protect forest reserves e.g. greening Ghana policy. • Availability of land to expand existing forest reserves and establish new ones • Availability of reserves across Northern Ghana.	• Over-exploitation of woodland for charcoal, timber, and firewood • Over-reliance on charcoal for domestic and informal-sector food and local beverage enterprises • Inability to prevent bush fires reduce the natural regeneration efforts of trees. • Inability to enforce national forest policies.	• Availability of funding opportunities under REDD+ • Payment for ecosystem services • Availability of carbon markets. • Availability of alternative energy sources, solar, liquefied petroleum gas, electricity • Financial support from global partnerships and collaborations • Reduction in global temperatures and extreme climate events.	• Annual bushfires • Excessive logging and charcoal production • Over-reliance on natural woodland for charcoal • Land tenure issues • High cost of petroleum products

Table 1.
Strength, weaknesses, opportunities and threats (SWOT) analysis of climate change adaptation and mitigation in Northern Ghana.

in the adaptation-mitigation nexus is to first consider mitigation in every adaptation measure. Effective adaptation must have a carefully thought-out mitigation measure but most of the adaptation measures are taken in anxiety missing out the consideration of the mitigation element. Adaptation and mitigation should not be decoupled in the fight against climate change and its impacts.

5. Conclusions

This chapter discussed the relationship between woodland and forest ecosystems, climate change adaptation, and mitigation in northern Ghana. The Savannah woodland and forest ecosystems have the *potential* to support adaptation activities and serve climate change mitigation needs. However, the need to adapt was over-emphasised, yet the capacity to adapt and mitigate climate change was downplayed. There is the need to understand that for every adaptation activity, there must be a corresponding mitigation strategy if the gap between adaptation and mitigation effects on the Savannah woodland and forest ecosystems in the rural areas of the country.

Author details

Raymond Aabeyir[1]*, Kenneth Peprah[2] and Gervase Kuuwaabong[1]

1 Department of Geography, SD Dombo, University of Business and Integrated Development Studies, Wa, Ghana

2 Department of Environment and Resource Studies, SD Dombo University of Business and Integrated Development Studies, Wa, Ghana

*Address all correspondence to: raabeyir@ubids.edu.gh

References

[1] Flottum K, editor. The Role of Language in the Climate Change Debate. New York: Taylor & Francis; 2017

[2] FES (Friedrich Ebert Stiftung) and GAWU (Ghana Agricultural Workers Union). Climate Change and its Impact on the Livelihood of Farmers and Agricultural Workers in Ghana. 2012

[3] Dessler E, Parson EA. The Science and Politics of Global Climate Change a Guide to the Debate. New York: Cambridge University Press; 2006

[4] Romm J. Climate change: What everyone needs to know. Oxford: Oxford University Press; 2016

[5] IPCC. Climate change widespread, rapid, and intensifying – IPCC. IPCC PRESS RELEASE, 2021/17/PR. 2021

[6] Fripp. Payments for Ecosystem Services (PES): A Practical Guide to Assessing the Feasibility of PES Projects. Indonesia: Center for International Forestry Research; 2014

[7] Giddens A. The Politics of Climate Change. 2nd ed. Cambridge: Polity Press; 2011

[8] Madhurima C, Banerjee A. Forest degradation and livelihood of local communities in India: A human rights approach. Journal of Horticulture and Forestry. 2013;**5**:122-129

[9] Purnomo H, Suyamto D, Abdullah L, Irawati RH. REDD+ actor analysis and political mapping: An Indonesian case study. The International Forestry Review. 2012;**14**(1):74-89

[10] Aabeyir R, Agyare WA. Woodland access arrangement for charcoal production and its influence on woodland degradation in Kintampo Municipality, Ghana. Scientific African. 2020;**10**:e00572

[11] O'Higgins RC. Savannah woodland degradation assessments in Ghana: Integrating ecological indicators with local perceptions. Earth & E-nvironment. 2007;**3**:246-281

[12] Forestry Commission of Ghana. Development of reference emissions levels and measurement, reporting and verification system in Ghana. Ghana's Forest MRV Project, FC/FCPF/MRV/REL/RFP/01/2013. 2013

[13] FAO. Global forest resources assessment 2000—Main Report. FAO Forestry Paper 140, Food and Agriculture Organization of the United Nations, Rome. 2001. p. 482

[14] Chidumayo EN, Gumbo DJ, editors. The dry forests and woodlands of Africa: Managing for products and services. Washington: Earthscan; 2010

[15] Quacou IE. Unsustainable management of forests in Ghana from 1900-2010. International Journal of Environmental Monitoring and Analysis. 2016;**4**(6):160-166. DOI: 10.11648/j.ijema.20160406.14

[16] Campbell B, Frost P, Goebel A, Standa-Gunda W, Mukamuri B, Veeman M. A conceptual model of woodland use and change in Zimbabwe. International Tree Crops Journal. 2000;**10**(4):347-366

[17] Sarfo-Mensah P, Oduro W. The dynamics of population change and land management in the savanna transition zone of Ghana. Population change and land management in Ghana. PLEC News

and Views New Series Number 3 October 2003. 2003

[18] Amanor K, Osei E, Gyampoh K, editors. Charcoal Burning in the Kintampo Districts: Policies, Environment and Livelihood Issues. DEAR Project, Institute of African Studies University of Ghana: Legon; 2005. Available from: http://www.nrsp.org/database/documents/2714.pdf

[19] Aabeyir R. Material and energy wastes associated with charcoal production. In: Leal FW, Azul AM, Brandli L, Lange SA, Wall T, editors. Affordable and Clean Energy. Encyclopedia of the UN Sustainable Development Goals. Cham: Springer; 2020. DOI: 10.1007/978-3-319-71057-0_145-1

[20] Brobbey LK, Hansen CP, Kyereh B, Pouliot M. The economic importance of charcoal to rural livelihoods: Evidence from a key charcoal-producing area in Ghana. Forest Policy and Economics. 2019;**101**(2019):19-31

[21] Nketiah KS, Asante J. Estimating national charcoal Production in Ghana. Ghana: Tropenbos; 2018

[22] Aabeyir R. Charcoal production and its implication for woodland degradation and climate change in the Forest-Savannah Transition Zone of Ghana: The case of Kintampo Municipality. A [PhD A Thesis] submitted to the Department of Civil Engineering, College of Engineering, KNUST, Kumasi. 2016

[23] Haynes K, Lassa J, Towers B. The influence of risk, gender and religion on child-centred disaster risk reduction. Children in a Changing Climate Working Paper 2. Brighton: Institute of Development Studies. 2010

Chapter 11

Global Change Drivers Impact on Soil Microbiota: Challenges for Maintaining Soil Ecosystem Services

Emoke Dalma Kovacs and Melinda Haydee Kovacs

Abstract

Global change refers to anthropogenic and climate pattern modification. The consequences of these changes are outstanding on aboveground biodiversity. Soil microbiota are key actors in soil processes, contributing significantly to numerous ecosystem services provided by soil. They are involved in the processes of nutrient cycling, organic matter decomposition, or pollutants degradation. Microorganisms are also able to synthesize volatile organic compounds that are secondary metabolites with multiple ecological roles and mechanisms of action—generally contributing to plant development. Changes in soil microbiota community could modify either negatively or positively their contribution in soil-provided ecosystem services through their involvement in soil functions that they mediate.

Keywords: microbiome, processes, soil functions, soil microbiota, soil ecosystem

1. Introduction

Global change is a hot topic of our days. Briefly, it could be defined as the sum of effects resulting from interactions between climate change and anthropogenic drivers. This phenomenon continues to grow in amplitude in the most part as a consequence of anthropogenic activities. Soil, a key environmental compartment that sustains life on earth and human development, it seems to be a fragile terrestrial ecosystem component in facing the challenges raised by global change. In its interplay with other environmental compartments under the pressure of global change, the soil could be at the same time both a contributor to as well a recipient of the impacts and factors of global change. However, at moment, under the challenge of global change drivers, the soil is considered as the least understood component due to its heterogeneity and complexity of it as properties, functioning, and provided services.

Now, global change is a phenomenon that happens, and its consequences on the ecosystem and delivered ecosystem services are predicted to continue even in future. As phenomenon, global change is the sum of end results between interactions of drivers and effects of changes in climate patterns and anthropogenic activities.

IntechOpen

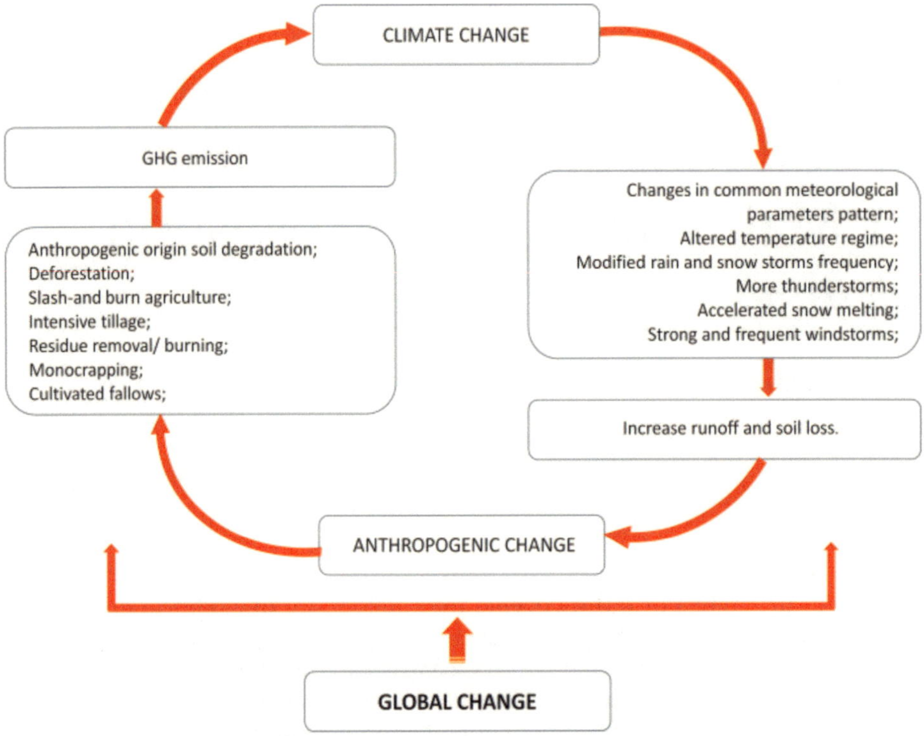

Figure 1.
Soil challenges under interactive effects between anthropogenic and climate change drivers.

Figure 1 presents soil faced challenges under the interactive effects between the main anthropogenic and climate change drivers.

Drivers are defined by Millennium Ecosystem Assessment as a natural or anthropogenic factor that induces a change in an ecosystem either directly or indirectly [1]. While *direct drivers* influence ecosystem processes without any doubt, *indirect drivers* usually generate an alteration on one or more direct drivers, thus acting more diffusely.

Ecosystem and delivered ecosystem service changes are usually caused by multiple interacting drivers (direct and/or indirect). They could interact at a multi-level scale, including spatial, temporal, and organizational scales. Therefore, global change drivers can impact ecosystems and consequently ecosystem services both over time, levels of organization, or intermittently [2].

Soil is a key resource for life on earth, acting as a provider of essential raw materials, promoter for human activities, and habitat and gene pool of biodiversity. These global phenomena impact the soil environment, changing its properties, and consequently related biota. The development of soil and its properties takes place most of the time in the presence of living organisms. Soil biodiversity is complex, including diverse micro, meso, and macrofauna and flora with a complex role in soil formation, functioning, and properties. Hereby, soil fauna diversity like invertebrates (earthworms, termites, etc.) modify soil physical and chemical properties [3], while grazers concentrate nutrients through dung middens [4]. Similarly, plants are the main source of organic carbon and also are geochemical pumps that return to soil bio-essential elements through litterfall and degradation [5, 6]. The amount and the distribution depth of organic carbon and essential elements are influenced by plant species [7]. Soil microbiota impact too soil's chemical and physical properties through the quality and quantity of organic compounds that they release, as well as through the biochemical reactions that they mediate [8].

Changes in soil properties due to global change is expected to impact biodiversity also. Modification or alteration in structure (exclusion or introduction of organisms or species), as well as in abundance of biota, could have also a reversible impact on the connected environment (property, functioning), ecosystem, and delivery of ecosystem services. However, soil ecosystem change is the after-effect of a large number of interactions between drivers. Whereas a considerable number of drivers are global, the current assemblage and interactions that bring about an ecosystem are more or less specific to a particular region or place. Although the influence on soil properties by higher-order plants [9, 10] and animals [11, 12] was well documented, a clear view on the potential influence of microbiota on soil properties is less profound. The reason is due to the difficulty of isolation and identification of species, for example., at moment only 1% of soil bacteria can be cultured. Similarly, there is minor knowledge of how microbiota structure and abundance could be influenced by global change drivers. Know-how on the relationship between global change drivers—soil microbiota—soil properties and functioning nexus are important when soil is accounted as a basic environmental compartment that serves us with various supporting, provisioning, and regulating ecosystem services, and when the way and the intensity of the response of soil system to global change are directly dependent to ecosystem dynamics.

Global change influences soil, and this in turn will impact global change further. Considering the time that is involved in soil formation, made it to be considered as a finite and non-renewable resource. This is the reason why is very important to understand how global change drivers put pressure on it in order to could conserve and protect its properties.

Ecosystem services refer to goods and/or services provided by an ecosystem. Soil functions are directly involved in the support and delivery of ecosystem services. Soil well-functioning depends on soil's physicochemical and biological properties. Changes in any of these properties could contribute either to decline or enhancement of provision with ecosystem services. Anthropogenic and climate change-related drivers are acknowledged to change soil physicochemical properties, which are supposed to influence soil biota. This could change also the soil biological properties. Alteration of soil properties could reduce or change soil functioning thus causing loss or alteration of provided ecosystem services.

2. Soil microbiota: key component of soil biodiversity and ecosystem

Soil microorganisms present a great diversity, although more of them are not cultured at moment. They are acknowledged as a key component of soil biodiversity due to their involvement in numerous and significant interactions in terrestrial ecosystems. Such interactions control soil's physical, chemical, and biological processes. Components of soil microbiota directly mediate and influence the stability and cycling of relevant elements and climate change. Microorganisms activities are related to the regulation of soil C sequestration and mineralization, nutrient cycling (N, P, etc.), and not finally with ecosystem productivity once that they facilitate the nutrient resource for higher components of biota (e.g., plants) through fast turnover [13]. Soil microbiota community components are also a relevant source of enzymes. They liaise soil potential for enzyme-mediated substrate catalysis [14, 15]. Microbiota components are indigenous to the environment and most of the time are capable to adapt to variable environmental conditions (temperature, redox potential, pH, moisture regime, and pressure) or to exist under oligotrophic conditions (low nutrients).

2.1 Soil microbiota community structure

Microbiota constituents could be grouped into three domain systems consisting *Archaea*, *Eukarya*, and *Bacteria*. *Bacteria* and *Archaea* are generally named as prokaryotes, while *Eukarya* as eukaryotes [16].

2.1.1 Bacteria

Bacteria, the most abundant microorganisms as a number of individuals (around 50 phyla) are free-living fewer complex organisms with great metabolic flexibility. Due to these features, they easily and promptly respond and adapt to changing environmental conditions. Bacteria can be grouped either by considering their cell envelope architecture (structural characteristic) or through their metabolism type (physiologic characteristic). Structurally, bacteria are grouped as Gram-positive (e.g., Bacillus, Clostridium, etc.) and Gram-negative (e.g., *Pseudomonas*, *Shewanella*). This difference mediates their survival in the environment.

Gram-negative bacteria cell envelope is a complex structure that allows them to interact with mineral surfaces and solutes from the environment. In that way, they obtain the required amount of nutrients for metabolism [17].

Gram-positive bacteria have a less complex cell envelope [18]. Their thick cell wall allows them to withstand challenging physical conditions of the soil environment [19].

Actinomycetes is a special group that now is classified as Gram-positive bacteria. These group differentiates by bacteria through their tendency to branch into small dimension filaments or hyphae that structurally resemble the hyphae of fungi [20]. They are widespread in soil environment and are recognized also as valuable antibiotic producers [21, 22]. Actinomycetes abundance increase with decomposed organic matter. However, they are strong pH-sensitive organisms, usually at pH below 5 pH units, their abundance decreases considerably. Contradictory with other bacteria species, their abundance increases once with soil depth. From ecological point of view, as growth strategy soil bacteria could be grouped as copiotrophs (e.g., *Betaproteobacteria*) and oligotrophs (e.g., *Acidobacteria*). Copiotrophs consume easily degradable organic C, while oligotrophs consume recalcitrant organic C. Although oligotrophs grow slowly but constantly, maximizing their yield under poor nutrient availability conditions. Copiotrophs to maximize their yield require high nutrient content. From bacteria metabolism point of view, considering the source of energy and C used for growth, Robert and Chenu [23] classified bacteria in several groups (chemoheterotroph, photoautotroph, photoheterotroph, etc.—see **Table 1**).

Therefore, according to their energy source, bacteria could be divided into four main groups (**Table 1**) as *phototroph* (energy is obtained from light—photosynthesis), *chemotroph* (oxidation of organic and inorganic chemicals), *autotroph* (carbon dioxide), *heterotroph or organotroph* (from organic compounds as glucose) [19]. Chemoheterotrophs (chemoorganotrophs) use organic compounds both as energy and as C source. Chemoautotrophs (chemolithotrophs) obtain energy from the oxidation of inorganic compounds and C from CO_2. Photoautotrophs get energy from light and

Metabolism type – examples	Metabolism	Carbon source	Energy source	Products
Chemoheterotroph	Respiration	Organic compounds	Organic compounds	
e.g., *Pseudomonas, Bacillus*	(aerobic)		(O_2)	CO_2, H_2O
e.g., *Micrococcus, Geobacter, Desulfovibrio*	(anaerobic)	e.g.,NO_3, Fe^{3+}, SO_4^{2-}		CO_2, NO_2^-, H_2O, N_2O, N_2, Fe^{2+}, S, S^{2-}
	Fermentation	Organic compounds	Organic compounds	
e.g., *Clostridium*	(anaerobic only)	Organic acids		CO_2, organic acids, alcohols
Chemoautotroph or Chemolithotroph	Chemolithotroph	CO_2	H_2, S^{2-}, NH_4^+, Fe^{2+}	H_2O, SO_4^{2-}, N^{2-}, Fe^{3+}
e.g., *Hydrogen bacteria*				
e.g., *Beggiatoa*	(aerobic)		(O_2)	
e.g., *Planctomycetes*	(anaerobic)		(NO_3)	
Photoautotroph	Photosynthesis	CO_2	Light + H_2O ($NADP^+$)	O_2
e.g., *Cyanobacteria*	(oxygenic)			
Bacteria including Purple sulfur bacteria (e.g., *Chromatium*); Purple non-sulfur bacteria (e.g., *Rhodospirillum*); Green non-sulfur bacteria (e.g., *Chloroflexus*); Heliobacteria (e.g., *Heliobacterium*)	(anoxygenic)	CO_2	Light + H_2S (bacterio-chlorophyll)	S^0
Photoheterotroph	Photoheterotrophy	Organic compounds	Light + H_2S (bacterio-chlorophyll)	S^0
e.g., many purple non-sulfur bacteria, purple sulfur bacteria but to a limited extent				

Table 1.
Bacteria classification based on their metabolism (modified after Pepper and Gentry, [19]).

fix C from CO_2. Photoheterotrophs obtain energy from light and C from organic compounds. Bacteria harvest energy either through respiration (aerobic or anaerobic process) or through fermentation (anaerobic process). *Archaebacteria* is a special heterogeneous group of bacteria that have the ability to leave even in extreme environmental conditions (e.g., environments with high sulfur or salt content). Usually, these organisms are classified either as *obligate anoxybiont* (total absence of oxygen in the media where they live) or *facultative anoxybiont* (they could survive in both aerobic and anaerobic environments, e.g., thermoacidophiles) [24, 25]. Obligate anoxynbionts include methanogen (produce CH_4 from CO_2) and halophile species (live in extremely salt environment). Although there are bacteria in soils that can have a pathogenic effect on plant biodiversity, most bacteria are recognized that have important functions that assure soil health and functioning through decomposing organic matter and contributing to producing nutrients available for other living micro and macroorganisms.

2.1.2 Archaea

Archaea, although could appear similar to bacteria they differ from them genetically and biochemically. They appear both in extreme and nonextreme environments. Extremophiles could survive in environment with extreme temperature (hot, cold), salinity, alkalinity, or acidity [19]. Major divisions of archaea are *Crenarchaeota*, usually thermophiles (live in high-temperature environment), and *Euryarchaeota* that include haloarchaeans (live in saline environments) and methanogens (live in anaerobic environment at low temperature) [26]. Their major functions related to soil are those connected to horizontal gene transfer between archaeans and bacteria, and nitrification process control.

2.1.3 Fungi

Fungi with great biomass are physically the largest group of eukaryotic microorganisms. They are regnant in soil environment with high adaptability to various conditions. Are valuable components of soil biodiversity due to their essential function as decomposers [27]. Considering their morphological description, fungi are grouped as molds, mushrooms, and yeasts. *Molds*, filamentous fungi, are found in many fungal phyla. *Mushrooms*, filamentous fungi, are part of Basidiomycota. They form the large fruiting bodies known too as mushroom. Both molds and mushrooms are very important decomposers of natural products [19]. They produce also extracellular substances that bind soil particles, forming stable soil aggregates, reducing, therefore, soil erosion. *Yeast*, are unicellular fungi with the ability to ferment under anaerobic conditions. Some components through symbiotic relationships with algae and cyanobacteria form lichens [28]. These secrete organic acids that help rocks and inorganic surfaces in degradation. Fungi have chemoheterotroph metabolism. This supports biosynthesis and energy production based on simple sugars, but they produce also secondary metabolites. These metabolites (e.g., exoenzymes) produced during the stationary phase of growth are acknowledged that help to reduce the competition for nutrients from other microorganisms, and some of them have antimicrobial properties. Exoenzymes break down complex polymers into simple C compounds for cells [29]. Based on that, fungi could be grouped as saprophytic and mycorrhizae.

Saprophytic fungi are important organic material degraders (e.g., dead plants and organisms), especially of complex polymers associated with them (e.g., cellulose and lignin from plants; chitin from insects) [30, 31]. They are also able to degrade chemical pollutants [32]. *Mycorrhizae* form a symbiotic relationship with a large number of plants. Through that relation, these fungi increase plant roots' absorptive area and prevent desiccation, as well increase nutrient uptake (especially phosphate) [33, 34]. Similarly, plants furnish sugar (obtained through photosynthesis) to fungi. Mycorrhizal fungi could be divided into ectomycorrhizal and endomycorrhizal fungi group [32].

2.1.4 Protozoa

Protozoa, are eukaryotic microorganisms with fundamental genetic differences between species. Protozoa species are usually heterotrophic organisms and consume bacteria, yeast, fungi, and algae [35]. Usually habit the top part of soil (up to 20 cm depth) and are concentrated near roots (due to availability of the high quantity of prey). They are involved in soil organic matter decomposition processes.

2.1.5 Algae

Algae, phototrophic organisms (metabolize in the presence of light for energy and CO_2 for C), are located at the top surface of the soil or very close to it (green algae and diatoms, which are heterotrophs as well photoautotrophs). They are strongly involved in soil formation processes through their metabolism. Once through photosynthesis, they introduce C to the soil, and secondly through metabolizing processes produce and release into soil carbonic acid and polysaccharides [36]. Released carbonic acid help in weathering surrounding mineral particles, while extracellular polysaccharides facilitate soil particle aggregation [37]. Soil algae are seasonally variable, higher abundance having in spring and fall period. In winter and summer, their development and abundance are suppressed considerably due to water-induced stress—soil moisture and/or desiccation [36].

2.2 Soil microbiota community structure and abundance: Role in soil processes

Soil, the base of terrestrial ecosystems, is inhabited by a large diversity of organisms. Microbiota inhabitants are key actors in several essential processes in soil. Through their diversity and varied metabolism mediate biochemical reactions and take part in multiple interactions and reactions in and between surrounding microhabitats. Through these, soil microbiota contributes to and liaises essential functions of soil ecosystems. Thus, soil microbiota significantly influences the soil ecosystem in its ability to provide ecosystem services.

2.2.1 Bacteria role in soil processes

Bacteria are recognized as primer decomposers of both organic matter and organic wastes. They change compounds and elements from inaccessible to usable forms for higher trophic components, thus significantly contributing to the cycling of essential elements and providing the nutritive resources required by below- and aboveground organisms. Geddes et al., [38] with others [39–42] revealed that *Rhizobium sp*. and *Bradyrhizobium sp*., bacteria present inside the host root system (leguminous plants root nodules) fix atmospheric N, the primary nutrient that influences plant growth and development. Arashida et al., [43] showed that *Pseudomonas sp*., *Bacillus sp*., *Azotobacter sp*., and *Azomonas sp*., bacteria from the proximity of plant root system, fix atmospheric N usable form as ammonia. Less frequent bacteria phyla in soils, such as *Verrucomicrobia*, is also involved in N fixation and associative activities [44]. Cyanobacteria are recognized as important improvers of C, N, and exopolymeric substances content in the soil. They release into soil amino acids, proteins, polysaccharides, carbohydrates, vitamins, and phytohormones as elicitor molecules that promote plant growth. Kumar et al., [45] (2013) also reported that cyanobacteria could facilitate plant resistance against both biotic and abiotic stresses. Actinobacteria species have also multiple roles in soil, such as fixation of atmospheric N (*Arthrobacter sp., Rhodococcus sp.*), minerals solubilization (*Ferrobacter sp.*), increase of antagonistic efficiency against different fungal root pathogens (*Streptomyces sp.*) as well as plant growth hormones production (*Frankia sp.*). They are also responsible for soil odor. Studies on dominant culturable bacteria, *Arthrobacter sp.*, *Streptomyces sp.*, *Pseudomonas sp.*, and *Bacillus sp.* revealed their involvement in nutrient cycling and biodegradation [46], degradation of recalcitrant organic compounds [47], and production of antibiotics [48] and biocontrol agents [49]. Functions in the soil of important autotrophic and heterotrophic bacteria are summarized in **Table 2** [19].

2.2.2 Archaea role in soil processes

Ren et al., [50] through studies performed on agricultural soils in that different management strategies and practices were applied demonstrate that archaea are serious contributors to soil processes such as nutrient cycling, ammonia oxidation, and minerals weathering. Archaea are involved also in carrying out N_2 fixation as well reduction of atmospheric N_2 to NH_4^+.

Bacteria	Characterization	Role and function in soil
Autotrophic bacteria		
Nitrosomonas	G-, aerobe	Convert $NH_4^+ \rightarrow NO_2^-$ (first step of nitrification)
Nitrobacter	G-, aerobe	Convert $NO_2^- \rightarrow NO_3^-$ (second step of nitrification)
Acidothiobacillus	G-, aerobe	Oxidize $S \rightarrow SO_4^{2-}$ (sulfur oxidation)
Acidothiobacillus denitrificans	G-, facultative anaerobe	Oxidize $S \rightarrow SO_4^{2-}$ (functions as a denitrifier)
Acidothiobacillus ferroxidants	G-, aerobe	Oxidize $Fe^{2+} \rightarrow Fe^{3+}$
Heterotrophic bacteria		
Actinomyceyes	G+, aerobe, filamentous	Produce geosmins and antibiotics
Bacillus	G+, aerobic, spore former	C cycling, production of insecticides and antibiotics
Clostridium	G+, anaerobe, spore former	C cycling, fermentation, toxin production
Methanotrophs	G-, aerobe	Methane oxidizers that can cometabolize trichloroethene using monooxygenase
Rhizobium	G-, aerobe	Fixes N symbiotically with legumes
Frankia	G+, aerobe	Fixes N symbiotically with nonlegumes
Agrobacterium	G-, aerobe	Pathogen for plants

Table 2.
Main autotroph and heterotroph bacteria roles in soil (adapted after Pepper and Gentry, [19]).

2.2.3 Fungi role in soil processes

Fungi perform several key functions in soil. They are considered important decomposers and mutualists. Saprophytic fungi, the main decomposers, degrade dead organic materials. These are transformed into fungal biomass, CO_2, and several small but essential molecules (e.g., amino acids). Hussain et al., [51] and Chaudhary et al., [52] reported that *Phoma sp*. and *Penicillium sp*., respectively, induce synthetic resistance against plant pathogens. Fungi such as *Trichoderma sp.* and *Phoma sp*. enhance biomass production, improve plant growth [53], and promote lateral root growth [54]. Mutualists are mycorrhizal fungi. Most time, these colonize plant roots. They favor phosphorous solubility and bring micro and macronutrients to plants. Arbuscular mycorrhiza fungi were reported that increase soil nutrient availability and improve nutrient acquisition by plants. Heidari and Karami [55] reported that mycorrhizas could enhance crop yield by fostering host resource uptake through sharing. They could minimize nutrient loss under extreme meteorological events also. Celik et al., [56] through their study assessed that arbuscular mycorrhiza fungi are important in both of forming stable soil aggregates as well in the improvement of water retention. Also, there are many studies that report their efficiency in the phytoremediation of polluted soils [50]. Yeast was summarized by Pepper and Gentry, [19] as consumers of bacteria and plant root exudates for the synthesis of plant protectants and many other useful and important compounds that enhance plant growth. Beeck et al., [27] highlighted that yeast produces significant extracellular polymeric substances. For that, yeast is often associated with soil structure formation and maintenance.

2.2.4 Algae role in soils

Recently, many studies have evidenced that algae are important in maintaining soil fertility. In their life cycle, they contribute to soil particle binding and facilitate soil erosion prevention [36]. It was reported also that algae are involved in rock weathering, thus they could be considered as wrapped up in soil structure building [57]. Algae increase soil water retention capacity, enhance submerged aeration through photosynthesis processes, and contribute in the reduction of soil nitrates through leaching or drainage processes [19]. After their life cycle, they contribute to soil nutrient resources with large amount of organic carbon.

2.2.5 Protozoa role in soil

Protozoa maintain microbial diversity and functional stability (microbial/bacterial equilibrium) through their nutrition (feeding and ingestion) and multitrophic interactions at that take place. Chen et al., [58] highlight in their paper protozoa positive influences on nutrient availability for plants, release of hormones for plant development, and biological control agents against organisms that could induce potential harmful diseases in plants. Ronn et al., [59] acknowledged that protozoa are also involved in the accumulation and stabilization of organic carbon in the soil.

3. Soil ecosystem services: Connection with soil microbiota

According to the implications that components of soil microbiota have in various soil functions, becomes obvious that soil belowground microbiota diversity is an important resource for maintaining the functioning of soil ecosystem and consequently of ecosystem services on which society depends. Soil microbiota through their metabolic pathways is directly involved in greenhouse gas removal, nutrient cycling, pathogens inactivation, and pollutants degradation. **Table 3** gives a summary of the

Service	Ecosystem services	Soil function	Role of microbiota
Support	Primary production	Support for terrestrial vegetation	Net and gross primary productivity is assured by soil microbiota through reactions in that transform organic and inorganic compounds into usable form for other organisms. Also, microbiota mobilizes nutrients from insoluble minerals to support plant growth.
	Soil formation and renewal	Soil formation processes	Microbiota speed up and modify soil physicochemical processes.
	Nutrient cycling	Storage, cycling, processing of nutrients and delivery to plants	Nutrient cycling among organic and inorganic pool is driven by soil bacteria, archaea, and fungi turnover. Fungi are effective in C and N storing in organic matter. AMF controls plant P nutrition. Saprophytic fungi generate more degradative enzymes, thus are important decomposers of recalcitrant plant litter. Bacteria higher turnover rate speed up gross mineralization and plant nutrient uptake. Nitrogen, phosphorus, and sulfur mineralization, nitrification, bioweathering of P minerals, and sulfur oxidation is assured by soil bacteria and fungi communities.
	Platform	Supporting structure for human activities	Microbes contribute to soil formation through elements cycling, and compounds and organic matter production. They are involved in minerals weathering. Microbial products are critical to soil aggregation. Improved soil structure makes it more habitable for living organisms.
Provision	Refuge	Habitat for resident and transient populations (terrestrial habitat)	Soil bacteria, archaea, and fungi contribute to soil formation and are the foundation of soil food webs, thereby underpinning the diversity of higher trophic levels.
	Water storage	Water retention and supply in the landscape	Soil microbiota are food and nutrient resources for higher trophic level organisms, such as plant roots, earthworms, and others. These contribute to macropores formation, which is related to hydrological processes (infiltration, drainage).

Service	Ecosystem services	Soil function	Role of microbiota
	Supply of food, fibers, biofuels, and wood (biomaterials)	Provision of plant growth and production	Soil microbes produce antimicrobial agents and enzymes useful for several biotechnological purposes. Microbiota species produce plant growth hormones, symbioses (mycorrhizal fungi and N_2 fixing bacteria), pathogen control, and degradation of stress ethylene (ACC deaminase-positive bacteria).
	Supply of raw materials of mineral origin	Provision of source materials	Soil microbes are involved in minerals weathering processes.
	Biodiversity and genetic resources	Source of unique biological materials and products (soil biota)	Soil bacteria, archaea, and fungi comprise a vast majority of the biological diversity on earth. Further, they are the foundation of soil food webs, thereby underpinning the diversity of higher trophic levels. Interactions among soil microbes and plants often determine plant diversity.
Regulation	Control of potential pests and pathogens	Population regulation (soil biota) to control pests, pathogens, and diseases	Soil bacteria, archaea, and fungi support plant growth by increasing nutrient availability, and by outcompeting invading pathogens through inter and intraspecific interactions (symbiosis, competition, host-prey association).
	Recycling and remediation action	Disposal and decomposition of residues and pollutants	Soils absorb and retain solutes and pollutants, avoiding their release into the water. Microbiota contributes to both the hydrophobicity and wettability of soils, impacting the ability of soils to filter contaminants.
	Water quality regulation	Filtration and buffering of water	Soil macropores are formed by plant roots, earthworms, and other soil biota, which may depend on soil microbes as food or for nutrients. Also, microbiota is involved in pollutant degradation. Nitrate respiring bacteria, fungal and bacterial contaminant degraders, metal oxidizing and reducing bacteria (e.g., sulfate oxidizers, *Geobacter metallireducans*) assure a dissimilatory nitrate reduction, co-metabolism and mineralization of organic contaminants, sulfate reduction and subsequent metal precipitation, metal respiration, and precipitation.

Service	Ecosystem services	Soil function	Role of microbiota
	Water regulation, and flood and drought control	Regulation of hydrological flows, buffering, and moderation of hydrological cycle	Soil pores through their capacity to retain and store quantities of water can mitigate and lessen the impacts of extreme climatic events. Soil microbiota are involved in soil pores characteristics defined as macropore formation and soil aggregate formation.
	Regulation of atmospheric GHG and climate regulation	Carbon sequestration and accumulation, regulation of the atmospheric chemical composition, and climate processes	Soil microbiota is involved in litter fragmentation and decomposition, and physical and chemical stabilization of residue carbon. By mineralizing soil carbon and nutrients, microbes are major determinants of the carbon storage capacity of soils. Denitrifying bacteria, fungi, and methane-producing and consuming bacteria regulate N_2O and CH_4 emissions from soil. Methane production is done by methanogens, while methane oxidation by methanotrophs. Chemoautotrophic nitrification, heterotrophic nitrification, denitrification, and co-denitrification are assured by nitrifying and denitrifying bacteria and fungi.
	Erosion control	Sediment retention	Soil microbiota promote plant growth through nutrient cycling and transformation in an available form for the plant, as well as through soil-root exchange enhancement, thus alleviating soil surface enhancement. Microbiota produces biological glues, and facilitates physical entanglement by roots and fungal hyphae.

Table 3.
Soil microbiota involvement in soil functions and processes assures diverse support, provision, and regulation services provision by the soil ecosystem (adapted after Aislabie and Deslippe, [60]).

main soil services description linked to soil microbiota. Therefore, to benefit from soil ecosystem services depends on soil biodiversity, which contributes to the capacity of soil to function. Soil functions assure and sustain plant and higher trophic levels of healthy development, as well as air and water quality. At moment, based soil functions in that soil microbiota are involved, acknowledged their potential implication in regulating, supporting, and provisioning services.

3.1 Regulating services

Regulating services, generated or intermediated by soil microbiota include in high proportion soil ecosystem processes related to climate regulation; water purification and cycle regulation; pest, pathogens, and diseases control; and bioremediation (e.g., degradation of organic pollutants). *Climate regulation* is mainly mediated by soil bacteria and fungi as they are involved in processes of C exchange between land and atmosphere. Soil as a global carbon sink relies on plants to fix atmospheric carbon and soil microbiota to convert that carbon into the soil. Consequently, soil microbiota excites carbon sequestration and storage through essential interactions with plants. Plant growth and plant biomass could be positively stimulated through nutrient solubilization and phytohormone production and regulation, processes mediated by soil bacteria and fungi. Moreover, soil microbiota are involved in soil structure formation and maintenance through aggregate formation, physically defend the decomposition of soil organic carbon. Conversely, soil microbiota are acknowledged in direct implication in soil organic matter decomposition, which leads to greenhouse gases emission into the atmosphere. Overall, soil microbiota plays an essential role in regulating the global carbon cycle. Soil carbon store and emission depend on land use and applied management practices. Therefore, soil climate regulation depends on that. Studies are divided into two, based on the potential effects of climate change on soil microbiota and their functions related to climate regulating services. The acceleration of heterotrophic microbiota activities rate is correlated positively with an increase in temperature. This was, first, considered to be a positive effect [61]. However, heterotrophic microbiota respiration under aerobic condition increases with temperature rise. Capek et al., [62] described that this will also increase soil organic carbon stock depletion. Loss of soil organic carbon was associated with significant climate feedback effects, such as rising CO_2 and CH_4 level in the atmosphere. In turn, Drigo et al., [63] showed that a higher CO_2 level damaged soil bacterial community structure. The level at which bacterial community is damaged depends on the cover plant type and nutrient availability in the soil. These depend most of the time on soil management practices. Moreover, several agricultural practices are associated with a higher CH_4 and N_2O emissions, both contributing further to climate change amplitude increase.

Water regulation and purification, implying soil microbiota, is assured through their degradation ability of various contaminants. Almansoory et al., [64] showed that *Serratia marcescens* can degrade total petroleum hydrocarbons from soil. *Pseudomonas sp*. degrade in soil herbicides as diuron. Song et al., [65] presented in their paper that *Rhizobium* bacteria degrade di-(2-ethylhexyl) phthalate. Soil bacteria could modify soil organic matter quality and quantity. This could change the water infiltration rate. Therefore, that could be also considered as an indirect effect of soil bacterial communities' influence on water regulation and purification services. *Diseases and pest regulation* by soil microorganisms could be assured through different mechanisms such as antagonism, competition, interference with pathogen signaling, or stimulation of host plant defenses [66]. Plants are often exposed to potential pathogens (e.g., *Erwinia*, *Pectobacterium*, *Pantoea*, *Acidovorax*, *Xanthomona*, etc.) that could cause leaf spots and blights, wilts, overgrowths, scabs, loss of fruits or even of species. Contrarily, the most efficient and dynamic bacteria that could suppress the development of diseases are those that belong to *Firmicutes, γ-* and *β- Proteobacteria*. Among fungi, *Ascomycota phylum* is the most representative.

Soil organic waste and xenobiotics biodegradation are assured by soil microbiota through processes such as transformation, mineralization, and stabilization. In order that these processes to take place favorable conditions are required. Generally, soil microbiota could use chemicals as substrate resources (energy, C, N, other nutrients, etc.) or could transform them by consecutive microbial enzymes or cofactors (co-metabolism) [66]. Microorganisms could adapt to contaminants; thus, the contaminant could promote their development. Microbiota in turn could tolerate or degrade contaminants (due to homeostatic capacity). The degradation capacity of microorganisms could be enhanced by supplemental nutrients/amendments addition, augmentation of degrading, or by promoting rhizosphere microbial degradation activity.

3.2 Supporting services

Supporting services are not directly used by humans, although they underpin soil ecosystem functions and processes on which humans depend. Soil microbiota are mainly involved in soil formation, nutrient cycling, water cycling, primary production, and habitat for biodiversity. All of these are related with supporting. Both bacteria and fungi are involved in supporting soil aboveground biodiversity through involvement in catabolic reactions throughout C and nutrients cycles are either break down, mineralized, or transformed. Bacteria that belong to *Actinobacteria*, *Proteobacteria*, and *Firmicutes* were reported to produce organic compounds that have influence on plant root system proliferation. Atmospheric N fixation is provided by free-living bacteria (*Azotobacter, Bacillus*, etc.). Higher available iron-chelating compounds are produced by bacterial species such as *Pseudomonas and Frankia*,. Dimkpa et al., [67] reported that species belonging to *Streptomyces* have the potentials for biofertilization, while *Agrobacterium sp*., *Bacillus sp*., and *Penicillium sp.,* possess phosphate-solubilizing capabilities. Recent studies show that plant rhizobacteria can smooth abiotic stresses related with global change drivers (drought, soil salt intrusion, extreme temperature, poor nutrient availability, contaminants, etc.).

3.3 Provisioning services

Provisioning services of soil ecosystems refers to goods that humans can use and benefit immediately. These are food, water, fiber, fuel, genetic resources, chemicals, medicines, and pharmaceuticals. Soil microorganisms are an important bioresource for several bioactive substances (antibiotics, biosurfactants, enzymes) [66]. They also mediate and facilitate several reactions and soil functions that the promote provision of mentioned soil goods. The complexity of linkages between soil functions (most mediated by communities of soil microbiota) and provided ecosystem services as well as their interrelations is given in **Figure 2**.

4. Challenges on soil ecosystem services: implications of global change drivers

Majority of ecosystem services required for humans' survival and development are resulted from soil processes and functions mainly mediated by biodiversity. This is due to involvement in soil biogeochemical and physicochemical processes, as well as in shaping the aboveground biodiversity and terrestrial ecosystem functioning through liaising nutrient cycles and turnover [48]. Once their structure can be influenced by environmental (soil physicochemical properties) and climatic conditions [68], they are considered as a relevant indicator of soil health and ecosystem sustainability. Thereby, soil biota is the main pillar that sustains and maintains the soil ecosystem. They are pivotal in supporting services such as soil organic matter decomposition, nutrient cycling, and organic and inorganic compounds degradation.

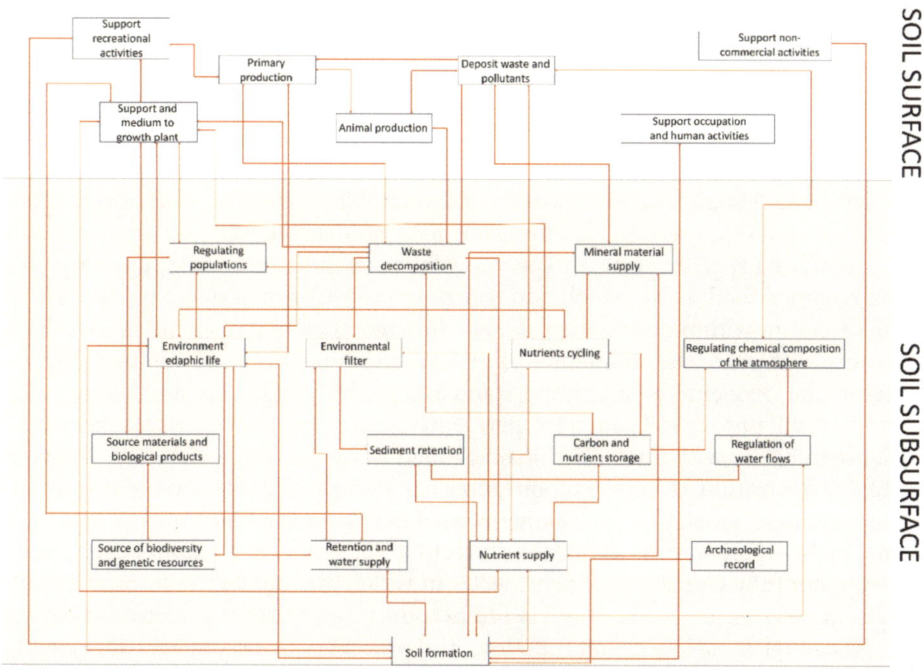

Figure 2.
Interrelationship between microbiota mediated and other soil functions and ecosystem services.

Through these, they contribute to soil fertility and quality. Microbiota sustains soil provisioning services by producing a wide range of organic compounds (e.g., polysaccharides, mucilage) that cohere soil particles. This influences soil supporting services as soil structure formation and development in time, through soil particle cementation. Further, soil structure stabilization is sustained by fungal filaments. Microbiota components are at the same time producers and consumers of soil organic carbon; thus, it is linked with supporting and regulating services of soil. Many studies underlay that soil microbiota could be a serious element in regulating services such as reducing atmospheric greenhouse gases. In this manner, they could limit climate change amplitude in future. Under favorable microhabitat conditions (optimal nutrient and carbon sources, available inorganic nutrients, proper acidity/alkalinity, temperature, aeration, moisture, etc.) microbial activity could considerably favoring and facilitating soil functions, and consequently the provision of soil ecosystem services at a higher amplitude. Niego et al., [69] reported that changes in soil microbiota diversity and abundance impact soil ecosystem processes. One of them is involvement in nutrient retention and cycling. These processes are related to the ability of soil microbiota components to break down organic matter and to release resources back into aboveground biota. Both plant litter decomposition as well N release back into aboveground plant diversity decrease with soil biodiversity shift in structure and abundance. Moreover, phosphorous, after increased rain events, could be lost through leaching. This is positively correlated with soil microbiota transformation and/or decline. Loss of soil biota become a great concern due to global change-related drivers induced pressures on it. Shift or loss of soil biota communities and abundance are thought to threaten soil ecosystem performance, further reducing the provision of ecosystem services [70]. Although it starts to become more clearer the importance of soil microbiota community structure and abundance in provision of soil ecosystem services, at the moment it is unclear how global change drivers will impact that. According to literature, until now most studies have focused on global change drivers' ecological consequences on aboveground biodiversity loss and less on belowground biodiversity. This is mostly attributed to the soils hidden diversity and heterogeneity. There are studies performed on land use and land management potential impact on soil microbiota. Usually, these connected soil microbiotas threaten with lower cycling of nutrients and other resources. Moreover, most studies focused on the effects and implications of a specific group or species. While soil microbiota components interact within complex food webs, which also contribute to changes in structure, abundance, and distribution within on trophic group or functional guild it is assumed that these may cause changes in the abundance, diversity, and functioning of another [70]. Therefore, become emergent to understand and get knowledge on how loss of soil microbiota will impact soil functions performance in providing ecosystem services.

Concurrent use with intensified management of the soil for agriculture, forestry, grazing, urbanization, mining, etc. purposes has serious outcomes on the provision of soil ecosystem services. Climate change-related events such as extreme meteorological events, higher temperature, and altered precipitation frequency amplify the effects. Although not fully elucidated which species in which manner has an impact on soil ecosystem services provision, it started to become clearer that soil microbiota components have a relevant role in soil functions and processes. Consequently, this could influence the provision of ecosystem services. It was reported that generally, microorganisms could manifest resistance, resilience, or extinction against some abiotic and biotic disturbance or stress. However, even in that cases the mechanisms are not completely understood. The impact amplitude seems to depend directly on the type of stress or disturbance, combinations of these, and their end influence on micro and macrohabitat properties.

4.1 Climate change impact on soil ecosystem services

It is acknowledged that soil biodiversity from all geoclimatic regions is affected by climate change. There are several reports that attest to the impact of various climate change drivers on several aboveground biodiversity components' health, functioning, and distribution. Considering literature, often reported that selected species have become extinct while others are endangered [71, 72]. However, both extinctions as well changes in health and functional status of species can alter seriously important ecological processes because they control, and mediate functions related to it. However, plants and animals respond in different ways at the pressures of climate change drivers. Usually, species have a tendency to cope with them. This is because of their evolutionary and ecological properties. Low-altitude inhabiting species extend their distribution at higher elevations while species from higher altitudes reorganize their relationships between community structure. Therefore, species range expansion poleward in latitude and upward in elevation is a common feature under changing climate properties. Generally, species are vulnerable due to their native habitat fragmentation and the negative effects of climate change drivers. They present individualistic responses to these challenges, but these will present a more pronounced and extended impact on the composition and functioning of future ecosystems [73]. Projected data reported in the literature consider macro-biodiversity. Information on how climate change impacts soil belowground microbiota, their involvement in ecosystem services provision as well how soil ecosystem will change, are scarce at present.

4.1.1 Atmosphere composition

Soils play a fundamental role in climate maintenance under favorable conditions, one of the major regulating services. Soil processes, mainly mediated by soil biodiversity, regulate climate through a balance of thermal and moisture exchange ratio, and greenhouse gases (H_2O, CO_2, CH_4, and N_2O) emission and retention [74]. Threatening of soil biodiversity can alter atmosphere quality and composition, which consequently will amplify climate change drivers further.

4.1.2 Temperature and precipitation regime

Degradation of soil ecosystem become frequent in many regions of the world. Changes in temperature, seasonality, and precipitation patterns have the potential to exacerbate soil ecosystem degradation. Many times, this results in lowering of ecosystem services' amount and efficiency. The high quality and quantity of ecosystem services are correlated with the health, complexity, and abundance of ecosystem species. However, species are limited in their adaptation to pressures such as temperature rise. Dow et al., [75] reported that an increase in average night temperature by 1° C decreased rice yield by 10%. Thus, changes in both temperature and precipitation regimes are often linked with food and other provisioning services decline. Alteration of evapotranspiration contribute also in the decline of primary productivity and food production [73]. These are supporting and provisioning services.

4.1.3 Extreme climate events

As changes in meteorological parameters could induce an alteration of the soil system's ability to translate the various ecosystem functions that support crop growth into provisioning services such as food, feed, and fiber, it becomes obvious that extreme climate events such as flood, drought, heat waves, and others, will have a more pronounced and dramatic effect.

4.2 Anthropogenic impact on soil ecosystem services

The prevalence of anthropogenic disturbance in the surrounding environment is obvious. These are highlighted through the soil physical and chemical properties decline. Soil properties decline manifest mainly through loss of its key constituents (clay, silt, soil organic carbon), diminution of water availability and holding capacity, abatement of key nutrients content, truncation of soil profile, and shallowing of topsoil depth as well through chemicals contamination and runoff. These deplete ecosystem C pool, enhance GHG reemission, and could induce anaerobiosis in soil layers. In part or in combination they could threaten soil microbiota through damage to their functioning, reduction in richness and abundance, or even in the extinction of species. The microbiology within soils supports a wide range of ecosystems underpinning the productive capacity and environmental sustainability of land use. Ample occupation of forests, grasslands, and wetlands for various purposes as the extension of living space and agricultural land resulted predominantly in the loss of biodiversity. This loss registered negative consequences on the soil chemical, biochemical, and physical properties. Ecosystem services are in high percent the end-result of interactions between plants, animals, and microorganisms from an ecosystem; biotic and abiotic properties of system; and human-engineered components of social-ecological systems. Ecosystem services provision, both in terms of quality and quantity, directly depends on land use type and use intensity, production system, and applied management [76]. These have an extensive impact on soil ecosystem services. Anthropogenic drivers and induced environmental stressors impact natural ecosystems through propagation of ecosystem functions that hamper ecosystem services provision [13, 30].

Land use change links human activities with ecological processes change that is closely connected with ecosystem services provision. Modification of land with aim of other use changes seriously soil ecosystem and its native biota abundance and spatial distribution. For example, in agriculture farms diversity (crops and/or livestock) is characterized mainly as "planned biodiversity." This always modifies soil ecological system structure and function, which end in failure of soil ecosystem ability to provide services. Therefore, the way of the use of soils has a decisive role on soil ecosystem services provision. Changes in natural land systems into agricultural or urban environment hampered soil ecosystem development, thereby reducing or stopping the provision of most support, provision, regulating and cultural services. Such alteration of services often amplifies soil issues such as erosion, poor fertility, desertification, and salinization [77]. Removal of forests for feedstock production for biofuels changed and damaged most services provided originally by the forest soil ecosystem. These are regulating services such as climate and water regulation through functions connected with carbon sequestration and accumulation, regulation of the atmospheric chemical composition and climate processes, as well functions linked to regulation of hydrological flows, buffering, and moderation of hydrological cycles. Provisioning services such as fiber, timber, genetic resources, and biochemicals are also reduced or totally shifted due to habitat modification and properties alteration. Such alterations extend their impact also on supporting and cultural services as well. Urban expansion is pronounced all around the world. This is also a driver associated with habitat loss and soil system degradation. Moreover, He et al., [78] and Wu et al., [79] reported that the continued expansion of urban areas will decrease significantly regional carbon storage, fostering thus climate change continuum and amplitude.

4.2.1 Land management

Intensive farming, mining, and industrial activities; as well as extended urbanization endangers local ecosystem services across landscapes. Soil erosion is a common feature of these activities. These reduce soil nutrient content which made microorganisms sensitive to this disturbance. Hereby, microbiota functions related to supporting services provision as nutrients element cycling and organic matter decomposition will decline [27, 32]. Until that moment farm diversity (crops and/or livestock) was characterized as "planned biodiversity." Although it comprised also "unplanned diversity," as weeds and pests, most of the time measures have been taken against them. Applied management practices were directed either to eliminate (e.g., pests) or promote (e.g., cultivated crops) populations or to enhance specific ecosystem processes (e.g., N fixation). Today, these practices and their effects at the macro-level made us to suppose that actions were felt also by associated species and functional diversity (e.g., soil belowground diversity), and consequently by their function performances in soil. The amplitude at which these were felt by soil belowground microbiota is not clearly acknowledged although there are evidence of their importance in several soil functions and processes performance. Crop diversification through crop rotations, cover crops, or intercropping enhanced regulation and sustainability of provided services. This was achieved through the facilitation of pest, weeds, and disease control by minimizing considerably or avoiding the use of agrochemicals. Furthermore, crop diversification stimulates soil microbial abundance and, in turn, soil biodiversity. Also, it supports carbon sequestration, which provides additional ecosystem services. Brussard et al., [80] stated in their paper that after abiotic disturbances or stresses, soil ecosystem and its inhabiting diversity could either restart succession from the stage to which it was set back or transcend into a new stability level. They assumed that after release from disturbance or stress, the reversion could take a long time. The required time usually could depend on spatial heterogeneity where recolonization must take place, and on microorganism restoration and dispersal abilities. However, studies in this sense are at early stage at this moment and the mechanisms and multilevel effects are not fully understood. Soil microbiota community resistance against stress and resilience from disturbance are important to be acknowledged for the proper management of biodiversity and provisions of linked ecosystem services. Crop production is associated with soil carbon content decline. This impacts soil functions as water and nutrient retention. The application of high amount of chemical N for intensive crops and vegetable cultivation caused an array of soil and connected ecosystem services alterations associated mainly with ecosystem pollution. These could be summarized in reactive N losses and greenhouse gas emissions. According to literature tilled agroecosystems without crop rotation decrease microbiota abundance and certain species richness. Opposite, agricultural practices such as crop-rotation, no-tillage and use of organic amendments enhance diversity abundance and community structure richness [81]. However, based on applied organic matter quality, the effects on microbiota could be either positive or negative. Drainage and irrigation, in proper manner, could enhance soil microbiota diversity and abundance. Tillage, a usual field operation practice modifies soil structure favoring agronomic processes (seed contact, root proliferation, water infiltration, etc.). Under conventional tillage, soil microbiota is altered structurally, morphologically, and functionally. This reduces microbial biomass, nutrient cycling, and enzymatic activities. Tillage reduces especially bacteria abundance and AMF diversity.

4.2.2 Chemicals and other extraneous elements addition to soil ecosystem

Pesticides and fertilizers can adversely impact selected soil biota species. This could result in the diminishment of their functionality, lower health status, or extinction. Such threat to soil biodiversity species will damage soil food webs, thereby underpinning the diversity of higher trophic levels. Moreover, modified interactions among soil species as well as a lower abundance or altered functioning of biota will significantly reduce supporting (nutrient cycling, soil formation, primary production) and regulating services (disease regulation pollination, climate, and water regulation). These consequently will negatively impact soil provisioning and cultural services. Municipal compost and animal slurries application to agricultural soils were considered as means to replace fertilizers used in conventional agricultural practices, and to reduce waste deposition in landfill sites. Acceptance of this practice was sustained also by other potential benefits such as the reduction of carbon emissions linked with mineral fertilizers production and use and with reduction of greenhouse gases emission through organic material decomposition and degradation in storing landfills. However, once with application of these management procedure on agricultural lands, new studies were performed assessing their impact. These evidenced several new safety issues linked to them. Most of these issues are related to the higher loads to the soil of potentially toxic chemicals, crops phytotoxicity, higher CH_4 and N_2O emission, as well as high potential in the introduction of alien and potential toxic microbiota components (e.g., pathogens). However, understanding and knowledge related to the impacts of applying organic waste materials to agricultural lands is developing at moment. There is information that introduction of alien species and/or pathogens in soil have the potential to change physical environment properties but is not clear how will this modify microbiota community structure and abundance that are involved in several key soil functions. EA published a report in 2008 [82] highlighted that organic waste material with higher Zn content shifted rhizobium bacteria abundance, which is responsible for nitrogen fixation into the soil. Considering agricultural and organic wastes, Garcia et al., [83] published that they could degrade the local ecosystem in multiple ways, reducing and altering therefore the provision of ecosystem services. They mentioned that swine manure ends up in nearby waters (rivers, streams, creeks, etc.) impacting fish diversity health and abundance, and water quality as well. However, more experimental and confirmatory data are still required in order to comprehend relations and connections between organic waste materials application to agricultural lands and changes in soil microbiota and ecosystem services, respectively.

5. Conclusions

Collectively, soil microbiota has key roles in soil ecosystem functioning processes, such as organic matter decomposition, nutrient cycling, and soil fertilizing. Through these, they assure soil with essential nutrients and elements with that soil could sustain the continuation and development of living organisms. As producer and consumer of soil organic carbon, they could sequester C into soil for a large period. They contribute to atmospheric greenhouse gas reduction and have the potential to limit the effects caused by greenhouse gases generated by climate change. Microbiota through their biochemical reactions assures in large part soil structure development and soil water retention capacity. With released polysaccharides and mucilage help to cement soil aggregates and consequently, reduce aggregates crumble when exposed to water. Summarizing all involvement and effects of soil microbiota components in soil processes and reactions, it could be concluded that most microbiota community components break down organic matter, recycle nutrients in soil, create humus, influence soil structure, fix nitrogen, promote plant development and growth as well protect them against pests and diseases. It is obvious that soil biodiversity is a critical element that assures soil ecosystem functioning and sustainability. Soil belowground microbiodiversity is as valuable as aboveground biodiversity, as they are being involved in many soils' biochemical and physicochemical processes. The effects of global change drivers on aboveground species are acknowledged through literature but knowledge on these how will impact soil microbiota and how a change in their structure and functioning will modify the sustainability of soil ecosystem functioning and ecosystem services delivery require additional research. In this context, exploring the influences of global change drivers on soil microbiota involved in key functions that contribute to and deliver ecosystem services is significant for understanding potential changes in regional soil ecosystem, ecology, and environment; enhancing soil ecosystem and ecology security; to promote a sustainable development maintaining in safe limits soil environment resources.

Acknowledgements

This work was supported by the Romanian Ministry of Research and Innovation through Institutional Performance-Projects for Financing Excellence in RDI [grant number 19PFE/2018 and 32PFE/ 2018].

Conflict of interest

The authors declare no conflict of interest.

Author details

Emoke Dalma Kovacs* and Melinda Haydee Kovacs
Research Institute for Analytical Instrumentation, INCDO-INOE 2000, Cluj-Napoca, Romania

*Address all correspondence to: dalmaemokekovacs@gmail.com

References

[1] MEA (Millennium Ecosystem Assessment). Ecosystem and Human Well-being. Volume 1. Current State and Trends. Washington D.C., USA: Island Press; 2005

[2] Nelson GC, Bennett E, Berhe AA, Cassman K, DeFries R, Dietz T, et al. Anthropogenic drivers of ecosystem change: An overview. Ecology and Society. 2006;**11**:29 http://www.ecologyandsociety.org/vol11/iss2/art29

[3] Sunnemann M, Siebert J, Reitz T, Schadler M, Yin R, Eisenhauer N. Combined effects of land-use type and climate change on soil microbial activity and invertebrate decomposer activity. Agriculture, Ecosystems and Environment. 2021;**318**:107490. DOI: 10.1016/j.agee.2021.107490

[4] Evans KS, Mamo M, Wingeyer A, Schacht WH, Eskridge KM, Bradshaw J, et al. Soil fauna accelerate dung pat decomposition and nutrient cycling in grassland soil. Rangeland Ecology & Management. 2019;**72**:667-677. DOI: 10.1016/j.rama.2019.01.008

[5] Kelsall M, Quirk T, Wilson C, Snedden GA. Sources and chemical stability of soil organic carbon in natural and created coastal marshes of Louisiana. Science Total Environment. 2023;**867**: 161415. DOI: 10.1016/j.scitotenv.2023.161415

[6] Canisares LP, Banet T, Rinehart B, McNear D, Poffenbarger H. Litter quality and living roots affected the formation of new mineral-associated organic carbon but did not affect total mineral-associated organic carbon in a short-term incubation. Geoderma. 2023;**430**:116302. DOI: 10.1016/j.geoderma.2022.116302

[7] Yang F, Zhong Y, Han G, Li X, Luo L, Cai X, et al. Effect of different vegetation restoration on soil organic carbon dynamics and fractions in the Rainy Zone of Western China. Journal of Environmental Management. 2023;**331**: 117296. DOI: 10.1016/j.jenvman.2023.117296

[8] Chapin FS, Matson PA, Vitousek PM. Principles of Terrestrial Ecosystem Ecology. Second ed. New York: Springer; 2011

[9] Jian T, Xia Y, He R, Zhang J. The influence of planting *Carex praeclara* and *Leymus secalinus* on soil properties and microbial community in a Zoige desertified alpine grassland. Global Ecological Conservation. 2022;**34**:e02002. DOI: 10.1016/j.gecco.2022.e02002

[10] Song Y, Song C, Shi F, Wang M, Ren J, Wang X, et al. Linking plant community composition with the soil C pool, N availability and enzyme activity in boreal peatlands of Northeast China. Applied Soil Ecology. 2019;**140**:144-154. DOI: 10.1016/j.apsoil.2019.04.019

[11] Frolla F, Aparicio V, Costa JL, Kruger H. Soil physical properties under different cattle stocking rates on Mollisols in the Buenos Aires Province. Geoderma Regional. 2018;**14**:e00177. DOI: 10.1016/j.geodrs.2018.e00177

[12] Traore S, Bottinelli N, Aroui H, Harit A, Jouquet P. Termite mounds impact soil hydrostructural properties in southern Indian tropical forests. Pedobiologia. 2019;**74**:1-6. DOI: 10.1016/j.pedobi.2019.02.003

[13] Dai W, Liu Y, Yao D, Wang N, Ye X, Cui Z, et al. Phylogenetic diversity of stochasticity-dominated predatory

myxobacterial community drives multi-nutrient cycling in typical farmland soils. Science Total Environment. 2023;**871**: 161680. DOI: 10.1016/j.scitotenv.2023.161680

[14] Acosta-Martinez V, Cruz L, Sotomayor-Ramirez D, Perez-Alegria L. Enzyme activities as affected by soil properties and land use in a tropical watershed. Applied Soil Ecology. 2007; **35**:35-45. DOI: 10.1016/j.apsoil.2006.05.012

[15] Wang X, Li Y, Wang L, Duan Y, Yao B, Chen Y, et al. Soil extracellular enzyme stoichiometry reflects microbial metabolic limitations in different desert types of northwestern China. Science Total Environment. 2023;**874**:162504. DOI: 10.1016/j.scitotenv.2023.162504

[16] Young IM, Crawford JW. Interactions and self-organization in the soil-microbe complex. Science. 2004; **304**:1634-1637. DOI: 10.1126/science.1097394

[17] Kim H, Lee J, Park J, Gho YS. Gram-negative and Gram positive bacterial extracellular vesicles. Seminars in Cell & Developmental Biology. 2015;**40**:97-104. DOI: 10.1016/j.semcdb.2015.02.006

[18] Siegel DS, Liu J, Ton-That H. Biogenesis of the Gram-positive bacterial cell envelope. Current Opinion in Microbiology. 2016;**34**:31-37. DOI: 10.1016/j.mib.2016.07.015

[19] Pepper IL, Gentry TJ. Microorganisms found in the environment. In: Pepper IL, Gerba CP, Gentry TJ, editors. Environmental Microbiology. 3rd ed. USA: Elsevier; 2015. pp. 9-45

[20] Zacchetti B, Wosten HAB, Claessen D. Multiscale heterogeneity in filamentous microbes. Biotechnology Advances. 2018;**36**:2138-2149. DOI: 10.1016/j.biotechadv.2018.10.002

[21] Bhatti AA, Haq S, Bhat RA. Actinomycetes benefaction role in soil and plant health. Microbial Pathogenesis. 2017;**111**:458-467. DOI: 10.1016/j.micpath.2017.09.036

[22] Shah AM, Rehman US, Hussain A, Mushtaq S, Rather AM, Shah A, et al. Antimicrobial investigation of selected soil actinomycetes isolated from unexplored regions of Kashmir Himalayas, India. Microbes Pathogens. 2017;**110**:93-99. DOI: 10.1016/j.micpath.2017.06.017

[23] Robert M, Chenu C. In: Biochemistry S, Stotsky G, Bollag JM, editors. Interactions between Soil Minerals and Microorganisms. Vol. 7. New York: Marcel Dekker; 1992. pp. 307-418

[24] Amend JP, Shock EL. Energetics of overall metabolic reactions of thermophilic and hyperthermophilic Archaea and Bacteria. FEMS Microbiology Reviews. 2001;**25**:175-243. DOI: 10.1111/j.1574-6976.2001.tb00576.x

[25] Li X, Li K, Wang Y, Huang Y, Yang H, Zhu P, et al. Diversity of lignocellulolytic functional genes and heterogeneity of thermophilic microbes during different wastes composting. Bioresource Technology. 2023;**372**: 128697. DOI: 10.1016/j.biortech.2023.128697

[26] Verhamme DT, Prosser JI, Nicol GW. Ammonia concentration determines differential growth of ammonia-oxidizing archaea and bacteria in soil microcosms. The ISME Journal. 2011;**5**:1067-1071. DOI: 10.1038/ismej.2010.191

[27] Beeck MOD, Persson P, Tunlid A. Fungal extracellular polymeric substance

matrices – Highly specialized microenvironments that allow fungi to control soil organic matter decomposition reactions. Soil Biology and Biochemistry. 2021;**159**: 108304. DOI: 10.1016/j.soilbio.2021.108304

[28] Henskens FL, Green ATG, Wilkins A. Cyanolichens can have both cyanobacteria and green algae in a common layer as major contributors to photosynthesis. Annals of Botany. 2012;**110**:555-563. DOI: 10.1093/aob/mcs108

[29] Helfrich M, Ludwig B, Thoms C, Gleixner G, Flessa H. The role of soil fungi and bacteria in plant litter decomposition and macroaggregate formation determined using phospholipid fatty acids. Applied Soil Ecology. 2015;**96**:261-264. DOI: 10.1016/j.apsoil.2015.08.023

[30] Huang C, Wu X, Liu X, Fang Y, Liu L, Wu C. Functional fungal communities dominate wood decomposition and are modified by wood traits in a subtropical forest. Science Total Environment. 2022;**806**: 151377. DOI: 10.1016/j.scitotenv.2021.151377

[31] Eichlerova I, Homolka L, Zifcakova L, Lisa L, Dobiasova P, Baldrian P. Enzymatic systems involved in decomposition reflects the ecology and taxonomy of saprotrophic fungi. Fungal Ecology. 2015;**13**:10-22. DOI: 10.1016/j.funeco.2014.08.002

[32] Talbot JM, Bruns TD, Smith DP, Branco S, Glassman SI, Erlandson S, et al. Independent roles of ectomycorrhizal and saprotrophic communities in soil organic matter decomposition. Soil Biology and Biochemistry. 2013;**57**:282-291. DOI: 10.1016/j.soilbio.2012.10.004

[33] Rozek K, Rola K, Blaszkowski J, Zubek S. Associations of root-inhabiting fungi with herbaceous plant species of temperate forests in relation to soil chemical properties. Science Total Environment. 2019;**649**:1573-1579. DOI: 10.1016/j.scitotenv.2018.08.350

[34] Mei L, Zhang P, Cui G, Yang X, Zhang T, Guo J. Arbuscular mycorrhizal fungi promote litter decomposition and alleviate nutrient limitations of soil microbes under warming and nitrogen application. Applied Soil Ecology. 2022; **171**:1043118. DOI: 10.1016/j.apsoil.2021.104318

[35] Altenburger A, Ekelund F, Jacobsen CS. Protozoa and their bacterial prey colonize sterile soil fast. Soil Biology and Biochemistry. 2010;**42**: 1636-1639. DOI: 10.1016/j.soilbio.2010.05.011

[36] Al-Maliki S, Ebreesum H. Changes in soil carbon mineralization, soil microbes, roots density and soil structure following the application of the arbuscular mycorrhizal fungi and green algae in the arid saline soil. Rhizosphere. 2020;**14**: 100203. DOI: 10.1016/j.rhisph.2020.100203

[37] Ghobashy MM, Mousa SAS, Siddiq A, Nasr HMD, Nady N, Atalla AA. Optimal the mechanical properties of bioplastic blend based algae-(lactic acid-starch) using gamma irradiation and their possibility to use as compostable and soil conditioner. Materials Today Communication. 2023;**34**:105472. DOI: 10.1016/j.mtcomm.2023.105472

[38] Geddes BA, Ryu MH, Mus F, Costas AG, Peters JW, Voigt CA, et al. Use of plant colonizing bacteria as chassis for transfer of N_2-fixation to cereals. Current Opinion in Biotechnology. 2015;**32**:216-222. DOI: 10.1016/j.copbio.2015.01.004

[39] Hsouna J, Gritli T, Ilahi H, Ellouze W, Mansouri M, Chihaoui S, et al. Genotypic and symbiotic diversity studies of rhizobia nodulating Acacia saligna in Tunisia reveal two novel symbiovars within the Rhizobium leguminosarum complex and Bradyrhizobium. Systematic and Applied Microbiology. 2022;**45**:126343. DOI: 10.1016/j.syapm.2022.126343

[40] Favero VO, Carvalho RH, Leite ABC, Santos DMT, Freitas KM, Boddey RM, et al. Bradyrhizobium strains from Brazilian tropical soils promote increases in nodulation, growth and nitrogen fixation in mung bean. Applied Soil Ecology. 2022;**175**:104461. DOI: 10.1016/j.apsoil.2022.104461

[41] Shameem RM, Sonali MIJ, Kumar PS, Rangasamy G, Gayathri KV, Parthasaraty V, et al. Nov., an efficient plant growth-promoting nitrogen-fixing bacteria isolated from rhizosphere soil. Environmental Research. 2023;**220**: 115200. DOI: 10.1016/j.envres.2022.115200

[42] Li Y, Yu H, Liu L, Liu Y, Huang L, Tan H. Transcriptomic and physiological analyses unravel the effect and mechanism of halosulfuron-methyl on the symbiosis between rhizobium and soybean. Ecotoxicology and Environmental Safety. 2022;**247**:114248. DOI: 10.1016/j.ecoenv.2022.114248

[43] Arashida H, Kugenuma T, Watanabe M, Maeda I. Nitrogen fixation in Rhodopseudomonas palustris co-cultured with Bacillus subtilis in the presence of air. Journal of Bioscience and Bioengineering. 2019;**127**:589-593. DOI: 10.1016/j.jbiosc.2018.10.010

[44] Bergman GT, Bates TS, Ellers KG, Lauber LC, Caporaso GJ, Walters WA, et al. The under-recognized dominance of Verrucomicrobia in soil bacterial communities. Soil Biology and Biochemistry. 2011;**43**:1450-1455. DOI: 10.1016/j.soilbio.2011.03.012

[45] Kumar M, Prasanna R, Bidyarani N, Babu B, Mishra BK, Kumar A, et al. Evaluating the plant growth promoting ability of thermotolerant bacteria and cyanobacteria and their interactions with seed spice crops. Scientia Horticulturae. 2013;**164**:94-101. DOI: 10.1016/j.scienta.2013.09.014

[46] Rashid MI, Mujawar LH, Shahzad T, Almeelbi T, Ismail MII, Oves M. Bacteria and fungi can contribute to nutrients bioavailability and aggregate formation in degraded soils. Microbiological Research. 2016;**183**:26-41. DOI: 10.1016/j.micres.2015.11.007

[47] Singh A, Srivastava N, Dubey SK. Molecular characterization and kinetics of isoprene degrading bacteria. Bioresource Technology. 2019;**278**:51-56. DOI: 10.1016/j.biortech.2019.01.057

[48] Xue Y, Zhao P, Quan C, Zhao Z, Gao W, Li J, et al. Cyanobacteria-derived peptide antibiotics discovered since 2000. Peptides. 2018;**107**:17-24. DOI: 10.1016/j.peptides.2018.08.002

[49] Patel P, Shah R, Joshi B, Ramar K, Natarajan A. Molecular identification and biocontrol activity of sugarcane rhizosphere bacteria against red rot pathogen Colletotrichum falcatum. Biotechnological Reports. 2019;**21**: e00317. DOI: 10.1016/j.btre.2019.e00317

[50] Ren L, Cai C, Zhang J, Yang Y, Wu G, Luo L, et al. Key environmental factors to variation of ammonia-oxidizing archaea community and potential ammonia oxidation rate during agricultural waste composting. Bioresource Technology. 2018;**270**: 278-285. DOI: 10.1016/j.biortech.2018.09.042

[51] Hussain H, Kock I, Al-Harrasi A, Al-Rawahi A, Abbas G, Green IR, et al. Antimicrobial chemical constituents from endophytic fungus Phoma sp. Asian Pacific Journal of Tropical Medicine. 2014;**7**:699-702. DOI: 10.1016/S1995-7645(14)60119-X

[52] Chaudhary S, Shankar A, Singh A, Prasad V. Usefulness of Penicillium in enhancing plants resistance to abiotic stresses: An overview. In: Gupta VK, Rodriguez-Couto S, editors. New and Future Developments in Microbial Biotechnology and Bioengineering: Penicillum System Properties and Applications. Amsterdam: Elsevier; 2018. pp. 277-284

[53] Martignoni MM, Garnier J, Zhang X, Rosa D, Kokkoris V, Tyson RC, et al. Co-inoculation with arbuscular mycorrhizal fungi differing in carbon sink strength induces a synergistic effect in plant growth. Journal of Theoretical Biology. 2021;**531**:110859. DOI: 10.1016/j.jtbi.2021.110859

[54] Ridout M, Houbraken J, Newcombe G. Zerotolerance of Penicillium and Phialocephala fungi, dominant taxa of fine lateral roots of woody plants in the intermountain Pacific Northwest, USA. Rhizosphere. 2017;**4**:94-103. DOI: 10.1016/j.rhisph.2017.09.004

[55] Heidari M, Karami V. Effects of different mycorrhiza species on grain yield, nutrient uptake and oil content of sunflower under water stress. Journal of the Saudi Society of Agricultural Sciences. 2014;(13):9-13. DOI: 10.1016/j.jssas.2012.12.002

[56] Celik I, Ortas I, Kilic S. Effects of compost, mycorrhiza, manure and fertilizer on some physical properties of a Chromoxerert soil. Soil and Tillage Research. 2004;**78**:59-67. DOI: 10.1016/j.still.2004.02.012

[57] Rodriguez-Caballero E, Aguilar MA, Castilla YC, Chamizo S, Aguilar FJ. Swelling of biocrusts upon wetting induces changes in surface micro-topography. Soil Biology and Biochemistry. 2015;**82**:107-111. DOI: 10.1016/j.soilbio.2014.12.010

[58] Chen X, Liu M, Hu F, Mao X, Li H. Contributions of soil micro-fauna (protozoa and nematodes) to rhizosphere ecological functions. Acta Ecologica Sinica. 2007;**27**:3132-3143. DOI: 10.1016/S1872-2032(07)60068-7

[59] Ronn MR, Griffiths BS, Young IM. Protozoa, nematodes and N-mineralization across a prescribed soil textural gradient. Pedobiologia. 2001;**45**:481-495. DOI: 10.1078/0031-4056-00101

[60] Aislabie J, Deslippe JR. Soil microbes and their contribution to soil services. In: Dymond JR, editor. Ecosystem Services in New Zealand – Conditions and Trends. Lincoln, New Zealand: Manaaki Whenua Press; 2013. pp. 143-161

[61] Butler OM, Lewis T, Rashti MR, Chen C. Energetic efficiency and temperature sensitivity of soil heterotrophic respiration varies with decadal-scale fire history in a wet sclerophyll forest. Soil Biology and Biochemistry. 2019;**134**:62-71. DOI: 10.1016/j.soilbio.2019.03.022

[62] Capek P, Starke R, Hofmockel KS, Bond-Lomberty B, Hess N. Apparent temperature sensitivity of soil respiration can result from temperature driven changes in microbial biomass. Soil Biology and Biochemistry. 2019;**135**:286-293. DOI: 10.1016/j.soilbio.2019.05.016

[63] Drigo B, Kowalchuk GA, Yergeau E, Bezemer TM, Boschker HTS, Van Veen JA. Impact of elevated carbon

dioxide on the rhizosphere communities of Carex arenaria and Festuca rubra. Global Change Biology. 2007;**13**: 2396-2410. DOI: 10.1111/j.1365-2486.2007.01445.x

[64] Almansoory AF, Hasan HA, Abdullah SRS, Idris M, Anuar N, Al-Adiwish W. Biosurfactant produced by the hydrocarbon-degrading bacteria: Characterization, activity and applications in removing TPH from contaminated soil. Environmental Technology and Innovation. 2019;**14**: 100347. DOI: 10.1016/j.eti.2019.100347

[65] Song M, Wang Y, Jiang L, Peng K, Wei Z, Li Y, et al. The complex interactions between novel DEHP-metabolising bacteria and the microbes in agricultural soils. Science Total Environment. 2019;**660**:733-740. DOI: 10.1016/j.scitotenv.2019.01.052

[66] Sacca ML, Caracciolo AB, Lenola MD, Grenni P. Ecosystem services provided by soil microorganisms. In: Lukac M, Grenni P, Gamboni M, editors. Soil Biological Communities and Ecosystem Resilience. Switzerland: Springer; 2017. pp. 9-24

[67] Dimkpa CO, Svatos A, Dabrowska P, Schmidt A, Boland W, Kothe E. Involvement of siderophores in the reduction of metal-induced inhibition of auxin synthesis in Streptomyces spp. Chemosphere. 2008;**74**:19-25. DOI: 10.1016/j.chemosphere.2008.09.079

[68] Sekaran U, McCoy C, Kumar S, Subramanian S. Soil microbial community structure and enzymatic activity responses to nitrogen management and landscape positions in switchgrass (*Panicum virgatum* L.). Global Change Biology. 2018;**11**:836-851. DOI: 10.1111/gcbb.12591

[69] Niego AGT, Rapior S, Thongklang N, Raspe O, Hyde KD, Mortimer P. Reviewing the contributions of macrofungi to forest ecosystem processes and services. Fungal Biology Reviews. 2023;**44**:100294. DOI: 10.1016/j.fbr.2022.11.002

[70] Geyer KM, Takacs-Vesbach CD, Gooseff MN, Barrett JE. Primary productivity as a control over soil microbial diversity along environmental gradients in a polar desert ecosystem. PeerJ. 2017;**5**:e3377. DOI: 10.7717/peerj.3377

[71] Wouyou HG, Lokonon BE, Idohou R, Zossou-Akete AG, Assogbadjo AE, Kakai RG. Predicting the potential impacts of climate change on the endangered Caesalpinia bonduc (L.) Roxb in Benin (West Africa). Heliyon. 2022;**8**:e09022. DOI: 10.1016/j.heliyon.2022.e09022

[72] Dimobe K, Ouedraogo K, Annighofer P, Kollmann J, Bayala J, Hof C, et al. Climate change aggravates anthropogenic threats of the endangered savanna tree Pterocarpus erinaceus (Fabaceae) in Burkina Faso. Journal for Nature Conservation. 2022;**70**: 126299. DOI: 10.1016/j.jnc.2022.126299

[73] Geest K, Sherbinin A, Kienberger S, Zommers Z, Sitati A, Roberts E, et al. The impacts of climate change on ecosystem services and resulting losses and damages to people and society. Loss and Damage from Climate Change. 2022; **2022**:221-236

[74] Schirpke U, Kohler M, Leitinger G, Fontana V, Tasser E, Tappeiner U. Future impacts of changing land-use and climate on ecosystem services of mountain grassland and their resilience. Ecosystem Service. 2017;**6**:79-94. DOI: 10.1016/j.ecoser.2017.06.008

[75] Dow K, Berkhout F, Preston B, Klein RJT, Midley G, Shaw R. Limits to adaptation. Nat. Climatic Change. 2013; **3**:305-307

[76] Baude M, Meyer B, Schindewolf M. Land use change in an agricultural landscape causing degradation of soil-based ecosystem services. Science Total Environment. 2019;**659**:1526-1536. DOI: 10.1016/j.scitotenv.2018.12.455

[77] Lang Y, Song W. Quantifying and mapping the responses of selected ecosystem services to projected land use changes. Ecological Indicators. 2019;**102**: 86-198. DOI: 10.1016/j.ecolind.2019.02.019

[78] He C, Zhang D, Huang Q. Assessing the potential impacts of urban expansion on regional carbon storage by linking the LUSD-urban and In VEST models. Environmental Modelling Software. 2016;**75**:44-58

[79] Wu Y, Tao Y, Yang G, Ou W, Pueppke S, Sun X, et al. Impact of land use change on multiple ecosystem services in the rapidly urbanizing Kunshan City of China: Past trajectories and future projections. Land Use Policy. 2019;**85**:419-427. DOI: 10.1016/j.landusepol.2019.04.022

[80] Brussard L, Ruiter PC, Brown GG. Soil biodiversity for agricultural sustainability. Agriculture, Ecosystems and Environment. 2007;**121**: 233-244

[81] Zhou YY, Wang JP. The effect of land-use types on composition of abundant and rare soil microbial communities in urban areas in Cyprus. Applied Ecology and Environmental Research. 2023;**21**(1):243-259

[82] Environmental Agency (EA). Using science to create a better place: Road Testing of "Trigger Values" for assessing site specific soil quality. Phase 1 – Metals, Science Report – SC050054SR1. Bristol: Environment Agency; 2008

[83] Garcia DJ, Lovett BM, You F. Predictive analysis of the industrial water-waste-energy system using an optimized grey approach: A case study in China. Journal of Cleaner Production. 2019;**228**:941-955. DOI: 10.1177/0958305X22109466

Chapter 12

Analyzing the Evolution of Land-Use Changes Related to Vegetation, in the Galicia Region, Spain: From 1990 to 2018

Sérgio Lousada and José Manuel Naranjo Gómez

Abstract

Considering the complex dynamics, patterns, and particularities that the Galicia region present—e.g., the fragility, shown to achieve sustainable development and growth—a study that analyzes the Land-Use related to the vegetation of this region is seen as pivotal to identifying barriers and opportunities for long-term sustainable development. Using GIS (Geographic Information Systems), the present chapter enables us to identify the dynamics and patterns of the evolution of the Land-Use Changes related to vegetation in the Galicia Region from 1990 to 2018 (years 1990, 2000, 2012, and 2018 using CORINE (Coordination of Information on the Environment) data). This study permits us to reinforce that the Land-Use Changes related to vegetation in the Galicia Region have undergone multiple changes—marked by increasing and decreasing periods. Also, can be considered a surveying baseline for the comparative analysis of similar works for different Land-Use Changes related to vegetation trends in Europe or worldwide. Land-Use Changes related to vegetation studies are reliable tools to evaluate the human activities and footprint of proposed strategies and policies in a territory. This chapter also enables us to understand that the main actors should design development policies to protect, preserve and conserve these incomparable landscapes, environments, ecosystems, and the region as a whole.

Keywords: Galicia region, GIS tools, land-use changes, regional studies, sustainable planning, territorial planning and management

1. Introduction

Institutions and citizens have been paying more attention to issues of sustainable development in the last few years [1, 2]. The international agendas have also specified a set of short-term measures and goals for reducing the impact of human activities on natural systems. Public bodies, on the other hand, do not always have the resources to formulate policies and plans capable of reacting to the increasing strains that the territories are subjected to [3].

Sustainable development is defined as "dissemination that meets current demands without jeopardizing future generations' ability to meet their own needs" [4]. This concept reflects the fact that sustainable development refers to a condition in which an input provides the best possible result without depleting natural resources. In accordance with the definition of sustainable development, there is a blueprint known as the Sustainable Development Goals (SDG) that directs people toward sustainable development [1].

The Sustainable Development Goals (SDGs) are a set of objectives that aim to create a more sustainable future by increasing wealth while also conserving the planet. These objectives place a premium on the long-term Outline Perspective Plan (OPP). The United Nations General Assembly launched the Sustainable Development Goals in 2015, and they are now implemented in all nations [1, 4].

As a result, achieving a balance between the economic, social, and environmental components is one of the most pressing concerns of our time. This equilibrium is particularly important in this setting because it directly affects both the acquisition and processing of natural resources. In the 1990s, the concept of sustainable development was first applied to mining planning and management [2, 5]. Over the last two decades, a lot of effort has gone into developing a sustainable approach to mining [2, 5].

Because of its potential to provide data that is accessible to a large audience, Geographic Information Systems (GIS) is an attractive tool for social workers. Administrators in the field of social work, for example, could utilize this technology to document the prerequisites for a new agency location. Furthermore, policymakers can offer the findings of a needs assessment or evaluative study, and academics can present the findings of a needs assessment or evaluative study [1, 6–8]. GIS can be traced back to a variety of technologies, processes, and procedures used in science, technology, and business, such as geodesy, mapping, geology, and seafaring; coordinate-time referencing of objects; processing and aggregation of photographic images from space for scientific and military purposes; and processing of geophysics and geodynamics data [1, 7]. GIS is defined by Burrough (1986) [8] as a set of tools for collecting, storing, retrieving, modifying, and displaying spatial data from the real world for a specific purpose.

As a result, rather than just presenting the results in tables, this study article uses maps to geo-visualize the data. GIS is not a new technology in today's world; it has been around for decades. It is also well-known for its capacity to provide a spatial-based solution [1].

GIS analyses based on the Corine Land Cover (CLC) database developed by the Copernicus program of the European Spatial Agency [3, 9] have been developed to determine a cognitive reference framework that shows the spread of land consumption at a national level and allows comparing the spread of this phenomenon among the various European countries.

Information on the Environment Coordination CORINE Land Cover (CLC) is a European effort that supports the collection and interpretation of geospatial data. It was initiated in 1985 in all nations of the European Community (EC). It was created with the following goals in mind: (a) obtain and synchronize interdisciplinary data on the state of the environment; (b) focus on priority areas in each EU country; (c) coordinate and coordinate data organization and management at the local and international levels; and (d) ensure data compatibility [10].

The CLC database is a tool for carrying out complicated geographical analyses based on various land use categories. As a result, the hierarchical structure of CLC classes has three levels. The first level of land use and land cover (artificial areas, agricultural areas, forest and semi-natural areas, wetlands, and water bodies) encompasses the five primary types of land use and land cover. There are fifteen departments on the second floor. Finally, the third level has 44 components that state that individual-level three classes' methodological scope is strictly defined [10, 11].

In context, the Geographic Information System (GIS) provides access to extensive land data sources and monitors land changes through high-resolution land cover assessments and change evaluations, particularly in urbanization regions [10, 12, 13]. Changes in human activities and urban ecological land cover can also be observed using these systems [13]. Furthermore, Urban Atlas (UA) has a wealth of other information, such as the classification of high-resolution satellite pictures (SPOT 2.5 m, ALOS 2.5 m, RapidEye 5 m), allowing for the separation of significant coverage classes. The lowest mapping unit is 0.25 hectares, which permits the development of land cover maps for only 305 large European cities with populations of more than 100,000 people and an estimated accuracy of 5 meters. Despite this, the UA only has 20 land cover classes, many fewer than the CLC [10].

Nowadays, Land-Use Changes studies are reliable tools to evaluate the human activities and footprint of proposed strategies and policies in a territory. The land is an important natural resource and a spatial carrier of human economic and social activities, and ecology. Land-use change reflects the impact of human activities on the natural environment, causing changes in surface structure (i.e., water bodies, climate, and ecology) and affecting the ecosystem service value [14]. The land is a non-renewable resource and while demand is constantly increasing, it is imperative to maintain a balance between demand and supply, needs and interests, or between contradictory uses, through Land-Use policies that achieve sustainable development and improve the quality of the environment [15, 16]. Very often, a poorly developed urban planning process leads to the changing of more natural land surfaces into artificial ones planned for human activities, therefore increasing social vulnerability. Therefore, the evaluation of the Land-Use Change process is important to the sustainable development of urban areas and to increase the resilience of territories and communities [10, 16]. On the other hand, Land-Use planning may also positively impact the environment by preserving natural resources, enhancing open space opportunities, or providing a significant reduction in traffic pollution [15].

Land-Use depends on numerous factors, including population, economic status, infrastructure, industrial activities, geographic conditions, land development policies, etc. [15, 17] and impacts numerous parameters, including flood risk, landslide probability, biodiversity, urban climate, hydrological processes, and pollution [15, 17–20].

Given the increasing number of disasters over recent years, one of the most efficient and accessible methods for reducing the pressure posed by natural or technological risks is reducing the vulnerability level of communities exposed to a particular hazard [21–24].

At all levels of government, there is a demand for instruments to enhance policy-making aimed at long-term planning. In this context, the Ecosystem Services (ES) approach [25, 26] provides a structured framework for developing more useful instruments for assessing environmental performance.

However, in the spatial planning process, the application of an ecosystem services approach to landscape analysis, sustainable planning, and decision making is largely inadequate [27, 28]. Many spatial landscape frameworks and environmental planning tools that incorporate the concept of ecosystem services have been created over the last decade [26, 29].

As a result, territorial planning and management strategy is a fundamental instrument for attributing wealth preconditions to the inhabitants, thereby fostering prosperity for future generations living on that territory, fostering the reduction of social imbalances and spatial inequalities, and serving as a stimulus for sustainable development [16, 30].

In the context of this study, the CLC data will be used to examine and evaluate the Land-Use Changes connected to vegetation in the Galicia Region between 1990 and 2018.

In this regard, we emphasize that the current study will contribute to science by enabling the collection of big data connected to Land-Use Changes associated to vegetation, as well as an overview of how they have evolved in the Galicia Region over the last three decades.

As a result of this research, we are able to give some principles and recommendations for future regional planning and management strategies and policies to be developed and implemented throughout the Galicia Region.

2. The Galicia Region: A brief overview

The Autonomous Community of Galicia (NUTS 2) is a Spanish region (NUTS 2) in the northwestern part of the country (**Figure 1**), with a total area of 29.574 km^2 and administratively divided into four provinces with a total of 313 municipalities and 3793 parishes, with an average population density of 91.35 hab/km^2 spread over more than 30,000 population centers, although the majority of its population is concentrated along the coast [31].

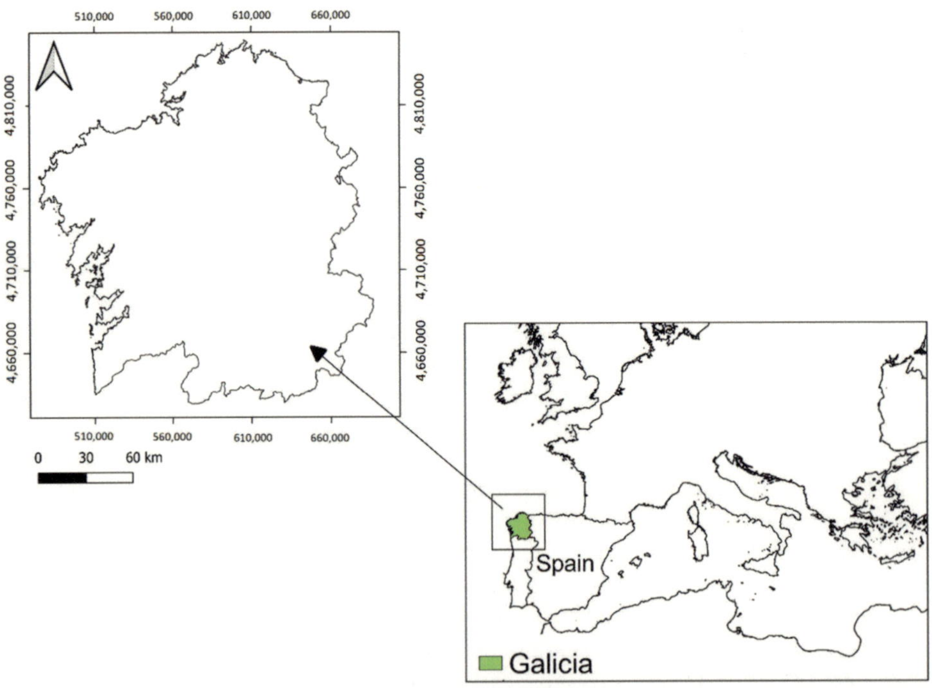

Figure 1.
Delimitation of the study area - Galicia region (Source: Authors by ESRI ArcGIS, 2020).

The altitude in Galicia ranges from sea level to nearly 2000 m and the topography includes plains as well as mountain areas with steep valleys [31, 32]. Galicia has two types of climate according to the Köppen–Geiger classification, the Csb (Mediterranean-Oceanica climate) and the Csa (Mediterranean climate) [33]. According to the Spanish official cartography, 69% of Galicia is covered by forestland (the forest terrain, according to the Galician forestry sector authorities, includes woodlands and shrublands) [31, 32]. The dominant tree species are three species of pine (Pinus pinaster, Pinus radiata, Pinus sylvestris), and two Eucalyptus species (Eucalyptus globulus and Eucalyptus nitens) and broadleaves (riparian species, Quercus robur, Quercus pyrenaica and Castanea sativa, among others). The 2015 analysis of the forestry sector indicates that 30% of the forestland was shrublands and rocky areas [32]. The forestry sector represents 3.5% of the Galician GDP and 50% of the timber cuts are Eucalyptus spp. followed by conifers [32]. Galician land forest is highly fragmented. According to official cadastral information, it is estimated that 162,188 ha are in cadastral parcels that are smaller than 0.5 ha [31, 32]; this accounts for approximately 40% of the land covered by the main productive tree species in Galicia.

Therefore, the study area is a region of extensive agroforestry tradition and high potential productivity [34], being approximately 61% (1.8 million ha) of its forest area territory [34]. With more than 1.4 million wooded hectares, and average growth of 12.3 million m^3 per year of wood, Galicia contributed just over 9.7 million m^3 in wood shorts in 2019 (almost half of the annual timber cuts in Spain), an annual rate of use increasable, under sustainability criteria, according to official statistics [31].

The Galicia Region has followed a path of dual productive specialization in forest and dairy production over the last half-century. It currently produces nearly half of the country's timber and 40% of its dairy. As a result of artificial plantations and spontaneous vegetation invasion [35], the area covered by trees and other woody vegetation rose dramatically, resulting in a major increase in the amount and continuity of biomass present on the terrain. As a result, the region is distinguished by a large percentage of forest area, accounting for about 60% of the territory's total area and 11% of Spain's total forest area [36].

Severe wildfires occur every few years due to sporadic, short but possibly strong periods of drought during the summer. In 2017, almost 62,000 ha were burned, the majority of which (approximately 42,000 ha) occurred in just a few days in early October [37]. From 1968 to 2012, there were 249,387 wildfires in the region [38], resulting in the burning of almost 8000 km2 (about a fourth of the total regional area) in the last 25 years (29.574 km^2). Different writers have identified a number of structural factors of fire igniting activity [39]. Traditional rural lifestyles are vanishing, as are tensions over land management and ownership, conflicts at the wildland-urban boundary, and socioeconomic conditions.

Property fragmentation is widely seen in the region as a significant impediment to the sustainable and economic management of forests and rangelands. According to current estimates, the region has almost 1.7 million proprietors (out of a total population of about 2.7 million) and over 11 million plots with an average size of 0.25 hectares [40]. Private owners own the majority of the land—they own more than two-thirds of the forest area—but the average size of a private holding is 1.5–2 ha per person [36]. Private properties, on the other hand, include common lands, which are legally recognized as a non-divisible kind of collective (albeit private) property. Community membership is limited by law, and it is open to everyone who lives in the same region as the community. As a result, communities are fluid entities: those who move in become owners, while those who leave lose their ownership rights. At 656,000 ha, common lands make up the final third (public property in the region is essentially non-existent), and are maintained by about 3000 local communities, with an average area of 200 ha per community [40].

Climatologically, Galicia Region has large differences between its coastal and inland areas. Average annual rainfall varies between 800 and 1000 mm in inland areas, and 1600 and 1900 in coastal areas. The annual mean temperature is 13°C, with remarkable differences between the coastal and continental temperatures; at the same elevation, in summer (winter), temperatures are on the order of 2°C higher (5°C lower) in the continental part. Thus, the lowest temperatures can be observed in the interior, where the highest mountains are located, with average minimum temperatures around 5°C. Summers are warm, particularly in the southeast of the area, with maximum temperatures exceeding 30°C [31, 41].

Galicia's territory is heterogeneous, having densely populated sections mixed in with more sparsely populated areas. Within them, there are some cities and headwaters of the region that act as focal points of activity [42]. Demographically and economically dynamic areas coexist with those characterized by a lack of vitality, but even within them, there are some cities and headwaters of the region that act as focal points of activity.

Rural depopulation is a serious issue in Spain, particularly in Galicia, where it is regarded as a demographic and territorial phenomenon. Indeed, since 2008, the population of the region has decreased by 9.2 percent. In terms of the foreign population, prior to the economic slump, the rise in immigrants helped to alleviate rural depopulation [42].

Other important factors contributing to rural depopulation include an aging population or low population density that prevents economic development. Age and gender disparities, on the other hand, may be to blame [43]. Rural masculinization happens at a young age as a result of largely female migration and a lack of equal productive and reproductive work, leading to a search for a higher educational level and career prospects in metropolitan regions [42]. Aging, geographical isolation, a lack of spatial integration with other surrounding places, bad connections and transportation difficulties, a lack of adequate social services, and lower levels of human capital and employment prospects are all disadvantages associated with low density. All of this unavoidably leads to a drop in the economy [42, 44].

Loss of human resources, lack of territorial growth, and incapacity to maintain commercial operations have not only economic, but also patrimonial and environmental consequences [45]. The loss of livestock and conventional agricultural uses, in terms of environmental effects, is a danger factor for natural environment protection. This is due to the fact that landscape changes are uncontrollable, and forest land management in rural regions is largely confined to individual plots [42].

In the Galician mountains, extensive livestock used to have an impact on the forest ecosystem, favoring mosaics and lowering fuel [46]. As a result, the fall in extensive animal husbandry in Spain (approximately 30% between 2004 and 2015) is seen as a contributing cause to forest fires [42].

The environmental implications of progressive abandonment of rural regions, such as soil loss and exposure to erosive phenomena over wide areas, can be deemed unfavorable in the medium term [43]. Furthermore, there is a lack of forest land management, which increases the risk of fire. Traditional burning activities and the use of fire in mountain management in Galicia have been linked to an increase in fire occurrence [42, 47].

3. Methodology

The data used was two layers of information. These are public and open and can be used to replicate this work in another work area. The analyzed area is the Galicia region, in Spain.

Firstly, land-use related to vegetation data were obtained. The European Space Agency (EEA) offers through the CORINE Land Cover (Coordination of Information-CLC) project a geodatabase using polygonal graphic features that evoke land uses throughout the European Union, for the years 1990, 2000, 2006, 2012 and 2018 [48].

The scale used is 1:100,000 in the Geodesic Reference System corresponding to the European Terrestrial Reference System 1989 (ETRS89) and the Mapping System is Universal Transverse Mercator (UTM), with the minimum cartographic unit (MCU) being equal to 25 hectares. The accuracy obtained has been increasing over the years, since in 1990 it was less than 50 meters, in 2000, 2006 and 2012 it was less than 25 meters, and finally, in 2018 it is less than 10 meters. Also, the information contained in these polygons is hierarchical in three levels of information (**Table 1**).

The second layer of information corresponds to the administrative delimitation of the Autonomous Community of Galicia. From the National Geographic Information Center in Spain (CNIG), as shown in **Figure 2**.

Level 1	Level 2	Level 3
1. Artificial surfaces	1.1. Urban fabric	1.1.1. Continous urban fabric
		1.1.2. Discontinuous urban fabric
	1.2. Industrial, commercial and transport	1.2.1.Industrial or commercial units
		1.2.2.Road and rail networks and associated land
		1.2.3.Port areas
		1.2.4.Airports
	1.3. Mine, dump and construction sites	1.3.1.Mineral extraction sites
		1.3.2.Dump sites
		1.3.3.Construction sites
	1.4. Artificial, non-agricultural vegetated areas	1.4.1.Green urban areas
		1.4.2.Sport and leisure facilities
2. Agricultural areas	2.1. Arable land	2.1.1. Non-irrigated arable land
		2.1.2. Permanently irrigated land
		2.1.3. Rice fields
	2.2. Permanent crops	2.2.1Vineyards
		2.2.2.Fruit trees and berry plantations
		2.2.3.Olive groves
	2.3. Pastures	2.3.1.Pastures
	2.4. Heterogeneous agricultural areas	2.4.1.Annual crops associated with permanent crops
		2.4.2.Complex cultivation
		2.4.3.Land occupied by agriculture

Level 1	Level 2	Level 3
3. Forests and semi-natural areas	3.1. Forests	3.1.1. Broad-leaved forest
		3.1.2.Coniferous forest
		3.1.3.Mixed forest
	3.2.Shrub and/or herbaceous vegetation association	3.2.1. Natural grassland
		3.2.2.Moors and heathland
		3.2.3.Scierophyllous vegetation
		3.2.4.Transitional woodland shrub
	3.3. Open spaces with little or no vegetation	3.3.1.Beaches, dunes, and plains
		3.3.2.Bare rock
		3.3.3.Sparsely vegetated areas
		3.3.4.Burnt areas
		3.3.5.Glaciers and perpetual snow
4. Wetlands	4.1. Inland wetlands	4.1.1.Inland marshes
		4.1.2.Peatbogs
	4.2. Coastal wetlands	4.2.1.Salt marshes
		4.2.2.Salines
		4.2.3.Intertidal flats
5. Water bodies	5.1. Inland waters	5.1.1.Water courses
		5.1.2.Water bodies
	5.2. Marine waters	5.2.1.Coastal lagoons
		5.2.2.Estuaries
		5.2.3.Sea and ocean

**For detailed information about the CLC Codes, the authors recommend the following source: www.eea.europa. eu/ publications/COR0-landcover, accessed on 30 November 2021.*

Table 1.
CORINE Land Cover nomenclature (Source: [48]).*

Subsequently, both layers of information were treated using ArcGIS 10.5 Geographic Information Systems (GIS) management software. Initially, all layers of information were transformed to the ETRS89-Azimuth Equiarea Coordinate Reference System of Lambert-2001, as this was adopted as official (ETRS-LAEA) [49]. Because ETRS-LAEA is based on the projection of equivalent areas in the territory. In this way, it serves as a reference for homogeneous units for all European countries. As a result, this coordinate system is used for the representation of analytical and statistical data.

Subsequently, the layer relating to the administrative divisions of the country, Spain has carried out a selection query through alphanumeric information and the Galicia region was selected. Subsequently, this single region was kept in a single layer of information. This layer of information was the limit of the scope of action of this work. The clip tool was then used, with Galicia's boundary as the reference layer. This procedure was used for each of the years studied (1990, 2000, 2006, 2012 and 2018). In this way, land uses were obtained, but only those that were included in the region. Subsequently, geometric measurement of the area of each of the polygons was performed in hectares. This resulted in the number of hectares of each polygon representative of land uses according to the CLC nomenclature.

Figure 2.
Delimitation of the study area - Galicia region (Source: Authors by ESRI ArcGIS, 2020).

Once this information was obtained, the alphanumeric information recorded in each of the tables for the years analyzed was exported to a database that was managed by the Microsoft Access database management program belonging to Microsoft Office 365 software.

Selection queries were made to the database using Structured Query Language (SQL) to select according to the CLC nomenclature, and then another grouping query was added to the previous query, also using SQL. Finally, the hectares for each land use were obtained for the years 1990, 2000, 2006, 2012 and 2018.

However, to take into account not only numerical but also geographical results, thematic maps were also obtained for each of the years. In this way, it was possible to identify where the greatest variation in land uses related to vegetation occurred and where there was predominant land uses related to vegetation.

To an easy understanding of the used methodology and case study selection criteria, a scheme has been developed (**Figure 3**).

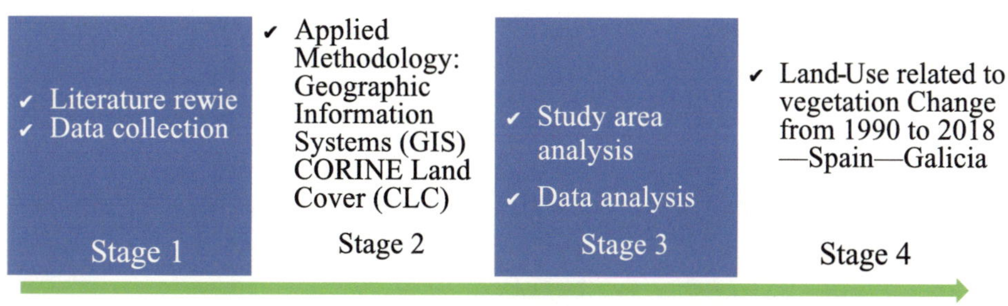

Figure 3.
Summary scheme of the used methodology and case study selection criteria (Source: Authors).

4. Results

The results come from the analysis of the land-use related to vegetation changes for the Galicia region in the years 1990, 2000, 2006, 2012 and 2018. The results will be exposed through the tables, and thematic cartography. This typology of results exposed allows for extracting the most relevant information and characterizing the evolution of land use based on the 15 land uses determined by CLC in level 2. The information is organized as presented in **Table 2**, in percentage.

In order to know what are the differences in area extension for every land use, the differences in percentage areas between years are calculated.

From the information in **Table 3**, it can be seen that the two greatest differences occur for land uses 2.4. and 3.1. between 2000 and 2006. Indeed, −12.88% for 2.4. corresponding to Heterogeneous agricultural areas and 3.1. corresponding to forests. For this reason, it was determined the percentage area for land uses, but in this case at the Level 3.

Table 4 shows that the highest percentage corresponds to 3.1.1. in 2006 whose value is 23.83% 2006, that is to say, about a quarter of Galicia is occupied by broad-leaved forest. Nonetheless, it is important to take into account that the percentage of this land use was very lower years before, 7.75% in 2000 and 7.63% in 1990. Consequently, a huge increase in the area occupied by broad-leaved forests occurred between 2000 and 2006. Although in a lower intensity and also between the same years, has occurred an increase of the land use 3.1.2. coniferous forest, from 2.78% in 2000 to 10.47% in 2006. On the contrary, a reduction of the area occupied occurred in the same years for the land uses 2.43. the land occupied by agriculture and 3.1.3. mixed forest.

Level 2	1990	2000	2006	2012	2018
1.1.	1.21%	1.25%	1.50%	1.52%	1.68%
1.2.	0.16%	0.22%	0.34%	0.37%	0.45%
1.3.	0.22%	0.32%	0.29%	0.32%	0.30%
1.4.	0.01%	0.02%	0.05%	0.05%	0.06%
2.1.	**0.00%**	**0.00%**	**1.61%**	**1.62%**	**2.50%**
2.2.	**0.22%**	**0.24%**	**0.26%**	**0.26%**	**0.18%**
2.3.	**0.00%**	**0.00%**	**0.16%**	**0.18%**	**0.28%**
2.4.	**41.07%**	**41.02%**	**28.14%**	**28.14%**	**27.53%**
3.1.	**31.27%**	**31.16%**	**39.81%**	**39.13%**	**38.33%**
3.2.	**23.99%**	**23.97%**	**25.78%**	**26.55%**	**26.85%**
3.3.	**1.05%**	**0.98%**	**1.32%**	**1.09%**	**1.09%**
4.1.	0.00%	0.00%	0.01%	0.01%	0.00%
4.2.	0.10%	0.10%	0.09%	0.09%	0.09%
5.1.	0.51%	0.55%	0.50%	0.53%	0.52%
5.2.	0.19%	0.18%	0.14%	0.14%	0.14%

**Values in bold corresponding to Land-Use Changes related to vegetation.*

Table 2.
Percentage of land uses according to level 2 of CLC nomenclature in the Galicia region (Source: authors).

Level 2	2000-1990	2006-2000	2012-2006	2018-2012
1.1.	0.04%	0.25%	0.02%	0.17%
1.2.	0.06%	0.12%	0.03%	0.08%
1.3.	0.10%	−0.03%	0.02%	−0.02%
1.4.	0.01%	0.03%	0.00%	0.01%
2.1.	0.00%	1.60%	0.01%	0.88%
2.2.	0.01%	0.02%	0.01%	−0.08%
2.3.	0.00%	0.16%	0.01%	0.10%
2.4.	−0.05%	**-12.88%**	0.00%	−0.61%
3.1.	−0.12%	**8.66%**	−0.68%	−0.80%
3.2.	−0.02%	1.81%	0.77%	0.30%
3.3.	−0.07%	0.34%	−0.23%	0.00%
4.1.	0.00%	0.01%	0.00%	−0.01%
4.2.	0.00%	−0.01%	0.00%	0.00%
5.1.	0.04%	−0.06%	0.03%	−0.01%
5.2.	0.00%	−0.04%	0.00%	0.00%

**Values in bold corresponding to Land-Use Changes related to vegetation - significant changes.*

Table 3.
Percentage difference of land uses according to level 2 of CLC nomenclature in the Galicia region (Source: authors).

Level 3	1990	2000	2006	2012	2018
2.4.1.	0.00%	0.00%	0.00%	0.00%	0.00%
2.4.2.	23.51%	23.57%	22.10%	22.10%	21.28%
2.4.3.	13.25%	**13.17%**	**5.77%**	5.76%	5.95%
3.1.1.	7.63%	**7.75%**	**23.83%**	23.68%	23.42%
3.1.2.	2.84%	**2.78%**	**10.47%**	10.02%	9.60%
3.1.3.	17.52%	17.38%	5.12%	5.06%	4.90%

**Values in bold corresponding to Land-Use Changes related to vegetation - significant changes.*

Table 4.
Percentage of land uses according to level 3 for 2.4. and 3.1. land uses according to CLC nomenclature in the Galicia region (Source: authors).

In this regard, to know the highest differences in the percentage area of the land uses, again it was calculated the difference between the years analyzed, but at the level 3 and specifically for the land uses classified into 2.4. and 3.1.

According to **Table 5**, it seems that the increase in certain kinds of land uses such as 3.1.1. broad-leaved forest and 3.1.2. coniferous forest is compensated by the decrease of other land uses like 2.4.2. complex cultivation, 2.4.3. the land occupied by agriculture and 3.1.3. mixed forest. Nonetheless, it is advisable to execute more exhaustive study research to know it.

Level 3	2000-1990	2006-2000	2012-2006	2018-2012
2.4.1.	0.00%	0.00%	0.00%	0.00%
2.4.2.	0.06%	**−1.47%**	0.00%	−0.82%
2.4.3.	−0.08%	**−7.40%**	0.00%	0.18%
3.1.1.	0.11%	**16.09%**	−0.16%	−0.26%
3.1.2.	−0.06%	**7.69%**	−0.45%	−0.42%
3.1.3.	−0.14%	**−12.25%**	−0.07%	−0.16%

**Values in bold corresponding to Land-Use Changes related to vegetation - significant changes.*

Table 5.
Percentage difference of land uses according to level 3 for 2.4. and 3.1. land uses according to CLC nomenclature in the Galicia region (Source: authors).

In addition, using ArcGIS 10.5 Geographic Information Systems (GIS) management software, it was possible to more accurately represent the location of each area (thematic cartography) – i.e., according to their respective CLC nomenclature and temporal variance, **Figures 4–10**.

Because between 2006 and 2000 has occurred the highest difference for land uses was classified into the groups 2.4. and 3.1. thematic cartography was made for this land use at level 3 and in these years.

Although **Figures 9** and **10** show the land uses for both years, it is really difficult to know where these changes have been produced. Because these land uses occupy most of the Galicia territory and they are very spread. However, it is possible to realize that

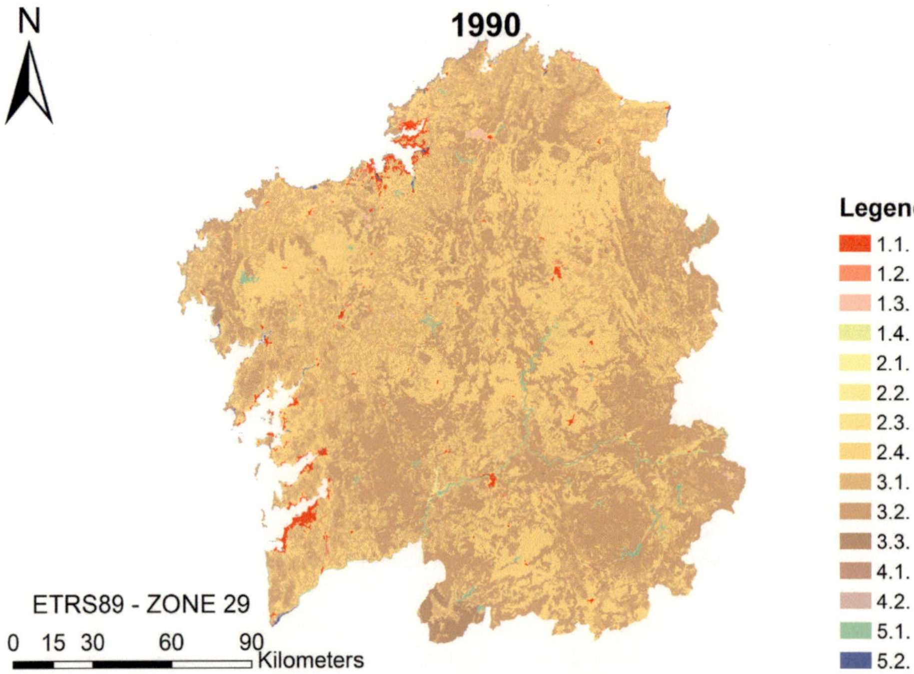

Figure 4.
Land uses according to level 2 of CLC nomenclature in the Galicia region in 1990 (Source: Authors by ESRI ArcGIS, 2020).

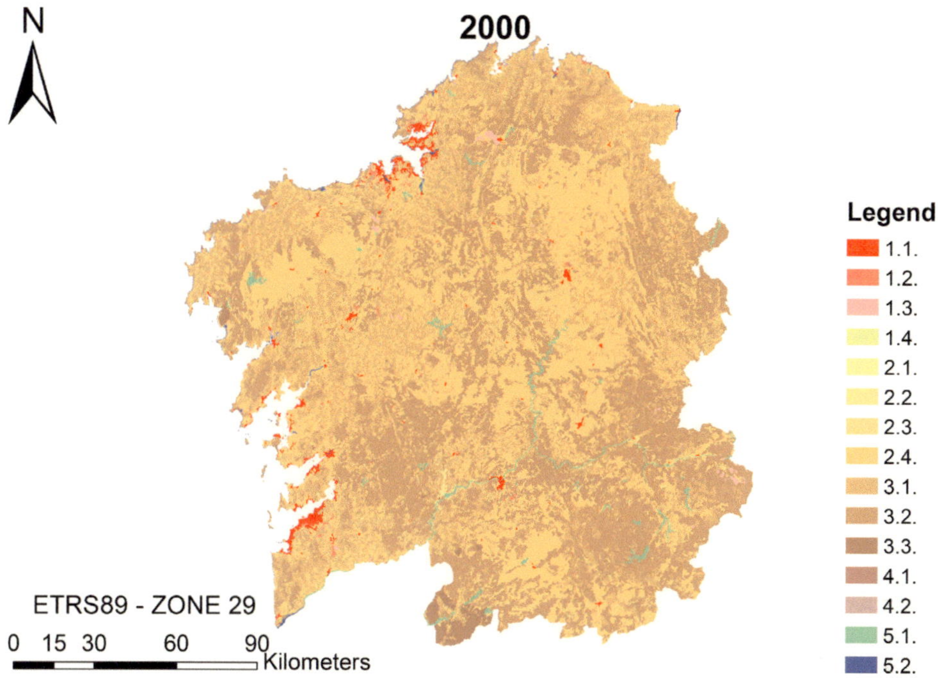

Figure 5.
Land uses according to level 2 of CLC nomenclature in the Galicia region in 2000 (Source: Authors by ESRI ArcGIS, 2020).

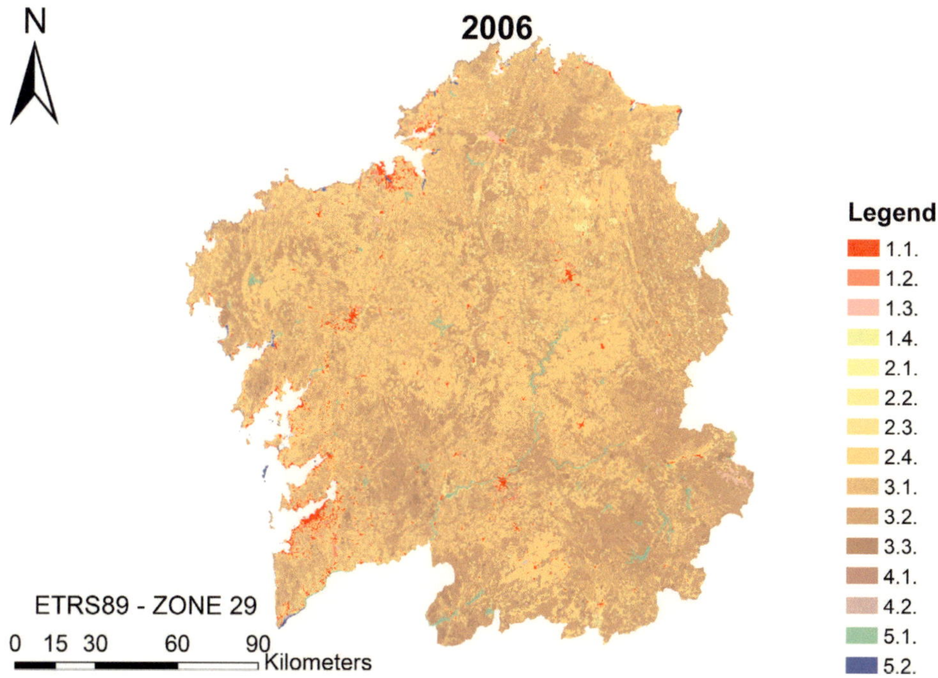

Figure 6.
Land uses according to level 2 of CLC nomenclature in the Galicia region in 2006 (Source: Authors by ESRI ArcGIS, 2020).

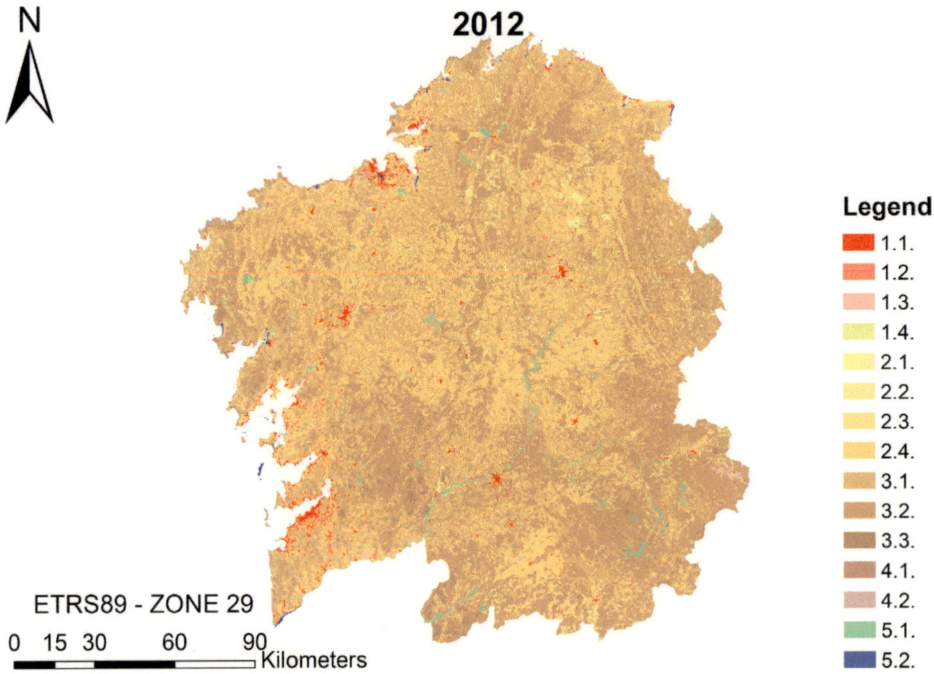

Figure 7.
Land uses according to level 2 of CLC nomenclature in the Galicia region in 2012 (Source: Authors by ESRI ArcGIS, 2020).

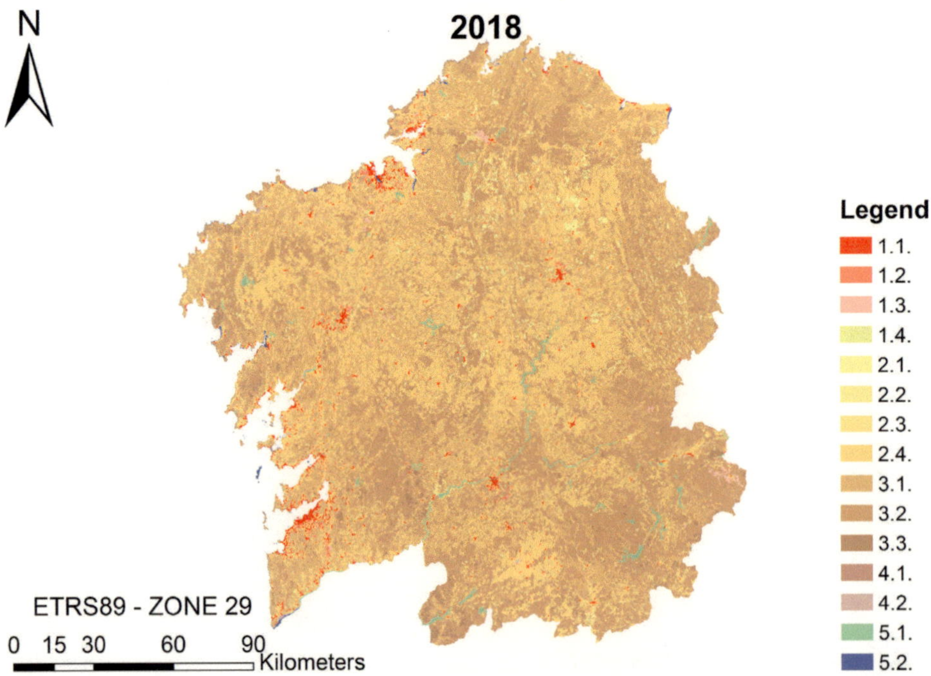

Figure 8.
Land uses according to level 2 of CLC nomenclature in the Galicia region in 2018 (Source: Authors by ESRI ArcGIS, 2020).

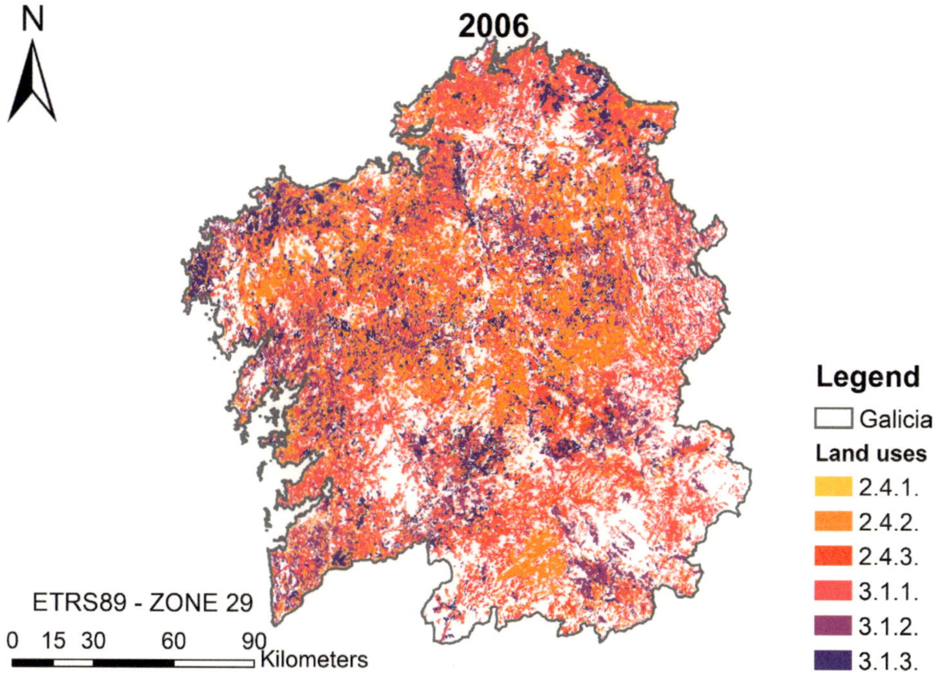

Figure 9.
Land uses according to level 3 for 2.4. and 3.1. land uses of CLC nomenclature in the Galicia region in 2006 (Source: Authors by ESRI ArcGIS, 2020).

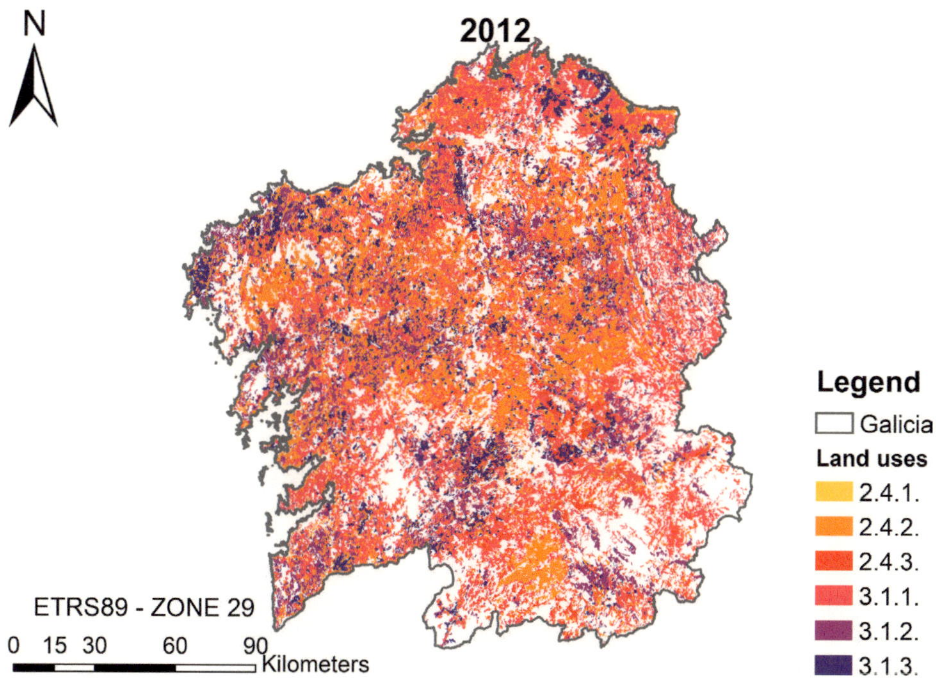

Figure 10.
Land uses according to level 3 for 2.4. and 3.1. land uses of CLC nomenclature in the Galicia region in 2012 (Source: Authors by ESRI ArcGIS, 2020).

in the north area 3.1.2. land uses disappear between the analyzed years. Besides, in the west area disappear land uses classified in 2.4. and 3.1., in favor of other land uses since more white areas appear. This effect is also observed in the southeast area.

5. Discussion and conclusions

In this section, we will address the results that come from the analysis of the land-use related to vegetation changes for the Galicia region in the years 1990, 2000, 2006, 2012 and 2018.

Therefore, the results presented through the tables and thematic cartography, in the previous section are related to the characterization of the evolution of land use based on the 44 uses of the soil determined by CLC. So, as we are analyzing the Land-Use Changes related to vegetation, we will give more importance to the CORINE Land Cover nomenclature associates, not neglecting the rest.

According to **Table 5**, it seems that the increase in certain kinds of land uses such as 3.1.1. broad-leaved forest and 3.1.2. coniferous forest is compensated by the decrease of other land uses like 2.4.2. complex cultivation, 2.4.3. the land occupied by agriculture and 3.1.3. mixed forest. Nonetheless, it is advisable to execute more exhaustive study research to know it.

The previously portrayed can be validated by the observation of the thematic cartography (**Figures 4–6**, **9** and **10**).

Corroborating what has already been portrayed concerning the Galicia Region, namely climatology, the rural depopulation, are factors that contribute to the increase of the Land-Use Changes related to vegetation, namely those related to the forest [31, 41, 42].

This temporal evolution – not only at the parish level but also at the municipality level – has been influenced by the land tenure regime and, as expected, by the land management carried out. Thus, forest ownership characterized by either solely private or solely public management showed a higher incidence of more productive forest types than mixed management [34].

In addition, the demographic aspects linked to these territorial units have contributed, directly or indirectly, to these forestry changes. So, densely populated areas have increased their forestland toward woodlands for timber production, although the environmental component of sustainable forest management requires a special weight given the strong urban and population pressure. On the other hand, the area of productive forestry did not increase in highly depopulated areas (unlike other forestry regions) because the economic incentives were insufficient to promote a future owner's interest [34].

The research of vegetation-related Land-Use Changes is critical for understanding regional trends and developments [50, 51]. It was feasible to discern changes in all CLC levels in the Galicia Region from 1990 to 2018 throughout this examination.

Thus, it was credible to establish that these Land-Use Changes related to vegetation suffered some changes, characterized by increasing and decreasing periods. Some of those decreasing values are disturbing and should have special attention by the government authorities to provide preservation and conservation of these unique Galician landscapes and environments.

The changes in the Land-Use related to vegetation could be understood as a direct manifestation of human activity over natural environments [52, 53]. Therefore, the natural factors and features—i.e., geomorphology, slope, relief, soil, and vegetation, among many others— are critical for the proper organization and distribution of the territory and their consequent land uses [52]. The lack of knowledge aligned with the existence of planning conducts to the destruction of the natural resources causing a relevant (negative) impact on the local communities [54].

Therefore, the study of the Land-Use Changes related to vegetation is seen as pivotal to understanding the dynamics and tendencies of these territories as well as to provide clues for the main actors to where the efforts toward sustainable development and growth should be placed.

In the final remarks, the Land-Use Changes related to vegetation could be understood as another tool for the knowledge of the territory—assessing the past and envisioning the future.

6. Limitations of the study and future research directions

Although this chapter provides some insight into the dynamics, trends, and specificities of Land-Use Changes associated to vegetation in the Galicia Region, more research is needed to uncover new variables and significant findings.

In these territories, regional policies and societal behaviors change frequently, necessitating close monitoring and new analyses of the directions and dynamics of Land-Use Changes associated to vegetation, as well as the management of sustainable development methods.

Furthermore, due to the employed CLC's minimum cartographic unit (25 hectares), some Land-Use in the Galicia Region could not be reflected in this study if these aspects were not identified. This problem would most likely be solved if newer versions of the CLC program were used, specifically the most recent one with a better resolution.

Future research on these regions may also combine cartography with protected natural spaces, their various figures, and Land-Use Changes related to vegetation through time.

Acknowledgements

This publication has been possible thanks to funding granted by the "Consejería de Economía, Ciencia y Agenda Digital" (Ministry of Economy, Science and Digital Agenda) of Extremadura govern, and by the European Regional Development Fund of the European Union through the reference grants GR21135, Research Group on Environment and Spatial Planning.

Conflict of interest

"The authors declare no conflict of interest."

Author details

Sérgio Lousada[1,2,3,4*] and José Manuel Naranjo Gómez[2,4,5]

1 Faculty of Exact Sciences and Engineering (FCEE), Department of Civil Engineering and Geology (DECG), University of Madeira (UMa), Funchal, Portugal

2 VALORIZA - Research Centre for Endogenous Resource Valorization, Portalegre, Portugal

3 RISCO - Civil Engineering Department of University of Aveiro, Aveiro, Portugal

4 CITUR - Madeira - Centre for Tourism Research, Development and Innovation, Madeira, Portugal

5 School of Agricultural Engineering, University of Extremadura, Badajoz, Spain

*Address all correspondence to: slousada@staff.uma.pt

References

[1] Yaakub NF, Masron T, Marzuki A, Soda R. GIS-based spatial correlation analysis: Sustainable development and two generations of demographic changes. Sustainability. 2022;**14**:1490. DOI: 10.3390/su14031490

[2] Assumma V, Bottero M, Caprioli C, Datola G, Mondini G. Evaluation of ecosystem services in mining basins: An application in the piedmont region (Italy). Sustainability. 2022;**14**:872. DOI: 10.3390/su14020872

[3] Botticini F, Auzins A, Lacoere P, Lewis O, Tiboni M. Land take and value capture: Towards more efficient land use. Sustainability. 2022;**14**:778. DOI: 10.3390/su14020778

[4] Mensah J. Sustainable development: Meaning, history, principles, pillars, and implications for human action: Literature review. Cogent Social Sciences. 2019;**5**:1653531. DOI: 10.1080/23311886.2019.1653531

[5] Bottero MC, Polo Pérez I, Taddia G, Lo RS. A geodatabase for supporting planning and management of mining activities: The case of Piedmont Region. Environment and Earth Science. 2020;**79**:83. DOI: 10.1007/s12665-020-8815-x

[6] Felke TP. Geographic information systems: Potential uses in social work education and practice. Journal of Evidence-Based Social Work. 2006;**3**:103-113. DOI: 10.1300/J394v03n03_08

[7] Andreev DV. The use of GIS technology in modern conditions. IOP Conference Series: Earth and Environmental Science. 2020;**421**:042001. DOI: 10.1088/1755-1315/421/4/042001

[8] Burrough PA. Principles of geographical information systems for land resources assessment. Geocarto International. 1986;**1**:54-54. DOI: 10.1080/10106048609354060

[9] Decoville A. Can the 2050 zero land take objective of the EU be reliably monitored? A Comparative Study. Journal of Land Use Science. 2016;**11**(3):331-349. DOI: 10.1080/1747423X.2014.994567

[10] Naranjo Gómez JM, Lousada S, Garrido Velarde J, Castanho RA, Loures L. Land-use changes in the canary archipelago using the CORINE data: A retrospective analysis. Landscape. 2020;**9**:232. DOI: 10.3390/land9070232

[11] Benedetti A, Picchiani M, Del Frate F. Sentinel-1 and sentinel-2 data fusion for urban change detection. IGARSS 2018. 2018 IEEE International Geoscience and Remote Sensing Symposium. 2018:1962-1965. DOI: 10.1109/IGARSS.2018.8517586

[12] Melchiorri M, Florczyk AJ, Freire S, Schiavina M, Pesaresi M, Kemper T. Unveiling 25 years of planetary urbanization with remote sensing: Perspectives from the global human settlement layer. Remote Sensing. 2018;**10**:768. DOI: 10.3390/rs10050768

[13] Washaya P, Balz T. Sar coherence change detection of urban areas affected by disasters using sentinel-1 imagery. The International Archives of the Photogrammetry, Remote Sensing and Spatial Information Sciences. 2018;**XLII–3**:1857-1861. DOI: 10.5194/isprs-archives-XLII-3-1857-2018

[14] Chen Q, Mao Y, Morrison AM. The influence of land use evolution on the visitor economy in wuhan from the perspective of ecological service value.

Landscape. 2022;**11**:1. DOI: 10.3390/land11010001

[15] Botezan CS, Radovici A, Ajtai I. The challenge of social vulnerability assessment in the context of land use changes for sustainable urban planning—case studies: Developing cities in Romania. Landscape. 2022;**11**:17. DOI: 10.3390/land11010017

[16] Castanho RA, Lousada S, Naranjo Gómez JM, Escórcio P, Cabezas J. Loures LF-P and L. dynamics of the land use changes and the associated barriers and opportunities for sustainable development on peripheral and insular territories: The madeira Island (Portugal). IntechOpen. 2018. DOI: 10.5772/intechopen.80827

[17] Grigorescu I, Mitrică B, Kucsicsa G, Popovici E-A, Dumitraşcu M, Cuculici R. Post-communist land use changes related to urban sprawl in the Romanian metropolitan areas. Human Geography – Journal of Studied Research Human Geography. 2012;**6**:35-46. DOI: 10.5719/hgeo.2012.61.35

[18] Caldas AM, Pissarra TCT, Costa RCA, Neto FCR, Zanata M, Parahyba RDBV, et al. Flood vulnerability, environmental land use conflicts, and conservation of soil and water: A study in the batatais SP municipality, Brazil. Water. 2018;**10**:1357. DOI: 10.3390/w10101357

[19] Zarzycki J, Korzeniak J, Perzanowska J. Impact of land use changes on the diversity and conservation status of the vegetation of mountain grasslands (Polish Carpathians). Landscape. 2022;**11**:252. DOI: 10.3390/land11020252

[20] Xie S, Zhang W, Zhao Y, Tong D. Extracting land use change patterns of rural town settlements with sequence alignment method. Landscape. 2022;**11**:313. DOI: 10.3390/land11020313

[21] Kundzewicz ZW, Kanae S, Seneviratne SI, Handmer J, Nicholls N, Peduzzi P, et al. Flood risk and climate change: Global and regional perspectives. Hydrological Sciences Journal. 2014;**59**:1-28. DOI: 10.1080/02626667.2013.857411

[22] Armaş I. Social vulnerability and seismic risk perception. Case study: The historic center of the Bucharest municipality/Romania. Natural Hazards. 2008;**47**:397-410. DOI: 10.1007/s11069-008-9229-3

[23] Barichivich J, Gloor E, Peylin P, Brienen RJW, Schöngart J, Espinoza JC, et al. Recent intensification of Amazon flooding extremes driven by strengthened walker circulation. Science Advances. 2018;**4**:eaat8785. DOI: 10.1126/sciadv.aat8785

[24] Vieira I, Barreto V, Figueira C, Lousada S, Prada S. The use of detention basins to reduce flash flood hazard in small and steep volcanic watersheds – A simulation from Madeira Island. Journal of Flood Risk Management. 2018;**11**:S930-S942. DOI: 10.1111/jfr3.12285

[25] Pilogallo A, Scorza F. Mapping regulation ecosystem services specialization in Italy. Journal of Urban Planning and Development. 2022;**148**:04021072. DOI: 10.1061/(ASCE)UP.1943-5444.0000801

[26] leBrasseur R. Mapping green infrastructure based on multifunctional ecosystem services: A sustainable planning framework for Utah's Wasatch front. Sustainability. 2022;**14**:825. DOI: 10.3390/su14020825

[27] Forkink A. Benefits and challenges of using an Assessment of Ecosystem Services approach in land-use planning. Journal of Environmental Planning and Management. 2017;**60**:2071-2084. DOI: 10.1080/09640568.2016.1273098

[28] Maes J, Jacobs S. Nature-based solutions for Europe's sustainable development. Conservation Letters. 2017;**10**:121-124. DOI: 10.1111/conl.12216

[29] Albert C, Aronson J, Fürst C, Opdam P. Integrating ecosystem services in landscape planning: requirements, approaches, and impacts. Landscape Ecology. 2014;**29**:1277-1285. DOI: 10.1007/s10980-014-0085-0

[30] Vulevic A, Macura D, Djordjevic D, Castanho RA. Assessing accessibility and transport infrastructure inequities in administrative units in Serbia's danube corridor based on multi-criteria analysis and Gis mapping tools. Transylvanian Review of Administrative Sciences. 2018;**14**:123-143. DOI: 10.24193/tras.53E.8

[31] López-Rodríguez G, Rodríguez-Vicente V, Marey-Pérez MF. Study of forest productivity in the occurrence of forest fires in Galicia (Spain). Sustainability. 2021;**13**:8472. DOI: 10.3390/su13158472

[32] Alonso L, Picos J, Armesto J. Forest land cover mapping at a regional scale using multi-temporal sentinel-2 imagery and RF models. Remote Sensing. 2021;**13**:2237. DOI: 10.3390/rs13122237

[33] Guitián MR, Rego PR. Clasificaciones climáticas aplicadas a Galicia: Revisión desde una perspectiva biogeográfica. Recursos Rurais Review of Institute Biodiversidade Agrar E Desenvolv Rural IBADER. 2007;**3**:31-53

[34] Marey Pérez M, Rodríguez Vicente V, Crecente MR. Using GIS to measure changes in the temporal and spatial dynamics of forestland: experiences from north-west Spain. International Journal of Forestry Research. 2006;**79**:409-423. DOI: 10.1093/forestry/cpl027

[35] Corbelle Rico EJ, Tubío Sánchez JM. Productivismo y abandono: dos caras de la transición forestal en Galicia (España), 1966-2009. Bosque Valdivia. 2018;**39**:457-467. DOI: 10.4067/S0717-92002018000300457

[36] Marey-Pérez M, Díaz-Varela E, Calvo-González A. Does higher owner participation increase conflicts over common land? An analysis of communal forests in Galicia (Spain). IForest - Biogeosciences For. 2015;**8**:533-543. DOI: 10.3832/ifor1060-008

[37] Chas-Amil M-L, García-Martínez E, Touza J. Iberian Peninsula October 2017 wildfires: Burned area and population exposure in Galicia (NW of Spain). International Journal of Disaster Risk Reduction. 2020;**48**:101623. DOI: 10.1016/j.ijdrr.2020.101623

[38] Boubeta M, Lombardía MJ, González-Manteiga W, Marey-Pérez MF, Boubeta M, Lombardía MJ, et al. Burned area prediction with semiparametric models. International Journal of Wildland Fire. 2016;**25**:669-678. DOI: 10.1071/WF15125

[39] Boubeta M, Lombardía MJ, Marey-Pérez M, Morales D, Boubeta M, Lombardía MJ, et al. Poisson mixed models for predicting number of fires. International Journal of Wildland Fire. 2019;**28**:237-253. DOI: 10.1071/WF17037

[40] Marey-Perez M, Loureiro X, Corbelle-Rico EJ, Fernández-Filgueira C. Different strategies for resilience to wildfires: The experience of collective land ownership in Galicia (Northwest Spain). Sustainability. 2021;**13**:4761. DOI: 10.3390/su13094761

[41] Chas-Amil ML, Prestemon JP, McClean CJ, Touza J. Human-ignited wildfire patterns and responses to policy shifts. Applied Geography.

2015;**56**:164-176. DOI: 10.1016/j.apgeog.2014.11.025

[42] de Diego J, Rúa A, Fernández M. Designing a model to display the relation between social vulnerability and anthropogenic risk of wildfires in Galicia, Spain. Urban Science. 2019;**3**:32. DOI: 10.3390/urbansci3010032

[43] Barreal J, Loureiro ML. Modelling spatial patterns and temporal trends of wildfires in Galicia (NW Spain). For Systems. 2015;**24**:e022-e022. DOI: 10.5424/fs/2015242-05713

[44] Bergstrand K, Mayer B, Brumback B, Zhang Y. Assessing the relationship between social vulnerability and community resilience to hazards. Social Indicators Research. 2015;**122**:391-409. DOI: 10.1007/s11205-014-0698-3

[45] Barreiro JB, Hermosilla T. Socio-geographic analysis of the causes of the 2006's wildfires in Galicia (Spain). For Systems. 2013;**22**:497-509. DOI: 10.5424/fs/2013223-04165

[46] Barreal J, Loureiro ML, Picos J. Estudio de la causalidad de los incendios forestales en Galicia. Economia Agraria y Recursos Naturales - Agricultural and Resource Economics. 2012;**12**(1):99-114. DOI: 10.7201/earn.2012.01.04

[47] Wigtil G, Hammer RB, Kline JD, Mockrin MH, Stewart SI, Roper D, et al. Places where wildfire potential and social vulnerability coincide in the coterminous United States. International Journal of Wildland Fire. 2016;**25**:896-908. DOI: 10.1071/WF15109

[48] CORINE Land Cover — European Environment Agency n.d. Publications Office of the European Union. Luxembourg. Available online: https://www.eea.europa.eu/publications/COR0-landcover [Accessed: April 28, 2022]

[49] Publications Office of the EU n.d. https://op.europa.eu/en/publication-detail/-/publication/96743011-0b4f-11ea-8c1f-01aa75ed71a1 [Accessed April 28, 2022]

[50] Gómez N, Manuel J. Impacts on the social cohesion of mainland Spain's future motorway and high-speed rail networks. Sustainability. 2016;**8**:624. DOI: 10.3390/su8070624

[51] Vulevic A, Castanho RA, Naranjo Gómez JM, Loures L, Cabezas J, Fernández-Pozo L, et al. Accessibility dynamics and regional cross-border cooperation (CBC) perspectives in the portuguese—Spanish borderland. Sustainability. 2020;**12**:1978. DOI: 10.3390/su12051978

[52] Gao P, Niu X, Wang B, Zheng Y. Restoration area based on GIS and RS of northern China. Scientific Reports. 2015;**5**:11038. DOI: 10.1038/srep11038

[53] Bertrand N, Vanpeene-Bruhier S. Periurban landscapes in mountain areas. Journal of Alpine Research | Revue de géographie alpine. 2007; **95**(4):69-80. DOI: 10.4000/rga.363

[54] Loures L. Land use: Assessing the past, envisioning the future. BoD – Books on Demand. 2019

Chapter 13

Impacts of Human Activities on the High Mountain Landscape of the Tatras (Example of the Border Area of the High and Belianske Tatras, Slovakia)

Veronika Piscová, Juraj Hreško, Michal Ševčík and Terézia Slobodová

Abstract

We summarize impacts of human activities on the alpine landscape at the border of the High and Belianske Tatras (Slovakia). The High Tatras, especially due to the glacial relief on the crystalline rocks and specific climatic conditions, represent the most attractive area of year-round tourism. The Belianske Tatras represent the limestone part of the mountain range, with rare communities, many endemics and glacial relics, and are among the rarest and most endangered mountains in Slovakia. In the past, this area was mainly affected by grazing, forest cutting and mining. Currently, the area is protected as the Tatra National Park, the Tatras Biosphere Reserve, by the Habitats Directive and the Birds Directive, tourism is the only human activity in the area. Due to tourism, the ridge trail of the Belianske Tatras has been closed since 1978 and one of the trails has been open since 1993. The current hiking, as the only activity in the area, is bearable, which was confirmed by experimental research. But hiking trails are threatened by many morphodynamic processes.

Keywords: high mountain landscape, the Tatras, land use change, human effects, trampling, synanthropisation, hiking trail

1. Introduction

The mountains are vital to life on Earth. Approximately 27% of the Earth's surface (40 million km^2) are covered by mountains [1], which possess at least one third of the species of the entire species diversity of terrestrial plants [2] and, at the same time, supply half of the human population with water [3, 4]. In the course of natural development, high mountains have become a refuge of many rare, endangered and endemic species and habitats [5]. Plant communities, differentiated both by height and species, very effectively capture water precipitation, snow, fog and ice. They regulate their runoff, allow uniform distribution of moisture throughout the soil horizon profile, and ensure a long-term balanced water regime [3]. For these reasons, plant communities in high altitudes have an irreplaceable role. These are mainly spring areas of watercourses, which are currently the last surface sources of clean water [4].

The alpine landscape in particular represents a unique biogeographical unit of the Earth. The territory covered by alpine vegetation is fragmented into several mountain regions [6, 7]. Alpine landscape can be found at all latitudes [8, 9]. It occupies 4 million km^2, which represents almost 3% of the Earth's land surface. Alpine vegetation hides a great variety of species around the world, including 8,000–10,000 species of vascular plants. Alpine ecosystems have a strong impact on humans. Around 10% of the world's population lives in high-altitude regions, and more than 40% of them depend in some way on the resources of these ecosystems, in particular drinking and irrigation water from high altitude basins [9].

The alpine landscape of Slovakia is understood as mountains with developed upper forest boundaries and higher vegetation zones: subalpine, alpine and subnival [10]. The subalpine vegetation zone follows the montane vegetation zone and ends with an upper limit of the continuous occurrence of shrubland at an altitude of approximately 1850 m above sea level [11]. The alpine zone follows the subalpine zone and extends to a height of about 2300 m above sea level. It consists of original, primary alpine meadows, which extend over the shrubland zone, the so-called alpine grasslands. The subnival zone is the highest vegetation zone of the Tatras at altitudes from 2300 m to the highest peaks. The vegetation is poor, more continuous vegetation cover does not exist, plants occupy rock cracks, walls and slits. In Slovakia, the alpine landscape occupies 320 km^2, which represents 0.7% of the country's territory [12]. The island character of the high mountains, their height and substratum ruggedness created suitable conditions for the creation of a varied mosaic of vegetation types with a number of naturally rare, relict and endemic plants. The Alpine landscape in Slovakia is found only in national parks, which mainly protect their ecosystems.

The unique alpine landscape with which humans have been connected since the past is represented by the smallest high mountains in the world, the Tatra Mountains. In the past, in the places where now exist the highest peaks, prehistoric seas were spreading, massive layers of sediments were deposited, mountain-forming processes were taking place and prehistoric animals were moving around the landscape. Today's appearance of the Tatras has been completed by processes in the last two million years. Mountain massifs elevated by alpine folding with remains of layers of sedimentary rocks formed mountain glaciers during probably four ice ages. They pushed huge volumes of rubble out of them and gave them the character of high mountains. The alternation of hot and cold periods, dry and humid periods in the Holocene was the key to the development of today's plant and animal kingdom. In the 11th century, they were surrounded by one large primary forest, and until the 14th century only isolated shepherds, treasure hunters and lumberjacks wandered into their valleys. Major changes occurred in the 14th – 17th centuries, when Wallachian colonization was directed to higher mountain areas. In the 18th – 19th centuries, most of the accessible forests were grubbed up for the needs of mining, metallurgy and construction. Following the shepherds and lumberjacks came researchers, tourists and climbers [13].

People perceived the rare beauty of the mountains and their uniqueness in the distant past. In order to preserve them, they declared the first protected areas. The aim was protection of their beauty, protection for religious and utilitarian reasons and protection of wildlife with original game for hunting. Later, biological, biogeographical and ecological aspects were pushed to the fore, such as protection of rare and endangered species and their habitats, protection of representative ecosystems, up to the systemically understood protection of natural ecosystems and original ecological processes [14].

However, a question has been hanging over national parks since the creation of the first ones – how to preserve the original nature from the emerging anthropic pressure and, at the same time, how to make the national park available for recreation and relaxation? This is a global issue applying to national parks around the world [14]. Protected areas are now considered as effective and promising instruments not only of a global strategy, but also of national strategies aimed at combating biodiversity loss [15, 16]. The mission of nature conservation areas is to preserve biodiversity and functioning natural ecosystems that serve as a refuge for many endangered plant and animal species and provide ecosystem services. However, they are often disturbed in most of the area of the intensively used landscape that surrounds them [14–16]. Despite great efforts to preserve the nature of the mountains, there are still significant changes in the environment. These changes are the result of climate change, deforestation or natural disasters, in many places in less developed countries in the world, also of mineral extraction, armed conflicts, poverty and hunger. In more advanced countries, as a result of modern times that have brought about the development of sports facilities, recreation and tourism, new roads, hotels and lifts have been built. This is a global trend. More than 50 million people visit mountains every year [17]. Many mountain towns around the world depend on the development of tourism. Catering and accommodation services for tourists who come to admire the mountains are developed in the villages.

The Tatras, despite their protection, they are threatened by an increasing number of tourists and increasing demands for the construction of infrastructure connected with services. In this chapter, we focus on the area located on the border of the High and Belianske Tatras, where the alpine landscape is characterized by various degrees of destruction, but, at the same time, almost undisturbed unique nature. We focus on the impacts of human activities on the high mountain landscape during the past and at present.

2. The border area of the High and Belianske Tatras (Slovakia)

The area of interest represents the border area between the High Tatras and the Belianske Tatras (**Figure 1**), which are parts of the Tatra Mountains. We chose it because of its rarity and uniqueness, but also because of the damage caused by human activities during the past and at present.

2.1 Characteristics of the area

The Tatra Mountains belong to the Alpine and Himalayan system and are a part of the extensive range of the Carpathian Mountains, spreading over the territories of Austria, the Czech Republic, Slovakia, Hungary, Poland, Ukraine, Romania and Serbia. The entire mountain range covers approximately 210,000 km^2, the length of the Carpathian arch reaches 1,500 km, the width ranges from approximately 12 to 350 km [18]. The highest part of the Carpathians are the Tatras, namely the High Tatras and Gerlachovský Peak with an altitude of 2654 m, which is their highest peak. Within the Carpathians, the study area belongs to the extensive mountain province of

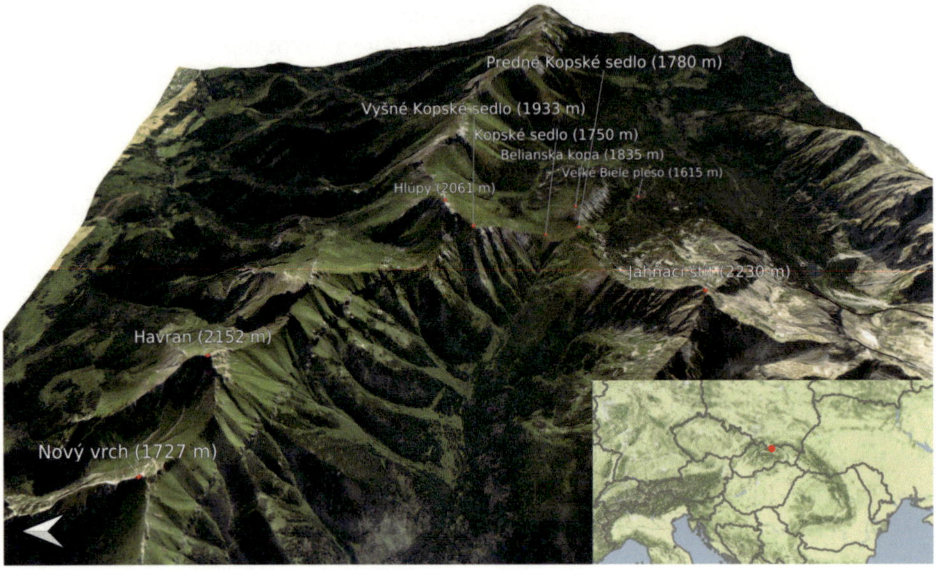

Figure 1.
Study area (Source: mapy.cz).

Figure 2.
The High Tatras – the tarn Veľké Biele pleso (1615 m MSL) (Piscová, 2021).

the Western Carpathians, situated in the western part of the Carpathian Mountains. The arc of the Western Carpathians extends to the territories of the Czech Republic (namely Moravia and Silesia), Slovakia, Poland (smaller northern part), Hungary (south-eastern part) and Austria (south-western parts). The total area of the Western Carpathians is about 70,000 km^2, the length of the mountains reaches 400 km. The course of the narrow Pieniny Klippen Belt divides the Western Carpathians into external and internal, which differ significantly in their geological history, geological composition and tectonic development [19].

The Tatras are considered to be a model of a high mountain range, because all typical alpine features can be found here in a relatively small area [20]. The environment which prevailed in the Tatras in the last Ice Age significantly marked the ductility of their surface. The relatively short warmer period of the Holocene (Ice Age, the younger Quarternary) was not enough to reshape these older forms of relief [21]. The High Tatras (**Figures 2** and **3**) are the highest mountain range of the Western Carpathians, formed by glacial activity. The Belianske Tatras (**Figure 4**) are considerably smaller than the High Tatras, it is the highest carbonate mountain range of the Western Carpathians, thanks to its Mesozoic sediments characterized by a characteristic gradation of mountain massifs [22].

Climatically, the Tatras fall into a cold area with a predominantly cold and cold mountain district. The area of the upper limit of the forest is characterized by an average annual air temperature of 2–4°C, an average July temperature of 10–12°C and an average annual rainfall of 900–1,200 mm. Above the border of the forest, the temperatures are even lower and precipitation is higher [23].

Figure 3.
The High Tatras, the summit Belianska kopa (1835 m MSL) (Hreško, 2005).

Figure 4.
The Belianske Tatras (Hreško, 2005).

A general characteristic of the soils of the High Tatras is an acidic to very acidic soil reaction. In the Alpine landscape of the Tatras we can find the following soil types: lithosols, rankers, rendzinas, cambisols and podsols [24].

There are two national parks in the whole territory of the Tatras: in the territory of Slovakia the Tatra National Park (TANAP), declared in 1948 with effect from 1 January 1949, with an area of 1,045 km^2 (of which the national park's own territory has 738 km^2, national park protection zone 307 km^2), and in Poland Tatrzański Park Narodowy, declared in 1954 with effect from 1 January 1955, with an area of 212 km^2. TANAP is the first and therefore the oldest national park in Slovakia. It was established by the Slovak National Council law No. 11/1948 Coll. on the Tatra National Park [25]. Its protection is currently provided for by Legislative Act No. 543/2002 Coll. on Nature and Landscape Protection [26], as amended.

Biosphere reserve Tatry (BR Tatras) entered into the world network of biosphere reserves on 15th February 1993 with an area of 113,251 ha. It is a bilateral biosphere reserve, which includes TANAP and its buffer zone (on the territory of the Slovak Republic) and Tatrzański Park Narodowy TPN (on the territory of Poland).

In addition to the protected areas at national level, the territory of the Tatras is also defined according to European legislation, in order to preserve the natural heritage important not only for the member state, but for the entire EU. These are the two directives: (1) Council Directive of the European Communities No. 79/409/EEC on the conservation of wild birds (Birds Directive) [27] and (2) Council Directive No. 92/43/EEC on the conservation of habitats and of wild fauna and flora (Habitats Directive) [28]. The Natura 2000 network consists of two types of areas: areas of European importance and protected bird areas. The area of European importance for Slovakia is defined by the Decree of the Ministry of Environment of the Slovak Republic No. 3/2004–5.1 of 14 July 2004 [29], which issues a national list of areas of European importance, and the protected bird area is established by the Decree of the Ministry of Environment No. 4/2011 Coll. [30], which declares the Protected Bird Area of the Tatras.

The studied area represents the boundary area of two geologically and geomorphologically distinct parts of the Tatras. The High Tatras, mainly due to its glacial relief, rock composition and specific climatic conditions, represent the most attractive area of year-round tourism. The Belianske Tatras represent one of the highest limestone mountains in Slovakia. With its habitats of rare communities and a number of endemics and glacial relics they are among the rarest and most endangered mountains in Slovakia. The studied area is located in the national nature reserve Belianske Tatras, which was declared to protect a territory with a great variety of species (even rare and endemic) and communities of fauna and flora, to protect the richness of glacial forms of relief on granites and mylonites as these geosystems are very unstable. In the past, especially due to the unbearably high number of visitors of this area, the rare ecosystems of the Belianske Tatras were damaged. In the study area, grazing, mining activities, general removal of shrubland and later unbearable hiking were taking place. The Belianske Tatras ridge trail has been closed to tourists since 1978. Since 1993, a part of the Monkova dolina valley has been accessible in one direction, since 2009 in two directions.

2.2 Rare ecosystems with many endemic and glacial relics

The Belianske Tatras represent one of the highest limestone mountains in Slovakia, with its habitats with rare communities and a number of endemics and glacial relics are among the rarest and most endangered mountains in Slovakia. The studied area is located in the Belianske Tatry National Natural Preserve (NNP), which was declared in 1991 by the Decree of the Slovak Commission for the Environment No. 166/1991 Coll. of 15.1.1991 on the State Nature Reserves and the Protected Sites in the TANAP [31]. The subject of NNP protection are habitats with rare communities and a number of endemics and glacial relics and the richness of glacial forms of relief on granites and mylonites, which are very unstable.

2.2.1 Development of vegetation since the last glaciation, current endemics and relics

After the retreat of glaciers, deep glacial valleys, morenas, lakes, rocky ridges with many peaks, towers and needles interwoven with tight saddles were created in the Tatras. After the end of the Ice Age (plesitocene), i.e. approximately 12,000 years ago, the vegetation of the Tatras began to form. On the outskirts of the Tatras, the vegetation in the glacier neighborhood had a tundra character with species of bush-like growth, such as *Dryas octopetala* (**Figure 5**) and *Salix reticulata* (**Figure 6**), which grow here to this day. Scattered tree vegetation consisted of shrub, pinus cembra, spruce, larch and birch [21].

In the preboreal (8300–6800 BC), the climate was colder on average by about 5°C than today and in the Tatras there was a dramatic change in the representation of tree and non-tree vegetation. Its ratio changed from 1:1 to 10:1. The upper limit of the forest was at a height of about 90–1000 m MSL, in addition to spruce it consisted of smaller areas of shrubs and stunted birch. Areas above 2000 m MSL were still covered by permanent snow and ice. In the boreal (6800–5500 BC), the average temperature increased by about 2°C more than at present. The upper limit of the forest was pushed to a height of 1700 m. In the forests predominated spruce, pine, fir, and towards the end of the period larches and broadleaves (lime, oak, birch). The Atlantic (5500–2500 BC) was a period with a relatively humid and warm climate, the temperature was higher by 3°C than today, precipitation was 60–70% more abudndant. The upper limit of the forest in the Tatras consisted of continuous pine-spruce forests with representation of larch and fir. On the southern side of the Tatras it reached a height of 1800 m MSL and on the northern side up to 1700 m MSL. Above the spruce stage, a strip of shrubland extending up to 2200 m MSL was formed. The pine remained more continuous only in cliff habitats. In the Subboreal (2500–800 BC), the temperature was slightly higher than it is now, but the air was drier, especially towards the end of

Figure 5.
Dryas octopetala (Piscová, 2008).

this period. In this period, we can assume the development of spruce forests, at lower altitudes beech and fir were also more abundantly represented. In mixed oak forests, oak prevailed over other plants, while elms almost disappeared completely. At the end of the Subboreal roughly the same zoning of vegetation with regard to the altitude that still exists today was formed. At the beginning of the Subatlantic (750 BC to the present day), the climate partially cooled down and was perhaps colder than it is today. The upper limit of the forest was lowered and vegetation elevational zones were stabilised as we know them today [21].

During the development of vegetation, the Tatras and their parts have become a specific territory for the occurrence of many endemics. Species such *Pinus cembra*, *Gentiana nivalis*, *Erigeron uniflorus*, *Carex lachenalii*, *Saussurea pygmaea*, *Artemisia eriantha* and *Ranunculus thora* [32] and a west-carpathian element *Luzula alpinopilosa* subsp. *obscura* [33] grow only in the Tatras. Only in the High Tatras there is *Cerastium uniflorum*, *Ranunculus pygmaeus*, *Ranunculus reptans*, *Armeria alpina* and *Juncus castaneus* [33]. Many species are found only in the Belianske Tatras [33], e.g., *Draba siliquosa*, *Draba fladnizensis*, *Draba pacheri*, *Petrocallis pyrenaica*, *Arctous alpina*, *Juncus triglumis*, *Kobresia simpliciuscula*, *Bellardiochloa variegata*, *Tofieldia pusilla*, *Carex atrofusca*. Glacial relics (remnants of the Ice Age) are represented by the species of *Dryas octopetala*, *Arctous alpina*, *Ranunculus reptans* [34]. Frequent fogs and a large amount of precipitation cause the occurrence of moist and humicolous species also in places that are exposed with occurrence of strong winds (*Arctous alpina*, *Carex atrofusca*, *Juncus triglumis*, *Pritzelago alpina*, *Pyrola carpatica*, *Saxifraga wahlenbergii*, *Tofieldia pusilla*), some of which in the Western Carpathians represent extremely rare postglacial relics.

Figure 6.
Salix reticulata (Piscová, 2008).

Several different plant growth forms have adapted to grow and reproduce under harsh environmental conditions [9, 35, 36]. As the altitude increases, the temperature decreases, the length of the vegetation season decreases and precipitation and humidity increase, which also causes the composition of the flora to change. These rare cushion-forming plants are one of the most conspicuous plants found in the most exposed alpine habitats [7]. Due to their low stature and compact form, cushion plants can modify environmental conditions creating particular microclimates within their canopies [9, 37], cushions are maintaining higher temperatures than their surrounding environment [38], reduce the wind speed by up to 90% [39], create their own humus and the diversity in the cushions is higher by 30–50% [37, 40–45], suggesting that cushions may influence the survival of other species [38]. Cushion plants also occur in the study area (**Figures 7** and **8**). The cushion plant form is not endemic to any single area or plant family. Cushion plants grow very slowly, with this inhibited growth there is increased longevity, with the largest cushions of some species reaching ages of up to 350–3000 years [46, 47].

2.2.2 Rare Communities of Plants

The alpine plant communities on the limestone subsoil of the Belianske and parts of the High Tatras with tourist trails are floristically extremely rich [48]. Their nomenclature is given in accordance with the work [49]. The surroundings of the trail leading from Tatranská Javorina (1000 m MSL) to Kopské sedlo (1750 m MSL) are lined with communities *Phleo alpini-Deschampsietum caespitosae* Krajina 1933)

Figure 7.
Cushion plants on the Hlúpy vrch (2061 m MSL) (Piscová, 2011).

Figure 8.
Cushion plants (Piscová, 2011).

Coldea 1983 (1424 m MSL)*Rhodiolo-Deschampsietum caespitosae* Krajina 1933 a *Vaccinio myrtilli-Calamagrostietum villosae* Sillinger 1933 (1450 m MSL); furthermore with the high-stem community*Geranio sylvatici-Calamagrostietum variae* (Sillinger 1932) Kliment et al. 2004 (1458 m MSL); the grass-herbal association *Ranunculo pseudomontani-Caricetum sempervirentis* (Krajina 1933) Dúbravcová ex Dúbravcová et Jarolímek (1508 m MSL) the community *Ranunculo pseudomontani-Caricetum sempervirentis* (Krajina 1933) Dúbravcová ex Dúbravcová et Jarolímek in which two other types of associations occurr, *Seslerietum tatrae* a *Rhodiolo-Deschampsietum caespitosae* Krajina 1933 (1642 m MSL). At an altitude of 1698 m MSL, we recorded a mosaic of communities *Festuco picturatae-Calamagrostietum villosae crepidetosum conyzifoliae*, *Seslerietum tatrae* Domin 1929 corr. Kliment et al. 2005, *Arenario tenellae-Caricetum firmae* (Br.-Bl. 1930) Šibík et al. 2004 and *Ranunculo pseudomontani-Caricetum sempervirentis* (Krajina 1933) Dúbravcová ex Dúbravcová et Jarolímek. In Kopské sedlo there is a community *Juncetum trifidi* Krajina 1933.

The trail leading from the Veľké Biele pleso (1615 m MSL) to the Kopské sedlo (1750 m MSL) is lined by communities *Ranunculo pseudomontani-Caricetum sempervirentis* (Krajina 1933) Dúbravcová ex Dúbravcová et Jarolímek nom. and *Festuco picturatae-Calamagrostietum villosae* Pawłowski in Pawłowski et al. 1928 corr. Kliment et al. 2004, at an altitude of 1642 m MSL communities of *Festuco picturatae-Calamagrostietum villosae* Pawłowski in Pawłowski et al. 1928 corr. Kliment et al. 2004,*Junco trifidi-Festucetum supinae* Krajina 1933 and *Festuco picturatae-Calamagrostietum villosae* Pawłowski in Pawłowski et al. 1928 corr. Kliment et al. 2004, then *Junco trifidi-Festucetum supinae* Krajina 1933 and *Ranunculo pseudomontani-Caricetum sempervirentis* (Krajina 1933) Dúbravcová ex Dúbravcová et Jarolímek nom. (1777 m MSL), in the Predné Kopské sedlo the community *Junco trifidi-Callunetum vulgaris* (Krajina 1933) Hatcher ex Šibík et al. 2007 (1778 m MSL) and the section between the Predné Kopské sedlo and Zadné Kopské sedlo is lined by communities *Juncetum trifidi* Krajina 1933 and *Rhodiolo-Deschampsietum caespitosae* Krajina 1933.

The hiking trail leading from the Široké sedlo (1825 m MSL) to the Kopské sedlo (1750 m MSL) passes through the communities *Ranunculo pseudomontani-Caricetum sempervirentis* (Krajina 1933) Dúbravcová ex Dúbravcová et Jarolímek nom., *Rhodiolo-Deschampsietum caespitosae* Krajina 1933 and *Seslerietum tatrae* Domin 1929 corr. Kliment et al. 2005, then at an altitude of 1831–1907 m MSL by the community *Seslerietum tatrae* Domin 1929 corr. Kliment et al. 2005. and at the altitude of 1919 m MSL passes through the arcto-alpine community of strongly blown ridges and edges, the association *Drabo siliquosae-Festucetum versicoloris* Petrík in Petrík et al. 2006 and the association *Seslerietum tatrae* Domin 1929 corr. Kliment et al. 2005. The trail continues through the community *Ranunculo pseudomontani-Caricetum sempervirentis* (Krajina 1933) Dúbravcová ex Dúbravcová et Jarolímek and at an altitude of 1927 m MSL, the community *Ranunculo pseudomontani-Caricetum sempervirentis*, however, this community is accompanied by association species *Seslerietum tatrae* Domin. 1929 corr. Kliment et al. 2005. From Vyšné Kopské sedlo (1933 m MSL) the trail continues to fall and passes through *the communities of Ranunculo pseudomontani-Caricetum sempervirentis* (Krajina 1933) Dúbravcová ex Dúbravcová et Jarolímek and *Seslerietum tatrae* Domin. 1929 corr. Kliment et al. 2005 (1907 MSL), then a mosaic of communities *Junco trifidi-Festucetum supinae* Krajina 1933, *Seslerietum tatrae* Domin 1929 corr. Kliment et al. 2005 and *Ranunculo pseudomontani-Caricetum sempervirentis* (Krajina 1933) Dúbravcová ex Dúbravcová et Jarolímek (1826 m MSL), which extend up to Kopské sedlo (1750 m MSL).

The closed trail leading from the Široké sedlo (1825 m MSL) to the Ždiarska vidla (2142 m MSL) is lined near the saddle by communities *Seslerietum tatrae* Domin 1929 corr. Kliment et al. 2005, *Junco trifidi-Festucetum supinae* Krajina 1933 and *Ranunculo pseudomontani-Caricetum sempervirentis* (Krajina 1933) Dúbravcová ex Dúbravcová et Jarolímek (2025 m MSL).

The closed trail leading from Vyšné Kopské sedlo (1933 m MSL) through a part of the Belianske Tatras, Front Meďodoly is lined near the saddle by the community *Festuco versicoloris-Oreochloetum distichae* Pawłowski et Stecki 1927 corr. Petrík et al. 2006 nom. Invers. propos. (2018 m MSL).

Only in the ridge positions of the Belianske Tatras are there Central European relic communities of strongly blown ridges and edges of the alpine to subnival zone *Oxytropido-Elynetalia* Oberdorfer ex Albrecht 1969 on strongly blown ridges and edges on neutral to slightly base substrates in the alpine zone. These are rare communities not only because of their limited occurrence, but mainly because of their floristic composition with a large number of Arctic-Alpine species. There are also Alpine grassy, cushion and shrubby communities of rock walls and fine skeletal rubble on mylonites in the alpine to subnival zones *Festucion versicoloris* Krajina 1933. These are the communities threatened all year round by alpine tourism. Rare is the occurrence of the species *Carex rupestris*, which more rarely penetrates the vegetation in the ridge part of the Belianske Tatras, into the union *Caricion firmae*. The main ridge of the Belianske Tatras (from Ždiarská Vidla to Predné Jatky) is also covered by the occurrence of the species *Elyna myosuroides*. An open, floristically rich community *Oxytropido carpaticae-Elynetum myosuroides* (Puşcaru et al. 1956) Coldea 1991 usually inhabits rocky edges or rock terraces. In the community *Drabo siliquosae-Festucetum versicoloris* Petrík in Petrík et al. 2006, there are a number of rare and phyto-geographically significant species such as *Bellardiochloa violacea*, *Draba fladnizensis*, *D. siliquosa*. Plants *Pyrolo carpaticae-Salicetum reticulatae* Petrík in Petrík et al. 2006 by their character represent the Arctic-Alpine tundra. On the wind-blown flat, only slightly inclined, sometimes almost flat places in ridge, less often in sub-ridge positions occurs the cushion-turf community *Festuco versicoloris-Oreochloetum distichae* Pawłowski et Stecki 1927 corr. Petrík et al. 2006 nom. Invers. propos. (2018 m MSL). The community *Silenetum acaulis* Country 1933 developed in the most extreme alpine locations where survival conditions are very limited. Growths of *Agrostio alpinae-Festucetum versicoloris* Pawłowski in Pawłowski et al. 1928 nom. Invers. propos. Inhabit terraces of almost vertical rock walls and rock ribs. The rubble habitats under the steep rock walls are inhabited by the communities of *Salicetum kitaibelianae* Krajina 1933.

2.2.3 The rarest animals living in the territory, endemic

The current state of animal distribution in the territory of the Tatra National Park is the result of long-term effects of natural and human factors. The Tatra fauna was particularly influenced by the cold periods (in the ice ages), from which the descendants of species inhabiting the northern taiga and tundra came [50]. The cold seasons were followed by warmer seasons with thermophilic species from eastern and south-eastern Europe. The Tatra fauna is therefore characterised by various geographical components, including mainly cosmopolitan, Palearctic, European (Euro-Siberian, Boreoalpine, Boreal, Samaric, Sudeten Carpathian) and endemic species.

A colourful metallic coloured *Carabus auronitens*, *Carabus fabricii* get into the subalpine and alpine zones. From the butterflies there are, for example, species of the genus *Erebia - Erebia pandrose* and *Erebia manta*, a glacial relic *Parnassius apollo*. Few species of amphibians get into the alpine zone, e.g. *Rana temporaria*, *Triturus alpestris* [50]. Only two reptile species, and even these only rarely, get into the alpine zone, namely *Lacerta vivipara* and *Vipera berus*. Regarding birds, alpine meadows and rocky habitats are inhabited by *Prunella collaris* and *Anthus spinoletta*, as well as *Phoenicurus ochruros*, *Oenanthe oenanthe* and*Lyrurus tetrix*. A frequent visitor to these locations is *Aquila chrysaetos*, *Falco tinnunculus* and *Corvus corax*.

The mammals in the subalpine and alpine zones, include relict species of the*Pitymys tatricus*, t*Microtus nivalis mirhanreini*, *Marmota marmota latirostris* and *Rupicapra rupicapra tatrica*, which are existentially bound only to these habitats. Sporadically, *Ursus arctos*, *Canis lupus*, *Lynx lynx*, *Vulpes vulpes* [50] get into the habitats of the alpine zone. Insectivores are represented by the species *Talpa europaea*, *Sorex araneus*, *Sorex minutus* and *Sorex alpinus*.

3. History of human activities in the territory and their consequences

The Tatras were surrounded by one large primary forest until the 11th century. Until the 14th century, only isolated shepherds, treasure hunters and lumberjacks [13] had strayed into the valleys. Later, the territory was influenced by pastoralism, mining, hunting and poaching, hiking, mountaineering and tourism. Thus, the scale and structure of the original landscape of the mountain range have been disrupted by human activity for centuries.

3.1 Pastoralism

Major changes occurred in the 14th–17th centuries, when Wallachian colonization was directed to higher mountain areas. The foothills of the Belianske Tatras were chronologically grazed as the first of the entire Tatras. Pastoralism used almost all vegetation cover, the average height of the upper limit of pastures reached up to 2000 m MSL and grazing took place even on very poorly accessible terrain [13]. Wooden huts of mountain sheep farming in the Belianske Tatras lay at an average height of almost 1500 m MSL. In the years 1891–1895 there was also a cheese factory north of the Belianska kopa. According to statistical data, the grazing culminated in the Belianske Tatras in 1803 [51]. At the time of the enactment of TANAP, landowners in the Belianske Tatras continued to have herds. The year 1955 was the last year of grazing in the Belianske Tatras [52].

The considerable consumption of wood at the salaše, but also the deliberate destruction of stands with the intention of expanding the grazing area contributed substantially to the disruption of the climatic upper limit of the forest (up to 200–300 m) and in many places the zone of shrubland almost completely disappeared. Meadow and pasture communities have become replacement communities after felled forests or burnt shrublands. The originally vast shrublands have been removed. On the secondary mountain grasslands (**Figure 9**) we may find *Calamagrostis villosa*, *Avenella flexuosa*, *Crepis conyzifolia*, *Trommsdorffia uniflora*, *Pulsatilla scherfelii*, *Homogyne alpina*, *Anemone narcissiflora* and others. However, fellings outside the peak subalpine zone paradoxically also contributed to greater landscape and biological diversity by creating pastures, larger meadows and polonyas in the forests [53].

The burning of shrublands in order to spread grazing seriously disturbed the original ecosystem, which led to erosion of the soil cover and violation of the water regime with all associated negative consequences (e.g. deterioration of the absorption ratios [53].

After the end of grazing, there was a change in the herbaceous vegetation. Favourable results were apparent within two years of the end of grazing [54].

Figure 9.
A territory grazed in the past (Piscová, 2013).

The number of synanthrope species decreased significantly, grass representation decreased and the number of vascular plants increased, secondary plant communities thus gradually disappeared, shrubland became naturally younger and succession progressed. However, on former pastures there are abandoned soils that are not able to produce and are difficult to regenerate, and there are, for example, old roads to sheepfolds left by shepherds, which are only slowly being overgrown. Negatives associated with the succession and overgrowth of former flowery mountainous grasslands with monotonous overgrowth of competitively very strong species of the genus *Calamagrostis*, oat (*Avenula* sp.) or the species of thistle (*Deschampsia cespitosa*), which usually cause irreversible changes in the composition of the original phytocenoses [55], have also been shown. However, negative effects that persist to this day include the inability to restore vegetation on soils damaged by wind and rain erosion, the formation of snow and stone avalanches [13].

3.2 Mining

During the Turkish invasion of Central Europe in the 16th century, mining also began to develop here. The development of mining led to the appearance of coal miners who burned wood into charcoal for furnaces in which ore was melted. Copper ore was mined in Kopské sedlo, silver was mined in Belianská kopa and gold in Jatky. The mining sites were still visible at the end of the 19th century. Gold was sought on Belianska kopa and on the ridge between Kopské sedlo and the Jahňací štít. Mines on Belianska kopa are documented by the charter of the town of Spišská Belá from 1585 [52]. Despite the enactment of the Tatra National Park in 1949, mining activity in the territory was definitively prohibited only by the Act of the Slovak National Council No. 287/199 Coll. SNR on State Nature Conservation [56].

The most obvious traces of mining were destroyed stands with consequences in the form of bad drainage conditions and losses of forest land. Therefore, a very significant element was that it changed the character of the region mainly due to the emergence of bare slopes. Nevertheless, since then, deliberately by foresters, but also by natural processes, these changes have softened or even disappeared [13].

3.3 Aromatic plant-based oil industry

In the alpine landscape, oilmen sought sustenance by the acquisition of medicinal oils and herbs and in their trade with them. In addition to balm, conifer oil and medicinal herbal extracts were used to produce other essences that were used in the treatment of the sick and in religious ceremonies [57]. An important location for oilmen in the Belianske Tatras was the Predné Meďodoly, where the so-called terpene and shrub oil factory was established [58]. It was located near the Biele plesá tarns. In 1890, the idea arose to develop a large-scale production of shrub oils. In 1897, the city of Kežmarok concluded a contract with the right to extract any quantity of shrub in the valleys of the Biela voda, the Zelené pleso tarn and the Biele plesá tarns and to distill aromatic oils from them on the spot. A working-class dormitory and a factory of medicinal 'oleum Pini pumilionis' shrub oils were built near the Biele plesá tarns. Due to a violation of the forest law of 1879, the contract was cancelled and the buildings were demolished. Pinus cembra oil "balsamus carpathicus" from the pinus cembra needles, twigs and cones [59] also played an important role in Spiš medicine.

The oil industry had a degrading effect mainly on shrub and pinus cembra stands.

3.4 Hunting and poaching

In the past, the owners of the Tatra mountainous territory, the kings and later the lower nobility, ordered their subjects to hunt in the area with the establishment of a mandatory game amount. The German prince, Prince Christian Kraft Hohenlohe, who bought a part of the Belianske Tatras in 1879, was one of the greatest hunters in the Tatras. He set up a game reserve here, in which people hunted until 1922 [60]. There were also poaching events in the Belianske Tatras, especially aimed at the marmot and chamois [57]. Poaching increased the most during the years 1919–1922. Due to extinction of the marmot in the Belianske Tatras, the restitution of marmots took place in 2009.

Hunting and poaching significantly affected the autochtonous fauna, which persisted even after the icing of past periods [61].

3.5 World Wars

The period of World War I is associated with a decrease in the number of grazed sheep in the Belianske Tatras, the number halved compared to 1803 [59]. During World War II, the German army captured the territory of the Belianske Tatras. Under the Hlúpy vrch, soldiers established a high-mountain firing position. In the massif of the Hlúpy vrch, or between the Hlúpy vrch (2060 m MSL) and the Zadné Jatky (2019 m MSL) there are preserved bunkers of the German army and trenches for artillery fire positions carved in the rock. The transport of military equipment, ammunition, food and necessary material to the ridge of the Belianske Tatras was provided from Tatranská Javorina through the Zadné Meďodoly by Soviet prisoners guarded by German soldiers [62]. Bunkers – artificial underground spaces were built in a mining way with the help of explosives documented in 1993–1995 by speleologists [63]. The limestones of which the bunker is formed are heavily cracked due to the method of construction (blasting). The Germans' stay on the ridge of the Belianske Tatras was reflected in the condition of the chamois. Zelina [51] states that the soldiers contributed to the strong decimation of the marmot population as well.

3.6 Tourism

In the past, coal miners, treasure hunters, hunters, but also domestic nobility, stepped into the valleys and on the peaks. In the dense Tatra forests, robbers and deserters, smugglers or serfs found their refuge. Only in the 16th century can we consider the discovery of the Tatras associated with the first attempts at tourist walks [61]. The first tourist hike in the study area was made by Kežmarok castle's lady Beata Laská in 1565 [59]. The mountains began to be visited by educators, students, and later especially by the public.

The construction of hiking trails in the study area dates from 1879. In 1898 the Veľké Biele Pleso tarn received a tourist connection from Tatranská Kotlina. The extension of this trail to Kopské sedlo was established in 1905. The hiking trail continues from Kopské sedlo through Zadné Medďodoly to Tatranská Javorina. In 1938, a new trail to Predné Meďodoly was built, because the original path was exposed to falling stones, landslides and snow avalanches [59, 64].

In 1922, the army, together with the Czechoslovak Tourist Club, built a cottage with two rooms near the Veľké Biele pleso tarn, which later became overnight accommodation for tourists. However, the Kežmarok hut burned down in 1974. The design of the new Kežmarok hut from 1985 was in the end rejected [65].

The Belianske Tatras in the past had the attribute of extremely popular mountains for tourists, they were easily accessible, undemanding and with a ridge without extreme height differences. Due to the natural environment and the view of the countryside, they were very actively used. The maximum traffic was also several thousand visitors per year [66]. The attack on this very sensitive territory in the form of an excessive number of tourists manifested itself quite strongly. Disturbed chamois migrated from the ridge to the forest, where they were threatened by beasts; golden eagles left their nests, the number of bears decreased, and the almost complete destruction of some botanical species such as *Leontopodium alpinum*, *Gentiana clusii* and *Aster alpinus* was also alarming [67, 68]. With an enormous number of tourists on the ridge, there were secondary negative phenomena, such as shortening of trails, trampling around them and pollution of the natural environment with garbage and excessive noise [67]. The TANAP Administration therefore announced on 1 July 1978 the complete and long-term closure of the area to the public with a view to restoring ecological equilibrium. Until 1983, even scientists could not enter the Belianske Tatras [69]. Pressure from the public and various interest groups to make the Belianske Tatras accessible are constantly increasing. A compromise between the interests of tourists and the State Nature Conservation of the Slovak Republic was the opening of the educational trail in the Monková dolina valley in 1993 (one-way, payment for entry). Since 2008, this educational trail has been open both ways free of charge. In the Tatras, the seasonal closure of trails along the Tatra huts from 1st November to 15th June applies to ordinary tourists.

4. Land-cover change in the area, with a particular focus on vegetation

In the Tatras, natural destructive factors have operated, operate and will always operate, but nature has dealt with them in the course of evolution and they do not pose a critical threat. It can eliminate their consequences very quickly. Man, and his activities have become the biggest negative factor in the course of history. Nature can cope with anthropogenic interventions with much more difficulties and must spend more energy maintaining the equilibrium or regenerating damaged components.

Land cover until the enactment of the national park was most intensively shaped in the studied area by shepherding and later by aromatic plant-based oil industry. Belianske Tatras were liberated from grazing and devastation by the SNC Act No. 11/1948 Coll. on the Tatra National Park [28], when, with effect from 1 January 1949, Belianske and High Tatras were declared our first national park – the Tatra National Park (By order of the SSR Government No. 12/1987 Coll. of 6 February 1987 [70], Western Tatras were also declared a part of TANAP). However, grazing was definitively abolished throughout TANAP only in 1954 [53].

Several studies from the territory claim [71, 72] that the change in land use since the enactment of the national park led to the spontaneous afforestation of land abandoned after the restriction of grazing (in order to protect nature), at the same time tourism was actively developing and local people were changing their orientation from agriculture to tourism.

The studied area falls under three cadastral territories: Tatranská Javorina, Tatranská Lomnica and Ždiar. In the cadastral areas, the spatial structure of the land cover for the period 1955–2010 [73] was analyzed. In all areas, there was a decline in the loss of coniferous forest, along with an increase in damaged forests, especially on the southern slopes in 1955–1986 and 1986–2010. In the area of Tatranská Javorina, a significant change was found in the proportions of alpine meadows and shrubland (*Pinus mugo*), especially on slopes facing south and south-west, by increasing the vegetation of the shrubland in places with higher altitudes, lower degree of slope and less sunlight, and in places with lower altitudes where radiation and the degree of slope are higher. In the area of Ždiar, an increased proportion of shrublands was observed in higher places, on moderate south-western and western slopes. In the period 1955–1968, there was an increase in shrubland (*P. mugo*) and a decrease in alpine meadows in locations with higher altitude and lower degree of slope. In the period 1968–2010, the increase of shrubland in relation to losses in coniferous forests prevailed. As in the area of Javorina, the study attributes the increments of shrubland to higher radiation and slopes at lower altitudes. Also, in the area of Tatranská Lomnica there was an increase in shrubland (*P. mugo*) for the period 1955–1968. There was more shrubland in places with higher altitudes and milder slopes (mainly in the south-west, west, north-west and north); in the period 1968–2010, increments of shrubland (*P. mugo*) were recorded more or less independently of other variables. In general, the increments of shrubland correspond to the declines of alpine meadows, the cover ofshrubland increased at higher altitudes: (1) mostly in sunny and less steep places and slopes and (2) at lower altitudes on sunny and steeper slopes. Decrease of shrubland in lower positions may be evidence of the spread of coniferous forests into higher altitudes, but none of the selected abiotic variables explain this change.

The results of the analysis show a slight upward shift of vegetation from 1956 to 2012. The most pronounced shift concerned shrubland (*P. mugo*). The spread of shrubland confirms the findings of several studies [71, 74, 75], which claim that the visible expansion of shrubland to higher altitudes is due to the abandonment of traditional land use, but also to better temperature conditions (longer seasonal growth, milder winters and shorter periods of snow cover) with sufficient water. On the other hand, the historical influence of man on vegetation in the Tatras is significantly limited by attempts to identify changes that are caused exclusively by climate change. According to some historical sources [76], the boundary of forest trees was lowered (especially in the Belianske Tatras), by 200 m on average, and by 350–400 m or more in some places. The declaration of the National Park (1949) and the subsequent ban on grazing (1954) were the main driving forces behind these changes in the Tatras.

5. Changes in the floristic composition of the peak vegetation in the territory since 1880

Plants are also sensitive to an increase in anthropogenic effects on the Earth's climate system [77]. Between 1957 and 1966, the number of species on mountain peaks in Europe increased by an average of 1.1. Since then, this trend has accelerated, so since 2006 and 2007, an average of 5.5 new species have moved to the highest mountain peak locations in a decade [78].

The same trend was observed in our study area, where, despite the general decline in species richness with increasing altitude, there was a clear percentage increase in the number of species between time periods (**Figure 10a, b**).

Ellenberg indicator values for the central-European flora [79] are routinely used to rapidly estimate site conditions from species composition, when measured

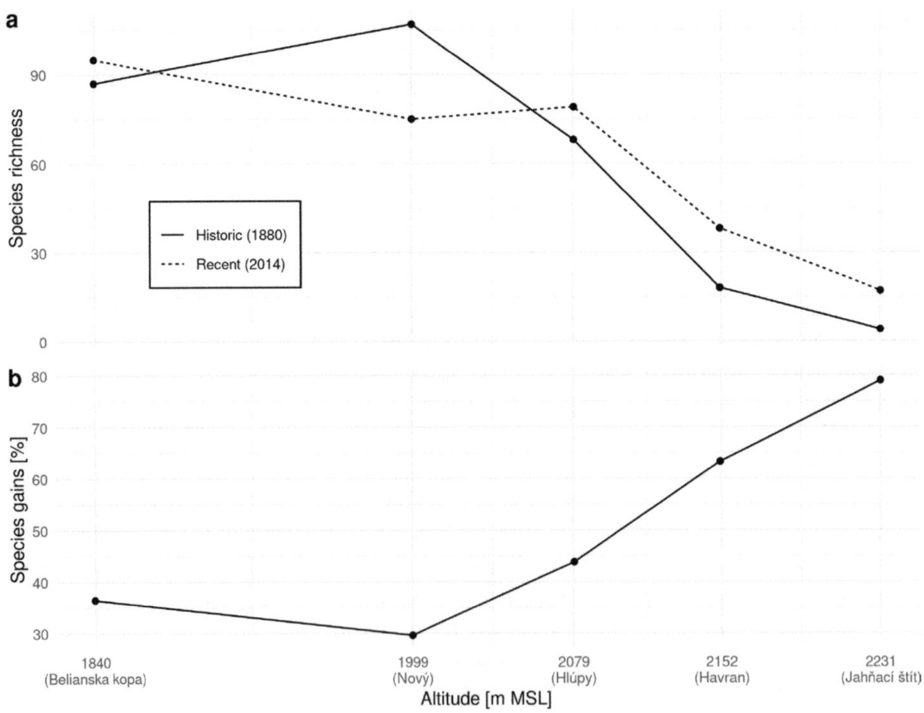

Figure 10.
a. Species richness in relation to altitude at different times; 10b. Percentages of species gain between the studied time periods (1880, 2014) in relation to altitude. Gain was calculated as: g/Stot where g is the number of species gained and Stot is the total number of unique species in both time periods.

values of environmental variables are not available [80, 81]. These indicators are estimates of species ecological optima along several main ecological gradients. Subsequent analyzes of Ellenberg's environmental indicators, using linear mixed modes (to pair design), showed significant differences between time periods in the light ($F_{1,390}$ = 5.14; p = 0.024), and soil reaction ($F_{1,392}$ = 6.17; p = 0.013) indicator. Despite its statistical significance, the simple effect sizes ($light_{mean\_diff}$ = −0.28; soil $reaction_{mean\_diff}$ = −0.65) were not significant enough to lead to more convincing conclusions.

6. Current land use and its bearing capacity for given activities

The current use of the studied area depends on the status of the national park. It can monitor and examine the dynamics of ecosystem development, its accessible part serves for the needs of education, interpretation, communication, tourism, recreation and the necessary infrastructure for the administration and guarding activities of the National Park Directorate.

However, the attractiveness of the Tatras, the smallest mountains in the world, is manifested by high tourist attendance in the long term. In the territory of TANAP there are about 600 km of marked trails that will lead tourists to the most interesting places. Hiking trails through the valleys allow ascents to the Tatra huts, some demanding and less demanding Tatra peaks, as well as passages through the Tatra saddles. The alpine landscape is thus under pressure from tourism and tourism-related activities.

6.1 Territory traffic

According to European Statistical Office [82], people living in Slovakia visit mountains the most of all European countries. Hiking is deeply rooted in Slovakia, with tourist traffic increasing every year. This trend is also confirmed by the data on the annual number of overnight stays in the High Tatras (**Figure 11a**).

We can see a similar short-term trend on one of the studied tourist routes Šalviový prameň (1213 m MSL) to Veľké Biele pleso (1615 m MSL) (**Figure 11b**), where we can see a steep increase, especially during the COVID pandemic restrictions (which also explains the rapid decline in the number of overnight stays in **Figure 11a** in 2020).

We believe that this increasing trend will continue, which may further affect the plant community composition and structure, even more.

6.2 Hiking and its impacts

A number of human activities in the alpine landscape began and ended over the years. However, one of them lasts almost 150 years and remains the only one to this day-hiking. The vulnerable territory of the studied area is affected not only by the bearable or unbearable number of tourists on the trails, but also by its location and the surface itself, which may constitute barriers for tourists.

6.2.1 Trampling impacts on vegetation

Another serious fact is trampling of the vegetation cover. Trampling is known to drive changes in plant community composition and structure [84–87]. Disturbance by trampling mainly affects vegetation directly by damaging plant tissues [88], and

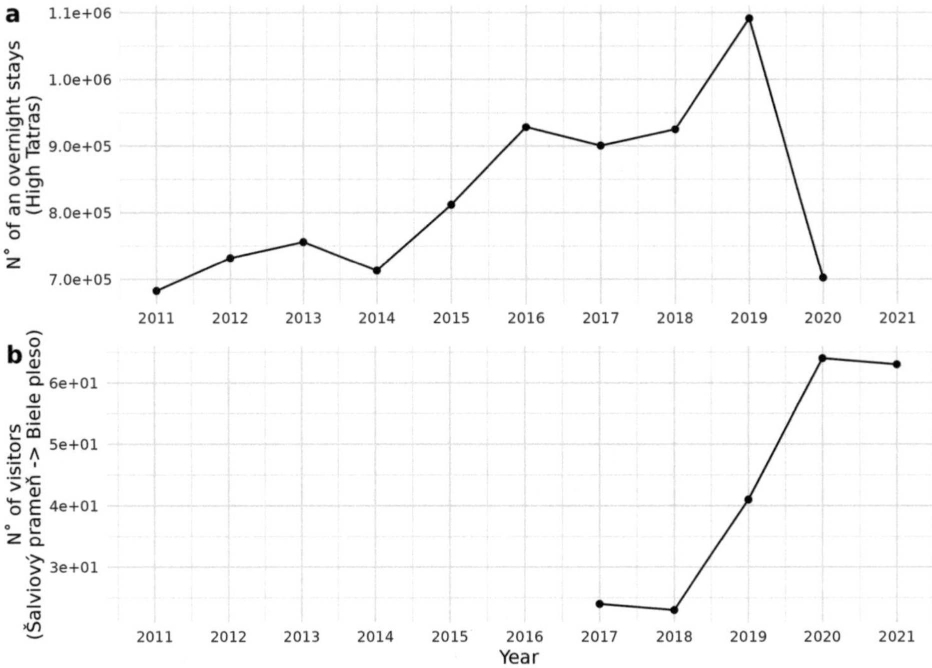

Figure 11.
a. Number of overnight stays in High Tatras area [83]; 11b. Number of visitors on the trail from Šalviový prameň (1213 m MSL) to Veľké Biele pleso (1615 m MSL), source: Správa TANAPu Tatranská Lomnica.

indirectly by modifications to soil structure [89], water regime [90], and nitrogen mineralization [91]. Other evidence indicates that the effects of trampling on soil compaction remain unclear [92–94], or at least are important only in areas of chronic disturbance (long-term effect) [95]. For single disturbance events, the direct effects of the damage to plant tissues are generally the most important [89]. Plants with similar ecological traits are estimated to respond to trampling in comparable ways [96]. Therefore, we have tried to find how the selected vegetation types resist trampling in three alpine communities: *Juncetum trifidi*, *Junco trifidi-Callunetum vulgaris* and *Seslerietum tatrae* using a standard short-term vegetation tracing protocol from Cole and Bayfiel [84]. However, we adjusted the design of trampled blocks and also changed the number of trampled areas according to the number of visitors to the site.

We based the evaluation of the resistance of species monitored on permanent surfaces on relative coverability. We based the calculation of the relative coverability on the sum of coverages of all types, which we preferred over a simple estimate of the total coverability [84]. **Figure 12** shows that there was a statistically significant interaction between trampling intensities and localities on the sum of the coverages ($F1.35, 12.12 = 45.6$, $p < 0.0001$). Therefore, the effect of the trampling intensities was analysed at each locality. P-values were adjusted using the Bonferroni multiple testing correction method. The effect of treatment was significant at all three localities (for Ks = $F1.09, 9.79 = 48.5$, $p < 0.0001$; PKs = $F1.01, 9.06 = 70.3$, $p < 0.0001$; VKs = $F1.1, 9.9 = 44.2$, $p < 0.0001$).

Our study [97] confirms earlier conclusions which stated that more resistant woody chamaephytes have less recovery abilities because of their woody habit. The statement that some communities are initially very prone to trampling due to the

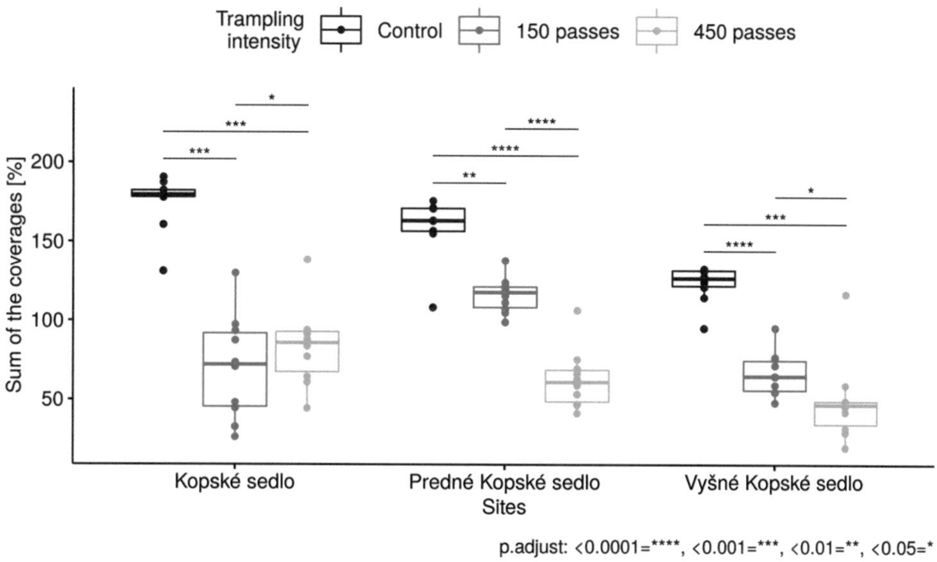

Figure 12.
Differences of sum coverages of all species between different trampling intensities for every locality [80].

high amount of sensitive herbal species was also confirmed. These plant associations are characterized by low, middle and high resistance to trampling, but hiking trails passing through communities can still be made available to tourists at a given traffic.

6.2.2 Synanthropisation

Synanthropisation is manifested by ecesia of the habitat-foreign plants [98]. The most common way of spreading such plants is the transport of diasporas from lower-lying habitats to higher altitudes along the routes of hiking trails. The ecological plasticity of these plants is a limiting factor for their maximum altitude of occurrence. Another way of spreading for habitat-foreign plants is their transport from high mountain higher altitudes to lower altitudes. This applies to species native to the Tatras (original). They stick to ecotopes at a more advanced stage of destruction as a result of hiking. They are plants of apophytic or facultative synanthropic species. Facultative synanthropic species also include species spreading by succession from the forest environment to places deprived of vegetation (grassy) cover by trampling. In terms of the impact of synanthropization on changes in species richness depending on the distance from hiking trails, we found significant differences. The results of the analysis of variance showed a gradual increase in species richness with increasing distance from the trails (**Figure 13a**; $F_{2,9}$ = 11.96; p = 0.003) by an average of 23% between distance categories. We also found differences in species richness between rest areas and their environs (**Figure 13b**; t_4 = −5.15; p = 0.007) by 34%. There are 5 synanthropic species in the area (*Plantago major*, *Plantago media*, *Prunella vulgaris*, *Taraxacum* sect. *Ruderalia*, *Tussilago farfara*) and 3 apophytic species (*Aegopodium podagraria*, *Chaerophyllum hirsutum*, *Urtica dioica*), which occur at a distance of up to 50 cm from the trails, while at the distance of up to 20 cm from the trail they reach the highest coverage of up to 5% and for 20–50 cm up to 2%.

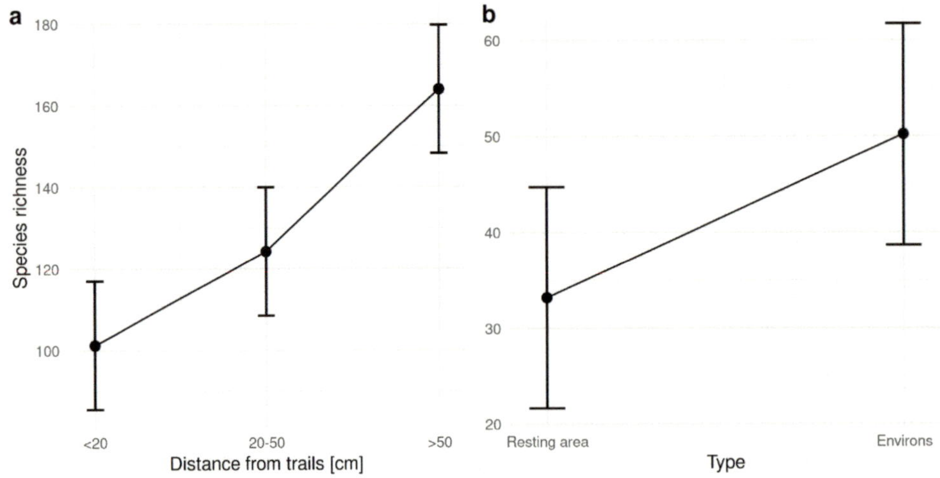

Figure 13.
a. Species richness differences in distances from hiking trails; 13b. Species richness differences in resting areas and their environs (whiskers represents 95% confidence intervals).

6.2.3 Branching, parallel trails and shortcuts of trails

When the surface of the trail is poorly treated, wind, precipitation and frost begin to affect the trail very intensively. As a result of intense erosive processes, the trail becomes

Figure 14.
Parallel trails in the Predné Kopské sedlo – the High Tatras (Piscová, 2011).

impassable for tourists and inappropriate also from the point of view of tourist safety, it constitutes an obstacle for tourists to be bypassed. This is how parallel replacement trails are created. In the studied area, the trails are accompanied by a number of parallel trails and shortcuts. The term parallel trail was originally used for a trail trodden by animals (e.g. chamois, cattle, etc.). Parallel trails were also created by trampling the vegetation cover when tourists walked off the trail. Parallel trails arise more often on small slopes, mainly in the grassland type of landscape, which allows tourists to deviate from the trail (**Figure 14** [55]). A trail with shortcuts is created mainly in places where the trail leads from saddles into valleys and has sharp turns – serpentines. New shortcuts as secondary trails have been created by undisciplined tourists close to the place of the largest curvature (bend) of the hiking trail to shorten their journey. The shortcuts are connected perpendicularly to the trails. They go in the direction of the greatest slope, thus conditioning the formation of erosion (**Figure 15**) [55].

6.2.4 Localisation of the hiking trail and morphodynamic processes threatening hiking trails

The location of a hiking trail should be in accordance with, or in the least contradiction with, the natural conditions of the territory. Otherwise, there is devastation not only of the trail itself, but also of its surroundings and the process of their regeneration is slow (**Figures 16** and **17**).

The subsoil in the studied area is susceptible to various forms of destruction by exogenous processes. Therefore, it is suitable only for a slightly concentrated, soft form of tourism.

Figure 15.
Shortcut on the hiking trail under the Vyšné Kopské sedlo – the Belianske Tatras (Piscová, 2011).

Figure 16.
Predné Meďodoly. Fading hiking trail, closed since 1978 (Piscová, 2010).

Figure 17.
Trail to Ždiarska vidla (2142 m MSL), closed since 1978 (Piscová, 2010).

Routes of hiking trails on the south-western and south-eastern slopes of the Belianske Tatras are threatened by many processes [99] associated with avalanches, nivation, surface runoff, coming off of soil and weathered cover, debris flows, rock

(a)

(b)

Figure 18.
Massive snow deposits covered the hiking trail after the avalanche event in April 2009 under the Ždiarska vidla (Hreško, 2009).

rushing and landslides (**Figure 18a, b**). Processes that arise in the immediate vicinity of the trails and are bound to their course – processes of nivation (dominant processes), erosive processes, gravitational descent of the weathered cover, slope-gravitational processes of the type of shallow landslides, eolithic processes in saddle and ridge positions.

Avalanches represent the process of movement of snow masses on mountain slopes, in avalanche troughs and in juvenile valleys. A special case of snow avalanches are "gliding avalanches", in which the soil-vegetation cover and the subsoil (including the trail) are destroyed almost in the entire width of the snow mass movement (**Figure 19**). An avalanche with a massive snow mass with a thickness of more than 2 meters and torn off fragments of rock substrate destroys vegetation and grinds the soil and weathered cover, which is documented by parallel grooves in the direction of the avalanche movement.

Another process endangering the trails associated with intense rainfall is debris flows. The limestone-dolomite subsoil of the exposed ridge of the Belianske Tatras provides a large amount of weathered, fragmentary material that forms the substantial mass of the flows. The most significant activation of debris streams occurred in 2016 (**Figure 20a, b**) at a precipitation intensity of more than 45 mm/h, which was confirmed by the analysis of the SHMI radar image.

The erosive effects of running water and concentrated surface runoff are mainly associated with torrential rains, which are involved in the fluviation of alpine ecosystems. Trails are in some places intersected by erosive grooves, which are formed in the gutters and on the bottoms of the juvenile valleys. Most often, the erosive effect is also manifested on the trails that represent local erosive bases (**Figure 21**) for the flowing surface water. This leads to significant destruction of trails and their deepening into more or less stable rock subsoil (**Figure 22**).

Figure 19.
The trail to Vyšné Kopské sedlo after a gliding avalanche (Hreško, 18 June 2009).

(a)

(b)

Figure 20.
Activated debris flows on the SW slope under the Hlúpy vrch after an intense downpour on 24 June 2016 (Hreško, 6 July 2009).

Nivation is a specific process, which is related to the long-term effect of snow fields on the hydric regime of the concave parts of slopes, especially if they are interrupted by a notch of a hiking trail. In the conditions of the SW slopes of the Belianske Tatras, we confirmed the occurrence of expansion cracks on the surface of the trails which were covered with remnants of snowfields for a longer time (**Figure 23a, b**).

Figure 21.
Fresh erosive groove near the trail under the avalanche trough of the Ždiarska vidla (Hreško, 6 July 2009).

The slow melting of the snow led to an increased retention of melt water, which reduced the stability of the soil-weathering layer and subsequently formed longitudinal expansion cracks. Such sections of trails have a high potential for slow coming off and descent. This is manifested by the sudden bends and dents of the trails, including their surroundings.

Nivation-eolic effects are concentrated mainly in the sub-ridge, ridge, sub-peak and saddle positions. The eolic-deflationary effect is based on the turbulent, backward effect of the flowing wind, which causes blowing off and removal of finer amounts of soil cover or even the nivation of the exposed weathered cover in the upper parts of leeward slopes, especially on the edges of the ridges (**Figures 24** and **25**). The

Figure 22.
Intensive deep erosion of the hiking trail under Kopské sedlo after an intense downpour in June 2016 (Hreško, 14 September 2016).

accumulation effect of both processes is destructive and does not allow a more successful process of vegetation succession. The eolic-nivation pads in the upper parts of the extremely steep slopes are often the source area of the initial debris flows, which, due to the smaller number of debris and fragments, may not reach the bottom of the slope, i.e. they remain "hanging" on the slope. The manifestations of wind erosion, more precisely the deflation of fine soil and weathering particles, focus mainly on the saddle and ridge positions of the Tatras. Another form conditioned by wind corrosion are eolic niches – pads with removed soil horizon of various shapes. Their edges are lined with overhangs, reinforced by root systems, which are intensively undermined by wind-blown particles. In cases of intense precipitation and snow melting, there is also a systematic washing out of niches and receding edges. Favourable conditions for the application of eolic deflation and corrosion are provided mainly by colourful shales

Figure 23.
Expansion cracks on the hiking trail after melting of the snow field on the SW slope of the Belianske Tatras (Hreško, 2009).

Figure 24.
Exemplary locations of the effect of eolic deflation of soil-weathered cover in the vicinity of Kopské sedlo (1750 m MSL) (Hreško, 2009).

Figure 25.
Vyšné Kopské sedlo, 1933 m MSL (Hreško, 2009).

and marls in the area of the Western and Belianske Tatras. The existing knowledge on the occurrence of eolic-deflationary forms is incomplete.

The slope gravity processes were activated in the form of shallow landslides of the soil-weathered cover as a result of intense rainfall or melting of snowfields (**Figure 26a, b**). In 2009, we found a landslide of the soil-weathered cover together with the vegetation cover of shrubland on the trail route in a shallow concave depression. The lithological properties of the geological subsoil were also applied to the formation of the landslide. In the area of interest, poorly reinforced, Mesozoic, hydrothermal altered limestone, lime, marly limestone, marl and shale are particularly susceptible to the formation of landslides.

6.2.5 Bearing capacity of trails in terms of their abiotic properties

The bearing capacity of the territory for the given traffic was addressed by several authors [100–104]. We have chosen the following parameter to evaluate the abiotic properties of trails and rest areas: gradient of the hiking trail, gradient of relief of terrain, rock resistance, soil types, surface coverage of hiking trails and the possibility of leaving the trail, according to the methodology [105].

According to our research [106], the trail leading from the Veľké Biele pleso tarn (1615 m MSL) to the Kopské sedlo (1750 m MSL) was characterized by medium bearing capacity in the lower sections and high bearing capacity in the higher parts. Therefore, in the vicinity of the trail we recommend to observe the erosion of the trail itself, the emergence of turf overhangs, in the event of trail impassability to strengthen its surface, we also recommend to intensify the patrol of nature guards in order to more closely monitor the frequency of attendance of the trail (rest area) and the movement of tourists outside the tourist-accessible places. The same applies to rest areas on the trail.

Figure 26.
Retrospective development of the slope along the tourist trail (1660 m MSL) below Hlúpy vrch (2061 m MSL) (Hreško, 11 October 2005, 2 July 2016).

In terms of abiotic properties of the trail and its surroundings, we can evaluate the bearing capacity of the trail from Tatranská Javorina (1000 m MSL) to Kopské sedlo (1750 m MSL) in the lower sections as high, in the higher sections as medium and in the final part before entering the Kopské sedlo as low. While in the lower parts we recommend to monitor the erosion of the trail itself, the emergence of turf overhangs, in the case of impassability of the trail to strengthen its surface, we also recommend to intensify the patrol of the nature guards due to a closer monitoring of the frequency of attendance of the trail (rest area) and the movement of tourists outside the tourists-accessible places. In the section before entering the saddle it would be necessary to reduce the frequency of attendance of the trail (rest area) by 50%, i.e., to leave the trail open only in one direction. Due to the entire length of the trail, it is possible to keep the current attendance.

The trail leading from the Široké sedlo (1825 m MSL) to the Kopské sedlo (1750 m MSL) is characteristic in terms of abiotic properties of the area through which it extends, medium and high bearing capacity (medium over high). Therefore, in the vicinity of the trail we recommend to observe the erosion of the trail itself, the emergence of turf overhangs, in the event of trail impassability to strengthen its surface, we also recommend to intensify the patrol of nature guards in order to more closely monitor the frequency of attendance of the trail (rest area) and the movement of tourists outside the tourist-accessible places.

7. Conclusions

The High Tatras mountains represent a unique alpine landscape with which humans have been connected since the past. The Belianske Tatras represent the limestone part of the mountain range. With rare communities and many endemics and glacial relicts, they are among the rarest and most endangered mountains of Slovakia. The High Tatras with glacial relief on crystalline basement and specific climatic conditions, represent the most attractive area of the Slovak high mountains. Tatras are the smallest mountains in the world. Like any other mountain range, the Tatras were first recognized by man, later he harvested wood, mineral richness, used grasslands as alpine pastures. Over time, however, one realized the uniqueness, rarity and value of the alpine landscape and began to protect it. At present, this area is protected as a national park, the Tatras Biosphere Reserve, the Habitats Directive and the Birds Directive. In the study area, all activities except recreation, sport, tourism, research, education and the construction of the necessary infrastructure are excluded.

With the enactment of the Tatra National Park in 1949, there were changes in the use of the territory of the Tatras. Since the Middle Ages, most villages in the Tatras had not been satisfied with the use of their mountain pastures and began to practise seasonal grazing on Tatra grasslands. After the end of this most intensive activity in the studied area, there has been a secondary succession towards greater stability. However, this process is not yet complete. In general, it can be stated from the results of the research that the former pastures gradually overgrow with forest cover and that the shrubland enters higher parts. The regeneration process is slow, as the recovery of this species in mountainous conditions is more difficult. Since the enactment of the national park (1949), the study area has gradually regenerated, but until now it has been mainly dealing with the surface grubbing-up of shrubland and grazing.

However, adverse changes in the studied area occurred even after the national park was enacted. They consist of visual disturbances, symptoms, as negative signs in the form of erosive manifestations that arose as a result of mass tourism on the ridge parts of the Belianske Tatras. The protection of the territory ensured the exclusion of those activities. Although there has been no development of tourism and infrastructure construction in the area, the Belianske Tatras ridge trail has been closed since 1978, one of the trails leading through the Monkova Valley has been open since 1993 in one way, and since 2008 again in two ways. The closed ridge trail regenerates here very slowly despite the long period of time. Most of the trails in the Belianske Tatras were inappropriately founded. In addition, the practice of guarding nature reveals numerous non-compliance with the entry ban. The most devastated parts of the trails here include places on long straight horizontal to diagonal traverses approximately in the middle parts of steep grass-herbal slopes, where disruption of the vegetation and soil-weathering layer created a local erosion base for water-gravity processes. All hiking trails in the area are endangered by many morphodynamic processes and the devastated parts of the trails form obstacles for tourists.

Tourist attendance has a number of direct and indirect impacts on the natural environment [107], so its monitoring is important. The situation in Slovakia is relatively non-specific in this case compared to other countries, as the systematic collection of tattendance data is missing. In most cases, the application of direct methods within Slovakia was associated only with the implementation of specific short-term projects [108] and so far there is no year-round continuous automatic monitoring of attendance, which cannot fully replace even several-day annual manual counting of visitors in the summer season in some areas of the Tatra Mountains [55].

From the point of view of sustainable development, it would make sense for the tourist closure of the ridge trail of the Belianske Tatras to continue with an appeal for preservation for future generations. Since uncontrolled tourism in a forbidden environment is more dangerous than a limited variant of ecotourism, the authors of the study [57] have reached the possibility of managing soft tourism in the territory, with supervision in the form of a guide. We recommend to monitor attendance and damaged habitats in the territory. A significant shortfall in mass tourism still remains minimal work with visitors to raise their environmental awareness as a means of reducing negative impacts on nature. What contributes to this is the lack of experience of municipalities, owners and administrators of protected areas in the regulation of tourism, as well as the lack of cooperation between the administrators of protected areas and tourism representatives.

Acknowledgements

This research was supported by Scientific Grant Agency of the Ministry of Education, Science, Research and Sport of the Slovak Republic and the Slovak Academy of Sciences, grants number VEGA 2/0018/19 Ecological Analyses of Landscape Acculturation in Slovakia since Early Prehistory until Today and VEGA 1/0546/21 Landscape Changes in the High Tatras Lake Basins, and with support from Project APVV-20-0108 Implementation of Agenda 2030 through biosphere reserves.

Author details

Veronika Piscová[1*], Juraj Hreško[2], Michal Ševčík[2] and Terézia Slobodová[2]

1 Institute of Landscape Ecology of Slovak Academy of Sciences, Nitra, Slovakia

2 Department of Ecology and Environmental Sciences, Constantine the Philosopher University in Nitra, Nitra, Slovakia

*Address all correspondence to: veronika.piscova@savba.sk

References

[1] Mallory GL. The reconnaissance of the mountain. In: Howard-Bury CK, editor. Mount Everest: The reconnaissance. London: Edward Arnold & CO; 1921. pp. 214-215

[2] Barthlott W, Lauer W, Placke A. Global distribution of species diversity in vascular plants: Towards a world map of phytodiversity. Erdkunde. 1996;**50**:317-327

[3] Messerli B, Ives JD, editors. Mountains of the world: A global priority. New York and Carnforth: Parthenon Publishing; 1997. p. 495

[4] Viviroli D, Weingartner R, Messerli B. Assessing the hydrological significance of the world's mountains. Mountain Research and Development Journal. 2003;**23**:32-40

[5] Körner C. Alpine ecosystems. In: LEVIN SA, editor. Encyclopedia of biodiversity. Academic Press: Waltham MA; 2013. pp. 148-157

[6] Bonnier G, Flahault C. Sur les modifications des végétaux. Annales des Sciences Naturelles. 1878;7:93-125

[7] Körner C. Alpine plant diversity: A global survey and functional interpretations. In: Chapin FS, Körner C, editors. Arctic and alpine biodiversity: Patterns, causes and ecosystem consequences: Ecological Studies. Berlin, Heidelberg, New York: Springer; 1995. pp. 45-62. DOI: 10.1007/978-3-642-78966-3_4

[8] Troll C. Klima und Pflanzenkleid der Erde in dreidimensionaler Sicht. Die Naturwissenschaften. 1961;**9**:332-348

[9] Körner C. Alpine plant life – Functional plant ecology of high mountain ecosystems. Heidelberg: Springer; 2003. p. 344. DOI: 10.1659/mrd.mm265.1

[10] Hreško, J. High-mountain landscape. Vysokohorská krajina [Internet]. 2011. Available from: http://147.213.211.222/sites/default/files/2011_2_057_058_editorial.pdf [Accessed: February 13, 2016].

[11] Dítě, D. Vegetation of the Tatra National Park 2: floristic characteristics of the territory. Rastlinstvo Tatranského národného parku 2: floristická charakteristika územia [Internet]. 2011. Available from: http://botany.cz/cs/tatry-2/ [Accessed: January 27, 2016].

[12] National biodiversity strategy of Slovakia [Internet]. 1997. Available from: https://www.cbd.int/doc/world/sk/sknbsap-01-p1-en.pdf [Accessed: February 12, 2016].

[13] Grivalský M. The use of the Belianske Tatras in the course of historical development [thesis]. Využívanie Belianskych Tatier v priebehu historického vývoja [diplomová práca]. Nitra, Slovakia: Constantine the Philosopher University in Nitra: Faculty of Natural Sciences; 2011

[14] Švajda J, Sabo P. Management of protected areas. Manažment chránených území. Banská Bystrica: Belanium; 2013. p. 128

[15] MEA. Ecosystems and human well-being: synthesis. Ekosystémy a lidský blahobyt: syntéza. Millennium ecosystem assessment. Prague: World Resource Institute, Czech ed. Centre for Environmental Issues: Charles University in Prague; 2005. p. 138

[16] CBD. About the Convention on Biological Diversity [Internet]. 2013.

Available from: http://www.cbd.int/convention [Accessed: September 30, 2013]

[17] Balsiger J, Debarbieux B. Should mountains (really) matter in science and policy? Environmental Science & Policy. 2015;**49**:1-7

[18] Karpaty.net. Carpathians. Karpaty [Internet]. 2022. Available from: https//: www.karpaty.net [Accessed: September 30, 2013]

[19] Mazúr E. Morphostructures of the Western Carpathians and their development. Morfoštruktúry Západných Karpát a ich vývoj. Acta Facultatis Naturalium Universitatis Comenianae. Geographical. 1979;**17**:21-34

[20] Nemčíková, M. Special nature and landscape protection in Slovakia. Osobitná ochrana prírody a krajiny na Slovensku [Internet]. 2008. Available from: http://dam.fpv.ukf.sk/mod/resource/view.php?inpopup=true&id=472 [Accessed: March 13, 2008]

[21] Collective of authors. Tatras - Nature. Tatry – Príroda. Prague: Publishing House M. Uhlíř - Baset; 2010. p. 637

[22] Lacika J, Urbánek J. New morphostructural division of Slovakia. Slovak Geological Magazine. 1998;**4**(1):17-28

[23] Lapin M, Faško P, Melo M, Šťastný P, Tomlain J. Climate regions. Klimatické oblasti. Map number 27. In: Atlas of the Country of the Slovak Republic. Banská Bystrica: Ministry of the Environment of the Slovak Republic, Bratislava; Slovak Environmental Agency; 2002. p. 95

[24] Koreň, M., Linkeš, V., Bublinec, E. Characteristics of soils. Charakteristika pôd. In Vološčuk, I. et al.: Tatra National Park. Biosphere reserve. Gradus, Martin; 1994. p. 86-105

[25] SNR Act No. 11/1948 Coll. on the Tatra National Park

[26] Act No. 543/2002 Coll. on nature and landscape protection.

[27] Council Directive of the European Communities No. 79/409/EEC on the conservation of wild birds (Birds Directive)

[28] Council Directive of the European Communities No. 92/43/EEC on the conservation of habitats and of wild fauna and flora (Habitats Directive)

[29] Decree of the Ministry of Environment of the Slovak Republic No. 3/2004-5.1 of 14 July 2004 issuing a national list of territories of European importance

[30] Decree of the Ministry of the Environment of the Slovak Republic No. No. 4/2011 Coll. which declares the Protected Bird Area of the Tatras

[31] Decree No 166/1991 Coll. Decree of the Slovak Commission for the Environment on state nature reserves and protected sites in the Tatra National Park

[32] Bertová L, editor. Flora of Slovakia IV/1. Flóra Slovenska IV/1. Bratislava: Veda; 1984. p. 432

[33] Futák J. Development of Tatra vegetation. Vývoj tatranského rastlinstva. In: Proceedings of works on the Tatra National Park. Martin: Osveta 1976;**17**:61-77

[34] Šibík J, Petrík A, Valachovič M, Dúbravcová, Z. Alpine vegetation of the class Carici rupestris Kobresietea bellardii Ohba 1974 in the Slovak part

of the Western Carpathians 2006. Vysokohorská vegetácia triedy Carici rupestrisKobresietea bellardii Ohba 1974 v slovenskej časti Západných Karpát [Internet]. 2006. Available from: https://ibotold.sav.sk/usr/Jozef/docs/SBS_2006_CK.pdf [Accessed: March 10, 2008]

[35] Bliss LC. Arctic and alpine plant life cycles. Annual Review in Ecology and Systems. 1971;**2**:405-438

[36] Billings WD. Adaptations and origins of alpine plants. Arctic and Alpine Research. 1974;**6**:129-142

[37] Cavieres LA, Badano EI, Sierra-Almeida A, Gómez-Gonzalez S, Molina-Montenegro MA. Positive interactions between alpine plant species and the nurse cushion plant Laretia acaulis do not increase with elevation in the Andes of central Chile. New Phytologist. 2006;**169**:59-69. DOI: 10.1111/j.1469-8137.2005.01573.x

[38] Cavieres LA, Badano EI, Sierra-Almeida A, Molina-Montenegro MA. Microclimatic Modifications of Cushion Plants and Their Consequences for Seedling Survival of Native and Non-native Herbaceous Species in the High Andes of Central Chile. Arctic, Antarctic, and Alpine Research. 2007;**39**:229-236

[39] Hager JA, Faggi AM. Observaciones sobre distribución y microclima de cojines enanos de la Isla de Creta y del noroeste de la Patagonia. Parodian. 1990;**6**:109-127

[40] Núñez C, Aizen M, Ezcurra C. Species associations and nurse plant effect in patches of high Andean vegetation. Journal of Vegetation Science. 1999;**10**:357-364

[41] Molina-Montenegro MA, Torres C, Parra MJ, Cavieres LA. Asociación de especies al cojín Azorella trifurcata (Gaertn.) Hook. (Apiaceae) en la zona andina de Chile central (37°S). Gayana Botanica. 2000;**57**:161-168

[42] Badano EI, Molina-Montenegro MA, Quiroz C, Cavieres LA. Efectos de la planta en cojín Oreopolus glacialis (Rubiaceae) sobre la riqueza y diversidad de especies en una comunidad alto-andina de Chile central. Revista Chilena de Historia Natural. 2002;**75**:757-765

[43] Arroyo MTK, Cavieres LA, Peñaloza A, Arroyo-Kalin MA. Positive interactions between the cushion plant Azorella monantha (Apiaceae) and alpine plant species in the Chilean Patagonian Andes. Plant Ecology. 2003;**169**:121-129

[44] Cavieres LA, Peñaloza A, Papic C, Tambutti M. Efecto nodriza del cojín Laretia acaulis (Umbelliferae) en la zona alto-andina de Chile Central. Revista Chilena de Historia Natural. 1998;**71**:337-347

[45] Cavieres LA, Arroyo MTK, Peñaloza A, Molina-Montenegro MA, Torres C. Nurse effect of Bolax gummifera cushion plants in the alpine vegetation of the Chilean Patagonian Andes. Journal of Vegetation Science. 2002;**13**:547-554

[46] Studano EI, Jones CG, Cavieres LA, Wright JP. Assessing impacts of ecosystem engineers on community organization: a general approach illustrated by effects of a high-Andean cushion plant. Oikos. 2006;**115**:369-385

[47] Ralph CP. Observations on Azorella compacta (Umbelliferae), a tropical Andean cushion plant. Biotropica. 1978;**10**(1):62-67

[48] Piscová V. Changes in the vegetation of the subalpine and alpine zones of the Tatras in selected locations affected by

man. Zmeny vegetácie subalpínskeho a alpínskeho stupňa Tatier na vybraných lokalitách ovplyvnených človekom [dissertation thesis]. Bratislava: Institute of Landscape Ecology SAS; 2009

[49] Klimentociation Živá planéta J, Valachovič M. (Ed.). Vegetation of Slovakia - Plant Communities of Slovakia, 4., Alpine Vegetation. Rastlinné spoločenstvá Slovenska, 4. Vysokohorská vegetácia. Bratislava: VEDA; 2007. p. 388

[50] TANAP Management: Fauna of the Tatra National Park. Fauna Tatranského národného parku [Internet]. 2008. Available from: http://spravatanap.sk/web/index.php/2012-08-24-09-58-43/fauna-tatranskeho-narodneho-parku-sk [Accessed: March 8, 2008]

[51] Zelina V. Movement of the marmot population in the Belianske Tatras. Pohyb populácie svišťov v Belianskych Tatrách. In: Proceedings of works on the Tatra National Park 8. Martin: Osveta; 1965. pp. 173-189

[52] Bohuš, I. Use of natural resources. Využívanie prírodných zdrojov. In: Vološčuk, I. et al.: Tatra National Park. Biosphere reserve. Martin: Gradus; 1994. p. 252-374

[53] Harvan L. How the pasture in the TANAP was solved. Ako sa vyriešila pastva v TANAP. In: Proceedings of works on the Tatra National Park 8. Martin: Osveta; 1965. pp. 231-253

[54] Šmarda J. Natural regeneration of grassland in Predné Meďodoly in the Belanské Tatras. Prirodzená regenerácia trávnych porastov v Predných Meďodoloch v Belanských Tatrách. In: Proceedings of works on the Tatra National Park 1. Martin: Osveta; 1957. pp. 57-62

[55] Piscová V, Hrnčiarová T, Hreško J, Dobrovodská M, Izakovičová Z, Izsóff M, et al. Use of the alpine landscape and its impact on changes in the environment (case study of Tatras and Low Tatras). Využívanie vysokohorskej krajiny a jeho dôsledky na zmenu prostredia (na príklade Tatier a Nízkych Tatier). Bratislava: Veda; 2018. p. 250

[56] Act 287/1994 Coll. of 23 August 1994 on nature and landscape protection.

[57] Sabo P et al. Let us save the high mountains of Slovakia. In: Zachráňme vysoké hory Slovenska. Piešťany: Civic Association Živá planéta; 2002. p. 132

[58] Midriak R. Morphogenesis of the surface of high mountains. Morfogenéza povrchu vysokých pohorí. Bratislava: Veda; 1983. p 516

[59] Bohuš I. Tatra valleys in reflection of time. Tatranské doliny v zrkadlení času. Tatranská Lomnica: Publishing House I & B; 2005. p 143

[60] Bohuš I. Who was who in the High Tatras. Kto bol kto vo Vysokých Tatrách. Vysoké Tatry. 1974;**6**(8):30

[61] Vysoketatry.sk: History. História [Internet]. 2022. Available from: https://www.vt.sk/hory/region/historia [Accessed: March 13, 2022]

[62] Pavlarčík S. Military monuments in the Tatra National Park – Bunker under the Hlúpy vrch. Vojenské pamiatky v Tatranskom národnom parku – Bunkre pod Hlúpym vrchom. Tatry. 2007;**46**(2):16-17

[63] Pavlarčík, S. Caverns of high mountain firing position under the Hlúpy vrch in the Belianske Tatras. Kaverny vysokohorského palebného postavenia pod Hl úpym vrchom v Belianskych Tatrách. Newsletter of the Slovak Speleological Society. Liptovský Mikuláš: Slovak Speleological Society; 1996;**27**(2):14-16

[64] Bohuš I. Tatra huts: Lighthouses in a sea of rocks and snow. Tatranské chaty: Majáky v mori skál a snehu. Tatranská Lomnica: Publishing House I & B; 2007. p. 140

[65] Kezmarskachata.sk: History. História [Internet]. 2022. Available from: https://www.kezmarskachata.sk/historia [Accessed: March 10, 2022]

[66] Spitzkopf, P. In the 13th chamber of TANAP. V trinástej komnate TANAP-u. Tatry. Lomnica: Správa TANAP, Tatranská; 2007;**46**(4):10-12

[67] Turošík J et al. Why the closure of the Belianske Tatras? Prečo uzáver Belianskych Tatier? Vysoké Tatry. 1978;**17**(4):2-3

[68] Paclová L, Pacl J. Has enhanced protection helped the Belianske Tatras? Pomohla sprísnená ochrana Belianskym Tatrám? Vysoké Tatry. 1982;**21**(6):9

[69] Kováčiková M. For the eyes of tourists, the beauty of the Belianske Tatras remains taboo – Human, the undesirable creature. Pre oči turistov zostáva krása Belianskych Tatier naďalej tabu – Človek, tvor nežiadúci. Vysoké Tatry. 1991;**30**(6):2-3

[70] SSR Government Ordinance No 12/1987 Coll. of 6 February 1987

[71] Boltižiar M. Structure of High Mountain Landscape of Tatra Mountains. In: Largescale Mapping, Analysis and Evaluation of Changes by Application of Data from Earth Remote Survey. Nitra, Slovakia: CPU, ILE, SAS; 2007

[72] Boltižiar M, Olah B. Land Use Changes of UNESCO Biosphere Reserves in the Slovak Carpathians Since the Late Eighteenth Century. In: Kozak J, editor. The Carpathians: Integrating Nature and Society Towards Sustainability, Environmental Science and Engineering. Berlin, Heidelberg: Springer; 2013. pp. 377-391. DOI: 10.1007/978-3-642-12725-0_27

[73] Solar J, Solar V. Land-cover change in the Tatra Mountains, with a particular focus on vegetation. Economics Monthly. 2020;**12**(1):15-26. DOI: 10.1553/eco.mont-12-1s15

[74] Švajda J, Solár J, Janiga M, Buliak M. Dwarf Pine Pinus mugo and Selected Abiotic Habitat Conditions in the Western Tatra Mountains. Mountain Research and Development. 2011;**31**(3):220-228. DOI: 10.1659/MRD-JOURNAL-D-09-00032.1

[75] Solár J, Janiga M. Long-term Changes in Dwarf Pine (Pinus mugo) Cover in the High Tatra Mountains Slovakia. Mountain Research and Development. 2013;**33**(1):51-62. DOI: 10.1659/MRD JOURNALD-12-00079.1

[76] Plesník P. Implication of Human Influence in the Area of Timber Line and Above in Tatra National Park Territory. Osveta, Martin: Treatises Concerning the Tatra National Park; 1078;**20**:67-91

[77] Grabherr G, Gottfried M, Pauli H. A Global Observation Research Initiative in Alpine Environments. Mountain Research and Development. 2000;**20**(2):190-191

[78] Steinbauer MJ, Grytnes JA, Jurasinski G, Kulonen A, Lenoir J, Pauli H, et al. Accelerated Increase in Plant Species Richness on Mountain Summits is Linked to Warming. Nature. 2018;**556**:231-234. DOI: 10.1038/s41586-018-0005-6

[79] Ellenberg H, Weber HE, Düll R, Wirth V, Werner W, Paulißen D. Zeigerwerte von Pflanzen in Mitteleuropa. Scr. Geobot. 1991;**18**:1-248

[80] Diekmann M. Species indicator values as an important tool in applied plant ecology – a review. Basic Applied Ecology. 2003;**4**:493-506

[81] Smart M, Tichý L, Dřevojan P, Lard J, Zelený D. Ellenberg-type indicator values for the Czech flora. Preslia. 2018;**90**: 83-103

[82] Eurostat. Trips to mountains by EU residents. Available from: https://ec.europa.eu/eurostat/en/web/products-eurostat-news/-/edn-20181211-1 [Accessed: March 14, 2008]

[83] Tatry.sk. Traffic statistics [Internet]. 2008. Available from: https://www.tatry.sk/infocentrum/dolezite-informacie/statistiky-navstevnosti [Accessed: March 14, 2008]

[84] Cole DN, Bayfield NG. Recretional trampling of vegetation: standard experimental procedures. Biological Conservation. 1993;**63**:209-215

[85] Gomez-Limon FJ, De Lucio JV. Recreational activities and loss of diversity in grasslands in Alta-Manzanares-Natural-Park, Spain. Biological Conservation. 1995;**74**:99-105

[86] Willard BE, Cooper DJ, Forbes BC. Natural regeneration of alpine tundra vegetation after human trampling: a 42-year data set from Rocky Mountain National Park, Colorado, USA. Arctic Antarctic and Alpine Research. 2007;**39**:177-183

[87] Forsberg O. Changes over time in remnant rural vegetation within built-up areas – recreational use, vegetation dynamics and conservation assessment. In: Acta Universitatis Agriculturale Sueciae [thesis]. Uppsala: Swedish University of Agricultural Sciences; 2010

[88] Bates GH. The vegetation of footpaths, sidewalks, cart-tracks and gateways. Journal of Ecology. 1935;**23**:470-487

[89] Roovers P, Gulinck H, Hermy M. Experimental assessment of initial revegetation on abandoned paths in temperate deciduous forest. Applied Vegetation Science. 2005;**8**:139-148

[90] Kozlowski TT. Soil compaction and growth of woody plants. Scandinavian Journal of Forest Research. 1999;**14**:596-619

[91] Breland TA, Hansen S. Nitrogen mineralization and microbial biomass as affected by soil compaction. Soil Biology & Biochemistry. 1996;**28**:655-663

[92] Cole DN. Effects of three seasons of experimental trampling on five montane forest communities and a grassland in Western Montana, USA. Biological Conservation. 1987;**40**:219-244

[93] Lei SA. Soil compaction from human trampling, biking, and off-roadmotor vehicle activity in a blackbrush (Coleogyne ramosissima) shrubland. Western North American Naturalist. 2004;**64**:125-130

[94] Amrein D, Rusterholz HP, Baur B. Disturbance of suburban Fagus forests by recreational activities: effects on soil characteristics, above-ground vegetation and seed bank. Applied Vegetation Science. 2005;**8**:175-182

[95] Kissling M, Hegetschweiler KT, Rusterhoz HP, Baur B. Short-term and long-term effects of human trampling on above-ground vegetation, soil density, soil organic matter and soil microbial processes in suburban beech forests. Applied Soil Ecology. 2009;**42**:303-314

[96] McIntyre S, Lavorel S, Tremont RM. Plant life-history attributes –their relationship to disturbance responses in herbaceous vegetation. Journal of Ecology. 1995;**83**:31-44

[97] Piscová V, Ševčík M, Hreško J, Petrovič F. Effects of a short-term trampling experiment on alpine vegetation in the Tatras, Slovakia. Sustainability. 2021;**13**(5):2750. DOI: 10.3390/su13052750

[98] Mšák L et al. Phytoindication of the tourist bearing capacity of the Tatra National Park. Fytoindikácia turistickej únosnosti Tatranského národného parku. In: Proceedings of works on the Tatra National Park. Osveta, Martin. 1990;**30**:123, 161

[99] Hreško J, Bugár G, Petrovič F. Changes of vegetation and soil cover in alpine zone due to anthropogenic and geomorphological processes. Landform Analysis. 2009;**10**:39-43

[100] Papánek F. Anthropic activity in the national park in terms of recreation and protection of nature and landscape. Antropické pôsobenie v národnom parku z hľadiska rekreácie a ochrany prírody a krajiny. Proceedings of works on the Tatra National Park. 1978;**20**:51-66

[101] Šoltésová A. Study of anthropic influences on the vegetation of the model territory of Tatranská Lomnica. Štúdium antropických vplyvov na vegetáciu modelového územia Tatranská Lomnica. In: Final Report of the State Research Task. Martin: Osveta; 1980

[102] Šoltés R. Anthropic influences on the bryoflora of the Tatranská Lomnica model territory. Antropické vplyvy na bryoflóru modelového územia Tatranská Lomnica. Proceedingsof works on the Taransky National Park. 1982;**23**:197-121

[103] Root M. Why is the trail on the ridge of the Belianske Tatras closed? Prečo je uzavretý chodník na hrebeni Belianskych Tatier? Tatras, State forests of TANAP, Tatranská Lomnica. 1995;**5**:2-3

[104] Barančoková M. Geoecological evaluation of the Belianske Tatras. Geoekologické hodnotenie Belianskych Tatier [dissertation thesis]. Bratislava: ILE SAS; 2006

[105] Hrnčiarová T, Altmanová M. Ecological evaluation of hiking trails in the central part of the Low Tatras. Ekologické hodnotenie turistických chodníkov v centrálnej časti Nízkych Tatier. In: Competitive work of young scientists. Bratislava: Institute of Experimental Biology and Ecology SAS; 1982

[106] Piscová V. Changes in the vegetation of the Tatras in selected locations affected by man. Zmeny vegetácie Tatier na vybraných lokalitách ovplyvnených človekom. Bratislava: VEDA – publishing SAS; 2011. p. 300

[107] Švajda J, Korony S, Brighton I, Esser S, Ciapala S. Trail impact monitoring in Rocky Mountain National Park, USA. Solid Earth. 2015;**7**:115-128

[108] Urban P, Škodaová M, Mezei A, Saxa A, Švajda J. Case studies on nature and landscape conservation and care of natural resources. Prípadové štúdie z ochrany prírody a krajiny a starostlivosti o prírodné zdroje. Banská Bystrica: FNS UMB in Banská Bystrica; 2015

Chapter 14

Phytoecological Study, Ethnobotanical and Dynamic of Dry Vegetation in the Ngazidja Island, Comoros Archipelago

Abdillahi Maoulida Mohamed, Frédéric Bioret and Vicent Boullet

Abstract

The dry vegetation of Ngazidja, Comoros Islands, is very rich in biological diversity. However, it is the most threatened and least known in this island. The purpose of this study is to provide ecological information about dry vegetation and to elaborate proposals for conservation. Data collection is ecological surveys, and ethnobotanical and socioeconomic surveys. Several floristic, and structural and soil parameters were processed and analyzed to characterize the sites. Five types of dry vegetal formations are identified: the dry forests of Lac-salé, Ngouni-Hamboda and Infoundihé-Chambouani, and the dry thickets of Singani, Domoni-Amboini, Hahaya. Eighteen surveys of 10 transects were undertaken. The forests are well stratified and the trees do not exceed 15 m high, with deciduous leaves. The density varies from 500 to 800 individuals per hectare in forest against 310 to 165 in thicket. A total of 103 species, belonging to 46 families, were recorded in these formations, including 70 trees and shrubs. Natural regeneration is only good for thickets and the dry forests of Lac salé and Infoundihé-Chambouani. Sustainable alternative solutions are proposed in order to improve the socioeconomic living conditions of the population and the conservation of biodiversity.

Keywords: dry vegetation dynamic, socio-economic, conservation biodiversity of ecosystems, Ngazidja, Comoros

1. Introduction

The recent geological nature of the Archipelago of the Comoros, their geographical position, the smallness of the territory and their multi-insularity give them a great originality that reflects the diversity of the landscapes and the richness of the vegetation, the fauna and the flora [1–3]. The flora of the Comoros has a great similarity with that of Madagascar. The archipelago is under a small influence of the African continent, while some elements of the flora could come from Asia [4]. The vegetation of the Comoros Islands is known for its richness in endemic plant species; however, it remains little studied. It is characterized by two types of vegetation: medium-high and high-altitude moist forests, and low-lying, deciduous, lowland forests [2, 5–7]. The proportion of land covered by forests would decrease from 6.6% in 1990 to 1.7% in 2012, with nearly 500 ha of forest being lost each year [8]. In Ngazidja, forest cover is estimated at 8658 ha of natural forest [9]. The dry forest and thicket are found along the South-East and North slopes as well as the West Coast. Remnants of poorly disturbed dry forests are reduced to small areas of a few hectares regularly grazed by livestock [7]. In addition to these threats, climate change does not spare them. Forest species in arid and semi-arid areas have reached their limit of adaptation to aridity and, therefore, are more vulnerable to increased drought than wetland forest species [10]. In 2006, Labat already said is urgent to identify these remnants of dry forests and thickets, because these types of vegetation are the most threatened. To our

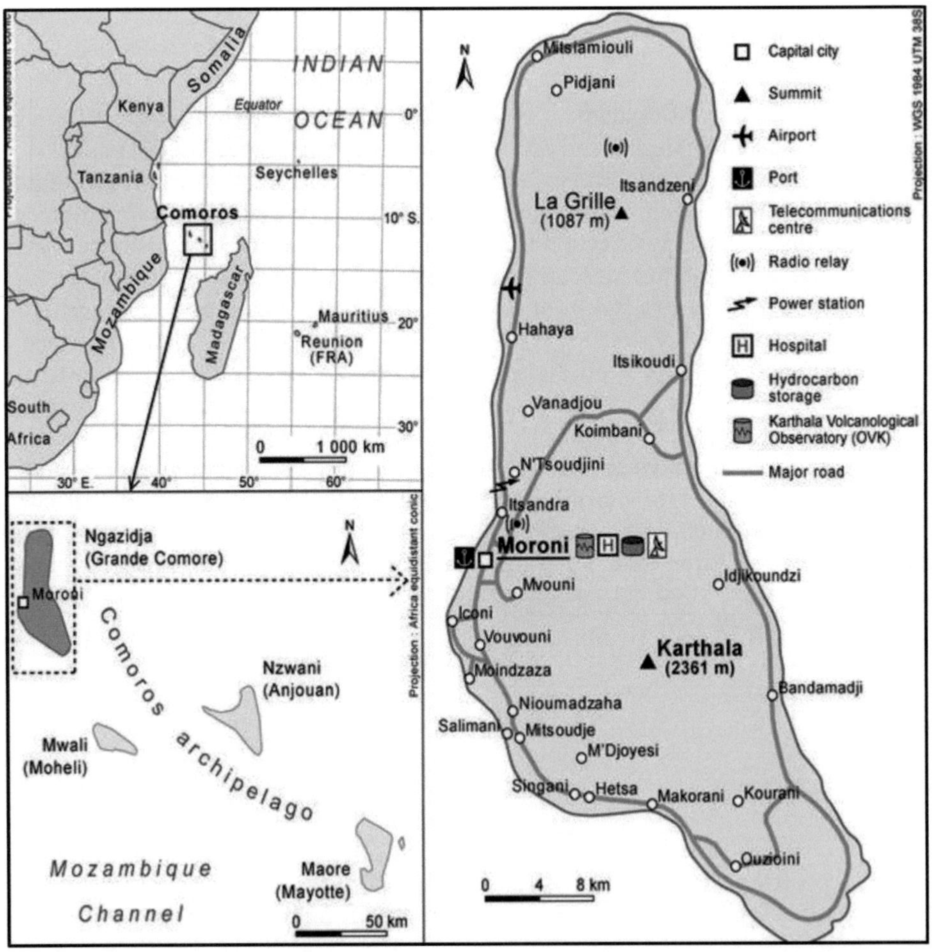

Figure 1.
Location of Ngazidja island, Comoros archipelago.

knowledge, no study has been undertaken on the ecological characteristics of the dry plant formations on Ngazidja, Comoros Archipelago (**Figure 1**). The present study aim at providing ecological information on the dry vegetation formations of Ngazidja Island and propose solutions for their conservation. The hypotheses are: (i) the dry plant formations are still very rich in endemic and indigenous species and are still threatened; (ii) These knowledge should facilitate the definition of ways of managing and conserving the natural and semi-natural habitats of dryland species, especially forests and thickets in a context of insularity but also to demographic and urban expansion socioeconomic dynamics. Our approach is based on the survey of the floristic and structural description, and the determination of threats and pressures.

2. Materials and methods

2.1 Identification of types of plant formations, choice and location of study sites

Identification of types of plant formations, choice and location of study sites. Surveys of the study areas were carried out between June and September 2019 and between January and May 2020. It consiste of locating the main well preserved dry vegetation allowing to have the maximum number of endemic or native species. The choice of study sites was based on bibliographic information, stratification of various maps and direct field observations. Some criteria were taken into consideration (**Figure 2**):

- the type of formation present in the site and its state of degradation;
- the accessibility of the site and the representativeness of the site formation.
- Among the visited sites, five types of dry vegetal formations were selected:
- dry forest of Lac-salé ("salt lake") is limited in the north-east by the village of Ivoini and in the south-east by the town of Bangoua-kouni;
- dry forest of Ngouni-Hamboda is limited in the north-east by the city of Moroni and in the south-east by the city of Ikoni;
- dry forest of Infoundihé-Chambouani is located in the south-west. It is bounded in the north-west by Infoundihé-Chambouani village and in the southwest by Dzahadjou and by the National Road towards Sima-Ambouani village;
- dry thicket of Singani is located in the south-west of this locality.
- dry thicket Domoni-Amboini-Hahaya is limited in the north-east by the locality of Ntsaouéni and in the south-east by the town of Hahaya.

Other very degraded dry plant formations (forests and thickets) were also visited in order to maximize the number of plant species.

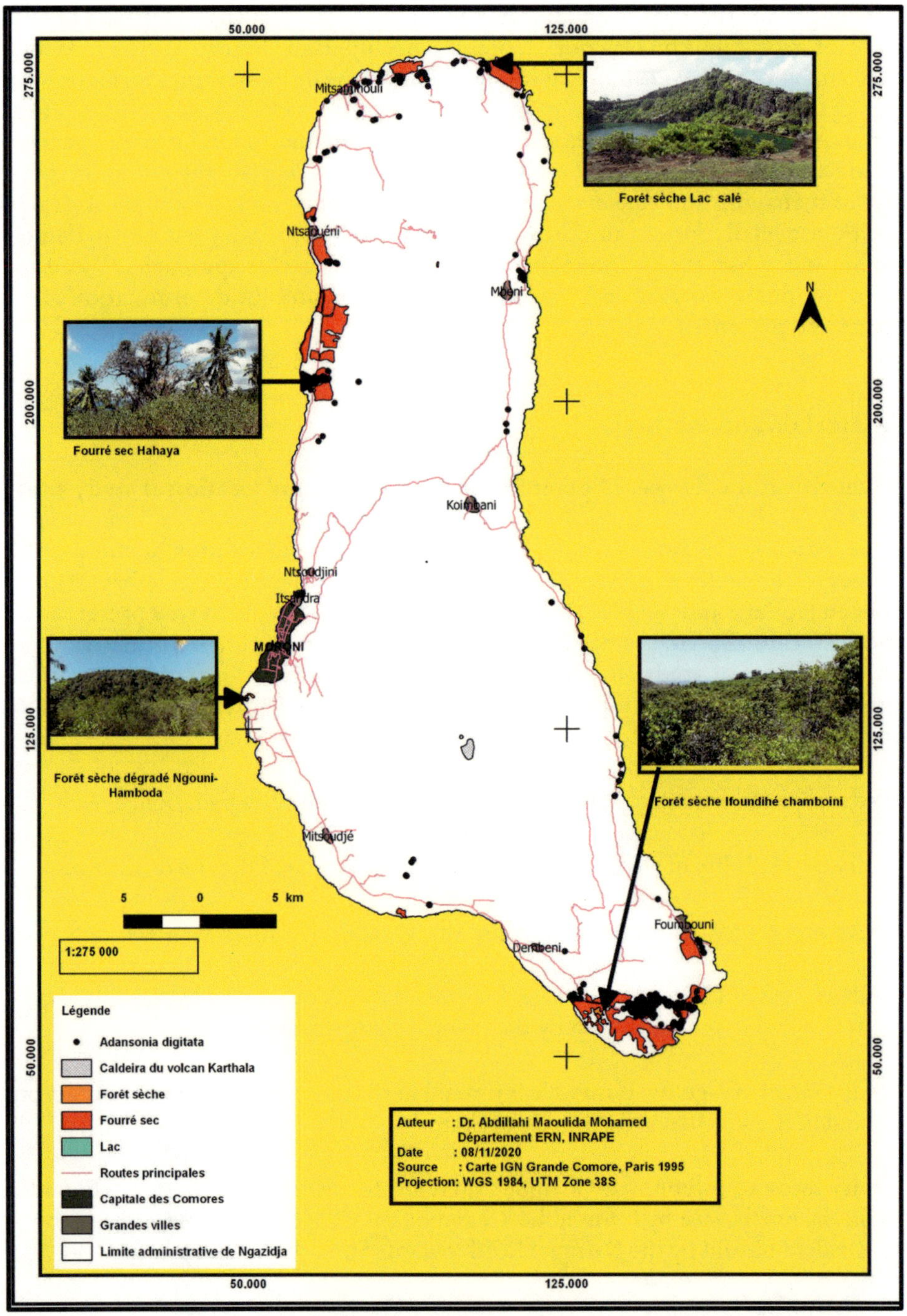

Figure 2.
Location of study sites and dry plant formations and the dry vegetation formations in Ngazidja Island.

2.2 Collection of data

The Phytoecological study was used. Surveys were conducted between October 2019 and January 2020 and between April and July 2020 to maximize the chances of harvesting easily identifiable, fertile specimens. The sampling was based on:

- the placeau of the Braun-Blanquet method, used to highlight the floristic diversity and the biological spectrum, according to [11] and based on the homogeneity criteria of [12];

- Gautier's linear survey [13] was used to describe the vertical structure and natural regeneration. For the horizontal structure, density was analyzed using the Quadrat Center to one point or QCP method [14];

Eighteen (18) surveys and 10 transects following a North–South gradient, perpendicular to the sea, were delineated. Part of species was identified onsite, another part was identified at the INRAPE herbarium and the rest with the flora of Madagascar and Comoros [15] with the databases samples of the herbarium of the Museum of Natural History of Paris [16] and the Herbarium of Mayotte on the basis of the National Botanical Conservatory of Mascaraing.

Species density was assessed by counting trees with a diameter at breast height greater than 10 cm. This DBH is established from the measurement of the circumference using a tape measure. The size of the trees was evaluated by the total height (Ht), measured using a clisimeter.

Physiomomic analysis, floristic diversity and similarity: Quantitative vegetation parameters (density, basal area, regeneration, species richness, biological spectra, maximum tree height) were used to characterize the types of vegetation.

For the study of natural regeneration, the Placeau method was used and consists in measuring the diameter of all the individuals of each species present in a 1 ha plot. Those iIndividuals are classified into two categories: regenerated species (DHP <10 cm or Nr) and seed species (DHP ≥ 10 cm or Ns). The regeneration potential is expressed by the number of regenerated individuals over the number of seed-bearing individuals, expressed by the regeneration rate (TR): TR (%) = (Nr/NS) X 100

- if the regeneration rate is less than 100%, the species is in difficulty of regeneration;

- if the regeneration rate is between 100% and 1000%, the regeneration is good;

- if this rate is greater than 1000%, the regeneration is very good.

Threat assessment on dry vegetation formations: Ethnobotanical surveys were conducted among local populations, 200 individus, in 14 localities bordering the study sites: Singani, Dzhadjou, Sima-Ambouani, Ifoundihé-chambouani, Ikoni, Moroni, Hahaya, Domoni-Ambouani, Domoni-Adjou, Ntsaouéni, Bangoua-kouni, Ouella, Ivoini and Ouémani. They understand the interactions of local communities with plant resources (different uses) as well as the threats that weigh on the habitats. Observations were also made in the field. The formula of [17] was used to evaluate the species utilization index in order to know the species that were widely used. The people concerned above all are those who appear to have a direct or indirect impact on dry forest or thicket: lumberjacks, wood traders, collectors of medicinal plants, peasants, carpenters, carpenters, farmers, head of non-governmental associations. These surveys were carried out collectively or individually, in the form of semi-direct interviews with semi-open, open or sometimes closed questions asked to men (65%) and women (35%) between 18 and 70 years old. The information is collected using a dictaphone.

3. Result

3.1 Types of dry vegetation formations identified

The types of dry vegetation formations, the natural and semi-natural identified during this study in Ngazidja island are forests, thickets and savannas (**Figure 2**):

- The dry forests were located in Ouemani-Mitsamihouli and on the ridges of the Lac-salé "salt lake" in Bangoi-kouni;
- In the region of Mbadjini (south-east) around Ikoni Hill (south-center);
- The thickets (shrubby or bushy) were located in the coastal part of the three islands, especially in the north-west part (Hahaya), in far north (Bangoi-kouni to Ivoini) and in north-east;
- Savannahs are foundi the eastern side of Karthala, in the Diboini Plateau and along all the coastal parts of this island (but at different stages of evolution according to their degradation). Our study concerned the ecological characterization of dry forests and thickets in Ngazidja: dry forest and thicket.

3.2 Physiognomic characteristics

Dry forests of Lac-salé and Ngouni-Hamboda (**Figure 3**) presente 3 strata, with some emergents with *Adansonia digitata* and Albizia sp. at a height of up to 20 m:

- A continuous lower stratum, with a recovery rate of 50% and an average height between 1 and 3 m. This stratum consisted in numerous grasses and shrubs such as *Acalypha indica* (Euphorbiaceae), *Oeceoclades lonchophylla* (Orchidaceae) (specie that is becoming increasingly rare in the north of the island), *Aloe alexandrei* and seedlings or young plants of the upper strata species *Erythroxylum platyclada, E. lanceum (Erythroxylaceae), Diospyros comorensis* (Ebenaceae), Ochna ciliata (Ochnaceae) and *Turraea wakefieldii* (Meliaceae).

Figure 3.
Dry forest of Hamboda with a mesophilic tendency.

- An average open shrub stratum with a recovery rate of 30 to 75% and a height of between 3 and 11 m. This stratum is composed of small trees such as *Terminala ulexoides* (Combretaceae), *Alangium salviifolium* (Alangiaceae), *Sorindeia madagascariensis* (Anacardiaceae), *Pyrostria bibracteata* (Rubiaceae), *Diospyros comorensis*, *Commiphora arafy* (Burseraceae) and *Phyllarthron comorense* (Bignoniaceae). This stratum was completely invaded by lianas such as, *Vanilla humblotii (Orchidaceae), Rhoicissus revoilii* and *Cissus quadrangularis* (Vitaceae).

- An upper stratum, moderately dense, with a recovery rate of 60%, varying in height from 10 to 15 m. The most abundant and dominant species in this stratum were *Ficus sycomorus* (Moraceae), *Albizia lebbeck* and *Albizia glaberima* (Mimosaceae) and *Drypetes thouarsii* et *D. comorensis* (Euphorbiaceae) (**Figure 4**) and *Ochna ciliata* (Ochnaceae). Various creepers such as *Rhoicissus revoilii* and *Saba comorensis* (Apocynaceae) were present in this stratum. The minimum area value of 300 m^2 corresponds to a cumulative number of species equal to 60.

In the Ifoundihé-Chambouani dry forest, three strata were identified with some emergents with *A. lebbeck* and *A. glaberima* and at a height of up to 15 m:

- A very open lower stratum with a recovery rate of 45% and whose height is less than 1 m. The species of this stratum are: *Erythroxylum lanceum* (Erythroxylaceae), numerous seedlings of *Ochna ciliata* (Ochnaceae) and numerous grasses such as *Tacca leontopetal*oides (Taccaceae) and *A. indica* (Euphorbiaceae).

- An average stratum with a recovery rate of 35% and a variable height between 2 and 4 m.

- The dominant species are *Sorindeia madagascariensis* (Anacardiaceae), *Turraea sericea* (Meliaceae) (**Figure 5**), *Terminalia ulexoides* (CombretaceaE), *Diospyros comorensis* and *Euclea racemosa* subsp. *schimperi* (Ebenaceae). This stratum is completely invaded by *Lantana camara*, *Saba comorensis, C. quadrangularis, Dioscorea comorensis* (Dioscoreaceae) and *Rhoicissus revoilii*.

Figure 4.
Drypetes comorensis (fruits).

Figure 5.
Turraea virens (flowers).

- An upper stratum of 6 to 12 m, discontinuous mainly composed of *Operculicarya gummifera* (Anacardiaceae), *F. sycomorus* (Moraceae) and *A. lebbeck* (Fabaceae). The minimum area value was 300 m^2. It corresponded to a number of 60 species.

Dry thickets of Singani (**Figure 6**) and Domoni-Ambouani-Hahaya have no stratification, with a very open canopy (10–20% recovery rate). It is a shrub vegetation 5 to 8 m high, which differed from other formations in the region only by reducing the size of the species and their foliage. Only baobabs and Albizia exceed 8 m and can reach 12 m. The minimum area of this thicket is 250m^2 and corresponds to 48 plant species. The dominant species are *Pyrostria bibracteata, Euclea racemosa subsp. schimperi, A. digitata, Vitex doniana, Albizia glaberrima, Phyllarthron comorense, F. sycomorus, Erythroxylum platycladum* (Erythroxylaceae), *Oeceoclades lonchophylla* and *Acampe pachyglossa* (Orchidaceae). The soil is covered by numerous seedlings of *Phyllarthron comorense, Euclea racemosa subsp. schimperi*.

Figure 6.
Acampe pachyglossa on tree of Vitex doniana, in dry Singani ticket.

3.3 Floristic characters

One hundred and three (103) species, belonging to 46 families, were recorded in the fives formations, including 70 trees and shrubs (**Table 1**). The floristic composition of dry Forests shows sixty-five (65) species in 54 genera and 35 families. The most represented families are respectively Rubiaceae, Euphorbiaceae, Fabaceae, Apocynaceae and Erythroxylaceae. *Euclea racemosa subsp. schimperi, A. digitata, Alangium salviifolium, Phyllarthron comorense, Vitex doniana, Albizia glaberrima, Sorindeia madagascariensis, Pyrostria bibracteata* and *Tamarindus indica* are the most abundant speces. The density varies from 500 to 800 plants per hectare. The different forms of biological adaptation observed in forest are foliar deciduousness in *A. lebbeck*, *Turraea sericea, A. digitata, Commiphora arafy*, pachycaulia in *A. digitata*, epiphytism in Aeranthes sp. and *Acampe pachyglossa*, Geophytism in *Dioscorea comorensis* and *T. leontopetaloides* and crassulescence in *C. quadrangularis*. The biological spectrum is represented by Mesophanerophytes (55%), Microphanerophytes (25%) and Nanophanerophytes (15%).

N°	Scientific name	Family	English name	Status
1	*Sorindeia madagascariensis* (Spreng.) DC.	Anacardiaceae	Grape mango	Rare and native to Comoros, S. Somalia, Mozambique, Madagascar.
2	*Carissa spinarum* L.	Apocynaceae	Conkerberry/ bush plum/ Simple-spined num-num	Rare and native to Comoros, Africa to Indo-China, Australia to New Caledonia.
3	*Ophiocolea comorensis* H. Perrier	Bignoniaceae	—	Endemic to Comoros
4	*Phyllarthron comorense* D.C	Bignoniaceae	—	Endemic to Comoros
5	*Commiphora arafy* H.Perrier	Burseraceae	—	Rare and native to Comoros, Madagascar.
6	*Alangium salviifolium* (L.f.) Wangerin	Cornaceae	Sage-leaved *alangium*	Rare and native to Comoros, E. Tropical Africa, Comoros, Indian Subcontinent.
7	*Dioscorea comorensis* R. Knuth	Dioscoreaceae	—	Endemic to Comoros
8	*Diospyros comorensis* Hiern.	Ebenaceae	—	Rare and native to Comoros, Madagascar.
9	*Euclea racemosa* subsp. *schimperi* (A.DC.) F.White	Ebenaceae	" the sea guarrie or dune guarrie "	Native to to Comoros, Egypt to S. Africa, S. Arabian Peninsula
10	*Erythroxylum lanceum* Bojer	Erythroxylaceae	—	native to Comoros, N. Madagascar.
11	*Turraea sericea* S.m	Meliaceae	—	Rare and native to Comoros, Madagascar.
12	*Turraea virens* L.	Meliaceae	Mozambique honeyssuchle-tree	Rare and native to Comoros, Madagascar.

N°	Scientific name	Family	English name	Status
13	*Ochna ciliate* Lam.	Ochnaceae	—	Rare and native to Comoros, Aldabra, Madagascar.
14	*Gomphia dependens* DC.	Ochnaceae	—	Rare and native to Comoros, Madagascar.
15	*Comoranthus obconicus* Knobl.	Oleaceae	—	Rare and native to Comoros, Madagascar.
16	*Chionanthus insularis*	Oleaceae	—	Rare and endemic to Comoros
17	*Oeceoclades lonchophylla* **(Rchb.f.) Garay & P.Taylor**	Orchidaceae	—	Rare and native to Comoros, Tanzania, Mozambique, KwaZulu-Natal in South Africa
18	*Pandanus maximus* Martelli	Pandanaceae	—	Endemic to Comoros
19	*Phyllanthus comorensis* Leandri	Phyllanthaceae	—	Rare and Endemic to Comoros
20	*Drypetes comorensis* **(Baill.) Pax & K.Hoffm.**	Putranjivaceae	—	Very Rare and Endemic to Comoros
21	*Drypetes thouarsii (Baill.) Leandri*	Putranjivaceae		Rare and Native to Madagascar
22	*Pyrostria bibracteata* (Baker) Cavaco	Rubiaceae	—	Native to Comoros, Aldabra, Kenya, Madagascar, Mozambique, Seychelles, Tanzania, Zimbabwe
23	*Vepris boiviniana* (Baill.) Mzira	Rutaceae	—	Rare and native to Comoros, Madagascar.
24	*Sterculia madagascariensis* R. Br.	Sterculioideae	—	Very Rare and Endemic to Comoros
25	*Aloe alexandrei* Ellert	Xanthorrhoeaceae	—	Rare and Endemic to Ngazidja -Comoros

Table 1.
List of endemic and/or rare species observed in the dry vegetation of Ngazidja.

Natural regeneration is good for the dry forests of Lac salé and Infoundihe-Chambouani (986.64% to 995.64%) and it is bad for the Ngouni-Hamboda dry forest (50.96%).

In the thicket, fifty-three (50) species in 45 genera and 30 families. The most represented families are Fabaceae and Asteraceae. At the generic level, Erythroxylum is the richest in species with three species. The density varies from 310 to 365 individuals per hectare. The different forms of biological adaptation observed in this dry thicket are pachycaulia in *A. digitata*, geophytism in *T. leontopetaloides*, epiphytism in *Angraecum eburneum*, *Acampe pachyglossa* and *Platycerium alcicorne*, and crassulescence in *C. quadrangularis*. The predominant biological types are microphanerophytes (50%) and nanophanerophytes (35%). Natural regeneration is good for thickets (315.18% to 383.67%).

3.4 Specific characteristics of the plant formations studied

After analyzing the relationships between the five types of vegetation formations and their floristic procession were established. A similarity between the vegetation formations, according to their dry forest and forested physiognomy and dissimilarity according to their floristic composition (**Table 1**) could be identified. Thus, the following vegetation formations were distinguished:

- The dry forest of Lac-salé with *Turraea virens*, *Comoranthus obconicus* (**Figure 7**), *A. digitata, A. alexandrei* (**Figure 8**) and *Drypetes comorensis;*
- The dry forest of Ifoundihé-Chambouani (**Figure 9**) with *Turraea sercicea*, *Commiphora araf* and *Chionanthus insularis;*
- The dry forest of Ngouni-Hamboda with a mesophilic tendency with *Diospyros comorensis, Carissa spinarum, Ophiocolea comorensis* and *Sterculia* madagascariensis;
- the dry thicket of Singani with Vitex doniana, Pandanus maximus, Pyrostria bibracteata and Euclea racemosa subsp. Schimperi;
- the dry thicket of Domoni-Ambouani-Hahaya with *Euclea racemosa* subsp. *schimperi*, *A. digitata, Erythroxylum lanceum* and *A. alexandrei*.

3.5 Pressures, threats and dynamic on dry vegetation

The threats affecting forests and dry thickets in the Comoros fall into two categories: anthropogenic threats and natural threats (**Figure 10**). Anthropogenic threats are mainly caused by the misuse of different resources from different vegetation formations:

Figure 7.
Comoranthus obconicus (fruits).

Figure 8.
Aloe alexandrei (flowers).

Figure 9.
Dry forest of Ifoundihé-Chambouani.

- Agriculture and urbanization: as the population increases, the need for arable land also increases. This case occurs in the study sites where several households complain of not having enough agricultural space, which leads them to clear the forest and thickets. Farmers continue to clear these areas, gradually practicing slash and burn cultivation (7%) and the search for space due to uncontrolled urbanization (32%).

- Overexploitation of forest wood for several uses: lumber, construction wood, fuel wood, animal fodder and medicinal plants. Moreover, although the thickets

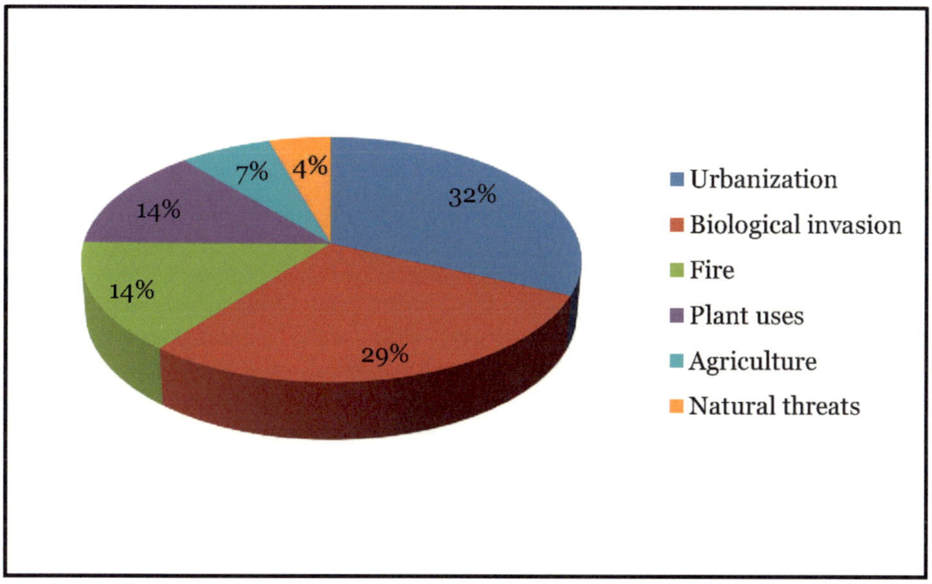

Figure 10.
Types of threats and their impacts.

of the Comoros are practically poor in tree species, riparian populations still use the few trees present in these thickets (*Phyllarthron comorense, A. lebbeck, Albizia glaberima, Ochna ciliata, Phyllanthus comorensis*). These uses vary from one locality to another. Thus, the most popular species used by the local population from the dry forest are *Alangium salviifolium, Phyllarthron comorense, Terminalia ulexoides, Drypetes comorensis, Phyllanthus comoriensis, Phoenix reclinata, F. sycomorus and Ochna ciliata.* The observations and information gathered on the ground show that plants in urban areas are often subject to felling, either for competing of space, risk of uprooted trees falling, or for cultural reasons (related to popular beliefs that baobabs and tamarinds are hosts to evil spirits) in some places.

- Breeding and stray animals: The threat to the dry thickets of Ngazidja is cattle. On a ground completely devoid of grass, especially in the dry season, cattle and goats roam in groups of 5 or 10 individuals in these formations, constantly grazing on grasses, seedlings of trees and shrubs, and lianas, thus preventing them from doing so. to regenerate properly.

- Invasive species: vegetation formations are also threatened by invasive species such as *A. lebbeck, Annona squamosa, Imperata cylindrica, Jatropha curcas*, *L. camara* and some lianas of small diameter (*Saba comorensis, C. quadrangularis and Rhoicissus revoilii*). However, *L. camara* is the most invasive because it forms clumps that block the regeneration of other species.

Threats of natural origin are volcanoes and cyclones. Only the habitats of Ngazidja are threatened by the Karthala volcano which is active. Many eruptions occurred and are noticed by long lava flows some of which are still very visible today. The last most devastating eruption was that of April 5, 1977 in Singani (Hambou), which destroyed many habitats. The Comoros Islands have experienced several cyclonic disturbances. The most recent cyclones affected in Ngazidja in 1984, 1985 and 2018–2019. They devastated homes, plantations and various dry ecosystems. These cyclones can cause windfall in vegetation formations, creating canopy openings [9]. All these actions have led to the reduction of almost 80% of the cover of dry plant formations in Ngazidja, in the last 40 years.

4. Discussion

The analysis results showed that the dry plant formations studied were different from each other, in terms of physiognomy, with species that characterized each formation. In fact, it consists of the dry forests of Lac-salé, Ngouni-Hamboda and Infoundihé-Chambouani, stratified forests and the two dry thickets of Singani and Domoni-Ambouani-Hahaya.

In terms of climate, the dry forest of Ifoundihé-Chambouani has several deciduous species while the dry forests of Ngouni-Hamboda and Lac-salé have few species that lose their foliage. Most species in these two last forests are usually evergreen for part of the year. Deciduous leaves condition the biological rhythm of a plant during the dry season to a slower state of life and it is one of the distinguishing criteria of tree formations of these dry forests. It is present in most trees of the upper and middle strata. Only trees from the mid and upper slopes lose their leaves during the dry season while those from the bottom do not lose their leaves during this season. For the Lac-salé, the lowland trees are found around Lake and thus directly receive moisture from this lake. This result is similar to what happens around Lake "Dziani Boundouni", on the island of Mwali [18].

At the national level, the dry forest facie of Ngazidja is characteristic of the dry ecosystem facies of Comoros archipelago similar at the dry forests found in Africa, Madagascar and the islands of the Indian Ocean, with true stratification and deciduous or semi-deciduous leaves. Because these dry forests have a canopy that can reach 15 m high with 20 m for the emergents like at Madagascar [19]. The abundance of small-diameter creepers and large-diameter thickets in dry forests is typical of dry vegetation communities in the Comoros [6].

The floristic composition of the dry forests is similar and as it is also the case for the floristic composition of the two thickets. Due to their proximity, species exchanges can occur between the Comoros Islands at the flora level [20]. Of the five vegetation formations studied, 106 species were identified including 70 trees and shrubs, which is very close to the score of 70 species of trees and shrubs found in the dry vegetation formations of Mayotte by [21, 22]. The dry forest Ngazidja has many floristic affinities with the "sub-humid" zone, semi-xerophile megatherm of Islands of Maoré (Mayotte) [23] and Mwali (Mohéli) [24] and particularly its particularly dry coastal fringe. Several species were found in the Maoré dry vegetation formations [25, 26] with over 90% of native species, some of which are endemic to the Comoros or Maoré (such as *Aloe mayottensis*).

Ethnobotanical surveys showed that trees and shrubs are also exploited for firewood. Generally, in the Comoros, wood remains the most used fuel as it is free: on Ngazidja, all ligneous plants are used as firewood although there are preferences for some tree species such as *Nuxia pseudodentata*. Currently, this case has gotten worse because of the high demography that this Island is experiencing. Thus, since forest species have become very difficult to find, it is the species of thicket (*A. lebbeck*, *Ochna ciliata*, *Pyrostria bibracteata*, *Euclea racemosa* subsp. *schimperi*) that are targeted.

Invasive species are also numerous in these formation: *A. lebbeck* (Fabaceae) lebbeck or lebbek tree or flea tree or frywood and *L. camara* (Verbenaceae) "common lantana" are considered among the highly invasives woody species of the Comoros Archipelago in dry vegetation [27, 28]. *L. camara* is very abundant and forms clumps in thickets. In Mayotte, this species is among the most threatening plant pests that invade any dry zone and forms dense bushes preventing any other species from establishing themselves [22, 23, 25].

In addition, most species are also completely invaded by lianas such as *Leptadenia madagascariensis and C. quadrangularis* (Vitaceae) veldt grape or adamant creeper or devil's backbone.

In Madagascar, nearly 23% of the 1003,000 ha of dry dense forests and 32% of the 1,444,000 ha of southern thorny scrub are considered degraded or secondary [29].

5. Conclusion

The study included Five types of dry vegetation formations in Ngazidja island, Comoros Archipelago: dry forests of Lac-salé, Ngouni-Hamboda and Infoundihé-Chambouani, and dry thickets of Singani and Domoni-Ambouani-Hahaya.

The physiognomy of the tree forests shows that they are well orgnized with deciduous leaves. These vegetation formations are different from each other with their own species:

- The dry forest of Lac-salé with *virens*, *Comoranthus obconicus, A. digitata, A. alexandrei* and *Drypetes comorensis;*

- The dry forest of Ifoundihé-Chambouani with *Turraea sericea*, *Commiphora arafy*, and;

- The dry forest of Ngouni-Hamboda with a mesophilic tendency with *Diospyros comorensis, Carissa spinarum, Ophiocolea comorensis* and *Sterculia* madagascariensis;

- the dry thicket of Singani with Vitex doniana, Pandanus maximus, Pyrostria bibracteata and Euclea racemosa subsp. Schimperi;

- the dry thicket of Domoni-ambouani-Hahaya with *Euclea racemosa* subsp. *schimperi*, *A. digitata, Erythroxylum lanceum* and *A. alexandrei*.

In all the studied formations, the overall natural regeneration is good. Mesophantophytes, nanophanerophytes and lianas of small diameter are the predominant biological types. The most common biological traits are foliar deciduousness, pachycaulia and aphylia.

In these plant formations, one hundred and three (103) species are distributed in forty-six (46) families, the most abundant of which are Fabaceae, Rubiaceae and Euphorbiaceae. The most dominant species are *Phyllarthron comorense, Diospyros comorensis, A. digitata, Euclea racemosa subsp. Schimperi*. Several endemic species were identified during this study, in these vegetation formations such as *Dioscorea comorensis, Diospyros comorensis, Phyllarthron comorense, Euclea racemosa subsp. Schimperi, Pyrostria bibracteata and Phyllanthus comoriensis*.

Some have just been observed for the first time in Ngazidja and are very rare and therefore threatened with extinction on the island such as *Turraea virens*, *Turraea*

N°	Scientific name	Family	English name	Status
1	*Sorindeia madagascariensis* (Spreng.) DC.	Anacardiaceae	Grape mango	Rare and native to Comoros, S. Somalia, Mozambique, Madagascar.
2	*Carissa spinarum* L.	Apocynaceae	Conkerberry/ bush plum/ Simple-spined num-num	Rare and native to Comoros, Africa to Indo-China, Australia to New Caledonia.
3	*Commiphora arafy* H. Perrier	Burseraceae	—	Rare and native to Comoros, Madagascar.
4	*Alangium salviifolium* (L.f.) Wangerin	Cornaceae	Sage-leaved *alangium*	Rare and native to Comoros, E. Tropical Africa, Comoros, Indian Subcontinent.
5	*Turraea sericea* S.m	Meliaceae	—	Rare and native to Comoros, Madagascar.
6	*Turraea virens* L.	Meliaceae	Mozambique honeyssuchle-tree	Rare and native to Comoros, Madagascar.
7	*Ochna ciliate* Lam.	Ochnaceae	—	Rare and native to Comoros, Aldabra, Madagascar.
8	*Gomphia dependens* DC.	Ochnaceae	—	Rare and native to Comoros, Madagascar.
9	*Comoranthus obconicus* Knobl.	Oleaceae	—	Rare and native to Comoros, Madagascar.
10	*Chionanthus insularis*	Oleaceae	—	Rare and endemic to Comoros
11	*Oeceoclades lonchophylla* **(Rchb.f.) Garay & P.Taylor**	Orchidaceae	—	Rare and native to Comoros, Tanzania, Mozambique, KwaZulu-Natal in South Africa
12	*Drypetes comorensis* **(Baill.) Pax & K.Hoffm.**	Putranjivaceae	—	Very Rare and Endemic to Comoros
13	*Drypetes thouarsii (Baill.) Leandri*	Putranjivaceae		Rare and Native to Madagascar
14	*Sterculia madagascariensis* R.Br.	Sterculioideae	—	Very Rare and Endemic to Comoros

Table 2.
List of species observed for the first time on the island of Ngazidja.

sericea, Comoranthus obconicus, Sterculia madagascariensis, *Alangium salviifolium, Commiphora arafy, Ouratea humblotii* and *Ochna ciliata* (**Table 2**).

Many of these species are widely used by the local population in various fields. The formations are subject to various pressures and threats of two types: anthropic and natural. The anthropogenic threat is the most important.

This article therefore aims at providing objective elements that can support conservation actions that could include:

- the delimitation or introduction of dry forests and thickets, in the strategy for the development of protected areas;

- the sensitization of national and local authorities and populations on the biological and ecological importance of dry forests and thickets;
- the ecological restoration of these ecosystems in areas degraded by local species;
- the evaluation of this ecosystem and development of management and conservation plans and steps to register in the International Union For The Conservation Of Nature Red List of Threatened ecosystem in Ngazidja island.

Acknowledgements

We would like to thank Center for International Cooperation in Agronomic Research for Development (CIRAD)-Madagascar, the " Sud Expert Plantes " 381 project for their financial support. Thank you also to all the people who supported us for the realization of this field study as well as the writing of this article, in particular the service of cooperation and cultural action (SCAC), service of the French Embassy in Moroni.

Author details

Abdillahi Maoulida Mohamed[1,2,3]*, Frédéric Bioret[3] and Vicent Boullet[3]

1 INRAPE, National Institute of Research in Agriculture Fisheries and Environment, Moroni, Comoros

2 Faculty of Sciences, University of Antananarivo, Antananarivo, Madagascar

3 Laboratory of Phytosociology, EA 7462 Geoarchitecture Territories, Urbanization, Biodiversity, Environment, UFR Sciences and Techniques, University of Western Brittany, France

*Address all correspondence to: abdillahimaoulida@yahoo.fr; amaoulidaabdillahieco1887@yahoo.com

References

[1] Voltskowa. Flora und fauna der Comoren. Vol. 3. Ostafrika: Reise, der Jahren; 1917. pp. 428-480

[2] Tilot V. Profil côtier de l'île de la Grande Comore. Comores: République Fédérale Islamique des Comores; 1998. p. 56

[3] Direction Générale de L'Environnement et des Forets. Seconde Communication Nationale sur les Changements Climatiques. Comores: Convention-Cadre des Nations Unies sur les Changements Climatiques (CCNUCC). Union des Comores; 2012. p. 159

[4] Battistini R. Paléogéographie et variété des milieux naturels à Madagascar et dans les îles voisines : quelques données de base pour l'étude biogéographique de la « region malgache ». In: Lourenço WR, editor. Biogéographie de Madagascar. Éditions de l'ORSTOM; 1996. pp. 1-17

[5] Boivin. Ancient crops provide first archaeological signature of the westward, Austronesian expansion. 2016. Available from: www.pnas.org. [Assessed September 30, 2016]

[6] Legris J. La Grande Comore: Climat et végétation. Institut française de Pondichéry. Travaux de la section scientifique et technique. 1969;**3**(5):1-28

[7] Adjanohoun EJ, Ake Assi L, Ahmed A, Eyme J, Guinko S, Kayonga A, et al. Contribution aux études ethnobotaniques et floristiques aux Comores. Paris: Agence de Coopération Culturelle et Technique; 1982. p. 216

[8] Keith P, Abdou A, Labat JN. Inventaire faunistique des rivières des Comores et inventaire botanique. Paris: Muséum d'Histoire Naturelle de Pris; 2006. p. 105

[9] FAO. Situation des forêts du monde. Rome: Département des forêts; 2012. p. 52

[10] Programme des Nations Unies pour l'Environnement. Atlas des Ressources Côtières de l'Afrique orientale. Comores: République Fédérale Islamique des Comores. République Fédérale Islamique des Comores; 2002. p. 54

[11] Lebrun J. Les formes biologiques dans les végétations tropicales. Bulletin des Sciences Botaniques. 1947;**15**:164-175

[12] Dalage A, Metaillie G. Dictionnaire de biogéographie. Paris: Centre Nationale pour la Recherche Scientifique; 2000. p. 579

[13] Gounot M. Méthode d'étude quantitative de la végétation. 1 ère édition ed. Paris: Masson; 1969. p. 224

[14] Gautier R, Birkinshaw C. Structure et flore de la forêt de sur pente d'Andranomena: Recherches pour le développement. Sciences Biologiques. 1998;**13**:15-29

[15] Brower JE, Von Ende CN. Field and Laboratory Methods for General Ecology. 3rd ed. United States of America: Brown publishers; 1990. p. 237

[16] Humbert H, Leroy J-F. Flore de Madagascar et des Comores. Paris: Muséum National d'Histoire Naturelle; 1936

[17] Lance K, Krenyien C, Raymond I. Extraction of Forest Products: Quantitative of Parc and Buffer Zone and Long Term Monitoring. Tananarive: Report to Park Delimitation Unit. WCS/PCDIM; 1994. p. 196

[18] Boullet V. Flore et habitats du littoral d'Itsamia (mohéli): Biodiversité,

conservation et restauration (version provisoire). Conservatoire Botanique National de Mascarin; 2009. p. 71

[19] Abdillahi MM, Cornu C, Raïma F, Charpentier M, Roger E, Rakouth B, et al. The baobabs of the Comoro Islands: Some biogeographical factors towards the protection and conservation of a neglected asset. Tropical Ecology. 2019;**6**:311-325

[20] Louette M, Meirte D, Jocque R. La faune terrestre des Comores. Centre Afrique: Musée Royal de l'Afrique Centrale; 2004. p. 455

[21] Pascal O. La végétation naturelle à Mayotte. Rapport interne. CTM/DAF/SEF. 1997. p. 90

[22] Pascal O. Plantes et forêts de Mayotte. France: Muséum national d'histoire naturelle. Institut d'écologie et de gestion de la biodiversité. Service du patrimoine naturel; 2002. p. 108

[23] Boullet V. Aperçu préliminaire de la végétation et des paysages végétaux de Mayotte (version provisoire). DAF Mayotte et Conservatoire Botanique National de Mascarin; 2005. p. 160

[24] Abdillahi MM, Frédéric B, Vincent B. Contribution to the management of natural forest plant resources in the Comoros: Prospects for conservation, local socio-economic development in protected areas. In: Advances in Sustainable Development and Management of Environmental and Natural Resources. 1st ed. Apple Academic Press; 2021. pp. 258-292

[25] Barthelat F, M'changama, M, Ali Sifari B. Atlas illustré de la flore protégée de Mayotte [Illustrated Atlas of the protected flora Mayotte]. 2006. Version B 06-06. Available from: http://daf.mayotte.agriculture.gouv.fr/. [Accessed: January 31, 2020]

[26] Amann C, Amann G, Arhel R, Guiot V, Marquet G. Plantes de Mayotte. Mayotte: Les Naturaliste de Mayotte; 2015. p. 370

[27] www.fao.org/dorep/007/j1922f/j1922f09.htm; Avril 2016

[28] www.fao.org.htm; Avril 2016

[29] Steven MG. Paysages naturels et biodiversité de Madagascar. UK: Biotope; 2008. p. 696

Chapter 15

Climate Change and Anthropogenic Impacts on the Ecosystem of the Transgressive Mud Coastal Region of Bight of Benin, Nigeria

Patrick O. Ayeku

Abstract

The transgressive mud coastal area of Bight of Benin is a muddy coastal complex that lies east of the Barrier/lagoon coast and stretches to the Benin River in the northwestern flank of the Niger Delta Nigeria. It constitutes a fragile buffer zone between the tranquil waters of the swamps and the menacing waves of the Atlantic Ocean. Extensive breaching of this narrow coastal plain results in massive incursion of the sea into the inland swamps with serious implications for national security and the economy. Climate change impacts from the results of meteorological information of the regions shows a gradual degradation in the past 30 years. Temperature, rainfall and humidity increase annually depict climate change, resulting from uncontrolled exploitation of natural resources is rapidly pushing the region towards ecological disasters. The ecosystem is very unique being the only transgressive mud coastal area of the Gulf of Guinea. The chapter describes the geomorphology, tidal hydrology, relief/drainage, topography, climate/meteorology, vegetation, economic characteristics, anthropogenic activities and their impacts on the ecosystem.

Keywords: vegetations, geomorphology, anthropogenic, tidal hydrology, topography

1. Introduction

The transgressive mud coastal zone is a fragile ecosystem, full of vast resources of minerals, foods, and energy but not without scenes of often conflicting uses. Presently, the uncontrolled and overexploitation of natural resources, vis a vis the climate change impacts pose a great threat to the economic prosperity and thereby resulting in environmental nightmares portending a lot of dangers for the generations to come [1]. This environment has witnessed a lot of degradation resulting from several years of oil exploitations and explorations. People's livelihood and their social well-being have often been affected adversely resulting from changes in the environment. Meanwhile, government over the years has dashed the hope of the inhabitants for protection. Bursting the petroleum pipeline by the people in an attempt to fight back, thereby disrupting the activities of the oil companies ended up compounding the challenges in their environments which eventually leads to oil pollution. Fishing is the major occupation of the predominant group of people (the Ilajes) in the coastal zone. Their settling pattern just like the Ijaws along the coast is in such a way that their houses were constructed with wood and were suspended on water.

This fragile ecosystem poses a delicate balance with the inhabitants in the area. The rate of environmental degradation in this region, as a result of anthropogenic pollution and climate change, is rapidly pushing the region towards ecological disasters. Uncontrolled reclamation of lands has been the last hope of the people due to lack of land to build settlements. Rivers, which are the people's only means of transportation, have been a nightmare, as a result of flood continual modifications of the rivers, thereby rendering them useless. Economically, human activities have been crippled. More also, shortage of land for development, flooding, siltation, occlusion and other environmental problems are associated with the hydrology and natural terrain of the area. Mangrove swamp forest (vegetation) reduces the impacts of floods, exacerbated by land subsidence, coastal erosion, and rising sea level. Also, the diurnal tidal movement modifies the floods which continuously impairs the river courses with significant impacts on the economy and human life patterns [2].

The transgressive Mud inter-tidal zone typically has a slope of 1:50, while beach elevation averages 3 m above the mean low water level. The coastline lies between Ajumo and the Benin river-estuary on the northwestern flank of the Niger delta [3]. The transgressive mud coastline runs from the northwest to southeast in the Ondo State coastline. About 90% of Nigeria's foreign exchange is derived from crude oil and gas, which is one of the major natural resources in this region. Other natural resources domiciled in this region include: fisheries, touristic resources, mangroves and forest.

2. Coastline geomorphology of the region

The Nigerian coastline is separated into four physiographic zones, as described by Fabiyi [4]. The sections include: the Strand Coast, the Niger Delta Coast, the Barrier-Lagoon Coast and the Transgressive Mud Coast. Each of the sections is associated with differing erosive activities, resulting from anthropogenic and natural factors as explained thus;

2.1 The Barrier-Lagoon coast

This complex is located in the Lagos State axis of Nigeria's coastline. It is predominantly made up of coarse sand beach, which allows easy drains of excess flood water and allows it to percolate. This easily flows back into the sea in few days.

2.2 The Niger Delta coast

The Niger Delta coast is predominated by fine beach sand and Mangrove forest. It is characterized by intense flooding around the communities and in some elevated areas, the vegetation is rain-fed deltaic. This is found along Delta, Akwa Ibom and Rivers States of Nigeria coastline.

2.3 The strand coast

It is the most eastern section of Nigeria coastline with a lot of vegetation (mostly Nypa Palms'- Nypa fruticans) which holds the fragile beach and acts as a barrier to ocean flooding. The beach is fronted by flat beaches and changes into a beach ridge plain. It is located at the Cross River axis of Nigeria coastline.

2.4 The transgressive mud beach or Mahin mud coast

This is located at the Ondo State axis of Nigeria coastline. It is composed of vegetated bluff, mud and flat marsh forming a transgressive mud beach. The area is characterized by a lot of floods trapped in the mud creeks, remaining there for days. Making it vulnerable to coastal flooding, thereby paralyzing activities in the affected communities. During the summer, it is left with mud cracks, formed from dried saturated mud (**Figure 1**).

Ondo State coastal plain (Transgressive mud coast) is narrow (about 700 m wide), unlike other coastal plains that have extensive mangrove swamps, lagoons and raised beaches. Freshwater swamp, lacustrine marshes and an intricate network of interconnected creeks backed the coastline by about 30–60 km wide freshwater marshes [5]. This, therefore, constitutes a fragile buffer zone between the Atlantic Ocean menacing wave and the tranquil freshwater swamp. Massive incursion of the sea into the inland swamps occurs as a result of extensive breaching into the coastal plain, leading

Figure 1.
Mud beach (A) and Mud crack, formed from dried saturated mud (B).

Figure 2.
Coastline recession in Ayetoro (A) and Awoye (B) communities resulting from massive incursion of the Atlantic Ocean.

to economic loss and threat to national security. One of the most important implications of massive incursion of the Atlantic Ocean is the coastline recession in Ayetoro and Awoye communities (**Figure 2**).

3. Geology of the region

It is believed that the mud beach coast evolved from the growth of the Niger delta into the Gulf of Guinea following the gradual retreat of the sea after a short-lived Paleocene transgression [1]. The major geological formations in the area include general alluvium, lagoonal marshes, abandoned beach ridges and coastal plains sand. The general alluvium comprises coarse, clayey, unsorted sands with clay lenses and occasional pebble beds which are lithologically indistinguishable from typical coastal plains sand strata [6]. These formations produce generally swampy soils on the nearly level coastal plains sand on alluvium, and very deep, well-drained soil, with very dark brown to dark brown surface sands from the nearly level coastal plains on coastal plain sand [7]. Elevation rises from about 1 m along the coastline to between 35 m (Igbokoda town) and 55 m (Okitipupa town) in the upland [8].

4. Soil of the region

The base of the sedimentary fill in Nigeria coastal area consists of unfossiliferrous sandstones and gravel weathered from the underlying pre-Cambrian basement [9].

Above the course materials are marine shales, sandstones and limestones of Santonian age, whose deposition was ended in parts of the Nigerian basin by folding, faulting, and basic igneous intrusion during the Santonian age. The next cycle of deposition began with the transgression that lasted into the Maestrichtian. The present Niger delta was initiated during regression that began in the early Eocene [10]. The soils underlying the Niger delta are generally characterized as soft, highly compressible, organic and inorganic silty clays overlying fine sands at great depths.

5. Vegetation of the region

The freshwater swamp forest is found in the inland freshwater areas between high forests and the mangrove swamp forest. The vegetation consists initially of species of reed (Phyragnites spp) and Papyrus (*Cyperus papyrus*). Their roots are submerged while the shoots stand above water and the trees which later cover the reed swamps form swamp forests which are poorer species than the forest on dryland. Palm-like raffia (*Raphia hookerii*) are characteristically present, so also are species of cane. The trees show layering and are still rooted with dense undergrowth of shrubs, and lianas where canopy is exposed to light. The eutrophic water (water that is fairly rich in mineral matter) in the area encourages silt soil to be formed which results from the accumulation of predominantly inorganic sediments in which the water is near the surface. There are floating grasses and other creeping plants along the edges of the creeks and rivers which are invaded either by water lettuce (*Pistia stratiotes*) or water hyacinths (*Eichhornia crassipes*).

The Mangrove Swamp Forest is characterized by evergreen trees and shrubs. The plants cover the sheltered muddy areas where land is rapidly encroaching on the sea in the estuarine and deltaic environments. Mangroves play an active role in building upland from the sea, obstructing currents, thus adding humus and raising the ground level seawards. The mangrove trees are essentially halophytes (adapted to saline habitats) receiving low saline water from rivers and higher saline water from the sea at different times of the day and seasonally. Hence, the predominant species in the brackish water zone are the red mangroves Rhizopora racemosa while other species of the red mangroves are found in the freshwater swamps inland. At the eastern margin of Awoye estuary, the vegetation consists mainly of red mangroves (*Rhizopora racemosa*) with white mangrove (*Avecinnia africana*) lining the inner edges of the creeks and along the rivers. Towards the seaside, Paspalum veginatum which after replacing felled mangroves becomes more dominant, sometimes excluding other species on the inner sand banks, with dense undergrowth of scrambling shrubs and trees, such as *Hibiscus tiliaceus*, Chrsobalanus orticulatis, etc. At the back swamp are raffia palms growing in the marshy areas, freshwater creek and along the rivers. The raffia palm is predominant in the area, it is readily available for construction purposes.

However, with active forest removal going on in the northern axis and extensive areas of marsh and mangrove forests being decimated in the southern parts, a large stretch of land along the coastline is now permanently inundated, especially in Ayetoro and Awoye areas. Thousands of peasant populations in so many rural communities are at the risk of a shortfall in their food security and means of livelihood [11]. Research and literature on the extent of degradation on the transgressive mud coast are lacking. Coastal erosion, canalization, inundation, the emergence of coastal grassland and rapid sedimentation of lagoons are some of the degradation processes identified. The increase in offshore exploration activities in the 70s led to more pronouncement of these degradation activities [12].

6. Mangroves ecosystem of the region

The most globally significant ecosystem between tropical rainforest and marine is mangrove. It represents one of the most productive natural ecosystems from the biological perspective. It is home to some unique endangered animal species. Economically, mangrove woods are used for construction works, useful chemical extraction (wood tar, tannin, alcohol, etc.) and furniture making, High quality charcoal and firewoods which are a good source of fuel are also sourced from mangroves. They also serve as nursery areas for marine animals and fish breeding arena.

In the environment, mangroves protect against storms (natural barrier against coastal erosion and tsunami), toxic substances, roots screening out debris, CO_2 absorption thereby reducing greenhouse gas and deposition of suspended sediments in water, creating mudflat for more mangrove. From the sociocultural perspective, it serves as a good source of food and medications for communities. Also, they are good environmental indicators for our climate.

The mangrove ecosystem in Nigeria is unique, being part of the Gulf of Guinea's large marine ecosystem, which is the largest in Africa and the third-largest in the world. It covers Lagos State, Ondo State and Niger Delta areas and occupies areas of about 10,000 km^2 in a 30–40 km wide belt [13]. Mangrove forests in Nigeria are found on the coast and stretch into the rivers and complex lagoons in several places. Spalding et al. [14] estimate the Nigerian mangrove to be about 10,500 km^2.

Plants that are usually associated with mangroves include buttonwood tree (Conocarpus), leather fern (*Acrosticum aureum*), hibiscus (*H. tiliaceus*), etc. Nypa fructicans from Singapore was introduced. Mangroves are harvested for consumption to meet the needs of the local communities [15] and are also being utilized as raw materials for industry [16]. Even though the demand has drastically declined, mangrove wood is still extensively used in Nigeria and the extraction of much higher volumes of wood is undertaken exclusively as selection harvesting under license agreements with the competent authorities [13].

Mangroves help protect the coastline from storm damage, wave action and erosion. They stabilize the elevation of land by accretion of sediment and also protect from damaging siltation seagrass beds and coral reefs. If the present level of mangroves depletion in Nigeria is not reduced, its coastal cities like Warri, Port Harcourt and Lagos will be drowned in the next couple of decades [17].

The area is generally inhospitable and difficult to develop. The area is inhabited mainly by fishermen/women and small farmers. The dense vegetation of mangrove forest found in this area has become a source of income generation, a reliable small-scale food processing and fuel wood for domestic usage [18].

7. Meteorology of the region

The meteorological station with long-time data record closest to the region is at Ondo town. Ondo town is about 85 km (north-south direction) from Ayetoro. Ondo town and the coastal communities fall under the same weather influence. Therefore the 30 years of meteorological data (1984–2014) used to characterize the area were based on Ondo town meteorological station and obtained through the central office of the Nigeria Meteorological Agency (NIMET) Lagos. A new meteorological station is now situated at Ayetoro, one of the communities but does not have long time data. In this station, air temperature and humidity, wind direction and rainfall are typically measured at 2 m above the surface on the muddy beach. Wind speeds are measured at 10 m height above the ground surface. Nigeria Meteorological Agency (NIMET) were contacted for information on the climatic data collected at Ondo from 1984 through 2014. The meteorological variables on which records were available were considered individually below;

7.1 Air temperature

The mean monthly temperature values at Ondo for the period of 30 years are presented in **Table 1**. The highest mean temperature was recorded in February (26.7 ± 0.5°C) while the lowest value was recorded in August (23.95 ± 0.4°C). With an annual mean air temperature of 26.86 ± 1.3°C, the year 2013 was the warmest year of the 30 years covered in the available record. The coldest year was 1991 with 25.2 ± 1.4°C annual mean value (**Table 2**). On the whole, a short mean air temperature range of 1.66°C was recorded.

7.2 Relative humidity

The monthly percentage relative humidity (% RH) from 1984 through 2014 was presented in **Table 1**. Relative humidity followed the same pattern as temperature. The lowest relative humidity for the months was recorded in January (68.16 ± 10.0%) while the highest value was recorded in August (90.6 ± 2.7%). The lowest annual relative humidity was recorded in 1984 (79.25 ± 10.0%), while the highest annual mean value of 87.42 ± 5.74% was recorded in 2014 (**Table 2**).

The high humidity experienced in this region makes the air to be close to saturation and thus with less capacity to store additional water. This tends to reduce the rate of evaporation despite high temperature and high energy input in the region. For humid conditions, the wind can only replace saturated air with slightly less saturated

Months	Rainfall (cm)	Relative humidity (%)	Mean Temp (oC)	Wind speed (m/s)	Wind class (Beaufort)
January	12.79 ± 20.1	68.16 ± 10	26.7 ± 0.5	1.09 ± 0.5	Light air
February	35.6 ± 35.3	70.6 ± 9.24	27.7 ± 1.1	1.45 ± 0.7	Light air
March	102.35 ± 53.2	78.58 ± 5.38	27.5 ± 0.8	1.9 ± 0.8	Light Breeze
April	165.15 ± 85.4	83.6 ± 1.5	27.1 ± 0.7	1.8 ± 0.7	Light breeze
May	180.49 ± 63.1	84.8 ± 4.7	26.5 ± 0.6	1.45 ± 0.6	Light air
June	229.88 ± 65.7	87.7 ± 2.3	25.45 ± 0.7	1.8 ± 1.0	Light breeze
July	257.5 ± 108.2	90.5 ± 1.7	24.29 ± 0.4	2.16 ± 0.9	Light breeze
August	183.6 ± 107.3	90.6 ± 2.7	23.95 ± 0.4	2.24 ± 0.9	Light breeze
September	280.1 ± 90.1	90.29 ± 1.2	24.6 ± 0.7	1.7 ± 0.7	Light breeze
October	179.1 ± 59.4	88.5 ± 1.9	25.38 ± 0.6	1.47 ± 0.9	Light air
November	54.8 ± 41.1	81.2 ± 3.5	26.8 ± 0.6	0.89 ± 0.5	Light air
December	10.4 ± 17.9	73.1 ± 6.6	26.8 ± 0.6	0.9 ± 0.7	Light air

Table 1.
Monthly meteorological data of the study area.

	Temperature (°C)	Relative humidity (%)	Wind speed (m/s)	Rain fall (cm)	Vapor pressure	Wind direction
Year	Annual mean	A. mean	A. mean	A. mean	A. mean	Dominant wind
1984	25.90 ± 1.63	79.25 ± 9.96	2.07 ± 0.62	123.17 ± 97.6	27.17 ± 2.17	SW
1985	25.78 ± 1.69	80.5 ± 10.72	2.22 ± 0.51	211.86 ± 171.4	27.26 ± 2.52	SW
1986	25.43 ± 1.4	83.08 ± 7.25	2.99 ± 0.9	130.55 ± 91.98	27.47 ± 2.69	SW
1987	26.38 ± 1.32	82.17 ± 6.58	2.18 ± 0.83	133.06 ± 118.2	28.77 ± 1.7	SW
1988	25.79 ± 1.39	82.92 ± 7.86	2.16 ± 1.02	141.22 ± 98.3	27.94 ± 2.36	SW
1989	25.69 ± 1.26	81.00 ± 13.79	1.63 ± 0.8	134.03 ± 107.3	26.65 ± 4.4	SW
1990	26.03 ± 1.7	83.00 ± 9.17	1.68 ± 0.67	113.12 ± 88.36	28.10 ± 1.57	SW
1991	25.20 ± 1.38	84.08 ± 6.93	1.56 ± 0.75	192.47 ± 161.96	27.98 ± 1.78	SW
1992	25.68 ± 1.68	80.25 ± 12.74	1.76 ± 0.62	126.68 ± 137.52	26.43 ± 3.72	SW
1993	25.78 ± 1.22	79.83 ± 10.95	1.73 ± 0.66	121.13 ± 98.7	26.64 ± 3.2	SW
1994	25.65 ± 1.15	81.08 ± 11.21	2.1 ± 0.87	150.00 ± 106.33	27.08 ± 2.94	SW
1995	25.98 ± 1.25	81.42 ± 10.04	1.92 ± 0.68	137.36 ± 123.9	27.67 ± 2.74	SW
1996	26.16 ± 1.26	84.17 ± 5.92	2.05 ± 0.77	146.69 ± 101.05	28.46 ± 1.12	SW
1997	26.01 ± 1.19	82.08 ± 11.04	1.49 ± 0.71	120.99 ± 94.4	27.56 ± 2.98	SW
1998	26.58 ± 1.6	79.58 ± 11.02	1.86 ± 1.1	125.75 ± 121	28.17 ± 2.9	SW
1999	25.89 ± 1.36	83.25 ± 7	1.88 ± 0.95	139.19 ± 112.3	28.05 ± 1.34	SW
2000	26.03 ± 1.56	80.42 ± 12.25	1.28 ± 0.48	129.61 ± 115.6	27.07 ± 3.22	SW
2001	25.98 ± 1.52	82.17 ± 9.25	0.88 ± 0.45	132.37 ± 117.8	27.85 ± 1.94	SW
2002	26.04 ± 1.36	80.83 ± 11.89	0.77 ± 0.39	139.89 ± 103	27.36 ± 3.22	SW
2003	26.14 ± 1.55	82.75 ± 6.72	0.96 ± 0.53	140.90 ± 157.4	28.42 ± 1.78	SW
2004	26.33 ± 1.52	82.08 ± 7.39	0.98 ± 0.78	134.68 ± 103.54	28.11 ± 1.49	SW
2005	26.3 ± 1.42	81.42 ± 9.98	1.17 ± 1.14	137.53 ± 121.13	28.32 ± 3.3	SW

	Temperature (°C)	Relative humidity (%)	Wind speed (m/s)	Rain fall (cm)	Vapor pressure	Wind direction
2006	26.43 ± 1.35	83.25 ± 7.52	1.19 ± 1.64	131.15 ± 86.23	28.80 ± 1.97	SW
2007	26.03 ± 1.26	80.58 ± 12.12	1.20 ± 0.46	133.08 ± 109.9	27.48 ± 3.69	SW
2008	26.17 ± 1.17	80.08 ± 12.2	0.98 ± 0.3	160.91 ± 139.52	27.24 ± 3.8	SW
2009	26.28 ± 1.17	84.25 ± 5.91	1.07 ± 0.22	135.11 ± 113.08	29.25 ± 1.42	SW
2010	26.68 ± 1.46	84.33 ± 6.5	1.54 ± 0.31	156.08 ± 133.78	30.08 ± 1.25	W
2011	26.41 ± 1.26	83.00 ± 9.38	1.36 ± 0.27	141.38 ± 119.33	28.88 ± 2.84	SW
2012	26.18 ± 1.25	85.92 ± 6.04	1.46 ± 00.28	132.63 ± 88.17	29.28 ± 2.01	W
2013	26.86 ± 1.34	86.00 ± 6.54	1.98 ± 0.31	143.83 ± 89.81	30.20 ± 2.06	W
2014	26.54 ± 1.47	87.42 ± 5.74	1.53 ± 0.26	173.97 ± 129.3	30.38 ± 1.48	S

Table 2.
Annual meteorological data of the study area.

air and remove heat energy [19]. It is noteworthy that the relative humidity in the area in 1984 (about 75%) has increased steadily over a period of 30 years to 90% in 2014.

7.3 Rainfall

The mean monthly rainfall from 1984 through 2014 is shown in **Table 1**. This area experiences a double maxima rainfall regime characterized by two high rainfall peaks. The rainy season begins around March (102.35 ± 53.3 cm) and attains a peak in June or July.

This first peak is followed by a short dry break in August (183.6 ± 107.3 cm), known as the August break usually lasting for about two to 3 weeks in August. This break is usually followed by the resumption of the rainy season and lasts to mid-October with a second peak usually in September. The period from late October through early March constitutes the dry season. The annual highest mean rainfall was recorded in 1985 (211.86 ± 171.39 cm) and the annual lowest mean was in 1990 (113.12 ± 88.36 cm) (**Table 2**). This coastal zone is characterized by high rainfall, with rainfall all the months of the year and annual variability of approximately 113–211 cm as indicated in the 30-year data record.

7.4 Wind speed over the region

The mean monthly wind speed, which is shown in **Table 1**, revealed a lower wind speed in the dry season (October to February) and which was classified as Light air with the range from 0.5 to 1.6 m/s (according to Beaufort wind speed classification). Also, the rainy season (March to September) experienced a higher wind speed (1.7–2.24 m/s) which was classified as Light breeze according to Beaufort wind speed classification. **Table 1** showed a downward progression of wind speed from 1984 through 2002 and an eventual steady increase from 2002 through 2014. Light breeze (2.99 ± 0.9 m/s) was recorded as the highest annual wind speed in 1986 while 0.77 ± 0.39 m/s (Light air) was recorded in 2002 as the lowest annual wind speed.

7.5 Wind direction

It is observed that the predominant wind in the study area during the period of study was the southwestern trade wind which originate from the Atlantic Ocean. The southwestern trade wind was predominant for 26 years of the annual record while western and southern trade winds were predominant only in 3 years and a year respectively. Moderate wind was also predominant in the study area.

8. Tidal hydrology

The daily tide data in Escravos for one annual cycle (2004) comprising about 1400 specific tide measurements (four readings per day for 1 year) were used to characterize the tidal hydrology of the area. The separation of tide phases into neap and spring tides was with regards to the moon phases.

Highlights of the tidal hydrology of the area based on tide measurements at Escravos bar (the closest tide station to the area which is about 10 km east of Awoye is presented in **Table 3**. The highest (or extreme) high water individual spring tide

Tide description	Abbreviation	n	Value ± s.d.
Extreme high water of spring tides	EHWS	1	1.77 m
Mean high water of spring tides	MHWS	370	1.50 ± 0.12 m
Average high tide level	AHTL	705	1.44 ± 0.15 m
Mean high water level of neap tides	MHWLN	335	1.37 ± 0.15 m
Mid tide level	MTL	1409	0.97 ± 0.49 m
Mean low water level of neap tides	MLWN	359	0.57 ± 0.16 m
Average low water level	ALWL	704	0.51 ± 0.18 m
Mean low water level of spring tide	MLWS	345	0.44 ± 0.17 m
Extreme low water level of spring tide	ELWS	1	0.14 m

Table 3.
Summary of the Tidal Hydrology of the region.

(EHWS) recorded over the one-year period was 1.77 m, and it was recorded 17th of April of the study year, a day after the full moon i.e. during the spring tide when the sun was passing the equator.

Mean high water of springs (MHWS) and mean low water of springs (MLWS) were 1.50 ± 0.12 m and 0.44 ± 0.17 m respectively, thus giving a mean range of spring tides values of 1.06 m (which is about 90% of the observed extreme range of springs tide). On the other hand, the mean high-water level of neap tides (MHWN) and low water level of neap tides (MLWN) were 1.37 ± 0.15 m and 0.57 ± 0.16 m, respectively. Giving a mean range of neap tides values of 0.8 m, which is about 55% of the observed extreme range of spring tides and about 80% of mean range of spring tides. Average high tide level (AHTL) and average low tide level (ALTL) were 1.44 ± 0.15 m and 0.51 ± 0.18 m respectively, while mid-tide level (MTL) for all recorded tides (n = 1409) was 0.97 ± 0.49 m (**Table 3** and **Figure 3**). The mean range of spring tides could be divided broadly into three commonly observed shore zones: the upper shore zone (the portion above the mean range of neap tides), the middle shore zone (corresponding to the mean range of neap tides), and the lower shore zone (corresponding to the portion below the mean range of neap tides). Two harmonic tide waves influence the tidal variation along the coastline of the area, one with a period of 12.5 hours and the other with a period of 25 hours [20]. The combination of these two harmonic tide waves usually produces two low tides and two high tides each day. The twice-daily (semidiurnal) tide of 12.5 hours predominates over the daily (diurnal) tide of 25 hours, generating a diurnal inequality, or mixed semidiurnal tides. This causes a difference in height between successive high and low waters. The result is two high waters and two low waters each day [21]. The tidal characteristics of the area have many features in common with those of the typical Atlantic coast with the domination of two unequal high water and two low waters occurring within 24 hours [22]. The mean neap range of 0.49 m recorded in the area is comparable with the British coasts amounting to between 0.45 and 0.55 m of the mean spring range [22]. This small neap range is typical of locations where tidal ranges are large. However, the overall mean range of about 1.1 m (MHWS-MLWS) in this area is lower than 1.7 m for Bonny bar and 2.0 m for Calabar in Nigeria.

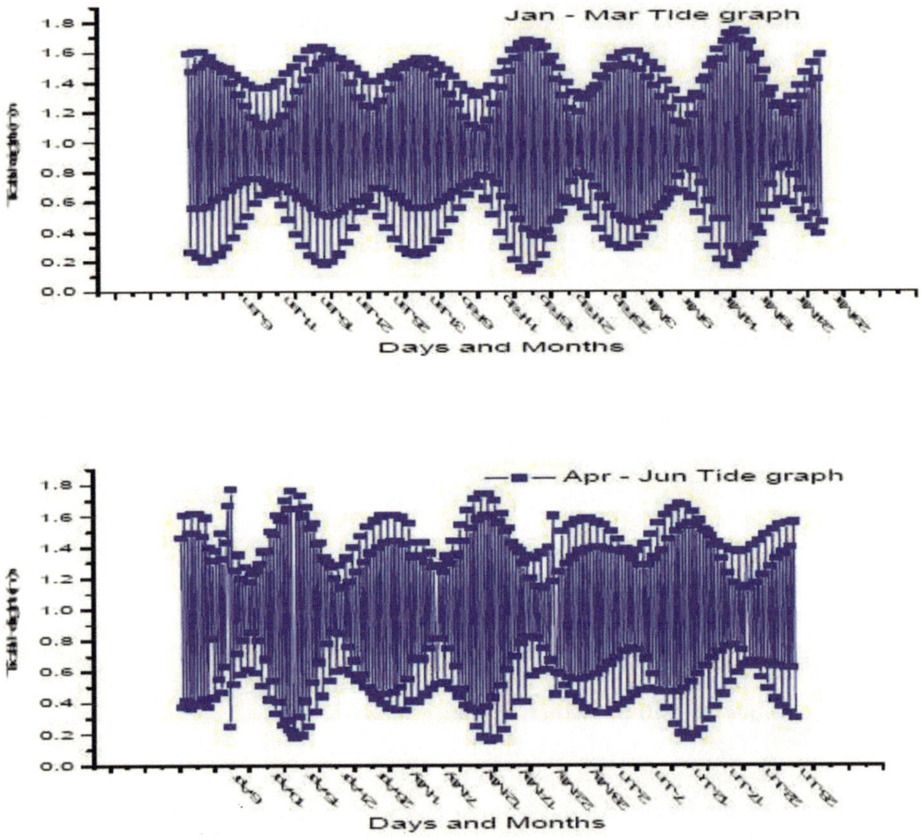

Figure 3.
Variations of the Tidal Hydrology of study region.

9. Relief and drainage

The area falls within the Atlantic system where most of the rivers are short, north-south flowing coastal rivers which follow more or less regular courses [23] and which drain into the sea. Rivers are being divided in a fairly simple line by western plains and ridges. The two major rivers, River Oluwa and River Ominla, display a drainage pattern which is dendritic but each river is parallel to the other and having a different tributaries. In the coastal plain area, river valley gradient is very low as it discharges its loads, leading to braided channel formation. Awoye and Abereke estuaries are the two major estuaries in the region. Abereke estuary located in the northern part of the area receives drainage from the Oluwa River and other major surrounding creeks. On the other hand, Awoye estuary located in the southernmost part of the area receives drainage from Ominla River, surrounding creeks and Benin River from Delta State.

10. Economic characteristics

The economy of the region is centred mainly on fishing, lumbering, farming, palm wine tapping, mat-weaving and petty trading. Fishing is the economic mainstay of the people in the region. The economic activities and the resultant occupation of the people have traditionally been related to the natural environment. This provides an opportunity for people in the riverine to be involved primarily in fishing, while people in the upland areas combine both farming and fishing with lumbering and carving of boats among other activities. A considerable number of people engage in traditional craft and modern processing industries. About 80% of the people in the area engage in fishing and that creates employment and generates a substantial income of about 90% of local GDP. Fishing activities in the area are carried out in two ways—freshwater fishing and ocean fishing.

11. Anthropogenic activities

11.1 Agriculture

Agriculture is the main source of livelihood for the people before the advent of crude-oil exploitation. Crude-oil exploitation results in conflicts and competition in the use of natural resources, which eventually leads to degradation of the environment. This affects the natural resources/livelihood (agriculture) of the host communities. Agriculture becomes less important in the polluted areas as it could not generate a reliable source of income for the communities. Fishing, Animal husbandry and Crop production are the major agricultural activities in the area. Agricultural wastes are discharged directly into the aquatic system [2].

11.2 Solid waste and sewage

The Ilajes (a dominant tribe in this region) are very enterprising people and one of the most dynamic in Nigeria. They are very good in aquatic skills and are able to adapt by conquering a harsh geographical environment, turning it to their advantage. Consequently, they were able to build communities like Ugbonla, Ayetoro, Zion Pepe, Awoye, Abereke, Araromi, Atijere, Ebute Ipare, Idiogba, Igbobi, Igbokoda, Igbolomi, Mahin, Mahintedo, Odun, Jirinwo, Odofado, Ago Nati, Ogogoro, Oloja, Ugbo, etc. The environment is dissected by several networks of river systems, which specifically make the coastal areas unfavorable for the development of road infrastructure that could serve as an engine for economic developments in the area. Many camps in the coastal area lack access to the hinterland except through hand-paddled canoes and some motorized boats. Some of these camps are directly on the Atlantic, especially where the tidal wave is relatively gentle. Therefore, all the communities in the area discharge their waste directly into the coast and the creeks.

11.3 Oil exploration

Oil exploration and exploitation started around 1977 by the then Gulf oil Company, presently called Chevron. This has metamorphosed into so many oil servicing and oil companies (such as; Express Petroleum and Gas Company/Conoco Energy Nigeria limited, Chevron-Texaco Nigeria limited, Global Pipeline, Consolidated Oil and Allied Energy, Agip Oil Nigeria Limited, Shell Petroleum Development Companies, etc) spreading vast installations and exploitation activities across the region [24]. The area is credited with about 14 oil fields, it contributes 12% of the country's crude oil production and reserves, with about 3.5 billion barrels of crude oil reserves.

The exploration, exploitation, and transportation of oil and gas in this region bring a serious problem with little or no economic development to the host communities by contributing a lot of pollutants to the ocean and the coastal zone. Some of these pollutants include spilling of hydrocarbons directly on the ocean and also those oil leakages from corroded pipelines, valves, production water effluents and ballast water discharges. Toxic chemicals from drilling fluids containing vessels and heavy metals (Vanadium, Plead, and Nickel) and other pollutants are being introduced from oil-field operations. All these pollutants are known to affect life forms [2].

12. Impacts of anthropogenic activities

12.1 Coastline recession

The major cause of coastline recession in this region is the wave attack of the clay ridge sediment, because of the absence of longshore current and weak nearshore littoral Guinea current. Tides and waves energy concentrated on the bare surface of the clay ridge, exposing it to direct wave impact because of the absence of the protective force of mangroves. A network of rills is developed on the plain due to wave backwash as it overruns the plain when the tide is high. Subsequent backwash and uprush of currents progressively widened and deepened the rills forming vertical heads U-shaped gullies. The walls of the gullies are terraced and are at the sub-tidal platforms level, becoming wave penetration avenue as the gully rapidly advancing into the coastal plain. The rate of gully head retreats measured the ranged from 5.7 m to 15.8 m annually [17].

Ground surface lowering and coastline recession result from the accumulation of ocean water in the depressions across the coastal plain. The energy uprush and backwash currents of the tides and waves affect the flood pools by increasing the soil water content. This relaxes the coherence of the unconsolidated soil, weakening the soil physicochemical interparticle bonds, thereby exposing the flood pool coastal plain to sheet wash erosion [5]. The coastal plains were eventually destroyed as the depressions gradually widened and deepened until the adjacent ones become incorporated and coalesced into an expanded subtidal platform.

Ebisemiju [17] reported coastline recession in the region (specifically Awoye) by about 3.31 km between 1974 and 1996 with annual rates varying between 31 m to 19 m in 1981. About 487 hectares of the coastline have been claimed into the Atlantic Ocean as a result of coastline recession within a short period of two decades (1973–1991) [17]. This has astronomically led to the loss of about 3000 hectares by 1996, reducing about 62% of the coastal plain and leading to the loss of arable land for animal husbandry and land for settlement in this narrow coastal plain. Presently, about 35 m of arable land is being loss into the Atlantic Ocean annually from this narrow coastal plain of the region.

12.2 Coastal erosion

Large-scale destruction of mangroves, canals and buildings are some of the direct impacts of erosion disasters in the region. These are evident in the ironwood stakes of abandoned houses found within the nearshore zones and intertidal platforms, indicating previous settlement.

A lot of inhabitants have been forced to abandon their houses and migrated inland to a safe location due to tidal floods and accelerated coastline recession. This makes the people relocate their communities at least once every 4 years. Today, the Awoye community (one of the settlements in the region) was originally about 3 km on the Atlantic Ocean from the present shoreline. Permanent structures could not be erected in the region because of the constant need to dismantle their houses and relocated them to a safer area due to the continuous threat of coastal erosion and tidal floods. Massive breaching of the coastline in one of the communities in the region (Ayetoro) may undoubtedly lead to a catastrophic event that will wipe out the entire community.

12.3 Seismic investigation

Seismic investigations (which involve explosive charges detonation below the ocean floor) were believed to have induced local subsidence. Ayetoro community claimed that it was Seismic investigations conducted in the area by an oil prospecting company, that generated shock waves which caused extensive damage to the major structures in the town, particularly the King's concrete palace. The claim was a memorandum sent by the community in 1981 to the NNPC (Nigeria National Petroleum Cooperation) but was refuted by the corporation. It is of note that explosion forces like that could result in the consolidation of subsurface sediment and liquefaction of surficial sediment. It could also lead to an increase in water depth and erodibility of sediment. Such activities could induce potential disturbance of the nearshore bottom leading to necessary conditions that promote coastal erosion. Increased water depth within the nearshore zone would enable larger waves to penetrate further inland than otherwise experienced. This would result in coastal erosion and the inundation of inland forests by seawater.

12.4 Oil spillage

Oil spillage has detrimental effects on both plants and animals. It is reported that oil spillage has caused constant threat to farmlands, crop plants, forest tree species and other vegetations in oil-producing areas in Nigeria [25]. There have been over 4000 oil spills in the Niger-Delta area of Nigeria since 1960. Toxicity of crude oil depends on its physical and chemical composition, the amount of the oil, the plant species and time of application as well as other environmental conditions [26].

Liver damages, infertility, disabilities, blindness, damages to fur and feather of birds and accidental poisoning are some of the direct effects of oil spillage in our ecosystem. It causes alterations in soil microbiological and physiochemical properties and affects soil fertility adversely, thereby having detrimental effects on the aquatic and terrestrial ecosystems.

13. Conclusion

The transgressive mud coastal ecosystem is full of vast resources of minerals, foods, and energy which has witnessed a lot of degradations as a result of several years of anthropogenic pollution (oil exploitations and explorations) and climate change (coastal recession). Meteorological information of the regions shows a gradual degradation in the past 30 years. Temperature, rainfall and humidity increase annually depict climate change, resulting from uncontrolled exploitation of natural resources is rapidly pushing the region towards ecological disasters.

Acknowledgements

The author is grateful to Prof. I. F. Adeniyi, Dr. A. O. Ajibare and Dr. L. T. Ogundele for the help rendered during the sampling period, laboratory analysis, statistical analysis and review of this paper.

Author details

Patrick O. Ayeku
Federal University Gusau, Zamfara State, Nigeria

*Address all correspondence to: patrickayeku@fugusau.edu.ng

References

[1] Wright JB, Hastings DA, Jones WB, William HR. Geology and Mineral Resources of West Africa. London: George Allen and Unwin; 1985

[2] Ibe, A. C. (1988). The Niger Delta and the global rise in sea level. In: Milliman, J. (ed.) Proceedings of the SCORE Workshop on Sea Level Rise and Subsidiary Coastal Areas, Bangkok, Oxford: Pergamon Press.

[3] Ibe AC. Marine erosion on a transgressive mud beach in western Niger delta. Geomorphology and Environmental Management. 1987;**12**:337-350

[4] Fabiyi OO. Mapping environmental sensitivity index of the Niger delta to oil spill: The policy, procedures and politics of oil spill response in Nigeria. In: Proceedings of MapAfrica, Held in Oliver Thambo Conference Centre. Johannesburg: South Africa; 2008

[5] Olorunlana FA. State of the environment in the Niger Deltal area of Ondo State. In: 1st Annual International Interdisciplinary Conference. Vol. 1. Azores, Portugal: AIIC; 2013. pp. 24-26

[6] Jones HA, Hockey RD. The Geology of Part of South-Western Nigeria. Geological Survey of Nigeria, Bulletin. 1964;**31**:12-18

[7] FDALR. Federal Department of Agriculture and Land Resources. Lagos, Nigeria: Annual Reveiw of Agriculture and Land Resources; 1985

[8] Iyun BF, Oke EA. Ecological and cultural barriers to treatment of childhood diarrhea in riverine areas of Ondo State, Nigeria. Social Science and Medicine. 2000;**50**:953-964

[9] Akpati BN. Geology of the Nigerian continental margin. In: Ajakaiye DE, Ojo SB, Daniyan MA, and Abatan AO. editors. Proceedings of the National Earthquakes in Nigeria. 1989. pp. 111-119

[10] Emery KO, Uchupi E, Philips J, Bowin C, Mascle J. Climate change impacts on marine ecosystems. AAPG Bullettin. 1974;**59**:2209-2265

[11] Akegbejo-Samson Y. Impact of Urban agriculture on water reuse and related activities on the rural population of the coastal settlements of Ondo State, Nigeria. African Journal of Food Agriculture Nutrition and Development. 2008;**8**(1):48-62

[12] Fasona MJ, Omojola A. Land cover change and land degradation in parts of the southwest coast of Nigeria. African Journal of Ecology. 2009;**47**(1):30-38

[13] Isebore CE, Awosika LF. Nigerian mangrove resources, status and management. In: Conservation and Sustainable Utilization of Mangrove Forests in Latin America and Africa Regions. 1993

[14] Spalding MD, Helen EF, Gerald RA, Davidson N, Ferdana ZA, Finlayson M, et al. Marine ecoregions of the world: A bioregionalization of coast and shelf areas. BioScience. 2007;**57**(7):573-583

[15] FAO. Mangrove M0061nagement in Thailand Malaysia and Indonesia. Vol. 4. Rome: Food and Agricultural Organisation Environment Paper; 1985

[16] Adegbehin JO, Nwoigbo I. Agroforestry practices in Nigeria. Agroforestry Systems. 1990;**10**(1):1-22

[17] Ebisemiju FS. Environment and Development Issues in the Niger Delta Area of Ondo State, Nigeria. Paper presented at the Stakeholders Forum/Workshop. Niger Delta Development Commission; 2001

[18] Awosika LF. The sub-Saharan Africa Coastal Zone: Assessment of Natural and Human Induced Changes. LOICZ Global Change Assessment and Synthesis of River Catchments-Coastal Sea Interaction; 2002

[19] IPCC. Climate Change 2007: Impacts adaptation and vulnerability. In:Parry ML , Canziani OF , Palutikof JP, Van der Linden PJ, and Hanson CE (ed.). Contribution of Working Group II to the Fourth Assessment. Report of the Intergovernmental Panel on Climate Change. Cambridge, UK: Cambridge University Press; 2007

[20] King CAM. Introduction to Physical and Biological Oceanography. London: The English Language Book Society and Edward Arnold (Publishers) Ltd.; 1975

[21] UNEP. Assessment of the State of Eutrophication in the Mediterranean Sea Map Technical Series. Vol. 106. Athens; 1996

[22] Lewis JR. The Ecology of Rocky Shores. The English University Press Ltd.; 1964

[23] Adebekun O. Atlas of the Federal Republic of Nigeria. Number 978-953–307-468-9. National Atlas Committee. INTECH; 1978

[24] Bayode OJ, Adewunmi EA, Odunwole S. Environmental implications of oil exploration and exploitation in the coastal region of Ondo State, Nigeria: A regional planning appraisal. Journal of Geography and Regional Planning. 2011;**4**(3):110-121

[25] Agbogidi OM. Response of Azolia africana Desv. and Salvinia nympphellula Desv. to the water soluble fraction of Odidi well crude oil. Journal of Science and Technology Resources. 2003;**2**(4):76-80

[26] Agbogidi OM, Nweke FU, Eshegbeyi OF. Effects of soil pollution by crude oil on seedling growth of *Leucaena leucocephala* (Lam. De Witt). Global Journal of Pure and Applied Science. 2005;**11**(4):453-456

Section 3

Human Influenced Ecosystems and Agriculture

Chapter 16

Vigor and Health of Urban Green Resources under Elevated O_3 in Far East Asia

Laiye Qu, Yannan Wang, Cong Shi, Xiaoke Wang, Noboru Masui, Thomas Rötzer, Toshihiro Watanabe and Takayoshi Koike

Abstract

Conservation of urban greens is an essential action for city residents, however, declining symptoms and/or traces in the annual ring of trees grown are found in parks and forest stands in a city as well as its suburb with a high level of ozone (O_3). Urban greens, including roof-green, provide comfortable conditions for the people and a moderate environment in a city. They are exposed to severe environments; heat, drought, air-pollutions, etc. even with intensive management of the people. How can we proceed with the conservation and wise use of urban greens? We should know the ecophysiological responses of urban trees to such a global environment as well as a local one. Defensive capacities of urban greens should be analyzed in terms of damages caused by biotic and abiotic stresses, and it is important to understand their interactions from the viewpoint of plant-insect/disease. There is a concern that some green areas are suffering from an outbreak of insects and diseases, reducing the vigor and health of urban greens. We discuss these based on specific examples, such as man-made forests, in cities in far east Asia for considering our approach to how to keep urban green resources.

Keywords: urban green, plant-insect interaction, ground-level Ozone (O_3), nitrogen dioxide (NO_2), heat island

1. Introduction

Air pollution has been considered as the most serious environmental problem for human health, associated with some million deaths worldwide per year. Among them, SOx was eliminated by desulfurization equipment in factories, however, nitrogen oxides ($NO_x = NO+NO_2$) have not well regulated. As NO_2 is one of the precursors of O_3, emission of NO_2 in Asia is estimated to be about four times higher than that of Europa and the U.S.A. [1–5]. Carbon dioxide (CO_2) increases global warming and increasing tropospheric O_3 are important global environmental issues today [6]. Cities have to cope with the challenges derived from poor air quality impacting the well-being of human health and citizen.

China has implemented various policies and measures to control air pollution and promote urban greening as well as forest rehabilitation since 1999 [7]. In 2004, China officially launched the construction of national forest cities, and gradually promoted and organized the construction of forest cities on a large scale across the country as proposed by the Chinese leader, Mr. JP Xi (2016). In 2019, forest city construction was included in Chinese Law as a legal guarantee for land greening and its protection. Urban greens, such as avenue trees and park trees, provide comfortable conditions for city residents, and a moderate environment in megacities in China [8] and in Japan [9, 10], and in small cities in Europe [11]. To efficiently reduce O_3 in cities, it is important to define suitable urban forest management, including proper species selection, with a focus on the reduction ability of O_3 via flux data [12, 13], and other air pollutants, such as PM [14–17], biogenic emission rates, allergenic effects [18] and maintenance requirements.

An epidemiological investigation is the most objective and practical method to determine the degree of O_3 damage by investigating the typical characteristics of O_3 damage to plants in the natural environment [19, 20]. The highest O_3 concentrations primarily occurred in July and August in northern China and the central part of Japan [21–23], and in September or April and May in southern China [23–26], in April–June in northern Japan [21]. Ozone causes cellular damage in plants, reducing stomatal control, lower CO_2 assimilation rates, and the occurrence of visible leaf injury [9, 27, 28]. These effects often accelerate senescence, diminish green leaf area and biomass, allocation of photosynthates to roots [29] and reduce photosynthetic capacity [28, 30, 31]. The investigation of visible injury of O_3 to the urban forest can not only judge whether the current concentration of O_3 in the air has reached the level harmful to the urban plants (e.g., [32]), but also determine the situation of urban exposure to O_3 pollution, and provide objective evidence for plant sensitivity evaluation [29].

Plants of green-roof are exposed to severe environments; heat, drought, PM, air-pollutions, etc. even with intensive management of the people [11, 33, 34]. There are many factors reducing the vigor and health of trees in avenue, parks, and suburban forests, etc. With the aggravation of O_3 pollution in urban air, the number of injured plant species in urban vegetation increased significantly, and the vitality of urban green resources was affected. Study shows that most plant functional types suffer a substantial decline in LAI as O_3 level increases [35]. Studies on the effects of O_3 on plants commonly used the following methods: manipulative exposure experiment: O_3-FACE) and epidemiological investigation (e.g., [9]). The effects of single or mixed pollutants on plant functional traits were studied through controlled experiments, such as OTC or artificial fumigation, to judge the resistance/sensitivity of plants to single pollutants [36] and multiple stresses.

We should know more about the negative effects of elevated O_3 on environmental conditions for plants as well as residents in big cities and surrounding areas through supporting service of ecological services; that is, plant primary productivities and nutrient cycling [37]. Moreover, we should consider biodiversity conservation, including plant-insect interactions for our future resources (e.g., pollination) as affected by elevated O_3 (e.g., [38–43]).

For accessing the methods of improving environmental conditions in cities and the vicinities, we state the role of urban green, declining symptoms, ecosystem vigor and health of big cities, especially under increasing ground-level O_3 and its dynamics. We showed the effects of environmental changes derived from elevated O_3 and other environmental factors on the heath and vigor of green infrastructure. Based on this evidence, we discussed a plausible understanding of the construction and maintenance of urban greening. Moreover, to attain the international Sustainable Development Goals (SDGs), the interactions in urban and suburban greens should be better understood to maintain the green infrastructure.

2. Role of urban green

2.1 A cool island

Urban greenness is essential for people's daily lives, while its contribution to air quality control is unclear. Green-roof is recently popular in big cities for moderating heat island phenomena through canopy transpiration, offering shade, and water management [44] even in cool climate regions. Some green-roof consist of vegetable garden for fun people and to inspire motivation of management. The mitigation effects of green spaces in urban areas are clearly observed as a role of the cool island [45, 46]. These imply the accumulation of a cold air mass in the park and its gravitational flow-out into the surrounding area. This practice of urban greenness is also found in Sapporo, northern Japan (**Figure 1**).

The expansion of an urban tree canopy is a commonly proposed nature-based solution to combat the excess urban heat-island problem. The influence of trees on urban climates via shading is driven by the morphological characteristics of trees, whereas tree transpiration is predominantly a physiological process dependent on environmental conditions and the built environment. The heterogeneous nature of urban landscapes, unique tree species (including vegetable crops) assemblages (**Figure 2**), and land management decisions make it difficult to predict the magnitude and direction of cooling by transpiration [48].

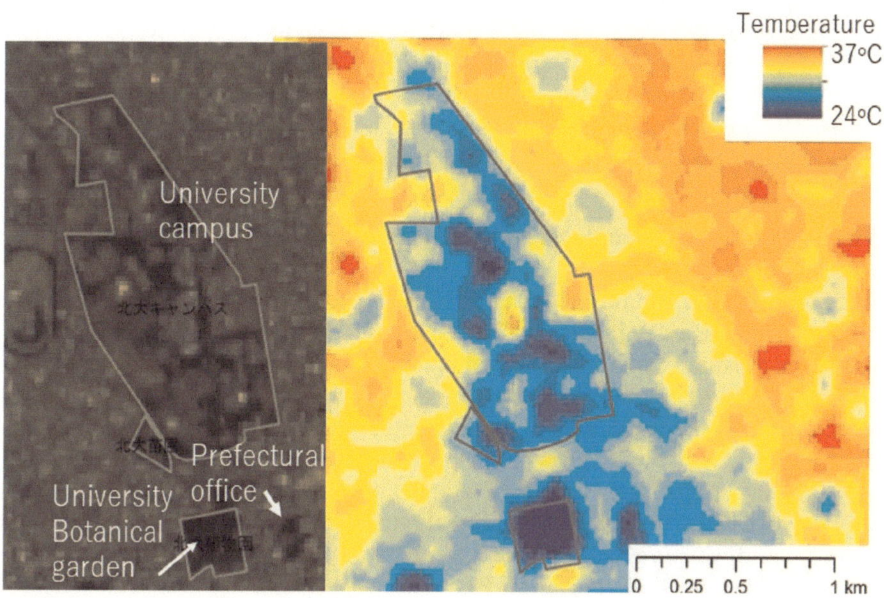

Figure 1.
Thermal image in Sapporo city center with green areas and their vicinity via the Landsat data (12 July 2015), offered Dr. H. Tani and the courtesy of Kyoritsu Publisher, Tokyo [47].

Figure 2.
Roof green with perennial plants and shrubs in Sapporo where we have heavy snow (A), and vegetable garden with small shade trees near the railway station in Tokyo (B) taken by T. Koike.

The Bowen ratios showed clearly the essential role of evaluation of transpiration capacity of urban tree canopy for moderating heat in a city. In this sense, we should keep the vigor and health of the green area in urban and suburban regions with high O_3 tolerance and also with O_3-sensitive plant species.

Ozone is a strong oxidative gas, and is not a substrate of plant physiological processes, but is an intensive stressor. We employ O_3-sensitive plants, such as morning glory (*Ipomoea nil*), as a biological indicator for avoiding O_3 stresses in a city [49]. As green area (trees and shrubs) has a high capacity for improvement of O_3 that is detected by flux monitoring [12, 50, 51], the green area has higher removal capacity (3.4 g m^{-2} $yr.^{-1}$ on average) than green roofs (2.9 g m^{-2} $yr.^{-1}$ as average removal rate), with lower installation and maintenance costs [13]. To overcome present gaps and uncertainties, they proposed a novel species air quality index (S-AQI) of suitability to air quality improvement for tree/shrub species. We recommend city planners to select species with an S-AQI > 8, that is, with high O_3 removal capacity, O_3-tolerant (e.g., [34, 52, 53]), non-emitting species of BVOC as a precoucer of O_3 [18], resistant to pests and diseases [38], tolerant to drought and non-allergenic plants. Consequently, green roofs can be used to supplement urban trees in improving air quality in cities.

2.2 The decline of urban and suburb green resources

With the effective promotion of national energy conservation and emission reduction measures, air pollution in China has been reduced [54].

Similar to most already-industrialized countries, China is now shifting away from SO_2-dominated to NOx- and O_3-dominated air pollution [55]. According to the bulletin on ecological and environmental conditions of China in 2019, the average O_3 concentration in China in 2019 was the highest in the past 5 years (148 $\mu g{\cdot}m^{-3}$), 20% higher than that in 2015, and the range of areas with high O_3 concentration was also expanding. Measurements from the national network for monitoring air pollutants show severe and worsening O_3 pollution in many areas, particularly the densely populated regions like Beijing-Tianjin-Hebei metropolitan area, YRD, PRD, etc. [56] (**Figure 3**). Beijing and its surrounding areas are typical composite pollution areas in North China, and the pollution degree has an obvious suburban gradient in somewhat [57–59], and plants have been threatened by high O_3 concentration [23, 60].

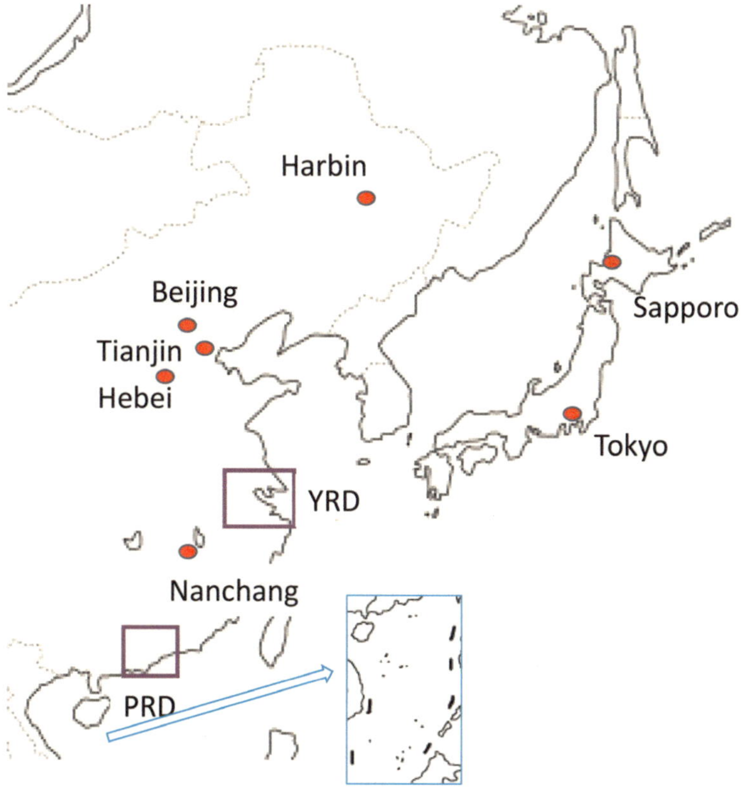

Figure 3.
Representative cities and districts in east Asia YRD Yangtze River Delta PRD: Pearl River Delta economic zone.

Although ground-level ozone (O_3) concentrations are expected to increase over the twenty-first century, especially in east Asia, with increasing NO_2 [4, 5].

2.2.1 Forest stand scale

In the O_3-sensitive beech forest as suggested by screening data of 18 tree seedling-scale by OTCs [34] (**Table 1**), we revealed earlier autumn senescence caused by elevated O_3 had already started, which was not found in the O_3-tolerant oak forests [12]. Ozone flux-based risk assessment was performed with the use of flux towers of carbon sequestration capacity in old temperate forests in Japan; pure forests of Siebold's beech vs. mixed evergreen and deciduous forests dominated by Konara oak (*Quercus serrata*).

Higher phytotoxic O_3-dose above a threshold of 0 uptake (POD-0) with higher canopy stomatal conductance (Gs) was observed in the beech forest than that in the oak forest. Light-saturated gross photosynthesis declined earlier in the late growing season with increasing POD-0.

The Tanzawa mountain range is located in the southwestern part of the region of Tokyo metropolis of Japan, and provides recreation activities to the residents of big cities. Dominate tree species are Siebold's beech forests are distributed in the high-elevation areas but have declined significantly. Recent ozone monitoring data suggest

O_3 Sensitivity	Leaf habit	Leaf growth	Species
High (sensitive to O_3)	Broad-leaved	Deciduous	*Populus maximowiczii, Populus nigra,* **Siebold's beech** (*Fagus crenata*), Japanese zelkova (*Zelkova serrata*)
		Evergreen	*Castanopsis sieboldii*
	Coniferous	Deciduous	Japanese larch (*Larix kaempferi*)
		Evergreen	Japanese red pine (*Pinus densiflora*)
Moderate	Broad-leaved	Deciduous	**Japanese white birch** (*Betula platyphylla* var. *japonica*),*Quercus serrata*
		Evergreen	*Quercus myrsinaefolia*, Camphora (*Cinamomum camphora)*
	Coniferous	Evergreen	Nikko fir (*Abies homolepis*)
Low (tolerant to O_3)	Broad-leaved	Deciduous	**Mizunara oak** (*Quercus mongolica* var. *crispula*)
		Evergreen	*Lithocarpus edulis, Machilus thunbergii*
	Coniferous	Evergreen	Japanese black pine (*Pinus thunbergii*), Sugi cedar (*Cryptomeria japonica*), Hinoki cypress (*Chamaecyparis obtuse*)

Adaptation from Yamaguchi et al. [34]. Bold means materials in Shi et al. [61] in section 4.

Table 1.
Classification of Japanese forest tree seedlings to O_{3-} sensitivity obtained from the OTC.

that high ozone concentrations (<100 ppb) may be a possible chronic cause of the loss of beech vitality started in 1980 days to date [62, 63].

2.2.2 Street landscape

Recently, efficient use of internet data offered an outline of the health and vigor of urban green in relation to air quality data (Air Quality Index: $PM_{2.5}$, PM_{10}, SO_2, NO_2, CO, and O_3) from 206 monitoring stations from 27 provincial capital cities in China [23–26]. Over 90% of air quality variation could be explained by socio-economics and geo-climates, suggesting that air quality control in China should first reduce efflux from social economics, while geo-climatic-oriented ventilation facilitation design is also critical. Pooled-data analysis at the national level showed that street-view greenness was responsible for 2.3% of the air quality variations in the summer and 3.6% in the winter; however, when separated into different regions, the explaining power increased to 16.2% at best. For different air quality components, greenness had the most significant associations with NO_2, CO, and O_3, and the street-view/bird-view ratio was the most powerful indicator of all greenness parameters. Large inter-city variations were observed in all the greenness parameters, and the weak associations between all street-view parameters and bird-eye greenspace percentage (21–73%) indicate their representatives of different aspects of green infrastructures.

Another assessment is shown in the traditional method of dendrochronology [64]. Urban tree growth is often affected by higher temperatures, reduced water availability, small and compacted planting pits like many avenue trees, as well as pollution inputs. Despite these reducing growth conditions, recent studies found a better growth of urban trees compared to trees at rural sites with high O_3, and enhanced growth of trees in recent times as shown in poplar [65]. Sakhalin fir (*Abies sachalinensis*) trees growing in urban vs. rural (or suburb) sites in Sapporo, northern

Japan analyzed the growth differences between growing sites and the effects of the pollution (e.g., NO_x, SO_2, and O_X, etc.) on tree growth (cf. **Table 1**). When exhaust gas of diesel cars circulates between urban areas and suburbs, the following chemical reactions occur: in the urban area; NO (from diesel cars) + $O_3 \rightleftharpoons$ (Ultraviolet ray) $\rightarrow NO_2$ [65]. Moreover, environmental conditions seem to positively affect tree growth, though with the exception of O_3, which had strong negative correlations with growth.

2.2.3 Foliar O_3 symptoms

Symptoms of O_3 injury had been observed at different levels in several plant species. The presence of foliar symptoms does not necessarily imply that there are significant effects on growth, yield or reproduction. Visible injury symptoms in bio-indicator plants are more perceptible and comprehensible for laymen than air pollution concentrations, which may help to enhance the environmental awareness to the general public. Therefore, it is an effective way to monitor the O_3 risk in urban and regional areas by investigating the typical characteristics of O_3 injury on plants as well as the street view [23–26]. These studies have identified sensitive species that could contribute to further investigated under controlled conditions.

The growth and physiological response of plants exposed to high concentrations of O_3 for a long time can reflect the degree of atmospheric O_3 pollution. Ozone levels in Beijing area were high enough to induce foliar symptoms by the year of 2021 at least, injury has been observing in different species [8, 60, 66, 67]. Visible injury was investigated on July and August in the main parks, mountainous and plain areas. The typical symptoms of the O_3 foliar injury are dark stipple, mottling and tip burn [66]. Some examples of typical symptom are illustrated as an atlas (**Figure 4**).

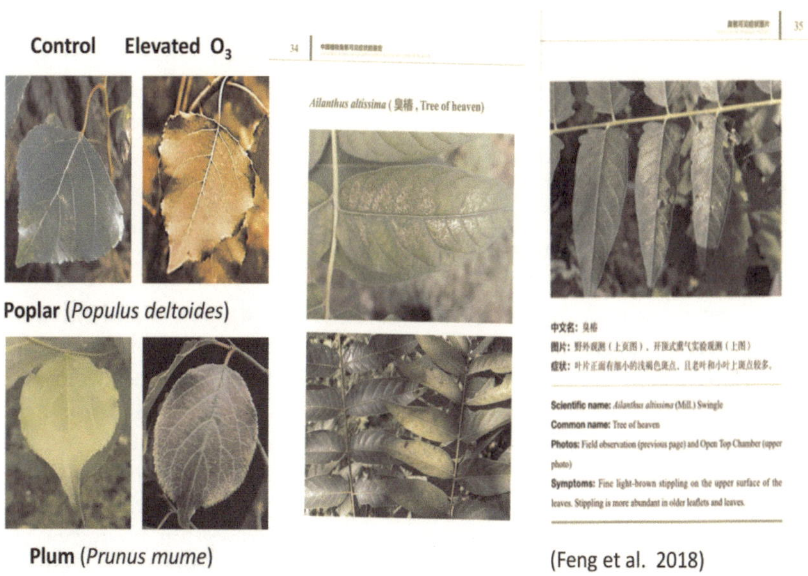

Figure 4.
Examples of damaged leaves of trees in a big city polluted with ozone. Photos of poplar and plum are cited from Nouchi [49], and tree of heaven is cited from Feng et al. [52, 53], the handbook for diagnosing O_3 damages of Chinese avenue and park trees.

Sensitive plants exposed to O_3 pollution usually show symptoms of visible leaf injury (ICP-Forest; [68]); most of them were surveyed on the basis of the Forest Health Expert Advisory System [69]. In recent years, foliar O_3 symptoms on plants were studied in Beijing, Hebei, Nanchang, and the surrounding areas of field investigations following the procedures for O_3 symptoms [8, 60, 66, 67, 70]; and [71].

Symptoms were more frequent in rural areas and mountains in northern Beijing, downwind from the city, and less frequent in city gardens [8]. *Ailanthus altissima* (Tree of heaven) is considered as the best bio-indicator tree for the Beijing area, because it is a native tree species with wide distribution, and symptoms of O_3 injury are easy to characterize. Similar investigations of O_3 damage symptoms on plant leaves were also carried out in Hebei Province. *Buxus megistophylla* was most frequently investigated and its injury symptoms of it were easy to be characterized, which will good to be used as an indicator species for ambient O_3 pollution in this region [71]. The other investigation in Nanchang city in Jiangxi province, which can represent subtropical urban areas in China, find that the O_3 concentration of suburban regions was significantly higher than in urban and exurban regions [70] (cf. **Figure 2**). The highest average O_3 concentration occurred in June at 40 ppb (AOT40 = 35.5 mg m^{-3} h^{-1}), well over the threshold considered (19.6 mg m^{-3} h^{-1}) to exert negative influences on the growth of wild plants. *Cerasus yedoensis*, *Phoebe shearer*, *P. bournei*, and *Litsea cubeba* might work as bio-indicators of ozone pollution in Nanchang [70].

Ozone stress in plants is recognized as a visible symptom of leaves at various levels of O_3 in the atmosphere [32, 52, 53, 72]. How can we obtain a bigger scale from individual tree scale to town scale? A typical example is recently reported in China with the use of big internet data [23–26]. They could identify the association between greenness and air pollution from a street view scale, which can favor urban greenness management and evaluation in other regions where street-view data are available. Urban forests play a vital role in terms of environmental quality related to particulate matter (PM including black carbon: BC) and studies on this area are increasing [73, 74]. The PM analyzes were divided into vertical and lateral directions at the stand scale [14]. The PM removal capacity of different vegetation types was usually in the following order: coniferous forest > evergreen forest > deciduous forest as related to foliar shape and longevity. Therefore, multi-scale analyzes on the effects of PM could help to better understand the roles of urban forests as a complex system.

3. Ozone and other environmental stress factors

The singly or combined effects of elevated O_3 on woody plants and surroundings were investigated using manipulative exposure experiment OTCs or O_3-FACE. Feng et al. [52, 53] integrated the results of 46 current O_3 fumigation experiments on 61 woody plants in OTCs in China. Net photosynthetic rate, growth index, and biomass of woody plants are affected by elevated O_3, showed varying degrees of reduction (**Figure 5**). Drought, high nutrients, temperature stress, and tropospheric O_3 pollution often co-occur in urban areas, adversely affecting urban plant health. Another experimental comprehensively analyzed the adsorption and absorption capacity of 537 species of plants in common urban forests in China to six kinds of air pollutants, namely, SO_2, NOx, freon (F), chlorine (Cl), O_3 and PM, six tree species with strong comprehensive tolerance to air pollutants were selected [75].

Identify possible indicators for O_3 sensitivity in woody plant species that could be useful for improving risk assessment and selecting appropriate species for urban

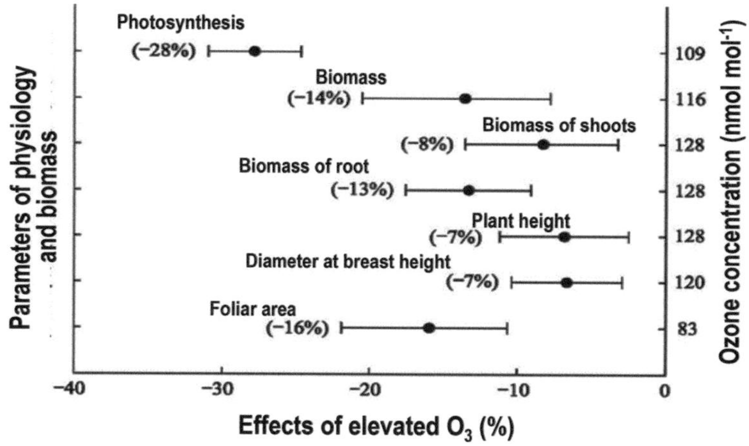

Figure 5.
Effects of elevated O_3 concentrations on net photosynthetic rate, growth index, and biomass of woody plants in China (Adapted from a meta-analysis by [52, 53]).

greening in areas with high O_3 levels. It seems feasible to plant tree species that are less sensitive to O_3 exposure, particularly in urban areas.

3.1 Heat

The combined impacts of air warming and O_3 on phenology and its functional traits are still not well illustrated for urban trees and roof green, including root crops (radish) on mini-farmland on buildings (cf. **Figure 2**) are still poorly understood. The reduction in the growth of radish (*Raphanus sativus* var. *sativus*) caused by elevated O_3 was accelerated by elevated temperature [72, 76]. This reduction discourages people's motivation for cultivation on a farm on rooftop (cf. **Figure 2**). Spring phenological and function in leaves of *Populus alba* and lian qiao (*Forsythia suspensa*) under ambient air (15.8°C, 35.7 ppb), increased air temperature (IT, ambient air temperature + 2°C) in a combination of two levels of O_3 (EO_3, ambient O_3 + 40 ppb), and their combined treatments (17.7°C, 74.5 ppb) [23–26]. Compared to EO_3, the combined treatment advanced the spring pheno-phase, increased growth, and induced a higher photosynthetic rate and antioxidative enzyme activities, which indicated that the positive effects of increased temperature alleviated the inhibition of growth induced by O_3. We expect that high temperatures open many stomata and increase the transpiration rate to a certain threshold. Therefore, it was expected that the amount of O_3 absorbed would also increase because the stomatal conductance would increase. In Italy, among three species with different water requirements (poplar>beech> oak), fast-growing plant species with high water requirements show more susceptible to O_3 and drought stress via the use of ^{13}C in leaves [77].

3.2 Nitrogen deposition

With the rapid industrial development and modern agricultural practices, increasing N deposition can cause nutrient imbalance. With remarkably high levels of NO_2 in east Asia [4, 5], O_3 could adversely affect the productivity of forest tree species, but risk assessments of O_3 impact were still limited [78]. We analyzed risk assessment of

O_3 on suburb forest tree species based on two previous studies via OTC [79], that is, the growth data in potted seedlings of Siebold's beech (*Fagus crenata*) and of three representative conifers, cypress (*Cryptomeria japonica*), pine (*Pinus densiflora*) and Japanese larch (*Larix kaempferi*). From deciduous leaf habit, N deposition increased in the O_3 sensitivity of beech while unchanged that of larch ([79], **Figure 6**). Based on the results, we conclude that the area with a high risk of O_3 impact does not necessarily correspond to the area with high O_3 exposure under different N levels.

Japanese larch and its hybrid larch (F_1; *Larix gmelinii var. japonica* × *Larix kaempferi*) are an important afforestation species in northeast Asia. We investigated whether N loading mitigates the negative O_3 impacts on two larch species [80]. Although N loading mitigated the negative effects of O_3 on Japanese larch, N loading did not mitigate O_3-induced inhibition of growth and photosynthetic capacity in the F_1. Elevated O_3 also reduced leaf nitrogen/phosphorus (N/P) ratio by elevated O_3, with significant effects in F_1, particularly under N loading. To avoid the negative effect of increasing O_3, we will plant Japanese larch but not F_1 at N rich site and/or accumulation of the N deposition site.

Larch species are associated with ectomycorrhizal (ECM) fungi, which play a critical role in nutrient acquisition for their hosts [4, 81, 82]). In this study, we investigated species richness and diversity of ECM fungi associated with a hybrid larch (F_1) and its parents (Dahurian larch; *L. gmelinii var. japonica*, Japanese larch; *L. kaempferi*), under simulated N deposition (0 and 100 kg ha^{-1} $yr.^{-1}$) with/without phosphorous (P) (0 and 50 kg ha^{-1} $yr.^{-1}$) planted in immature volcanic ash (Vitric Andosols) with low nutrient availability. F_1 showed heterosis in relative biomass, which was most apparent under high N treatments. Except for Dahurian larch, effects of the nutrient addition to ECM fungal community in F_1 were intermediate. F_1 was tolerant to high N loading, which was due to a consistent, relatively high association with *Suillus* sp. and *Hebeloma* sp. [81].

A recent trial was carried out for analyzing the combined effects of O_3 (low and two times ambient O_3), elevated CO_2 (ambient vs. 700 ppm), and 3 levels of soil N supply on Siebold's beech seedlings grown in climatized chambers [83]. They found that elevated CO_2 ameliorated O_3-induced reductions in photosynthetic activity, whereas the negative effects of O_3 on photosynthetic traits were enhanced by soil

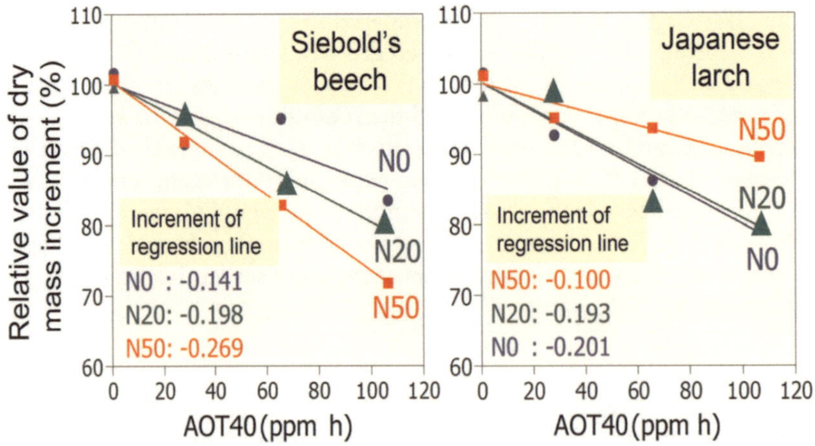

Figure 6.
Contrasting sensitivity of Siebold's beech vs. Japanese larch to elevated O_3 (as shown in AOT40) with increasing nitrogen loading to the soil (Re-illustrated from [79]).

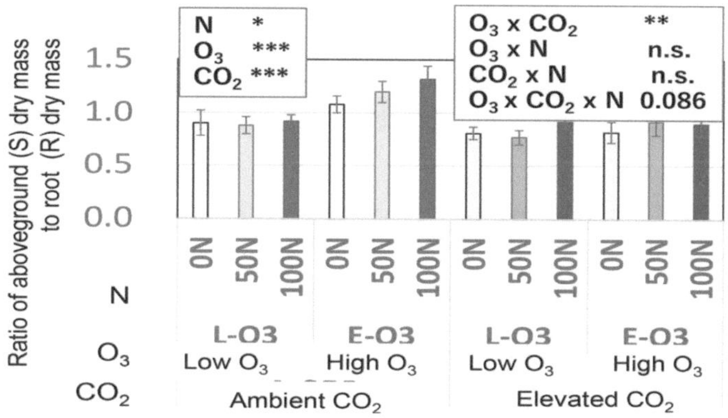

Figure 7.
*Ratio of aboveground mass (S) to root mass (S/R) of Siebold's beech seedlings raised under different O_3, CO_2, and three levels of N roading (Adapted from [83]). *$p < 0.05$, **$p < 0.01$, ***$p < 0.001$, n.s. not significant, the actual p-values are shown when $0.05 \leq p < 0.10$.*

nitrogen supply. Although we observed two- or three-factor interactions of gas and soil N in leaf photosynthetic traits, the shoot-to-root dry mass ratio (S/R) was the only parameter for which a significant interaction was detected (**Figure 7**) among seedlings. O_3 caused a significant increase in S/R under ambient CO_2, whereas no similar effects were observed under elevated CO_2.

4. Biological stresses on urban green

Elevated ground-level O_3 induces adverse effects in plants. Vigor and health of plants under elevated O_3 are discussed through biological stresses: (i) the composition and diversity of plant communities by affecting key physiological traits [38, 84, 85]; (ii) foliar chemistry and the emission of volatiles (mainly BVOC) [43], thereby affecting plant–plant communication, plant-insect interactions, and the composition of insect communities [38, 42, 43, 85]; and (iii) plant–soil-microbe interactions (e.g., [81]) and the composition of soil communities by disrupting plant litterfall [61] and altering root exudation, soil enzymatic activities, decomposition, and nutrient cycling [86, 87]. The community composition of soil microbes is consequently changed, and alpha diversity is often reduced. The effects depend on the environment and vary across space and time.

4.1 Plants' defense capacity under elevated ozone with elevated CO_2

Most carbon-based defense chemicals are synthesized from photosynthates [88]. As found in birch [89, 90] and cauliflower [20], we expect plant defense will increase at elevated (E) CO_2 and decrease at elevated (E) O_3. Aspen-FACE in the northcentral U.S.A. provided a combination of CO_2 (560 ppm) and O_3 (x ambient 1.5 times) [91] and showed performance of forest tent caterpillar (*Malacosoma disstria*) larvae. ECO_2 reduced N and increased tremulacin (a glycoside in poplar plants) levels, whereas EO_3 increased early season N and reduced tremulacin as compared to controls. With respect to insects, ECO_2 had almost no effect on larval performance. Larval performance improved in EO_3, but this response was negated by the addition of ECO_2

(i.e., CO_2 + O_3 treatment). The tent caterpillars will have the greatest impact on aspen under current CO_2 and EO_3, due to increases in insect performance and reduced in tree growth, whereas the insect will have the least impact on aspen under high CO_2 and low O_3 levels, due to moderate changes in insect performance and enhances tree growth.

4.2 Outbreak of insect and disease, a role of symbiosis

Defense capacities of plants are mostly originated from photosynthates, as a result, defense traits (leaves, branches, bark, etc.) of plants under elevated O_3 are lower. Consequently, trees and shrubs under elevated O_3 are attacked by diseases, insects, and wild animals [38, 62]. Outbreaks of sawfly and repeated sawfly attacks are fatal for the weakened beech trees. Another indirect biotic factor is the increased population of sika deer (*Cervus nippon*), which eat up ground vegetation and destroys related community balance. Sika deer is not fond of eating plants containing high levels of C-based defense chemicals (such as phenolics and lignin) [92]. Therefore, individuals whose defense levels have decreased due to O_3 are grazed.

The factors affecting beech decline in the Tanzawa mountains are complicated, however, further scientific research activities in various fields are required to understand the phenomena and to recover the beech forest vegetation [62, 63].

Colonization and species abundance of ECM fungi were revealed in 2-year-old hybrid larch (F_1) under the combination of elevated CO_2 and O_3 in OTC (O_3 < 6 nmol mol^{-1}, EO_3: 60 nmol/mol; ambient and elevated CO_2 [380 vs. 600 mmol mol^{-1}]). After two growing seasons, ECM colonization and root biomass increased under elevated CO_2 [81]. Additionally, O_3 impaired ECM colonization and species richness, and reduced stem biomass. Concentrations of aluminum (Al), iron (Fe), molybdenum (Mo), and phosphorous (P) in needles were reduced by O_3, while potassium (K) and magnesium (Mg) in the roots increased. No effects of combined fumigation were observed in any parameters except the P concentration in needles. The tolerance of F_1 to O_3 might potentially be related to a shift in ECM community structure. Special ECM to larch (*Suillus* sp.) could keep infection even under EO_3, so we can expect *Suillus* sp. as a candidate for improving larch rhizosphere [81].

4.2.1 Oak wilt disease

Deciduous oak dieback in Japan has been known since the 1930s [93], but it has been spreading to the whole Japanese island, mainly facing to sea of Japan side where N deposition, as well as transboundary O_3 levels, was high [94]. N deposition [83] and O_3 bring an imbalance in S/R ratio [29] and physiological homeostasis. The symbiotic ambrosia fungus *Raffaelea quercivora* is the causal agent of oak dieback and is vectored by oak beetle (*Platypus quercivorus* [Murayama]). This is the first example of an ambrosia beetle fungus that kills vigorous trees. Future global warming will possibly accelerate the overlapping of the distributions of *P. quercivorus* and most *Quercus* sp. with the result that oak dieback will become more serious in Japan.

The Meiji Jingu (shrine) is an important green resource in a mega-city, Tokyo, Japan, and planned natural regeneration smoothly continued until around 1990, however, invasive palm seedlings started to prevent the regeneration of various

Figure 8.
A: Inside view of tthe Meiji Shrine forest (courtesy of Ms. Harumi Ejiri-Noda). a: Insect trap with "continuous rotho" attached to the oak stem at holes for collecting adult beetle, b: Vinyl film inhibiting shatter-proof wrapped at damaged tree stem,c: Bark beetle (Platypus quercivorus), B: Dutch elm disease in the campus of Hokkaido University in Sapporo. Allow indicates declining individual elm tree infected with beetle. a: Trace under bark attacked by the beetle, b: Elm bark beetle (Scolytu esurient).

broadleaved trees. Progress of "Heat Island phenomenon" may have enabled the overwintering capacity of invasive palm [95]. Since its seed dispersal depends on birds, seedlings of palm are found in many urban forests. Moreover, palm leaves are O_3 tolerant, while O_3 suppresses root growth [77], but the moist forest floor seems to allow them to continue to grow. However, mature oaks are declining with intensive infection of wilt disease (**Figure 8A**). How can we manage the serious damages? The Health and vigor of oaks seem to be declining even though most oaks have relative O_3 tolerance (cf. **Table 1**) but their root growth is usually suppressed. We hardly regulate the progress of oak wilt disease.

4.2.2 Die-back of elm tree

A sudden increased mortality of *Ulmus davidiana* var. *japonica* (Japanese elm) trees occurred during 2014–2016 in Sapporo, northern Japan [96]. The estimated damaged tree age ranged between 36 and 186 years. Elm bark beetle (*Scolytus esuriens*) was regarded as a key insect for the spread of the disease. Fungi isolated from elm bark beetle and the wood of the galleries were identified as *Ophiostoma ulmi* and *Ophiostoma novo-ulmi*. Most declining trees with wilting branches were present around these beetle-attacked trees (**Figure 8B**). Therefore, Dutch elm disease caused this occurrence of Japanese elm dieback. This declining may be related to declining of tree vigor and health under degradation of the atmospheric environment with increasing nitrogen deposition and O_3 [80].

Higher leaf N content and lower plant defense (low condensed tannin) content in N loading and lower condensed tannin content in elevated O_3 were observed with O_3-FACE, suggesting that both N loading and elevated O_3 decreased the leaf defense and that N loading further enhanced the leaf quality as food resource of insect herbivores [19, 20, 89, 90]. Visible foliar injury caused by N loading might directly induce the reduction of the number of elm sawfly individuals.

Although elevated O_3 suppressed the plant defense capacity in leaves as found in white birch [41, 42, 97] and poplar [43], a significantly lower number of elm leaf beetle was observed in elevated O_3. Why did the number of leaf beetle in leaves in elevated O_3 decrease with lowered chemical defense?

4.2.3 Insect-plant interaction: O_3 as an environmental disturber

Plant–insect interactions are basic components of biodiversity conservation. Ground-level O_3 usually suppresses not only plant activities and disrupts interaction webs among plants-insect [41]. O_3 mixing ratios in suburbs are usually higher than in the center of cities and may reduce photosynthetic productivity, as suggested by Gregg et al. [65] for poplar and Moser-Reischl et al. [64] for fir. As a result, carbon-based defense capacities of plants may be suppressed by elevated O_3 more in the suburbs than in the urban region. Contrary to this expectation, grazing damage by leaf beetles (found in birch, alder, elm, and poplar: [43]) has been severe in some urban centers in comparison with the suburbs. To explain differences in grazing damages between urban areas and suburbs, the disruption of olfactory communication signals by elevated O_3 via changes in plant-regulated biogenic volatile organic compounds (BVOC) and long-chain fatty acids with double bonds are considered (e.g., [41, 43]). Ozone-disrupted BVOCs of plants should be considered to explain insect herbivory activities in urban and suburban systems.

5. Construction and maintenance of urban greening

The green space in the city not only serves as a cool island but also provides relaxation. However, even with management, there is always a risk of decline due to air pollution, high temperatures, high vapor deficit, nutrient imbalance, etc. due to narrow planting holes and restriction of the root system. "Meiji Jingu" forest was designed to be maintained by natural regeneration created in the center of Tokyo metropolis [98], but the invasion of palms by the heat island and the decline of oak wilt disease are imminent (**Figure 8A**, **B**). Even if the roadside trees and rooftop greening can be managed sufficiently, the viewpoint of nutrient cycling is important for park trees [99]. Additionally, we should introduce a moderating method of application of ethylene-di-urea (EDU). This is considered a chemical that offers protection to the treated plants against O_3 [100–102].

5.1 Nutrient cycling

Effects of tree species on mineral soil, litter, and root properties are found to be inconsistent, and understanding general cross-site patterns and the possible mechanism is important for enhancing the forest ecological service through proper tree species selection. Leaf litterfall nutrient concentrations and their ratios are a common indicator of site nutrient status and a critical component of many ecosystems [87]. Concentrations of N and P in the leaf litter are related to foliar concentrations, but they are reduced by nutrient resorption during senescence as affected by O_3 [61, 84]. However, few studies have assessed how the timing of litter collection affects estimates of nutrient concentration [61, 84]. The emphasis on sampling senesced leaf tissue at a single point in time leads to biased estimates of nutrient concentrations, stoichiometry, and litterfall and resorption fluxes, especially for P than for N in birch, beech and maple at the Hubbard Brook Forest. In northeast China, mineral conditions of four tree species; larch (*Larix gmelinii*), pine (*Pinus sylvestris* var. *mongolica*), poplar (*Populus* spp.), and elm (*Ulmus pumila*) were compared in several minerals, the component of SOC and soil N concentrations. Elm could capture more mineral SOC and nutrients, poplar induced mineral soil P depletion, and pine litter was of more recalcitrance for decomposition [103].

The effects of elevated O_3 in different soil conditions (brown forest, volcanic ash, and serpentine soil) on foliar elements stoichiometry were investigated in *Betula platyphylla* var. *japonica* (white birch), (*Quercus mongolica* var. *crispula* (oak) and Siebold's beech (*Fagus crenata*) with a O_3-FACE [61]. Soil nutrients have distinct impacts on retranslocation rate of K, Fe, and P. A negative correlation between foliar N and the metal elements was found in white birch. From the differences of foliar contents as well as their retranslocation rate, Siebold's beech with determinate shoot growth pattern was rather more sensitive to O_3 stress on foliar contents, meanwhile oak was possibly susceptible to O_3 on dynamics of immobile elements. Mn and K can become indices in assessing the O_3 and soil effects.

Decomposition is directly related to nutrient cycling [104]. Litter decay dynamics of paper birch (*Betula papyrifera*) were assessed at the Aspen-FACE in the northcentral USA. Leaf litter was decomposed for 12 months under factorial combinations of ambient (360 ppm) vs. elevated CO_2 (560 ppm), crossed with 36 vs. 55 $nLO_3\ L^{-1}$. *In situ*, litterbags methods revealed that CO_2 enrichment regardless of O_3 produced poorer quality litter (high C/N, lignin/N, and condensed tannins) than did ambient CO_2 (low C/N, lignin/N, and condensed tannins). Substrate quality differences were reflected in the mass loss rates of litter (k-values), which were high for litter generated under ambient CO_2 (0.89 $yr.^{-1}$) and low for litter generated under elevated CO_2 (0.67 $yr.^{-1}$), which suggests regulating the storage of fixed C and the release of CO_2 from northern forest ecosystems. Additionally, revegetation should be needed in NE China where we have a limited amount of rainfall (<500 mm $yr.^{-1}$) after the thinning or harvest of the "restoring agricultural land forest" [7]. We should carefully select larch species, such as *Larix gmelnii* under salt stress and EO_3 [105].

5.2 EDU application

Increasing evidence on the antiozonate efficacy of EDU against the phytotoxic action of O_3 is becoming more readily available [100, 101]. EDU is a very promising antiozonant with its antiozonate action being observed when applied to roots in concentrations of 275.7 to 374.3 mg L^{-1}. The effect of ambient O_3 on visible foliar injury, growth, and biomass in field-grown poplar cuttings of an Oxford clone sensitive to O_3 irrigated with EDU or water for 3 years. Protective effects of EDU on O_3 visible injury were found but no increase in stem height and diameter was found [102]. EDU was more effective in some cultivars compared with others although this remains inexplicable [33]. Additionally, the biochemical mechanism of its antiozonate activity is still unclear [78].

6. Concluding remarks and future perspective

Urban vegetation or green infrastructure, as a cost-effective and nature-based approach, aids in meeting clean air standards, which should be taken into account by policy-makers. Nowadays, complex atmospheric pollution seriously threatens the vitality of suburb trees as well as urban trees (e.g., [23]). Continuous attention should be paid to the long-term changes of O_3 and related impacts on urban greening. On the basis of "matching site with trees," more attention should be paid to the relationship between plants and urban planting small environment [33]. Highly ranked tree species that can inclusively resist O_3 and other pollutants which will be the priority target of urban greening.

Recent high air pollution in many cities indicates the urgent need for policy action and for urban development based on local air quality management; the prospects of improving urban air quality through proper design and protection of vegetation within local planning strategies [106], such as selecting of non-emission species of BVOC as a precursor of O_3. In addition, we should make more effort to reduce the emission of CO, NOx, VOC, etc. as O_3 precursors [22, 64]. Elevated ground-level O_3 can adversely affect plants and inhibit their growth and productivity, threatening ecological health via pollination [38, 40, 41] food security, etc. Therefore, it is important to develop ways to protect plants against O_3-induced negative effects [13, 107]. Of course, we should choose tree species of non-emitter of BVOC as a precursor of O_3 as shown by screening experiments of tree species [18]. It would be effective to create a space where the wind can easily pass through the city plan to discharge O_3 [13, 46]. In addition, we expect that new studies should be green chemistry for the continuous support of green infrastructure for city residents.

Acknowledgements

We thank many precious studies in O_3-related research in the world for improving our knowledge. The last author is grateful for offering much information offered by Prof. E. Agathokleous, Dr. M. Watanabe, Dr. M. Yamaguchi, Prof. WJ. Wang, Dr. Y. Mizuuchi, Prof. T. Izuta, and Dr. T. Watanabe. Financial support in part by the National Key Research and Development Program of China (2017YFE0127700; LY. Qu), JST (No. JPMJSC18HB: representative researcher, M. Watanabe of TUAT and T. Watanabe of HU), and the AUDI environmental foundation through TUM, Germany.

Abbreviations

AOT40	Accumulated ozone exposure over a threshold of 40 ppb
BVOC	biogenic volatile organic compound
CO	Carbon oxide or Carbonic oxide
ECM	Ectomycorrhizal fungi
ECO_2	Elevated CO_2
EO_3	Elevated O_3
EPA	U.S.A. air pollutant emission
FACE	Free-Air CO_2 enrichment
Gs	stomatal conductance
ICP	International cooperative program on effects of air pollution on natural vegetation and crops
IT	Increased air temperature
LAI	Leaf area index (leaf area per unit land area; $m^2\ m^{-2}$)
LRTAP	National emissions reported to the convention on long-range transboundary air pollution
LandSat	Land Satellite
NO_2	Nitrogen dioxide
O_3	Ozone
O_3-FACE	Ozone in a free-air controlled environment

OTC	Open top chamber
PM	Particle matter
POD-0	Phytotoxic O_3-dose above a threshold of 0 uptake
PRD	Pearl River Delta economic zone
S-AQI	Species air quality index
S/R ratio	shoot to root dry mass ratio
SDGs	Sustainable Development Goals
SOC	Soil organic carbon
SOx	Sulfur oxide complex
YRD	Yangtze River Delta

Author details

Laiye Qu[1,2], Yannan Wang[3], Cong Shi[4], Xiaoke Wang[1], Noboru Masui[5], Thomas Rötzer[6], Toshihiro Watanabe[7] and Takayoshi Koike[1,7]*

1 Urban and Regional Ecology, Research Center for Eco-Environmental Science, Beijing, China

2 University of Chinese Academy of Sciences, Beijing, China

3 Jiangxi Academy of Forestry, Nanchang, China

4 School of Environmental Science and Engineering, Tiangong University, Tianjin, China

5 Graduate School of Agriculture, Hokkaido University, Sapporo, Japan

6 Forest Growth and Yield Science, School of Life Sciences, Weihenstephan, Technical University of Munich, Freising, Germany

7 Research Faculty of Agriculture, Hokkaido University, Sapporo, Japan

*Address all correspondence to: lyqu@rcees.ac.cn; tkoike@for.agr.hokudai.ac.jp

References

[1] EPA-US: Air Pollutant Emissions Trends Data. Available from: https://www.epa.gov/air-emissions-inventories/air-pollutant-emissions-trends-data)

[2] Kurokawa J, Ohara T. Long-term historical trends in air pollutant emissions in Asia: Regional emission inventory in Asia (REAS) version 3. Atmospheric Chemistry and Physics. 2020;**20**:12761-12793. DOI: 10.5194/acp-20-12761-2020

[3] LRTAP. National emissions reported to the convention on long-range transboundary air pollution. Available from: https://www.eea.europa.eu/data-and-maps/data/national-emissions-reported-to-the-convention-on-long-range-transboundary-air-pollution-lrtap-convention-15

[4] Qu LY, Wang YN, Masyagina O, Kitaoka S, Fujita S, Kita K, et al. Larch: A promising deciduous conifer as an eco-environmental resource. In: Cristina A. editor. Conifers - Recent Advances Gonçalves. London: IntechOpen, 2022b. Available from: https://www.intechopen.com/chapters/80260

[5] Qu LY, Wang XN, Mao QZ, Agathokleous E, Choi DS, Tamai Y, et al. Responses of ectomycorrhizal diversity of larch and its hybrid seedlings and saplings to elevated CO_2, O_3, and high nitrogen loading. Eurasian Journal of Forest Research. 2022a;**22**:23-27. DOI: 10.14943/EJFR

[6] De Marco A, Sicard P, Feng ZZ, Agathokleous E, Alonso R, Araminiene V, et al. Strategic roadmap to assess forest vulnerability under air pollution and climate change. Global Change Biology. 2022. DOI: 10.1111/gcb.1627

[7] Zhang PC, Shao GF, Zhao G, Le Master DC, Parker GR, Dunning JB Jr, et al. China's forest policy for the 21st century. Science. 2000;**288**:2135-2136. DOI: 10.1126/science.288.5474.2135

[8] Feng ZZ, Sun JS, Wan WX, Hu EZ, Calatay V. Evidence of widespread ozone-induced visible injury on plants in Beijing China. Environmental Pollution. 2014;**193**:296-301. DOI: 10.1016/j.envpol.2014.06.004

[9] Izuta T. Air Pollution Impacts on Plants in East Asia. Springer: Tokyo. 2017. p. 322. eBook ISBN: 978-4-431-56438-6

[10] Mizuuchi Y, Ueda H, editors. Shinto Shrines as Social and Environmental Infrastructure. Vol. 83. Journal of the Japanese Institute of Landscape Architecture; 2020. pp. 236-292 (in Japanese)

[11] Saitanis CJ, Katsaras DH, Riga-Karandinos AN, Lekk DB. Evaluation of ozone phytotoxicity in the greater area of a typical Mediterranean small city (Volos) and in the nearby forest (Pelion Mt.), Central Greece. Bulletin Environmental Contamination Toxicology. 2004;**72**:1268-1277. DOI: 10.1007/s00128-0

[12] Kitao M, Yasuda Y, Kominami Y, Yamanoi K, Komatsu K, Miyama T, et al. Increased phytotoxic O_3 dose accelerates autumn senescence in an O_3-sensitive beech forest even under the present-level O_3. Scientific Reports. 2015;**6**:32549. DOI: 10.1038/srep32549

[13] Sicard P, Agathokleous E, Araminiene V, Carrari E, Hoshika Y, De Marco A, et al. Should we see urban trees as effective solutions to reduce

increasing ozone levels in cities? Environmental Pollution. 2018;**243**: 163-176. DOI: 10.1016/j.envpol.2018.08.049

[14] Han DH, Shen HL, Duan WB, Chen LX. A review on particulate matter removal capacity by urban forests at different scales. Urban Forestry & Urban Greening. 2019;**48**:126565. DOI: 10.1016/j.ufug.2019.126565

[15] Yamaguchi M, Otani Y, Li P, Nagao H, Lenggoro IW, Ishida A, et al. Effects of long-term exposure to ammonium sulfate particles on growth and gas exchange rates of *Fagus crenata, Castanopsis sieboldii*, *Larix kaempferi* and *Cryptomeria japonica* seedlings. Atmospheric Environment. 2014;**97**: 493-500. DOI: 10.1016/j.atmosenv.2014.01.023

[16] Yamaguchi M. Experimental Studies on the effects of fine particulate matter ($PM_{2.5}$) and ozone on plants. Journal of Japanese Society of Atmospheric Environment. 2021;**56**:25-33 (in Japanese with English summary)

[17] Yamaji K, Ohara T, Uno I, Kurokawa J, Pochanart P, Akimmoto H. Future prediction of surface ozone over east Asia using Models-3 community multiscale air quality modeling system and regional emission inventory in Asia. Journal of Geophysical Research Atmospheres. 2008;**113**(D8). DOI: 10.1029/2007JD008663

[18] Tani A, Mochizuki T. Exchanges of volatile organic compounds between terrestrial ecosystems and the atmosphere. Journal of Agricultural Meteorology. 2021;**77**:66-80. DOI: 10.2480/agrmet.D-2

[19] Agathokleous E, Belz RG, Calatayud V, De Marco A, Hoshika Y, Kitao M, et al. Predicting the effect of ozone on vegetation via linear non-threshold (LNT), threshold and hormetic dose-response models. Science of the Total Environment. 2019a;**649**:61-74. DOI: 10.1016/J.SCITOTENV.2018.08.264

[20] Agathokleous E, WaiLi Y, Ntatsi G, Konno K, Saitanis CJ, Kitao M, et al. Effects of ozone and ammonium sulfate on cauliflower: Emphasis on the interaction between plants and insect herbivores. Science of the Total Environment. 2019b;**659**:995-1007. DOI: 10.1016/j.scitotenv.2018.12.388

[21] Hatakeyama S. Gaseous species. In: Izuta T, editor. Air Pollution Impacts on Plants in East Asia. Tokyo: Springer; 2017. pp. 3-20. DOI: 10.1007/978-4-431-56438-6_1

[22] Tatsumi K. Rice yield reductions due to ozone exposure and the roles of VOCs and NOx in ozone production in Japan. Journal of Agriculture Meteorology. 2022;**78**:89-100. DOI: 10.2280/agrmet.D-21-00051

[23] Wang WJ, Parrish DD, Wang SW, Bao FX, Ni RJ, Li X, et al. Long-term trend of ozone pollution in China during 2014-2020: Distinct seasonal and spatial characteristics. Atmospheric Chemistry and Physis. 2022c; open access. DOI: 10.5194/acp-2022-123

[24] Wang CC, Ren ZB, Dong YL, Zhang P, Guo YJ, Wang WJ, et al. Efficient cooling of cities at global scale using urban green space to mitigate urban heat island effects in different climatic regions. Urban Forestry & Urban Greening. 2022a;**74**:127635. DOI: 10.1016/j.ufug.2022.127635

[25] Wang WJ, Tian PL, Zhang JH, Agathokleous E, Xiao L, Koike T, et al. Big data-based urban greenness in Chinese megalopolises and possible contribution to air quality control.

Science of the Total Environment. 2022b;**824**:153834. DOI: 10.1016/j.scitotenv.2022.153834

[26] Wang WJ, Xu S, Li B, Chen W, Li Y, He XY, et al. Responses of spring leaf phenological and functional traits of two urban tree species to air warming and/or elevated ozone. Plant Physiology and Biochemistry. 2022d;**179**:158-167. DOI: 10.1016/j.plaphy.2022.03.015

[27] Ainsworth EA. Understanding and improving global crop response to ozone pollution. Plant Journal. 2017;**90**: 886-897. DOI: 10.1111/tpj.13298

[28] Jolivet Y, Bagard M, Cabané M, Vaultier M-N, Gandin A, Afif D, et al. Deciphering the ozone-induced changes in cellular processes: A prerequisite for ozone risk assessment at the tree and forest levels. Annals of Forest Science. 2016;**73**:923-943. DOI: 10.1007/s13595-016-0580-3

[29] Agathokleous E, Saitanis CJ, Wang XN, Watanabe M, Koike T. A review study on past 40 years of research on effects of tropospheric O_3 on belowground structure, functioning, and processes of trees: A linkage with potential ecological implications. Water, Air Soil and Pollution. 2016;**227**:33. DOI: 10.1007/s11270-015-2715-9

[30] Koike T, Watanabe M, Hoshika Y, Kitao M, Matsumura H, Funada R, et al. Effects of Ozone on forest ecosystems in East and Southeast Asia. Elsevier Developments in Environmental Science. 2013;**13**:371-390. DOI: 10.1016/B978-0-08-098349-3.00017-7, 10.1016/j.ecolind.2019.04.017

[31] Sitch S, Cox PM, Collins WJ, Huntingford C. Indirect radiative forcing of climate change through ozone effects on the land-carbon sink. Nature. 2007; **448**:791-794. DOI: 10.1038/nature06059

[32] Gao F, Calatayud LV, García-Breijo F, Armiñana RJ, Feng ZZ. Effects of elevated ozone on physiological, anatomical and ultrastructural characteristics of four common urban tree species in China. Ecological Indicators. 2016;**67**:367-379. DOI: 10.1016/j.ecolind.2016.03.012

[33] Saitanis CJ, Agathokleous E. Exogenous application of chemicals for protecting plants against ambient ozone pollution: What should come next? Current Opinion in Environmental Science & Health. 2021;**19**:100215. DOI: 10.1016/j.coesh.2020.10.003

[34] Yamaguchi M, Watanabe M, Matsumura H, Kohno Y, Izuta T. Experimental Studies on the effects of ozone on growth and photosynthetic activity of Japanese forest tree species. Asian Journal of Atmospheric Environment. 2011;**5**:65-78. DOI: 10.5572/ajae.2011.5.2.065

[35] Zhou SS, Tai APK, Sun S, Sadiq M, Heald CL, Geddes JA. Coupling between surface ozone and leaf area index in a chemical transport model: Strength of feedback and implications for ozone air quality and vegetation health. Atmospheric Chemistry and Physics. 2018;**18**:14133-14148. DOI: 10.5194/acp-18-14133-2018

[36] Li P, Calatayud V, Gao F, Uddling J, Feng ZZ. Differences in ozone sensitivity among woody species are related to leaf morphology and antioxidant levels. Tree Physiology. 2016;**36**:1105-1116. DOI: 10.1093/treephys/tpw042

[37] ME. Guide to the Millennium Assessment Reports. 2000. Available from: https://www.millennium assessment.org/en/index.html

[38] Agathokleous E, Feng ZZ, Oksanen E, Sicard P, Wang Q, Saitanis CJ, et al.

Ozone affects plant, insect, and soil microbial communities: A threat to terrestrial ecosystems and biodiversity. Science Advances. 2020;**6**(eabc1176): 1-17. DOI: 10.1126/sciadv.abc1176

[39] Agathokleous E, Kitao M, Kinose Y. A review study on ozone phytotoxicity metrics for setting critical levels in Asia. Asian Journal of Atmospheric Environment. 2018;**12**:1-16. DOI: 10.5572/Ajae.2018.12.1.001

[40] Fuentes JD, Roulston TH, Zenker J. Ozone impedes the ability of a herbivore to find its host. Environmental Research Letters. 2013;**8**:014048. DOI: 10.1088/1748-9326/8/1/014048

[41] Masui N, Agathokleous E, Mochizuki T, Tani A, Matsuura H, Koike T. Ozone disrupts the communication between plants and insects in urban and suburban areas: An updated insight on plant volatiles. Journal of Forestry Research. 2021a;**32**:1337-1349. DOI: 10.1007/s11676-020-01287-4

[42] Masui N, Agathokleous E, Tani A, Matsuura H, Koike T. Plant-insect communication in urban forests: Similarities of plant volatile compositions among tree species (host vs. non-host trees) for alder leaf beetle *Agelastica coerulea*. Environmental Research. 2021b;**204**:11996. DOI: 10.1016/j.envres.2021.111996

[43] Masui N, Koike T, Kitaoka T, Watanabe Y, Satoh F, Watanabe T. Outbreak of insects and disease of two poplar cuttings grown under free air elevated Ozone. Tree and Forest Health. 2022;26:26-27. (in Japanese). Available from: http://www.thrs.jp/journal/journal_index.html

[44] Oberndorfer E, Lundholm J, Bass B, Coffman RR, Doshi H, Dunnett N, et al. Green roofs as urban ecosystems: Ecological structures, functions, and services. Bioscience. 2007;**57**:823-833. DOI: 10.1641/B571005

[45] Narita K, Sugawara H. Cold-air Seeping-out Phenomena in an Urban Green Space. Journal of Geography. 2011;**120**:411-425 (In Japanese with English summary). DOI: 10.5026/JGEOGRAPHY.120.411

[46] Ortía MA, Casanelles-Abella J, Chiron F, Deguines N, Hallikma T, Jaksi P, et al. Negative relationship between woody species density and urban green spaces in seven European cities. Urban Forestry & Urban Greening. 2022; **127650**. DOI: 10.1016/j.ufug.2022.127650

[47] Koike T, Kitao M, Ichie T, Watanabe M. Physiological Ecology of Woody plants. Kyoritsu Publisher: Tokyo. p. 262. ISBN: 9784320058125. (in Japanese)

[48] Winbourne JB, Jones TS, Garvey SM, Harrison JL, Wang L, Li D, et al. Tree transpiration and urban temperatures: Current understanding, implications, and future research directions. Bioscience. 2020;**70**:576-588. DOI: 10.1093/biosci/biaa055

[49] Nouchi I. Atlas of Visible Injuries of Representative Tree Species. National Institute of Agro-Environmental Studies. 2002, (in Japanese), ISBN: 10: 4842500794

[50] Nowak DJ, Crane DE, Stevens JC. Air pollution removal by urban trees and shrubs in the United States. Urban Forestry & Urban Greening. 2006;**4**: 115-123. DOI: 10.1016/j.ufug.2006.01.007

[51] Wada R, Ueyama M, Tani A, Mochizuki T, Miyazaki Y, Kawamura K, et al. Observation of vertical profiles of NO, O3, and VOCs to estimate their sources and sinks by inverse modeling in a Japanese larch forest. Journal of

Agricultural Meteorology. 2020;**76**:1-10. DOI: 10.2480/agrmet.D-18-00029

[52] Feng ZZ, Li P, Yuan XY, Gao F, Jiang LJ, Dai LL. Progress in ecological and environmental effects of ground-level O_3 in China. Acta Ecologica Sinica. 2018a; **38**:1530-1541. DOI: 10.5846/stxb201704100618

[53] Feng ZZ, Peng JL, Calatayud LV, Tang HY. Identification of ozone visible injury in Chinese plants. China Environment Publishing Group, Available from: http://www.cesp.com.cn/ (in Chinese with English summary). 2018b. ISBN: 978-7-5111-3704-3

[54] Li K, Jacob DJ, Liao H, Shen L, Zhang Q, Bates KH. Anthropogenic drivers of 2013-2017 trends in summer surface ozone in China. Proceedings of the National Academy of Sciences of the United States of America. 2019;**116**:422-427. DOI: 10.1073/pnas.1812168116

[55] Zeng Y, Cao Y, Qiao X, Seyler BC, Tang Y. Air pollution reduction in China: Recent success but great challenge for the future. Science of Total Environment. 2019;**663**:329-337. DOI: 10.1016/j.scitotenv.2019.01.262

[56] Xu XB. Recent advances in studies of ozone pollution and impacts in China: A short review. Current Opinion in Environmental Science & Health. 2021; **19**. DOI: 10.1016/j.coesh.2020.100225

[57] Liu XW, Xu XB, Lin WL. Variation characteristics of surface O_3 in Beijing and its surrounding area. China Environmental Science. 2010;**30**:946-953 Corpus ID: 131153021

[58] Wang J, Zhang HX, Wang XK, Ouyang ZY, Mou YJ. Study on air pollutants in three representative regions of Beijing. Environmental Chemistry. 2011;**30**:2047-2053

[59] Zhao PS, Dong F, Yang YD, He D, Zhao XJ, Zhang WZ, et al. Characteristics of carbonaceous aerosol in the region of Beijing, Tianjin, and Hebei, China. Atmospheric Environment. 2013;**71**:389-398. DOI: 10.1016/j.atmosenv.2013.02.010

[60] Wan WX, Manning WJ, Wang XK, Zhang HX, Sun X, Zhang QQ. Ozone and ozone injury on plants in and around Beijing, China. Environmental Pollution. 2014;**191**:215-222. DOI: 10.1016/j.envpol.2014.02.035

[61] Shi C, Watanabe T, Koike T. Leaf stoichiometry of deciduous tree species in different soils exposed to free-air O_3 enrichment over two growing seasons. Environmental and Experimental Botany. 2017;**138**:148-163. DOI: 10.1016/j.envexpbot.2017.03.012

[62] Kohno Y. Decline of *Fagus crenata* in the Tanzawa Mountains, Japan. In: Izuta T, editor. Air Pollution Impacts on Plants in East Asia. 2017. pp. 151-162. DOI: 10.1007/978-4-431-56438-6_10

[63] Taniwaki T, Aihara K, Saito H, Yamane M. Factors and its interactive effects on beech forests decline in the Tanzawa Mountains. Bulletin of the Kanagawa Prefecture Natural Environment Conservation Center. 2016;**14**:1-12 (in Japanese), ISSN 1349-2500

[64] Moser-Reischl A, Rötzer T, Biber P, Ulbricht M, Uhl E, Qu L, et al. Growth of *Abies sachalinensis* along an urban gradient affected by environmental pollution in Sapporo, Japan. Forests. 2019;**10**:707. DOI: 10.3390/f10080707

[65] Gregg JW, Clive G, Jones CJ, Dawson TE. Urbanization effects on tree growth in the vicinity of New York City. Nature. 2003;**424**(6945):183-187. DOI: 10.1038/nature01728

[66] Wan WX, Xia YJ, Zhang HX, Wang J, Wang XK. The ambient ozone pollution and foliar injury of the sensitive woody plants in Beijing exurban region. Acta Ecologica Sinica. 2013;**33**:1098-1105. DOI: 10.5846/stxb201203220388

[67] Zhang HX, Sun X, Yao YH, Wan WX, Xiao Y, Sun BF, et al. Ground-level ozone distribution pattern in summer of Beijing and its foliar injury effect upon plants. Acta Ecological Sinica. 2014;**34**: 4756-4765. DOI: 10.5846/stxb201306261781 in Chinese with English summary

[68] Schaub M, Calatayud V, Ferretti M, Brunialti G, Lövblad G, Krause G, et al. Assessment of ozone injury. In Manual on methods and criteria for harmonized sampling, assessment, monitoring and analysis of the effects of air pollution on forests. 2016. p. 21. ICP Forests. Available from: https://www.dora.lib4ri.ch/wsl/islandora/object/wsl:11916

[69] Innes J L, Skelly I M, Schaub M. Ozone and broadleaved species. A Guide to the Identification of Ozone-induced Foliar Injury. Bern Switzerland: Paul Haupt Publishers, 2001. ISBN: 10: 3258063842

[70] Cheng XY, Luo CW, Liu QJ, Meng SW, Zhou YS, Zhou H. Tropospheric ozone distribution and injury on leaves of sensitive woody plants in Nanchang city, China. Plant Science Journal. 2016; **34**:211-219. DOI: 10.11913/PSJ.2095-0837.2016.20211

[71] Wan WX, Zhang S, Li J, Sun X, Guan ZG, Yu XH, et al. Regional differences of urban ozone pollution and its damage to plants in Hebei province. Ecology and Environmental Sciences. 2021;**30**: 2185-2194. DOI: 10.16258/j.cnki.1674-5906.2021.11.009

[72] Izuta T, Miyake H, Totsuka T. Evaluation of air-polluted environment based on the growth of radish plants cultivated in small-sized open-top chambers. Environmental Sciences. 1993;2:25-37. Available from: https://cir.nii.ac.jp/crid/1573950400042023680

[73] Grantz DA, Zinsmeister D, Burkhardt J. Ambient aerosol increases minimum leaf conductance and alters the aperture–flux relationship as stomata respond to vapor pressure deficit (VPD). New Phytologist. 2018;**219**:275-286. DOI: 10.1111/nph.15102

[74] Takahashi K, Ohta A, Sase H, Murao N, Takada K, Yamaguchi M, et al. Seasonal variations in the amount of black carbon particles deposited on the leaf surfaces of nine Japanese urban greening tree species and their related factors. International Journal of Phytoremediation. 2020. DOI: 10.1080/15226514.2022.2072808

[75] Li P, Wei YY, Feng ZZ. Preliminary screening for the urban forest against combined air pollution. Environmental Science. 2020;**41**:4495-4503. DOI: 10.13227/j.hjkx.202004038

[76] Nakashima K, Nishi Y, Kawada S, Yamaguchi M. Assessment of ambient air pollution in Nagasaki city, Japan, based on ozone impacts on growth of radish (*Raphanus sativus* L.) using the Open-Top Chamber Method. Journal of Japanese Society of Atmospheric Environment. 2018;**53**:186-193. DOI: 10.11298/taiki.53.186

[77] Paoletti E, Hoshika Y, Arab L, Martinia S, Cotrozzi L, Weber D, et al. Date palm responses to a chronic, realistic ozone exposure in a FACE experiment. Environmental Research. 2021;**195**:110868. DOI: 10.1016/j.envres.2021.110868

[78] Agathokleous E, Feng ZZ, Saitanis CJ. Effects of ozone on forests. In: Akimoto H, Tanimoto H, editors. Handbook of Air Quality and Climate Change. Springer Nature Singapore Pte Ltd.; 2022. DOI: 10.1007/978-981-15-2527-8_24-1

[79] Watanabe M, Yamaguchi M, Matsumura H, Kohno Y, Koike T, Izuta T. A case study of risk assessment of ozone impact on forest tree species in Japan. Asian Journal of Atmospheric Environment. 2011;**5**:205-215. DOI: 10.5572/ajae.2011.5.4.205

[80] Sugai T, Okamoto S, Agathokleous E, Masui N, Satoh F, Koike T. Leaf defense capacity of Japanese elm (*Ulmus davidiana* var. *japonica*) seedlings subjected to a nitrogen loading and insect herbivore dynamics in a free air ozone-enriched environment. Environmental Science and Pollution Research. 2020;**27**:3350-3360. DOI: 10.1007/s11356-019-06918-w

[81] Wang XN, Qu LY, Mao QZ, Watanabe M, Hoshika Y, Koyama A, et al. Ectomycorrhizal colonization and growth of the hybrid larch F_1 under elevated CO_2 and O_3. Environmental Pollution. 2015;**197**:116-126. DOI: 10.1016/j.envpol.2014.11.031

[82] Wang Y, Xu S, Zhang W, Li Y, Wang N, He X, et al. Responses of growth, photosynthesis and related physiological characteristics in leaves of *Acer ginnala* Maxim. to increasing air temperature and/or elevated O_3. Plant Biology. 2021;**23**:221-231. DOI: 10.1111/plb.13240

[83] Watanabe M, Li J, Matsumoto M, Aoki T, Ariura R, Fuse T, et al. Growth and photosynthetic responses to ozone of Siebold's beech seedlings grown under elevated CO_2 and soil nitrogen supply. Environmental Pollution. 2022;**304**: 119233. DOI: 10.1016/j.envpol. 2022. 119233

[84] Shi C, Eguchi V, Meng FK, Watanabe T, Satoh F, Koike T. Retranslocation of foliar nutrients of deciduous tree seedlings in different soil condition under free-air O_3 enrichment. iForest - Biogeosciences and Forestry. 2016;**9**:835-841. DOI: 10.3832/ifor1889-009

[85] Wang XN, Agathokleous E, Qu LY, Watanabe M, Koike T. Effects of CO_2 and/or O_3 on the interaction between root of woody plants and ectomycorrhizae. Journal of Agriculture Meteorology. 2016;**72**:95-105. DOI: 10.2480/agrmet.D-14-00045

[86] Schoonhoven LM, van Loon JJA, Dicke M. Insect–Plant Biology. 2nd ed.. 2005, Oxford University Press. p. 440. Available from: https://baloun.entu.cas.cz/~cizek/EkologieLesaPrednaska/InsectPlantInteractions/InsectPlantBiology.pdf

[87] See CR, Yanai RD, Fahey TJ. Shifting N and P concentrations and stoichiometry during autumn litterfall: Implications for ecosystem monitoring. Ecological Indicators. 2019;**103**:488-492. DOI: 10.1016/j.ecolind.2019.04.017

[88] Schoonhoven LM, Jermy T, van Loon JJA. Insect–Plant Biology: From Physiology to Evolution. 1st ed. Chapman & Hall; 1998. p. 409

[89] Abu ElEla SA, Agathokleous E, Ghazawy NA, Amin TR, ElSayed WM, Koike T. Enzyme activity modification in adult beetles (*Agelastica coerulea*) inhabiting birch trees in an ozone-enriched atmosphere. Environmental Science and Pollution Research. 2018a; **25**:32675-32683. DOI: 10.1007/s11356-018-3243-0

[90] Abu ElEla SA, Agathokleous E, Koike T. Growth and nutrition of *Agelastica coerulea* (Coleoptera: Chrysomelidae) larvae changed when fed with leaves obtained from an O_3-enriched atmosphere. Environmental Science and Pollution Research. 2018b; **25**:13186-13194. DOI: 10.1007/s11356-018-1683-1

[91] Lindroth LR. Impacts of elevated atmospheric CO_2 and O_3 on forests: Phytochemistry, trophic interactions, and ecosystem dynamics. Journal of Chemical Ecology. 2010;**36**:2-21. DOI: 10.1007/s10886-009-9731-4

[92] Agetsuma N, Agetsuma-Yanagihara Y, Takafumi H, Nakaji T. Plant constituents affecting food selection by sika deer. The Journal of Wildlife Management. 2019;**83**:669-678. DOI: 10.1002/jwmg.21615

[93] Kamata N, Esaki K, Kato K, Igeta Y, Wada K. Potential impact of global warming on deciduous oak dieback caused by ambrosia fungus *Raffaelea* sp. carried by ambrosia beetle *Platypus quercivorus* (Coleoptera: Platypodidae) in Japan. Bulletin of Entomological Research. 2002;**92**:119-126. DOI: 10.1079/BER2002158

[94] Kitada T, Okamura K, Nakanishi H, Mori H. Production and transport of ozone in local flows over Central Japan-comparison of numerical calculation with airborne observation. In: Gryning SE, Batchvarova E, editors. Air Pollution Modeling and Its Application XIII. Boston, MA: Springer; 2000. pp. 95-106. DOI: 10.1007/978-1-4615-4153-0_10

[95] Hagiwara S. Current Condition of The Shirogane-forest. 2010. Available from: https://www.kahaku.go.jp/research/researcher/my_research/meguro/hagiwara/index.html

[96] Miyamoto T, Masuya H, Koizumi A, Yamaguchi T, Ishihara M, Yamaoka Y, et al. A report of dieback and mortality of elm trees suspected of Dutch elm disease in Hokkaido, Japan. Journal of Forest Research. 2019;**24**:396-400. DOI: 10.1080/13416979.2019.1679942

[97] Agathokleous E, Sakikawa T, Abu ElEla SA, Mochizuki T, Nakamura M, Watanabe M, et al. Ozone alters the feeding behavior of the leaf beetle *Agelastica coerulea* (Coleoptera: Chrysomelidae) into leaves of Japanese white birch (*Betula platyphylla* var. *japonica*). Environmental Science and Pollution Research. 2017;**24**:17577-17583. DOI: 10.1007/s11356-017-9369-7

[98] Mizuuchi Y, Nakamura KW. Landscape assessment of a 100-year-old sacred forest within a shrine using geotagged visitor employed photography. Journal of Forest Research. 2021;**26**:267-277. DOI: 10.1080/13416979.2021.1892251

[99] Chapin III FS, Matson PA, Mooney HA, Vitousek PM. Principles of Terrestrial Ecosystem Ecology. 2002, Springer Verlag, SPIN 1086. p. 392. ISBN: 978-1-4419-9503-2

[100] Agathokleous E, Koike T, Watanabe M, Hoshika Y, Saitanis CJ. Ethylene-di-urea (EDU), an effective phytoprotectant against O_3 deleterious effects and a valuable research tool. Journal of Agriculture Meteorology. 2015a;71: 185-195. Available from: https://www.jstage.jst.go.jp/article/agrmet/71/3/71_D-14-00017/_pdf/-char/en

[101] Agathokleous E, Koike T, Saitanis JC, Watanabe M, Satoh F, Hoshika Y. Ethylenediurea (EDU) as a protectant of plants against O_3. Eurasian Journal of Forest Research. 2015b;18: 37-50, Available from: https://eprints.lib.

hokudai.ac.jp/dspace/bitstream/2115/60324/1/102-Agathokleous-2.pdf

[102] Hoshika Y, Pecori F, Conese I, Bardelli T, Marchi E, Manning WJ, et al. Effects of a three-year exposure to ambient ozone on biomass allocation in poplar using ethylenediurea. Environmental Pollution. 2013;**180**: 299-303. DOI: 10.1016/j.envpol.2013.05.041

[103] Wei CH, Xiao L, Shen G, Wang H, Wang WJ. Effects of tree species on mineral soil C, N, and P, litter and root chemical compositions: Cross-sites comparisons and their relationship decoupling in Northeast China. Trees. 2021;**35**:1971-1992. DOI: 10.1007/s00468

[104] Kopper BJ, Lindroth RJ. Effects of elevated carbon dioxide and ozone on the phytochemistry of aspen and performance of an herbivore. Oecologia. 2003;**134**:95-103. DOI: 10.1007/s00442-002-1090-6

[105] Wang YN. Ecophysiological study on the effect of elevated O_3 and salinity stress on the growth and photosynthesis of three larch species. [Master thesis] of Graduate School of Agriculture, Hokkaido University. p. 29. DOI: 10.3389/ffgc.2019.00053

[106] Badach J, Dymnicka M, Baranowski A. Urban vegetation in air quality management: A review and policy framework. Sustainability. 2020;**12**:1258. DOI: 10.3390/su12031258

[107] Agrawal SB, Agrawal M, Singh A. Tropospheric Ozone: A Hazard for Vegetation and Human Health. Cambridge Scholars Publishing. 2021, ISBN (10): 1-5275-7057-6. p. 646

Chapter 17

Agroforestry: An Approach for Sustainability and Climate Mitigation

Ricardo O. Russo

Abstract

Agroforestry Systems (AFS), or the association of trees with crops (or animals), is a strategy for land management and use that allows production within the sustainable development: (a) environmentally (production environmentally harmonic); (b) technically (integrating existing resources on the farm); (c) economically (increase in production), and (d) socially (equality of duties and opportunities, quality of life of the family group). As an intentional integration of trees or shrubs with crop and animal production, this practice makes environmental, economic, and social benefits to farmers. Given that there is a set of definitions, rather than a single definition of Agroforestry (AF) and AFS, it is justified to explore the historical evolution and the minimum coincidences of criteria to define them and apply them in the recovery of degraded areas. Knowing how to classify AFS allows us to indicate which type or group of AFS is suitable for a particular area with its characteristics. The greatest benefit that AFS can bring to degraded or sloping areas lies in their ability to combine soil conservation with productive functions. In other words, AF is arborizing agriculture and animal production to obtain more benefits including climate change adaptation and mitigation by ecosystem services.

Keywords: agroforestry, agroforestry systems, silvopastoral systems, land use, tree biomass, climate change mitigation, carbon sequestration, live fences, shade trees

1. Introduction

Agroforestry systems (AFS) date back to the Mayan civilization, from 600 to 300 BC, with an apogee estimated to have lasted until 300 or 900 AC. this culture developed in the region of humid forests, but it is claimed that its agrarian system would have developed in the highlands of Guatemala until reaching the Yucatan jungle, where they practiced a pre-Hispanic style of agriculture adapted to forest management, which may well be called agroforestry [1–3]. The Mayan were poly-farmers; so that, they can be considered a culture with knowledge of land use and forest management; they used to practice a shifting cultivation system, which implies rotation of land use with periods of farming and resting the soil, and sometimes the selective logging leaving some useful trees. They farmed in small fields or clearings in the forest, and from the neighboring forest they took medicine, food, and building materials. This whole system of management of the natural forest and itinerant agriculture was based on the knowledge of the phenological cycle of certain trees. They also practiced horticulture and fruit growing in a multi-story system [4, 5].

It is estimated that in Latin America (LA) the AFS reaches an area between 200 and 357 million ha, including 14–26 million ha in Central America (CA), the most prominent are the commercial Silvopastoral Systems and the AFS of perennial crops under shade including coffee and cocoa plantations [6]. Although, these figures may have changed today given that the SPS has increased due to climate change mitigation actions and the AFS with coffee and cocoa may have decreased. An updated LA inventory of agroforestry areas would be valuable to land planners, resource managers, and decision-makers. This limits the amount of data that can be useful for multi-scale efforts.

A conceptual controversy may arise about whether agroforestry is a forestry activity or an agricultural one. Agroforestry as a concept should not be confused with other related terms, such as forest farming, which covers all the effects of forests and trees on the environment and agriculture, particularly the related socio-economic aspects. So not any kind of random combination of forest, fruit trees, ornamental trees, or service trees with crops or pastures is defined as an agroforestry system. It is also required that their combination be intentional, carried out systematically, and to produce various types of products; the system is the result of an important interaction, both ecological and economical between various types of crops; and that the system maintains or, as far as possible, improves the productive capacity of the land. There are three essential conditions to define an AFS: (1) at least two plant species interacting biologically; (2) at least one of the plant species is a tree or woody perennial; and (3) at least one of the plant species is managed for crop production (annual or perennial) or forage [2, 3, 7].

2. Historical development

One of the first documents on agroforestry in CA (CA) was possibly that of Cook in 1901 [8] who recognized several beneficial effects of shade trees, particularly legumes, on coffee plantations. Later, Holdridge in 1951 [9] described the use of *Alnus acuminata* (alder) associated with grasslands in the highlands of Costa Rica (CR). This type of land use system was also described by Budowski in 1957 [10], who reported the success of *Cupressus lusitánica* as a windbreaker, in the highlands of dairy regions and *Cordia alliodora* as a shade tree in grasslands in humid lowlands, both in CR.

According to Holdridge [11], there are three major basic land uses agricultural, grazing, and forestry, and while other human activities occupy land (such as for industrial purposes, urban developments, and transportation infrastructure), they do not directly use the soil resources in the sense of the three major uses. Agroforestry activity arises when one of the main uses, agriculture or grazing overlaps with forestry. The mixture of species with different requirements also allows an enhancement of the interception of radiation by vertical stratification of the components and better use of horizontal space [12].

Combe in 1979 [13] identified three main fields of hypotheses related to AFS within the framework of economics, ecology, and forestry.

Economic hypothesis: it is assumed that AFSs allow obtaining net income higher per unit area in the long term than the possible income with each isolated component.

Ecological hypothesis: it is assumed that trees in an AFS contribute to the conservation of the environment and particularly of the soil, especially when the induced combination represents a simulation of the types of vegetation that would occur in natural successions. In addition to the effects on the soil, important impacts on the microclimate, the fauna, and other factors that affect the biological balance are assumed.

Silvicultural hypothesis: it is assumed that the trees in an AFS can and should be managed according to the principles of classical forestry, always considering the particular requirements of the associated crops. Adequate silvicultural treatment is an indispensable condition for achieving and optimizing the positive economic and ecological results exposed in the previous hypotheses. In CA, there was a historical process, which had its beginnings with the definition of Combe and Budowski [7], presented that year in the First Workshop of Agroforestry Systems held in Turrialba, CR that can be summarized as follows:

It is the set of land use and management techniques that involves the combination of trees with crops (annual and/or perennial), with animals, or with both at the same time, in a plot, either simultaneously or successively, to obtain advantages of the combination.

These combinations can be simultaneous or staggered in time and space, and their objective is to optimize the production of the system and ensure sustained performance [7, 14].

With the creation of the International Centre for Research in Agroforestry-ICRAF (Currently World Agroforestry Centre) in Nairobi, Kenya in 1977/78, a space for discussion and analysis of agroforestry issues was established. Within this framework of internal debates, the initial ideas were refined, and a definition was agreed upon in which the criterion of "deliberate association" and that of "significant ecological and/or economic interactions between its components" was highlighted [12, 15]. In the decade of the 80s, there was agreement that agroforestry is a modality of integrated land use that seeks greater production, especially under conditions of marginal land or low level of inputs in the same area, and some cases of AFS in CA are exemplified, such as the coffee plantations or shaded cocoa plantations of *Erythrina* and *Cordia* and in turn the concept of "agroforestry practices" is introduced as operational aspects of an AFS, for example, the pruning of the trees of the system [16, 17]. Nair's definition [18] summarizes the concept as follows: "*Agroforestry is a land-use system in which woody species are grown intentionally in combination with crops or cattle on the same land, either simultaneously or in a sequence. The objective is to increase the total productivity of plants and/or animals in a sustainable manner, especially under levels of low technical inputs and in marginal lands. It involves the social and ecological integration of trees and crops*". Simultaneously, in CR, was conceptualized a complementary definition includes requirements or conditions: "*Agroforestry is a form of land use for multiple crops in which some fundamental conditions are met: (1) There are at least two components that interact biologically; (2) At least one of the components is a perennial woody; and (3) At least two species are managed for "agricultural" purposes in the broad sense of the word*" [2].

Initially, most studies in agroforestry were descriptive from a biophysical point of view, in addition, it was accepted that agroforestry was a new name for a set of old practices; but much attention was paid to socio-economic aspects [18, 19], which have been widely discussed by other authors [20], and include a great diversity of products such as wood, foliage, fruits, resins, fuel and fodder; and numerous environmental services (climatic, hydrological, soil, ecological) and human (ethical and esthetic).

Most definitions highlight the interactions among plant or animal components and their local environment and the spatial and temporal patterns of productive activities. Furthermore, open the possibility of considering and planning the social relations of production, and the interactions between communities and the outside world. Most of these aspects have been contemplated by Montagnini et al. [21, 22] in their comprehensive books on agroforestry.

3. Interactions among components in AFS

The functioning and adaptability of AFS depend on a dynamic relationship between plant species (a woody component with annual or perennials crops) and their abiotic environment (soil and water), as well as physical and chemical interactions in the environment (rainfall, temperature). These interactions and processes are of great importance for the long-term sustainability of the system. While the interactions are complex and interrelated simultaneously, they can be simplified from the point of view of the biological relationship between the two basic populations of an AFS, the woody component, and a crop; they may benefit or damage each other; or in other cases, the relationship may be neutral, all this depending on species and density of the tree component, the type of shade it produces according to the type of, type of canopy, tree crown, its branching habit, all of which have a fundamental role in AFS. After all, and since an AFS is an agroecosystem, which according to Hart [23], is an ecosystem that includes an agricultural or livestock productive component (crop populations, domestic animals, or both), an AFS can be syncretical defined as an agroecosystem with a woody perennial or tree component (**Figure 1**).

The effects of the woody component (trees, shrubs, palms, and bamboos) of an agroforestry system on soil and crops of an agroforestry system on soil and crops are very important because AF can increase farm productivity in several ways; first, the total output per unit area of tree/crop/livestock combination is greater than any single component alone; second, crops and livestock protected from the damaging effects of wind are more productive; and third, new products make the financial operations of

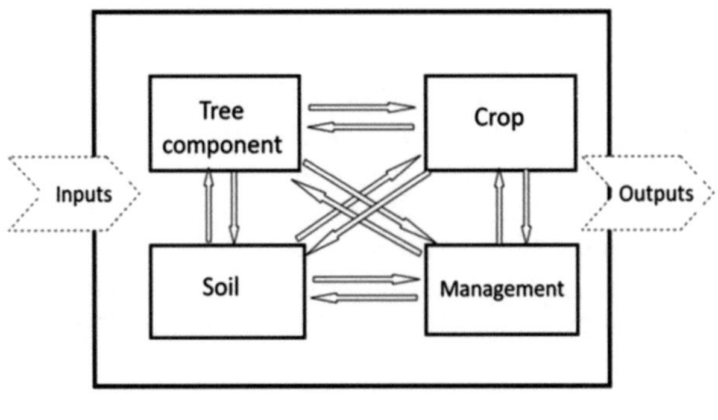

Figure 1.
Interaction among components of an agroforestry system, with limits, inputs, outputs, components, and interaction among components. Inputs are solar radiation, rainfall, fertilizers, and money invested in the system. Outputs are agricultural products, wood, and firewood from the tree component and ecosystem services. Management is what the farmer does with the components and inputs.

Effects of the woody component	
On soil	**In crops**
The leaf litter is the source of organic matter	Shade avoids excessive exposure to radiation
Nutrient supply	Intercept and mitigate wind
Improves soil structure	Attenuate the impact of rains
Controls erosion	Shade reduces air temperature
Favor water infiltration	Increase relative humidity
Limit runoff	Reduces weed dispersion
Reduces soil temperature	The positive effect of shadow

Table 1.
Effects of the woody component (trees, shrubs, palms, and bamboos) of an agroforestry system on soil and crops.

a small agricultural enterprise more diverse. These effects are shown both on the soil and in crops and are outlined in **Table 1**.

4. Canopy effects in agroforestry systems

The canopy is a set of crowns and branches of the trees; it is like a filter that intercepts the photosynthetically active radiation (PAR) or light that reaches the associated crops under the canopy and modifies it in quantity and quality. This interception projects a shadow, with physical effects (light/shadow, absorption efficiency, spectral modification of the transmitted light), and physiological actions are also triggered, such as photocontrol of germination, elongation of internodes, leaf expansion, and the development of the photosynthetic structure in the associated crops (**Figure 2**).

The canopy is characterized by having a structure and a floristic composition that can be managed, thus regulating the amount of shade depending on the crop's needs and the farmer's objectives. To measure the density of the agroforestry canopy, the Leaf Area Index (LAI) can be used, which represents the sum of all the existing leaf areas in a soil area. The LAI is an indicator of the canopy's ability to intercept solar radiation and predict the type of shade it produces dense, medium, or light shadow. The type of shade that the canopy produces can also be expressed in the percentage of coverage of the cups, in expressions such as 50% shade; although it is not necessarily an accurate indicator because the shadow is a dynamic process that moves on the floor of the AFS as the sun makes its apparent movement on the horizon. The position, shape, and accumulation of tree shadows, in different places and at different dates and times of an agroforestry plot, can be calculated with *software* designed in CATIE called *ShadeMotion*, which requires supplying the number of trees, location, shape, size, and density of foliage of the trees; as well as, the size of the land, degree of slope, and geographical latitude where the plot is located [24–26]. In response to shade, most plants produce less dry matter, retain photosynthates in the shoot at the expense of root growth, develop longer internodes and petioles, and produce larger and thinner leaves. The net photosynthesis (NP) of the crop increases with the Leaf Area Index (LAI) but reaches a ceiling when LAI is around 3 and could be plotted like the adjunct one attached. LAI is defined as the relationship between the sum of green leaf areas of

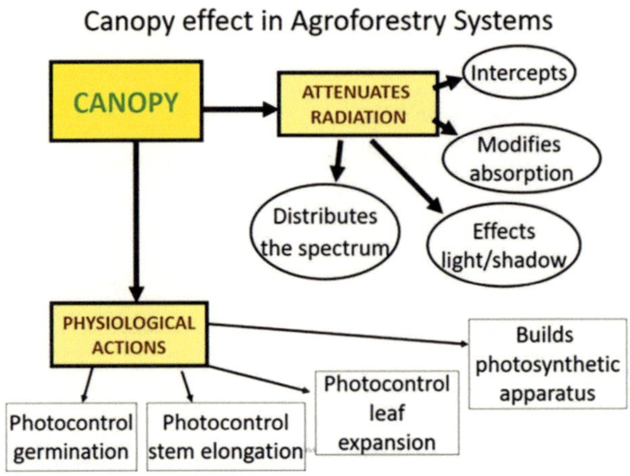

Figure 2.
Effect of the canopy on agroforestry systems.

the canopy of shade trees per unit ground surface area (LAI = Σ leaf area/ground area, m^2/m^2; in broadleaf canopies).

5. Classification and characterization of agroforestry systems

Classifying agroforestry systems including their environmental and site variants took a long time, without reaching a consensus or a global classification. In this context, ICRAF conducted a global inventory of AFS between 1982 and 1987, the results of which resulted in a classification scheme that is generally accepted today [16–18]. This inventory was designed to collect, synthesize, and disseminate information on existing AFS in developing countries. As a result of it, Nair in 1993 concluded that "*irrespective of the sociocultural differences in different geographical regions, the major types of agroforestry systems are structurally similar in areas with similar ecological conditions. Thus, agroecological regions can be taken as a basis for the design of agroforestry systems*." [18]. This project also made it possible to generate a list of the main herbaceous and woody perennial plants reported as components of existing systems and their main uses in different regions. Among the first classifications [7, 16], AFS were grouped into (a) sequential, (b) simultaneous; and (c) linear systems, according to the sequence of the tree component and the crop, and by the type of accompanying crop (annual or perennial). Sequential AFS include *shifting cultivation* and *Taungya* systems (annual crops combined in a forest plantation). In simultaneous AFS, all those combinations are grouped at the same time and placed trees with crops (annual or perennial), or with pastures; while on the live fences, hedges, and windbreaker curtains are grouped. Nair [18] made a grouping of agroforestry classifications into four groups: (1) For its structure; (2) For its functions; (3) Ecological; and (4) Based on socio-economic criteria; although, the first two have prevailed. Knowing the existing classifications allow identifying those AFS most appropriate to recover degraded areas through their restoration. However, when an AFS is used to stop deforestation and the recovery of degraded forest areas, the results may not be satisfactory unless forecasts are taken for the social welfare of the people involved. The known

Classification of traditional AFS	Type of AF system	Example of agroforestry systems
Sequential	Shifting agriculture	The traditional agriculture of cutting and burning trees practiced since ancient times.
	Taungya Systems	A temporary combination of a forest plantation during its initial phase, with the production of annual crops until the shade of the canopy allows it.
Simultaneous	Trees with annual crops	*Alley cropping*, rows of a woody nitrogen-fixing plant are associated with an annual crop.
	Trees with perennial crops	Growing coffee or cocoa under shade trees such as *Erythrina poeppigiana* y/o *Cordia alliodora*.
	Agroforests	Management of secondary forests, in association with one or more tree species of economic utility. Systems Quesungual o Kuxum-Rum.
	Silvopastoral systems	Association of trees with pastures and livestock. Grazing in forest and fruit plantations.
	Mixed home gardens.	Characterized by their complexity, are multi-specific, combine various forms of life and maintain production throughout the year.
Linear systems or in alignment	Live fences	Fences with live poles to which the wire is fixed and periodically pruned.
	Live hedges	A row of tree species was established at very close distances.
	Windbreak curtains	Multiple rows of tree species are planted perpendicular to the direction of prevailing winds.

Table 2.
Examples of traditional Agroforestry Systems in CA. Source: modified from Combe y Budowski [7] and Nair [16].

agroforestry classifications are hierarchical and arbitrary because the objective is defined by the user, and organized by components (trees, crops, livestock), temporal arrangements (sequential or simultaneous), and spatial arrangements, among others. Since there is a relationship between the concepts of the definition and the construction of classification, it is important to consider other aspects such as management, forestry, planting densities, establishment and maintenance costs, environmental services provided, and forest production associated with AFS, to avoid ambiguity when classifying it. Some examples of traditional AFS in CA are presented in **Table 2**.

6. Sequential agroforestry systems

Sequential AFS occurs at a site where there is a chronological succession between a period with annual crops and another with a forest component; that is, annual crops and regeneration of the natural forest or tree plantations follow each other over time. This category includes modalities of migratory agriculture with fallow management and taungya systems, where annual crops are made interspersed between rows of trees in the stage of establishment of a forest plantation until the foliage of the trees is developed (**Figure 3**).

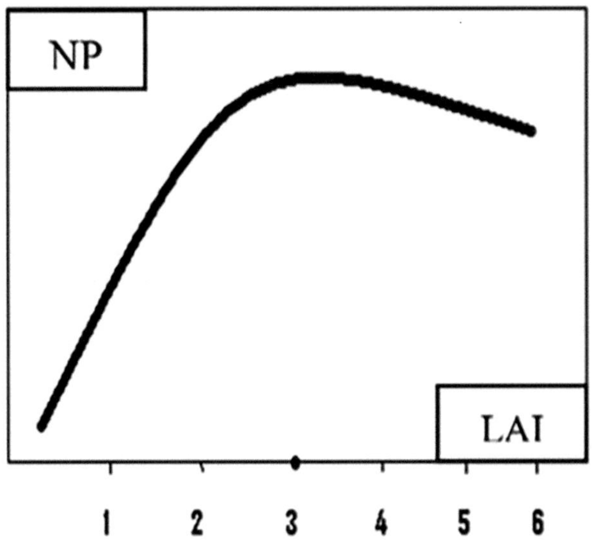

Figure 3.
Relation between Net photosynthesis (NP) and Leaf Area Index (LAI). (Russo, R. Agroforestry Course).

6.1 Shifting agriculture

The migratory agriculture also called itinerant, nomadic, or "*shifting cultivation*" is possibly the oldest of the agricultural systems and consists of the slash and burn of natural vegetation with the option of clearing the land to cultivate. On the other hand, it has been an important source of subsistence for rural populations in the tropics. Its application has varied according to the site and local conditions, but several practices are almost universal; among them is the rotation of cultivation sites or milpas (rotation of trees and crops), the cleaning of land by burning (slash-grave-burn in Mexico), the exclusion of chemical fertilizers, the exclusive use of manual labor, planting by hand and short periods of cultivation alternated with long periods of fallow. The system was developed in conditions of low population density, oriented towards subsistence, with a high concurrence of forests and simultaneous production of several crops with different harvest times. Fertility is restored through a long fallow period, and during the first production season little or no weeding is needed [27]. In addition, slash, grave, and burn fallows serve as habitats for wildlife, corridors between patches of forest and as shields against edge effects such as extreme temperatures, desiccation, and fires [28].

6.2 *Taungya* system

The *Taungya* system (TS) is a reforestation method that allows the temporary combination of a forest plantation in its establishment phase with the production of short-cycle crops, such as maize and beans, or horticultural crops (**Figure 4**). Under certain conditions, the TS works better than pure reforestation, since there is an intermediate use of the land in agriculture, which if it did not occur could proliferate weeds that compete with the plantation [17]. The word *taungya*, which means "hillside terrain" is Burmese (Burma, today Myanmar from where the system migrated in 1870). In India, that same practice was called "kumri". In Java, the TS was used to plant 40,000 ha

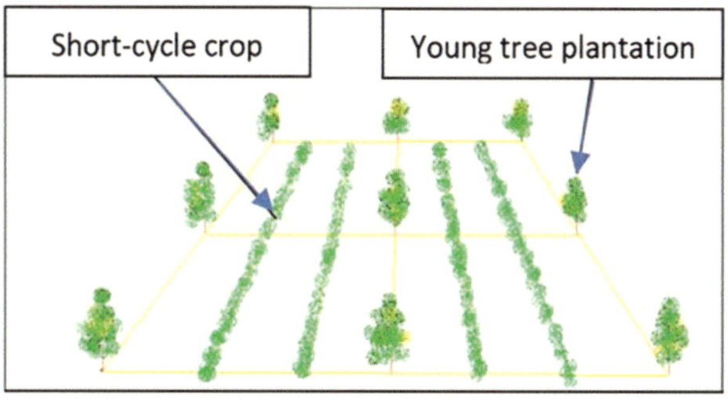

Figure 4.
Taungya system: a young tree plantation with a short-cycle crop between rows (Russo, Agroforestry Course).

of *Tectona grandis* (teak) in the late 1800s; in 1920 there was 190,000 ha, in 1952 312,000 ha and currently exceeding 700,000 ha [29, 30]. Annual crop yields in *taungya* combinations are usually lower than in pure crops, but they produce income that covers planting costs and can be considered an added value to reforestation, which would not otherwise be obtained. To be successful, TS must be applied in places where there is a need for land; soils are suitable for producing food crops with reasonable temporary yields, without causing excessive soil deterioration; tree species in demand are of proven adaptability; there is a peasant population, and local staff is trained to operate the system [27]. The TS has been very successful to establish forest plantations under different conditions. In the dry areas of what is now Myanmar, with 450–1100 mm of annual rainfall, communal forests were established using taungya systems for 2–3 years to provide firewood from *Acacia* spp., *Albizia lebbek*, *Senna siamea*, *Dalbergia sissoo*, *Melia azedarach*, *Prosopis* spp. and *Eucalyptus* spp.; and to produce pulpwood. An example is the *Gmelina arborea* plantations in Jari in southern Brazil, established through the *taungya* system; rice and beans have been harvested for 2 years in plantations of *Cunninghamia lanceolata*; rice, cotton, and corn with *Eucalyptus.* In addition, among the ecosystem benefits that TS provides, it is a restoration tool, combining afforestation actions with agricultural activity during the early stages of tree establishment, which represents an economic and social benefit, while preventing the establishment of weeds and contributing organic matter to the soil with crop stubble (**Figure** 5).

Figure 5.
Taungya system: Left: Young plantation of Eucalyptus deglupta (eucalyptus) and Araucaria hunsteinii (klinki) intercropped with Zea mays (freshly harvested corn) between rows, in Guácimo, CR; and, Right: New plantation of Swietenia macrophyla (mahogany) intercropped with Eryngium foetidum (coriander) between rows, in Turrialba, CR. Photos: Rolando Camacho.

7. Simultaneous agroforestry systems

Simultaneous AFSs occur at a site where there is a simultaneous and continuous combination of an agricultural component with a forestry component, whether timber, fruit, multi-use, or service trees. These AFSs include all kinds of tree associations with annual or perennial crops, mixed home gardens, agroforests, and silvopastoral systems.

7.1 Trees with annual crops

7.1.1 Alley cropping

Alley cultivation is a simultaneous AFS of trees with annual crops, consisting of rows of trees, usually 4–6 m apart between rows x 2 m between trees, interspersed with annual crops between rows of planting (**Figure 6**). Trees are pruned before planting and branches are left in alleys to incorporate organic matter into the soil and in turn suppress weeds. In this AFS, conveniently, the trees are of nitrogen-fixing species (Fabaceae such as *Erythrina* spp., *Gliricidia sepium,* and *Leucaena leucocephala*), mainly in soils of low fertility, where the nitrogen content (N) is low. The main intention of alley cultivation is the recycling of nutrients to maintain or increase crop yields through soil improvement, weed control, and erosion control.

Reminding that the main mechanisms of gain of N in the soil are: (a) N contributed with the rainfall; (b) N from non-symbiotic fixation; (c) N from symbiotic fixation; (d) N provided by organic fertilizers; and e) N from the mineralization process from fresh remains (vegetable and animal), in this case, the fallen leaves and branches from the trees. Therefore, this is a production system that adapts well to low fertility soils in degraded areas and to dry and semi-arid areas, since it favors the restoration of fertility and physical conditions of the soil. In addition, producers can obtain from trees other products such as poles, firewood, fodder, green manure, and atmospheric nitrogen fixation. The latter has current importance because the action of reducing nitrogen fertilization is a way to reduce nitrous oxide emissions into the atmosphere, so it is considered a way to mitigate climate change. In areas, with steep slopes, the

Figure 6.
Left: Alley Cropping of banana (Musa paradisiaca) with rows of Moringa oleifera, Caribbean region of CR; and Right: Organic banana system, rows of Musa AAA subgroup Cavendish) intercropped with a nitrogen-fixing tree (Erythrina berteroana), a timber tree (Cordia alliodora), and cocoa (Theobroma cacao), Caribbean region of CR. Photos: R. Russo.

rows of trees can be established in contour lines as a living barrier for water conservation and to deter erosion. In addition, they are a way of conserving the soil that does not require physical conservation structures.

7.1.2 Crops under the cover of forest curtains

This category includes any form of short-cycle agricultural monocultures or polycultures such as corn, beans, onions, celery, lettuce, tomatoes, coriander, and other horticultural species, in association with windbreaker curtain-like trees in windy areas, multiramified live fences, or rows of trees in contour lines in hillside areas. All these alternatives that integrate crops with the planting of trees are a form of conservation and restoration of degraded areas, which contribute to conserve biodiversity and water resources. Crops undercover, forest.

7.1.3 Trees with perennial crops

The simultaneous association of trees and perennial crops is a common practice in CA. The most prominent examples are coffee and cocoa crops under shade. The beneficial effects of shade trees, particularly *Fabaceae*, on coffee were recognized and described at the beginning of the last century [8]. Cocoa, unlike coffee, adapts to fertile inshore sites (from 0 to 700 meters above sea level); while coffee is a crop in higher areas. These AFSs with shading trees are more sustainable alternative to perennial monocultures because they give added value in terms of diversifying production, providing habitat for greater biodiversity, favoring soil conservation, and serving as protection of water resources. All these elements are important when contemplating the recovery of degraded areas.

7.1.4 Coffee and cocoa plantations under shade

These systems simultaneously combine trees with perennial crops, such as *Coffea arabica*. The main crop is interspersed with the trees that contribute with environmental services, additional products, soil improvement, microclimate beneficial to the crop, and serves as a tutor or support for vine crops such as *Piper nigrum* or *Vanilla planifolia*.

Coffee (*C. arabica*): Shaded coffee is perhaps the oldest and most important crop, as it is estimated that it covers seven hundred thousand hectares in CA, of which more than 80% are AFS and most of it is grown by small-scale farmers on farms no larger than 5 ha [31–33]. Permanent shade trees can be timber (*C. alliodora*, *Cedrela odorata, Swietenia macrophyla, Dalbergia retusa*, *Tabebuia donnell-smithii*, *Schizolobium parahybum, Grevillea robusta*, also, *Terminalia amazonia*, *Gmelina arborea, Eucalyptus* spp., among others; also fruit trees (*Citrus* spp., *Inga edulis, I. vera, Persea americana*, *Macadamia* spp., *Psidium guajava*), or multipurpose trees (*Erythrina poeppigiana*, *E. fusca*, *Gliricidia sepium*; *Leucaena leucocephala*), *among others* (**Figure 7**) Also, the temporary shade of banana and plantain (*Musa* spp.) is in the early stages of the establishment of the system. In all cases, shade trees play an important role, in light regulation by the various layers or strata of crown trees. Aspects such as planting densities of shade trees, and regulation of shade by pruning or pollarding of branches have sound importance; given that coffee cultivated under excessive shade produces fewer coffee grains and increases in production can be favored with the management

Figure 7.
Left: A coffee plantation (Coffea arabica) shaded with a nitrogen-fixing tree (Erythrina poeppigiana) just totally pollarded, in Tres Ríos, CR; and Right: View of a coffee crop associated with Eucalyptus deglupta located in Juan Viñas, CR.

of the shade, through pruning of the trees, which allows air circulation and greater penetration of light [34]. A coffee AFS with shade trees of *E. poeppigiana* annually pollarded (pollarding is the pruning of all branches at a certain height of the tree trunk) located in Tres Rios, CR and a SAF of coffee associated with *Eucalyptus deglupta* located in Juan Viñas, CR, are presented in **Figure 7**.

Cocoa (*Theobroma cacao*) farming in CA is practiced by small farmers extremely poor (indigenous people, Afro-Caribbean, and mestizo), living in remote zones. Cocoa is cultivated at 100–800 m altitude in small plots (1.2 ha/farm) with low yields: 75–150 kg/ha/year in zones with frosty pod rot (*Moniliophthora roreri*) and with poor management; and 200–350 kg/ha/year where there is frosty pod rot and minimal management (**Table 1**). Cocoa trees are typically spaced at 4 × 4 m (625 plants/ha) in most countries. Most farmers have two or more cocoa plots per farm. Cocoa trees are 4–6 m tall and are associated with shade trees at a density in the range of 85–166 trees/ha. Most shade trees are planted, and some species are selected from the natural regeneration. Shade trees are used for timber (*Cordia alliodora, Cedrela odorata*), fruit (*Musa spp., Citrus spp*., avocado (*Persea americana*), coconut (*Cocus nucifera*), peach palm (*Bactris gasipaes*), mango (*Mangifera indica*), and shade providers (*Inga* spp., *Gliricidia sepium*). Shade tree canopies usually have three vertical strata (low <10 m, medium 10–20 m tall, and high >20 m tall) [35, 36].

7.1.5 Agroforests

The concept of agroforests, despite being traditional systems, has been incorporated more recently into the definitions and classifications of FAS. Agroforests are areas with a predominance of trees and shrubs or communities that resemble forests, where there are plots or clearings with agricultural practices along with structures typical of natural forests due to their floristic composition and their multiple stratifications. An example is the so-called *Quesungual* agroforestry system (SAQ), or *Kuxur Rum*, an agroforestry modality that was practiced ancestrally by the Mayan cultures and is still carried out in some regions. This system is reported to be practiced in countries such as Guatemala, El Salvador, Honduras, and Nicaragua. It consists of pruning selected trees in an area of natural forest (usually secondary) up to the middle of the trunk, without damaging the roots, and planting corn in the clearings or gaps with greater solar radiation [37]. It should be noted that in the case of Honduras, after Hurricane

Figure 8.
Cocoa (Theobroma cacao) agroforest in rehabilitation with a shade of numerous tree species, Changuinola, Panama.

Mitch, in the areas where SAQ was practiced, such intense damage was not observed. Studies carried out in southeastern Honduras, on hillside land with slopes greater than 12% (approximately 80% of the country's area); have shown this system reduces the vulnerability to climate change of smallholder subsistence farmers, and that it has great potential to improve livelihoods and help to adapt climate changes on tropical slopes; it is a good option to mitigate greenhouse gases, and in turn, it performs other services for a better sustainable agricultural use [38]. Other examples are cocoa agroforests in the area of the binational Sixaola River basin between CR and Panama (**Figure 8**).

7.1.6 Silvopastoral systems

Silvopastoral Systems (SPS) are agroecosystems in which a tree component is deliberately associated with an herbaceous one (natural or improved pastures) and a livestock production component (domestic animals) in the same site so that there are biological interactions between both to maximize the land use. In other words, they temporarily and spatially combine the maintenance of pastures (natural or cultivated ones) with livestock production activities, along with tree species. To this the silvopastoral practices can be added, in which the woody component does not need to be in the same site as the animal component because forage can be transported; such as the case of forage banks or living fences, which are pruned, and the forage produced by the pruning is supplied to confined animals [39, 40]. The limits, inputs, outputs, components, and interactions are shown in **Figure 9**. Silvopastoral systems found most frequently in CA are: (a) trees in pasture lands, including grazing in secondary forests and fallows; (b) grazing in forest and fruit plantations; (c) living fences; (d) perimeter shelterwood; and (e) fodder banks or crop and utilization of forage trees and shrubs. Tree species identified in pasture lands in CA are diverse and are according to the characteristics of vegetation, climate, and altitude of each region. In most cases, the trees are from natural regeneration and have been allowed to grow in densities that do not affect pasture growth, in a range from 10 to 70 trees per hectare but can reach up to 100 trees; with a basal area (BA) ranging from 1 to 7 m^2/ha, although some authors mention that is possible to have up to 200 trees/ha [42]. Among the most frequent tree species found in animal production systems is *Cordia alliodora, Cedrela odorata, Enterolobium cyclocarpum, Pithecolobium saman,*

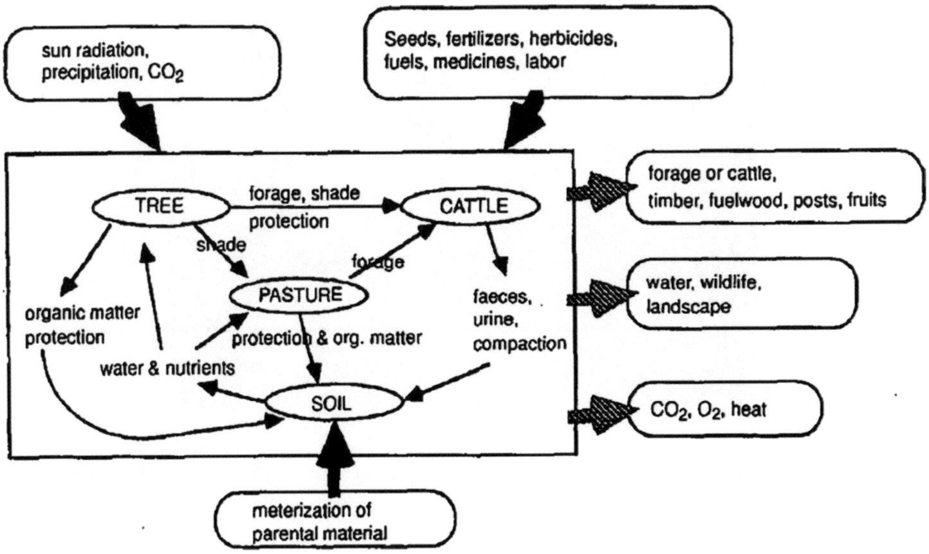

Figure 9.
Interactions in a silvopastoral system. Source: taken from [41].

Guazuma ulmifolia, Tabebuia rosea, Pterocarpus rohrii, *Pentaclethra macroloba, Mangifera indica*, *Gliricidia sepium* in the lowlands (up to 600 masl); and *Citrus* spp., *Platymiscium dimorphadrum, Persea americana, Inga* spp., *Psidium guajava*, *Bursera simaruba, Brosimum alicastrum, Alnus acuminata;* in midlands and uplands (600–2000 masl).

8. Linear agroforestry systems

Linear or alignment systems are tree plantations in rows of one, two, or more rows, such as live fences, hedges, living barriers, tree, and shrub lines (timber, fruit, multi-use), and windbreak curtains, usually associated with an agricultural crop or grasslands. They are useful especially on small farms because they offer many opportunities for the production of goods and services of interest to the farmer and are one of the most commonly promoted agroforestry technologies in forest and agroforestry extension and development programs in CA [43, 44].

8.1 Live fences

A living fence is made up of live poles in a row of trees or shrubs that delimit a property, or you can divide or parcel it out internally. Depending on the species used, it can produce firewood, wood, fodder, flowers for honey, fruits, and poles among others. They are very common in the countries of CA and one of their most important functions is the delimitation of farms, or paddocks (**Figure 10**). The most commonly used tree species as living poles are *Gliricidia sepium, Bursera simarouba, Erythrina berteroana, E. costaricensis, Spondias* spp., *Mangifera indica, Ficus isophlebia, Pochote quinata; Delonix regia*, *M. indica*, and *Simaruba glauca*, among others. The fences can be established with a single species (mono-specific), or with more than two species (multi-specific) [43–45].

Figure 10.
Left: Mono-specific living fence with Erythrina berteroana poles in Sarapiquí, CR; Right: Multi-specific living fence in Bonanza, Nicaragua.

8.2 Windbreaker curtains

A windbreaker is a linear tree plantation, which forms a barrier, to mitigate the negative effects of winds and regulate microclimate conditions, which consists of spinning multiple lines of trees, established perpendicular to the direction of the prevailing winds. They are included in the AFS when they are associated with an agricultural or livestock production system. The trees are planted in several parallel rows, and the protection depends on the height of the curtain and the compactness of the tree crown to stop the wind. It is generally accepted that a windbreaker curtain provides services and benefits to agricultural establishments. In addition, a well-managed curtain also produces timber and fuel wood. Several aspects must be considered for the design of a curtain, among them are (a) Orientation; (b) Distance between trees and between rows of trees; (c) Density; and (d) Height of trees. The height of the curtain trees is the most important factor to consider in your design, as it determines the area it protects. The maximum wind mitigation distance of a curtain varies between 15 and 20 times the height of the trees. That is, if a curtain is 10 m high, it will protect up to a distance of 150–200 m. For instance, in León, Nicaragua, curtains of three strata and five lines of *Eucalyptus camaldulensis, Leucaena leucocephala,* and

Figure 11.
Left: Windbreaker curtain of Cupressus lusitanica (cypress) associated with horticultural crops in Ochomogo, CR. Right: Agroforestry landscape of windbreak curtains associated with horticultural crops and pastures, Cartago, CR.

Tecoma stans were established to protect the soil in cotton (*Gossypium hirsutum*) fields during the dry season [46]. In CR, the cases of curtains of *Cupressus lusitánica*, from Mexico are known in high altitude areas of the Central Valley in the provinces of Heredia and Cartago (personal observations] (**Figure 11**).

9. Attributes and characteristics sought in agroforestry systems

There are desirable attributes and characteristics of AFS in different aspects: (a) As for the selection of the woody species, keep in mind that it is easy to establish and grow quickly; (b) Regarding the architecture, phenology, and compatibility of the woody species with the associated crops, it is desirable that they make little competition for water and nutrients; that they have an open and narrow crown with small leaves; that they have a strong root system and as far as possible deep; no allelopathic effects; with branches and stems that are not brittle and that do not host pests or diseases. (c) Regarding the management and physiology of the woody component, it is desirable that: they tolerate full exposure to the sun; have self-pruning of branches; tolerate frequent pruning; regrow easily; be easy to handle (without thorns or stinging latex); and that if possible, fix nitrogen. (d) As for ecological functions, they are desirable attributes that have functional biodiversity and promote biological control; provide habitats for avifauna and other non-harmful animals; encourage soil conservation and fertility and maintain foliage in the dry season. The main three groups of attributes and characteristics (**Table 3**) that farmers expect from an AFS are (a) productivity; (b) sustainability; and (c) adaptability [47–49].

Attributes	Desirable characteristics of agroforestry systems
Productivity	The system produces goods, merchandise, and services required by producers
Sustainability	Maintains or increases productivity over time: producing while preserving and conserving producing
Adoptability	It is accepted, even under socioeconomic and biophysical constraints prevailing locally

Table 3.
Desirable attributes and characteristics of agroforestry systems. Source: Modified from [47–49].

10. Current trends in agroforestry

At present agroforestry has become a significant issue in scientific research because the human face new challenges to ensure food security and climate change mitigation. The research interest in the field has boosted, and about 139 countries have been involved with the research in the field of AF and connected topics. These publications cover 66 subject categories and a great diverse research theme. The most used keywords in AF research have been changed from "Intercropping", "Alley cropping", and "Multipurpose trees" to "Carbon sequestration", "Ecosystem service", and "Climate change" [50, 51]. Other topics like Small-island agroforestry in climate change and sustainable development goals have been developed [52]. Also, extensive analysis and proposals to face the challenges of the new millennium by first-line researchers, covering topics of Biodiversity Conservation, and Food Sovereignty, Climate Change have been brought together in a work of vast value for researchers and students [53].

11. Conclusions

Agroforestry systems are a viable option to reduce land degradation and generate income for rural families. However, due to the cost structure and the return period, technical and financial assistance (payment for environmental services) should be considered for the adoption and empowerment of these systems to be successful in the long term. CAn farmers are familiar with a set of traditional AFS, including shaded coffee, shaded cocoa, silvopastoral systems (SSP), and row trees.

The different modalities of the AFS allow the diversification of family farming, the sale of surplus production, and the efficient use of the natural resources of the farm (water, land, biodiversity, energy); factors that are linked to the degree of development of the peasant economy and that would allow more comprehensive productive, food, and nutritional schemes. Due to the similarities in their structure, energy flows, and nutrient cycles with natural forest ecosystems, AFS is considered to be an alternative for ecologically sustainable use for climatic zones where natural vegetation is a forest.

Agroforestry systems, whether traditional or innovative, allow the development of strategies for the maintenance of productivity based on the regulation of nutrient recirculation through the choice of species, planting densities, and the management of canopy shade on crops through pruning. All this makes it possible to maximize income and minimize the loss of nutrients from the soil.

Although the advantages of the tree component (trees, shrubs, palms, and bamboos) are always highlighted, there can also be negative effects on crops and soil when planting density and shade are excessive and when the choice of species is not the most appropriate.

There are ancestral agroforestry modalities (*Quesungual* or *Kuxum-Rum*) that are very appropriate for tropical areas with a dry season and the recovery of degraded areas. Indigenous peoples and local communities (IPLC) own or manage a considerable area of existing forests in the CA region; consequently, they are related to agroforestry practices comprising subsistence crops such as maize, beans, bananas, plantain, and cocoa, managed through low-impact concepts and combined in agroforestry systems; where multiple crops are mixed with timber trees, and with permanent crops such as cocoa, they offer a different vision of what agroforestry systems and the ancestral management of the natural forest should be since they develop a sustainable production in which the soil is never left uncovered. After all, Agroforestry is a form of productive restoration of degraded areas because it improves soil fertility, increases resilience to climate change, and provides alternative sources of income to local people.

Agroforestry is part of the concept of the Nationally Appropriate Mitigation Action (NAMA) mechanism, which is based on a combination of public and market incentives for the implementation of greenhouse gas (GHG) mitigation measures. An example is the NAMA for the coffee sector of CR, which constitutes a broad platform for coordination and participation of the sector together with governmental, non-governmental, and international cooperation entities, covering an area of more than 90,000 hectares and 50,000 producers, for the improvement of competitiveness (cost savings and diversification of the coffee agroforestry system), and seeks at the same time the differentiation of the sector maintaining its access to markets and contributing to a low emission economy.

In a brief summary of the above, Guatemala has experiences and achievements in community forest management, with more than 20% of forests managed communally or municipally (380,000 hectares managed sustainably by community concessions in Petén); in Panama, 54% of forests and carbon are in indigenous territories and indigenous peoples organized under the National Coordinator of Indigenous Peoples of Panama (COONAPIP); Nicaragua has interesting approaches in the Autonomous Regions (21 titled territories with more than 3.6 million hectares, which are more than 62% of the forests in North and South Atlantic Autonomous Regions (RAAN and RAAS); in Honduras, more than 400,000 ha are in the hands of communities since the Forestry Law of 2007, there is titling of seven territories and 760,000 ha in the Mosquitía; while in CR, indigenous peoples, who constitute 2% of the population with 12% of the forests in the country, have Indigenous Development Associations (ADI) and from these rights the Payment of Environmental Services (PES) was established in indigenous territories with institutions consolidated by the National Forest Financing Fund (FONAFIFO).

Finally, Agroforestry is a possible alternative to receiving payment for the environmental services (PES) they produce. In the case of CR and Guatemala, there exist formal PES programs that incentive agroforestry; promote the incorporation of trees in agroecosystems; as an alternative for the recovery of forest cover, income generation, and also as a means for the reduction of greenhouse gas emissions. Honduras and Panama provide environmental services in their legislation, and Dominican Republic is in the process of formally implementing PES. In the case of mixed crops involving timber trees, it will undertake to increase and/or reorder the number of trees and reduce the impact of the crop on soils and waters and that its activity coincides with the capacity of land use; in addition, they could constitute an opportunity to strengthen the processes of conservation, sustainable use and poverty reduction in the CA region.

Acknowledgements

I dedicate this review to the memory of Dr. Gerardo Budowski (1925–2014) with whom I was introduced to the concepts of agroforestry; also, to my colleagues in my initial experience (1981–1986) at The Tropical Agricultural Research and Higher Education Center (CATIE), where I took the first steps into this multi-disciplinary field that combines agriculture with forestry and livestock activities.

Conflict of interest

The author declares that the literature research was conducted in the absence of any commercial or financial relationships that could be construed as a potential conflict of interest, and without external funding.

Comments

The apparent limitation of the study would be that the author cited mainly old literature, but the author felt that the pioneering work of the researchers who laid the foundations of the AF both in the New and Old World could not fail to be recognized. However, it is recognized the mutual importance of both the pioneering and current researchers of these sustainable cultivation technologies.

DOI: http://dx.doi.org/10.5772/intechopen.105406

Author details

Ricardo O. Russo
University of Costa Rica, San José, Costa Rica

*Address all correspondence to: ricardo.russo@ucr.ac.cr

References

[1] Hernandez XE, Levi S, Arias RL. Hacia una evaluaci6n de los recursos naturales renovables bajo el sistema roza-tumba-quema en México. In: Lund HG, Caballero-Deloya M, Vilareal-Canton R, editors. Land and Resource Evaluation for Natural Planning in the Tropics: Proceedings of the International Conference and Workshop, Jan 25-31, 1987; Chetumal, Mexico. Washington, D.C.: USDA Forest Service; 1987. pp. 330-340

[2] Somarriba E. Revisiting the past: An essay on agroforestry definition. Agroforestry Systems. 1992;**19**:233-240

[3] Budowski G. The socio-economic effects of forest management on the lives of people living in the area. In: The Case of Central America and Some Caribbean Countries. Turrialba, Costa Rica: CATIE; 1981. pp. 16-17

[4] Gligo N, Morello J. Notas sobre la historia ecológica de América Latina. In: Sunkel O, Gligo N, editors. Estilos de Desarrollo y Medio Ambiente en América Latina. México, D.F: Fondo de Cultura Económica; 1980. pp. 112-148

[5] Bene JG, Beall HW, Côté A. Trees, food, and people: Land management in the tropics. International Development Research Centre. 1977:52

[6] Somarriba E, Beer J, Orihuela JA, Andrade H, Cerda R, DeClerck F, et al. Mainstreaming agroforestry in Latin America. In: Nair PKR, Garrity DP, editors. Agroforestry: The Future of Global Land Use. The Netherlands: Springer; 2012. pp. 429-453

[7] Combe J, Budowski G. Classification of agroforestry techniques. In: de las Salas G, editor. Proceedings of the Workshop on Agroforestry Systems in Latin America. Turrialba, Costa Rica: CATIE; 1979. pp. 17-47

[8] Cook OF. Shade in Coffee Culture. Washington, D.C.: U.S. Department of Agriculture, Division of Botany, Bulletin No. 25; 1901. p. 116

[9] Holdridge LR. El jaúl, Alnus acuminata, para los arbolados de las fincas en Costa Rica. Caribbean Forester. 1951;**12**:53-57

[10] Budowski G. Quelques aspects de la situation forestière au Costa Rica. Bois & Forets Des Tropiques. 1957;**55**:3-8

[11] Holdridge LR. Life Zone Ecology. San José, Costa Rica: Tropical Science Center; 1966. p. 149

[12] Malézieux M, Crozat Y, Dupraz C, et al. Mixing plant species in cropping systems: Concepts, tools, and models. A review. Agronomy for Sustainable Development. 2009;**29**(1):43-62

[13] Combe J. Concepto sobre la investigación de técnicas agroforestales en el CATIE. Turrialba, Costa Rica: CATIE; 1979. p. 20

[14] Lundgren B. Introduction. Agroforestry Systems. 1982;**1**(1):3-6

[15] Lundgren BO, Raintree JB. Sustained agroforestry. In: Nestel B, editor. Agricultural Research for Development: Potentials and Challenges in Asia, ISNAR. Vol. 1982. The Hague: International Service for National Agricultural Research; 1982. pp. 37-49

[16] Nair PKR. Classification of agroforestry systems. Agroforestry Systems. 1985;**3**:97-128

[17] Combe J. Agroforestry techniques in tropical countries: Potential and limitations. Agroforestry Systems. 1982;**1**:13-27

[18] Nair PKR. Definition, and concepts of agroforestry: Community forestry, Farm forestry, and social forestry. Chapter 3. In: Nair PKR, editor. An Introduction to Agroforestry. Dordrecht, The Netherlands: Kluwer Academic Publishers; 1993. pp. 13-16

[19] Mercer DE, Miller RP. Socioeconomic research in agroforestry: Progress, prospects, priorities. Agroforestry Systems. 1998;**38**:177-193

[20] Krishnamurthy L, Ávila M. Agroforestería básica. Serie textos básicos para la formación ambiental. México D.F: PNUMA-ORPALC-RFAALC.; 1999. p. 340

[21] Montagnini F, Somarriba E, Murgueitio E, et al., editors. Sistemas Agroforestales. Funciones Productivas, Socioeconómicas y Ambientales. Cali, Colombia/Turrialba, Costa Rica: Editorial CIPAV/CATIE; 2015. p. 454

[22] Ashton MS, Montagnini F, editors. The Silvicultural Basis For Agroforestry Systems. Boca Raton,Florida: CRC Press; 2000. p. 278

[23] Hart RD. Agroecosistemas: Conceptos básicos. Turrialba, Costa Rica: CATIE; 1985. p. 159

[24] Quesada F, Somarriba E, Malek M. ShadeMotion 3.0: Software para calcular la cantidad de horas de sombra que proyectan un conjunto de árboles sobre un terreno. s/f. Turrialba: C.R.: CATIE. p. 31

[25] Somarriba E. Estimación visual de la sombra en cafetales y cacaotales. Agroforestería en las Américas. 2002;**2002**(35/36):86-94

[26] Somarriba E, Zamora R, Barrantes J, et al. ShadeMotion: The Analysis of Tree Shade Patterns. Turrialba, C.R: CATIE; 2020. p. 50

[27] Wadsworth FH. Forest Production for Tropical America. Agriculture Handbook 710. Washington, DC: U.S. Forest Service; 1997. p. 563

[28] Ferguson BG, Griffith DM. Tecnología agrícola y conservación biológica en El Petén, Guatemala (Agricultural technology and biological conservation in Petén, Guatemala). Manejo Integrado de Plagas y Agroecología (Costa Rica). 2004;(72):72-85

[29] Krishnapillay B. Silviculture and management of teak plantations. Unasylva. 2000;**201**(51):14-21

[30] Roshetko JM, Rohadi D, Perdana A, et al. Teak agroforestry systems for livelihood enhancement, industrial timber production, and environmental rehabilitation. Forests, Trees, and Livelihoods. 2013;**22**(4):241-256

[31] Riyandoko ME, Roshetko JM. Guidelines for establishing coffee agroforestry systems. Bogor, Indonesia: World Agroforestry Centre (ICRAF) Southeast Asia Regional Program; 2017

[32] FEWS-NET (Famine Early Warning Systems Network). América Central: Informe de café. In: Country/Region Special Report. 2018. Washington, DC: FEWS NET; 2018

[33] Anta Fonseca S. El café de sombra: un ejemplo de pago de servicios ambientales para proteger la biodiversidad. Coyoacán, México D.F, Instituto Nacional de Ecología; 2007

[34] Farfán-Valencia F, Baute-Balcázar JE. Efecto de la distribución espacial del

sombrío de especies leguminosas sobre la producción de café. Cenicafé. 2010;**61**(1):35-45

[35] Somarriba E, Beer J. Cocoa-based agroforestry production systems. In: Shade Grown Cocoa Workshop. Panama: Smithsonian Institution; 1998

[36] Somarriba E, Villalobos M, Orozco-Aguilar L. Cocoa in Central American. GRO-Cocoa No. 14. 2008. pp. 3-7

[37] Gamboa H, Gómez W, Ibrahim M. Sistema agroforestal Quesungual:una buena práctica de adaptación al cambio climático. Informe Técnico/CATIE: Serie Técnica; 2009. no. 337

[38] Mendoza AA, Manrique E, Mora JL. Potencial del Sistema Agroforestal Quesungual (SAQ) en tierras de ladera en el sudeste de Honduras: Reduciendo la vulnerabilidad al cambio climático de pequeños agricultores de subsistencia. III Workshop REMEDIA. Abril 2014. Valencia. pp. 53-55

[39] Russo RO. Agrosilvopastoral systems: A practical approach toward sustainable agriculture. Journal of Sustainable Agriculture. 1996;7(4):3-17

[40] Russo RO. Reflexiones sobre los sistemas silvopastoriles. Pastos y Forrajes. 2015;**38**(2):157-161

[41] Russo RO. Agrosilvopastoral Systems: A Practical Approach Toward Sustainable Agriculture. Journal of Sustainable Agriculture. 1996;7(4):5-17

[42] Uribe F, Zuluaga A, Valencia L, Murgueitio E, Zapata A, Solarte L, et al. Establecimiento y manejo de sistemas silvopastoriles. Manual 1, Proyecto Ganadería Colombiana Sostenible. Bogotá, Colombia: GEF, BANCO MUNDIAL, FEDEGAN, CIPAV, FONDO ACCION, TNC; 2011. p. 78

[43] Budowski G, Russo RO. Live fence posts in costa rica: A compilation of the farmer's beliefs and technologies. Journal of Sustainable Agriculture. 1993;**3**(2):1993

[44] Budowski G. Living fences in tropical America: A widespread agroforestry practice. In: Gholz HL, editor. Agroforestry: Realities, Possibilities, and Potentials. Dordrecht. The Netherlands: Martinus Nijhoff Publishers; 1987. pp. 169-178

[45] Harvey CA, Villanueva C, Villacís J, et al. Contribution of live fences too the ecological integrity of agricultural landscapes. Agriculture, Ecosystems, and Environment. 2005;**111**:200-230

[46] Mendieta López M, Rocha Molina LR. Sistemas Agroforestales. Managua, Nicaragua: Universidad Nacional Agraria; 2007. p. 115

[47] Beer J. Advantages, disadvantages, and desirable characteristics of shade trees for coffee, cacao, and tea. Agroforestry Systems. 1987;**5**:3-13

[48] Murgueitio E. Silvopastoral systems in the neotropics. Colombia: Fundación CIPAV; 2005

[49] Torquebiau EF. A renewed perspective on agroforestry concepts and classification. Life Sciences. 2000;**323**:1009-1017

[50] Liu W, Yao S, Wang J, Liu M. Trends and features of agroforestry research-based on bibliometric analysis. Sustainability. 2019;**11**(3473):1-15

[51] Liu W, Wang J, Li C, Chen B, Sun Y. Using bibliometric analysis to understand the recent progress in agroecosystem services research. Ecological Economics. 2019;**156**:293-305

[52] van Noordwijk M. Small-island agroforestry in an era of climate change and sustainable development goals. In: van Noordwijk M, editor. Sustainable Development Through Trees On Farms: Agroforestry in its Fifth Decade. Bogor, Indonesia: World Agroforestry (ICRAF) Southeast Asia Regional Program; 2019. pp. 233-247

[53] Montagnini F, editor. Integrating Landscapes: Agroforestry for Biodiversity Conservation and Food Sovereignty. Springer; 2017. p. 501

Chapter 18

Agroforestry

Gyan Shri Kaushal and Rajiv Umrao

Abstract

Agroforestry has the double benefit of making grain and wood from bushes alongside crop advent from a solitary actual property parcel. It is beneficial to broaden teak, crease, sesame, and eucalyptus for wood, lumber for apparatuses, crease, bamboo and teak, bamboo for paper and mash, and karanj for biofuel. Manure bushes are for land recovery, soil wellbeing, and food security; herbal product bushes for sustenance and pay; feed bushes that in addition broaden smallholder domesticated animals' advent; lumber and gasoline wooden bushes for secure residence and energy; healing bushes to war infection, especially in which there's no drug store; and bushes that produce gums, tars, or plastic items. Ranchers having substantial land maintenance commonly take in this path of motion of tree planting. Significant tree species for block planting are poplar, eucalyptus, and deck. This vegetable offers a top-notch wellspring of high-protein dairy farm animal grain. The range is separated from the leaves, blossoms, bark, seeds, and so on from palas blossoms to orange, from one purple to any other, from one blue to any other, and so forth getting colors. If you are taking a gander on the historic backdrop of horticultural ranger carrier in India, we are those with the maximum profile of rehearsing rural ranger carrier. This assists with saving the dampness withinside the dust for pretty a while. So, we ought to make use of strategies like ranger carrier, and natural product cultivation. Raising animals at the homestead, cows, bison, goats, and hens bring fulfilment to the ranch. The agroforestry framework has the selection to study munching with the aid of using the reception of feed tree species with suitable grasses. Safeguarding systems and streets from floating snow, funding price range in domesticated animals' advent with the aid of using lessening wind chills, safeguarding crops, giving untamed existence territory, disposing of climatic carbon dioxide and handing over oxygen, lowering breeze pace, and as a consequence proscribing breeze disintegration and particulate depend withinside the air, diminishing commotion contamination, and moderating heady fragrance from focused domesticated animals' activities.

Keywords: agroforestry, vegetation, ecosystem, trees in agroforestry, benefits of agroforestry

1. Introduction

Picking basically dry land or decrepit land for forestry is helpful. These principally incorporate plantations and slopes, the side of the road close to homesteads, medium and shallow terrains, and regions with afforestation along the banks of a canal. These manors assist with staying away from soil disintegration. Agroforestry has the double advantage of delivering feed and wood from trees alongside crop creation from a solitary land parcel. It is valuable to develop teak, crease, sesame and eucalyptus for lumber, wood for instruments, crease, bamboo and teak, bamboo for paper and mash, and karanj for biofuel. Agroforestry structures art a range of growing scenes sharing for all plans and motive woody plant life utilized in blend with creatures contacting and yield creation [1]. Shorter regularity than monoculture development, multifunctional help with group and money organization plus the valuable effect of parts. These structures consolidate shelterbelts, crunch no man's land, fruitful fields, blooming backwoods, forest area gardens, worked on ignored land, the forest got together with crop creation, widely appealing yield advancement, and common green establishments. Agroforestry is to incorporate viable woody yields into green activities to line a monetarily, socially, and ecologically important development [2]. Agroforestry is alluring production, further developing movement agribusiness, increasing regular soil matter levels, set climatic nitrogen, reuse supplements, modifying microclimate, and upgrading the structure's benefit ceaselessly in regards to the possibility of reasonable creation [3]. Forest areas are seen as critical structures because of the regular organizations that they give, similar to carbon accumulation and get, oxygen creation, great enormity, and climate for arranged animal species; moreover, forests are indirect providers of the water framework for metropolitan life, fauna, helpful plants, and, when the whole thing is declared in done, huge parts of untamed life [4].

2. History of agroforestry

Establishing crop-related trees in close association with one another is a notable culture that applies to ranchers. Following the historical backdrop of Imnon ranger service, King eliminated forlorn timberlands to the Centre of the time in Europe, sliced, developed food crops for different periods in coordinated spaces, and already trees. Endless planting was a definitive practice. This "rural framework" is currently not normal in Europe, yet before the end of the last century, it was generally executed in European nations and was carried out during that period. Notwithstanding, before the end of the nineteenth century, the making of ranger service and rural ranches was a significant objective of dynamic Imnon ranger service. Above all else, stress revision is not expected. In 1806, Tibeto-Burman U. Dish Hle of the Tonze Forest in Thararrawaddy, as well as the Asian nation (Burma) at the station of the National Empire, said, "Taungya," awarded on account of the branch office, Dietrich Brandis. There are reports that "this is likely the most practical method for establishing teak." From this beginning, the application steadily spread. It was presented in South Africa in 1887, which was an Asian country at that point, and in 1890 it moved to the heart and geology of the Indian pioneer country [5].

The administering reasoning of the other Taungya framework was to recognize woods cultivated each time an icon or landless laborer was harmed. To complete organic undertakings, laborers can develop the land between the sapling lines for the creation of produce. This is a rearrangement of a framework with various subtleties in various nations and districts, because of the woodland chief's obligation to the timberland site; the essential objective of the examination of these blended frameworks was to affirm that:

- Almost no injury to woods species happened.
- The development level of tree species was not unreasonably repressed by rivalry with crops.

- Decide the ideal time and grouping for establishing one of the trees or harvests to guarantee endurance and ascent of the tree crop.

- Woodland species that had the option to endure rivalry with farming species are determined.

- Decide the ideal establishing stretch for the development of the following arboreal harvest.

So, the examination led was attempted for science by foresters. It appears to be the foresters directing the examination never imagined the framework as being fit for building a major commitment to a rural turn of events, or its true capacity as a land—the executive's framework [5]. If you take a gander at the historical backdrop of horticultural ranger service in India, we are the ones with the most seasoned record of rehearsing rural ranger service. All through District, most farming ranger service frameworks have been drilled since Neolithic sheep. Indians respect Ashok Vatika, a nursery made out of plants and fruiting trees in the Hindu awe-inspiring Ramayana, to act as an illustration of a rural ranger service framework by an academic administrator of attendant preparation. Indeed, even today, there are a few ceremonies related to tree and horticultural ranches in Asian nations. Following different public drives since the 1970s, the Indian government has driven further examination in the field of rural ranger service. Horticulture is additionally significant because of social and otherworldly practices. A few specialists brought up that numerous administration regulations believed them to nurture partners thwarting the improvement of ranger service [6]. The Indian National Agricultural Forestry Policy might be a far-reaching strategy system intended to further develop occupations in farming by amplifying horticultural efficiency to alleviate worldwide environmental change. The Government of India sent off this arrangement in February 2014 at the World Congress on Command Agricultural Forestry in Big City. India has turned into a significant world forerunner in taking on partner rural ranger service strategies. The arrangement intends to unite trees, harvests, and animals on a similar land to further develop efficiency and ecological attributes. It was composed to battle restricted rural efficiency brought about by the proceeded reduction in the helping of farmers' area possession brought about by fast development and occasional decrease in agrarian action. One more part of nursing, rural, and ranger service necessities was the absence of backwoods and trees in Asian nations. As per the 2019 Asian Country Forest Survey, the nation has 8.7 million ha of timberland trees, representing 24.56% of the nation all out the geological region. Nonetheless, under the public wood's strategy, a third (33.3%) of the nation is covered by woodland trees [6].

3. Where agroforestry fits

It envelops a wide scope of working trees that are developed on ranches and in-country scenes and incorporates the age of science-based tree enterprise opening doors that can be significant later on. Among these are compost trees for land recovery, soil wellbeing, and food security; organic product trees for nourishment and pay; feed trees that further develop smallholder domesticated animals' creation; lumber and fuel wood trees for haven and energy; therapeutic trees to battle sickness, particularly where there is no drug store; and trees that produce gums, saps, or plastic items [7].

An expected 1.2 billion rustic individuals currently practice agroforestry on their ranches and in their networks and rely on its items [8]. Their tree-based endeavors assist with guaranteeing food and health security, increment their pay and resources, and assist with taking care of their territory and the executive's issues.

These difficulties are:

1. To assist with destroying hunger through supportive of unfortunate food creation frameworks in distraught regions in light of agroforestry strategies for soil ripeness renewal and land recovery.

2. Lessen country destitution through market-driven, privately driven tree development frameworks that create pay and assemble resources.

3. Advance the wellbeing and sustenance of the rustic poor through agroforestry frameworks.

4. Ration biodiversity through coordinated protection advancement arrangements in light of agroforestry innovations, innovative foundations, and better approaches.

5. Safeguard watershed administrations through agroforestry-based arrangements that empower poor people to be compensated for their provision of these administrations.

6. Empower the provincial poor to adjust to environmental change, and to profit from arising carbon markets, through tree development.

7. Construct human and institutional limits in agroforestry innovative work.

4. The need for agroforestry

Ranger service is one such land of the board framework. In this, the accessible assets are completely used by expanding the advantageous trees in horticulture alongside occasional and lasting yields. It is gainful to do afforestation deliberately, for example, in a square or rectangular establishing technique. Various techniques for afforestation are chosen by the sort of land, precipitation, and nearby requirements. (i) Neglected lands, (ii) Desolate terrains, and (iii) Lands with desolate terrains are reasonable for ranger service. From this the rancher can expand his pay through feed, kindling, wood for farming carries out, a safe house, and products of the soil scale backwoods produce. Mounting the anthropological individuals requires an enormous sort of dinner and wood, which in the end delivers unnecessary strain on cultivable land and backwoods in India. In India, the assembling cap limit of wood for lumber is restricted to roughly 0.7 m^3/ha/year in assessment to the range regular area of around 2.1 m^3/ha/year achieving oversize original around of the call for and supply. Agroforestry is believed to be one of the practical substitutes, has been empowered the cap to hold onto land debasement, sequestering carbon and get better net page efficiency through connections among wood, soil, yields, and animals, and therefore fix the environmental factors and decorate the useful [9].

5. Perspectives on agroforestry

The wood delivered from it does not break down for a long time. Grain yields can likewise be filled in the open space between the trees as the trees get greater. The grass may likewise be accessible for the creatures to eat. The dirt surface is not really great for planting. Light and weak land can likewise be utilized for this reason. Once spent, there is a compelling reason needed to spend consistently. Likewise, different harvests can be filled in this piece for up to 4–5 years. You and the creatures can likewise get great shade. Trees block the air. This assists with saving the dampness in the dirt for quite a while. So, we ought to utilize techniques like ranger service, and organic product cultivation. Raising animals on the homestead, cows, bison, goats, and hens carry success to the ranch. Today our relatives go to the city for work as there is no work on the homestead. In the city, they face incalculable challenges. In the event that they can get work in rustic regions, they will not need to go to the city and deal with this multitude of issues. Animal cultivating is called protein cultivating. The ranch is so named on account of its high protein content in the items like meat and milk. Also, domesticated animals and dairy items should be effectively and financially showcased nearby. Such circumstances are helpful for this sort of cultivation. The sort of ranger service in which plants are created as nourishment for creatures is, for example, Grub, grain, and animals' feed.

6. Methodology

6.1 Agroforestry

Occasional harvests are planted in a succession of trees at an efficient and proper distance. Soybean beasts, wheat, and oilseeds can be developed alongside greens, neem, mahogany, eucalyptus, and bamboo. Conceal lenient harvests like turmeric, ginger, and restorative plants ought to be planted when the shade of trees increments. In tall and straight developing woods, teak, eucalyptus, cypress, and silver oak are planted on the bank around the plantation. This shields the plantation from solid breezes and hot breezes.

6.2 Agricultural garden

In this strategy, occasional or perpetual harvests can be filled in organic product trees for a couple of years. Organic product crops and intercrops are chosen by water and soil accessibility e.g. tamarind, Ber, Amla, drumstick, custard apple. Ringer and Blackberry can be filled in light and loamy soils. In the cultivation area, crops like mango, chickpeas, guava, and pomegranate are developed and vegetables or blossoms are filled in them.

6.3 Forests

In this technique, grub crops like guinea grass, rhododendron grass, para grass, napier, dinanath, anjan, uneven, pavana, stylo, ranmug, and fenugreek grass alongside babul, Anjan, Neem, Kadulimb, Hadaga, Shevari, Maharukh, Tuti, and Babul plants are planted deliberately or scattered. This framework is done on light, shallow, or sloppy neglected lands, which builds the fruitfulness of the dirt alongside the grub.

6.4 Arable land

These incorporate some superior grain crops alongside customary cereals, Napier mixtures, fenugreek or maize grub trees with subabhul or lumber giving trees, teak, mahogany, and so forth in a three-layered framework.

6.5 Intensive planting

In this, financially significant trees are planted at brief distances, for example, in high thickness. Its primary objective is to make unrefined components accessible for paper creation, biofuels, building posts, and power age.

7. Selection of trees in agroforestry

7.1 Suitable tree for different lands

Various techniques for afforestation are chosen by the kind of land, precipitation, and nearby requirements. Ripe, desolate, and fruitless grounds are reasonable for ranger service. It gives feed, kindling, and wood for farming executes, haven and foods grown from the ground items.

1. In uneven terrains—Neem, Subabhul, Eucalyptus, Acacia, Saundad.

2. In saline soils—Eucalyptus, Wadi Acacia, Neem, Acacia, Sisu, Karanj, Khair.

3. In light and shallow terrains—Acacia Anjan, Subabul, Sisu, Cirrus, Shevari, Neem.

4. In saline soils—Eucalyptus, Wadi Acacia, Neem, Acacia, Sisu, Karanj, Khair.

5. In wetland lands—Karanj, Giripushpa, Suru, Bhendi, Shevari, Nilgiri.

Woods balance the air, environment, and climate. However, this woodland asset is being obliterated. The pace of environmental change is expanding quickly because of the huge contamination, expanding populace and urbanization that go with deforestation. This has prompted an expansion in the occurrence of catastrophic events, for example, temperature changes, increasing ocean levels, extreme and unseasonal downpours, floods, dry seasons, softening of ice, and consumption of the ozone layer.

7.2 Scope of agroforestry

- Woods cover in India is around 20.60% of its absolute topographical areas of 67.71 million ha, against the best inclusion of 33.33%.

- In India, the timberland cover in bumpy areas is just 35.85% against the best inclusion of 66.66%.

- The per capita accessibility of forest in India is just 0.08 ha against the world normal of 0.64 ha, though it is of 1.07 ha. In created nations, agroforestry is the most ideal choice to overcome any barrier among wanted and accessible backwoods inclusion in the country.

- Public Wasteland Development Board assessed that 123 million ha of land is lying as no man's land in India [10]. Thus, appropriate agroforestry practices could recover these regions for reasonable improvement other than 50 million ha debased land because of mining action into creation framework:

- The land holding is decreasing step by step because of segment tension with expanded requests for feed, fuel, little wood, and other minor items produced using the forest. In this way, the strain on regular woodland could be enormously decreased taking on appropriate agroforestry rehearses nearby lands.

- The agroforestry plots normally create nonstop income which is not plausible in arable land and permits expansion of ranch exercises and utilizes natural assets.

- There is an incredible interest in fuel or kindling, especially in the country of India which could be met through the reception of agroforestry rehearses. Around 87% of the yearly gathered wood in India is utilized as kindling, and fuel wood. In addition, 60–80 million tons of dry cow fertilizer (identical to 300–400 million tons of newly gathered excrement) are used as fuel which might have been used as manure for crop creation.

- Continued brushing by creatures past the conveying limit of the land barely passes on any vegetation to endure except if uncommonly secured. The agroforestry framework has the choice to examine munching by the reception of feed tree species with appropriate grasses.

- Agroforestry has the most noteworthy possibilities in the age of provincial work because of the incredible variety of items with moderately less venture and that too with untalented country local areas.

- Agroforestry gives a more noteworthy degree to the reception of preservation practices and supports cooperative watershed investigation ways to deal with the board of scenes containing blended possessions, vegetation types, and land uses. Accordingly, agroforestry rehearses the battle against hunger, and insufficient asylum and can check ecological debasement in a more prominent manner [11].

7.3 Benefits of agroforestry

7.3.1 Increase forest production through forest conservation

As of now, ranger service can possibly turn out feasible revenue to the ranchers as well as adapt to the catastrophes of environmental change. Ranger service gives advantages like grub, organic products, vegetables, kindling, soil surface, the executives of corrupted and decrepit terrains, and wind insurance. What's more, ranger service gives unrefined substances to the paper and pressed wood businesses. To give impulse to ranger service in the current situation, the Government of India executed the National Agroforestry Policy in 2014 [12] and under this, multi-reason trees are being proliferated and advanced in agribusiness. Under this, National Forestry and

Bamboo Mission were begun in 2017. Wood is utilized in building, furniture, paper, kindling, fuel, coal, toys, and so on. Alongside wood, timberlands give you honey, food, organic products, blossoms, creature feed, strands, bamboo, stick, therapeutic plants, sweet-smelling plants, tannins, normal colors, leaves, gum, excellent wood carvings, wood, kach, khirsal. The bark, roots, seeds, polish, silk, flavors, plant pesticides, and so forth are acquired from the backwoods. Organic products like Jambul, Chinch, Fanas, Kokam, Karwand, Alu, Amla, Tadgole, Charoli, Mango, Toran, and so forth are acquired from the woodlands; curry leaves, takla, bharangi, cher, pev, karanda, white musli, math, ghol, kuda, and so on. Vegetables; Kevada, sandalwood, and cinnamon are gathered from fragrant plants. Sivan, Dhaman, Kanchan, Apta, Shewari, Tuti, Asana, Kinjal, Ain, Bamboo, and so forth for creature feed. The leaves of the species are utilized. Around 30% of the complete animals rely upon grass got from woodlands. In summer, the leaves of these trees are valuable as nourishment for goats, sheep, and creatures.

7.3.2 Supply of raw materials to industries

Individuals in the backwoods region gather honey from the woodland region, products of Ringi-Rita trees for the cleanser industry, Shikakai, Wawding, Bibba, PalasPhule, Gulvel, Arjunsal, Hirda, Behda, Amla organic products. Gum is gathered from Ain, Khair, Biwla, Kandol, Neem, and Acacia. Interest in this speciality has developed altogether because of ongoing corporate embarrassments. The woodland is a plant that gives a normal tone. The variety is separated from the leaves, blossoms, bark, seeds, and so forth from palas blossoms to orange, from one red to another, from one blue to another, and so on. Get colours. Interest in this speciality has developed altogether because of ongoing corporate outrages. Different articles are made by pulling ropes, strings, and strings from various pieces of the tree. From bamboo strings to mats, mats, brushes, bushels, sheets, and so on. Many items are made. There are 145 assortments of bamboo found in India. 5,000,000 tons of bamboo is collected in India consistently. Viticulture, paper making, feed, and so on. Bamboo is utilized for some reasons. Tendu leaves, Palas, Kanchan, Muchkund, Banana, Chavai, and so forth are utilized to make Drona, Patravali, Vidi, Buke, and houses.

7.3.3 Storage of forest produces

It is important to embrace traditional techniques while gathering, drying, and putting away timberland to produce natural products, tree bark, and so forth. A few organic products ought to be left on the trees while gathering. While stripping, strip in a similar heading. Subsequent to collecting the wood products, they should be dried well; any other way they can get rotten and ruined. While putting away backwoods produce, keep it in an all-around ventilated place, in a sack or in a container. Wood's produce ought not to be kept on the ground. Enormous scope planting of woodland is expected to yield trees. This makes it conceivable to set up numerous businesses in country regions.

7.3.4 Availability of legumes

A few ancestral social orders make living by selling foods grown from the ground. Wild organic products are profoundly therapeutic, containing an elevated degree of nutrients, proteins, normal sugars, and so on. These organic products are of the incredible interest in the market, for example, Aloo, Bhedas, Jambul, Atki, Anjan,

Chinch, Charoli, Amla, Devhara, Bor, Kokam, Yellow Kokam, and so forth, found in woodlands. Vegetables and a few tubers are tracked down just in timberlands. These vegetables and onions contain numerous illnesses. Takla Bhaji is utilized to scrub the stomach; Bharangi Bhaji is utilized to eliminate tooting. Greenery, Pev, Shevaga, Curry, White Kuda, White Musli, Karanda, Cartoli, and so on are utilized for vegetables. Since these timberland items are normal in nature, they contain no fake synthetic compounds and are beneficial for wellbeing [13].

7.3.5 Agroforestry for improved air and water quality

Agroforestry practices, for example, windbreaks and shelterbelts are promoted as having various advantages. These advantages incorporate really safeguarding structures and streets from floating snow, investment funds in domesticated animals' creation by lessening wind chills, safeguarding crops, giving untamed life territory, eliminating climatic carbon dioxide and delivering oxygen, decreasing breeze speed and accordingly restricting breeze disintegration and particulate matter in the air, diminishing commotion contamination, and moderating scent from concentrated domesticated animals' activities, among others. As of late, interest in involving shelterbelts as an expected way to deal with managing domesticated animals' scents has gotten impressive consideration. Vegetative supports can channel airstreams of particulates by eliminating residue, gas, and microbial constituents. In their essentials survey on this point with definite situation to pig smell Tyndall and Colletti [14]. Agroforestry rehearses are additionally a demonstrated technique to give clean water. In customary farming frameworks, not exactly 50% of the applied N and phosphorous manure is taken up by crops. Subsequently, the overabundance of manure is washed away from farming fields through surface overflow or drained into the subsurface water supply, along these lines debasing water sources and diminishing water quality [15]. Riparian cradles assist with cleaning spillover water by diminishing the speed of overflow, along these lines advancing penetration, dregs affidavit, and supplement maintenance. Supports likewise decrease supplement development into groundwater by taking up the abundance supplements. A few examinations have shown that agroforestry vegetative supports decrease nonpoint source contamination from column crop horticulture [16]. For instance, [17] archived expanded supplement evacuation proficiency when trees were consolidated into a riparian support strip put on the boundary of agronomic field plots in Iowa. Trees with profound establishing frameworks in agroforestry frameworks can likewise further develop groundwater quality by filling in as a "wellbeing net" by which overabundance supplements that have been drained underneath the establishing zone of agronomic harvests are taken up by tree roots. These supplements are then reused once more into the framework through root turnover and litterfall, expanding the supplement use proficiency of the framework [18]. Trees additionally have a more drawn-out developing season than most agronomic harvests, which increments supplement endlessly use productivity in an agroforestry framework by catching supplements when the trimming season. A couple of studies have revealed the security net job of trees established in both the tropical [19] and calm locales [20]. For instance, in a walnut cotton rear entryway editing framework in northwest Florida, [18] revealed a 72% decrease in nitrate-N at a profundity of 0.9 m contrasted with monoculture cotton. In a silvopastoral framework in Florida, the USA, [20] observed soil phosphorus fixations in pastures with and without a 20-year-old cut (*Pinus elliottii*) pine trees and reasoned that silvopastoral affiliations improved soil supplement maintenance and decreased supplement transport in surface and subsurface water. In general, the ebb and flow proof recommends that agroforestry frameworks could assume a significant part in moderating water quality issues emerging from concentrated farming practices. In spite of their advantageous impacts, the creators saw that shallower soil profundity and the resulting diminished accessibility of soil water were reasonable restricting root advancement and development of the poplar trees.

8. Broadly trees are grown on farmlands as under

1. **Scattered trees**—Sub-mountain locale is having dispersed trees on farmland. The ranchers hold the neighborhood trees like khair, kikar, dhak, shisham, mango, amla, phulai, and beri on their fields any place they end up recovering normally. These plants give fuel, feed during the lean period, and furthermore help in soil and dampness protection. In the undulating plain district, where a long-lasting water system source is accessible ranchers additionally embrace block planting of poplar at more extensive dispersing and between developing the agrarian harvest (**Figure 1**).

2. **Boundary planting**—This training is appropriate to all classes of ranchers, particularly the little ranchers who can develop trees and supplement their agrarian pay without forfeiting any land from rural activities. These single or twofold columns of trees on field bunds and water system channels develop rapidly and prosper well as they get water and manure applied to crops and have no rivalry for light and crown advancement. Limit line of trees ought to ideally be planted in the north-south course to limit the antagonistic impact of shade on connecting crops. It is obviously true that the misfortune in crop yield is much of the time remunerated by the pay created from collecting

Figure 1.
Scattered trees.

Figure 2.
Boundary planting.

Figure 3.
Block planting.

trees toward the finish of the revolution period. Reasonable homestead ranger service tree species are eucalyptus, poplar, subabul, mulberry, dek, toon, shisham, and so forth (**Figure 2**).

3. **Block planting**—This is an act of developing trees in a rectangular or square establishing design all through the field. Ranchers having huge land holding for the most part embrace this course of action of tree planting. Significant tree species for block planting are poplar and eucalyptus. Numerous agrarian harvests (wheat, potato, mustard, berseem, turmeric, sugarcane, and so forth) can be productively raised intermixed with block estates of trees (**Figure 3**).

9. Potential agroforestry trees

A tree animal group is thought of as appropriate for agroforestry in the event that it is having the greater part of the accompanying qualities: quickly developing nature, straight trunk, clear bole, fewer branches, restricted crown, the self-pruning mature beneficial outcome of litterfall and its disintegration on soil and with crops, great attractiveness, and so on. Remembering the above qualities, some tree species that could be possibly filled in agroforestry in the state are momentarily talked about underneath:

1. *Gmelina arborea* (Goomar teak, White teak and local name Gamhar) is a quickly developing deciduous tree with a straight trunk that happens normally all through a larger piece of India at elevations up to 1200 m. It leans towards soggy rich openly depleted soils with 750–4500 mm precipitation. It stays hindered on dry, sandy or unfortunate soils. It blossoms from February to March when the tree is pretty much leafless while organic products mature from May to June. This tree is really great for apiculture as its plentiful blossoms produce bountiful nectar and quality honey. It is a light demander and reasonably ices strong. Its lumber is utilized in developments, furniture, sports, instruments and counterfeit appendages. Once prepared, it is a consistent wood and decently impervious to rot and ranges from exceptionally impervious to respectably impervious to termites. It is proliferated through seeds as well as cuttings. Pivot for pulpwood is around 8 years and that of sawn wood is around 10 years under positive circumstances (**Figure 4**).

Figure 4.
Gmelina arborea *trees.*

Figure 5.
Toona ciliata.

2. *Toona ciliata* (Burma toon, Indian cedar, Indian mahogany and local name Mahanaim, Tun) is a huge deciduous tree. It fills well in damp areas up to an elevation of 1500 m with a mean yearly precipitation of 750–4000 mm. It leans towards very much depleted, profound, rich soil and does not excel on wet, minimal, or unfortunate sandy ones. Plants are spread through seeds. In block estate, trees foster a tall tidy bole up to a stature of 9–12 m which suits well for agroforestry. It is an ice strong and great coppice. The leaves are once in a while hacked for feed. White fragrant blossoms show up in spring. This is a significant species utilized for apiculture. The lumber is red in variety, simple to work, and profoundly esteemed. It is utilized broadly for furniture, wood framing, development, boat building, bureau making, matchboxes, beautiful compressed wood and facade outside utilizes, mouldings, and so on (**Figure 5**).

3. *Melia composita* (Ghora neem) is a quickly developing deciduous tree. It develops well at a height up to 1800 m and means yearly precipitation of 750–2500 mm. It is a light demander and great coppice. It is a potential agroforestry tree animal type for square and limited ranches in Punjab. It fills well in very depleted soil with moderate soil richness. It can likewise come up in soils with saltiness up to 4 ds/m. The tree blossoms in Jan-April and the natural product ages from Nov to Feb in the following year. It is spread through seeds and cuttings. Pivot for pressed wood is around 8 years and for sawn wood around 12 years. Wood is modestly hard that can be utilized for divider sheets, entryway boards, furniture, agrarian carries out, match industry, paper mash, and ground surface. It is viewed as the best-unrefined substance for compressed wood, for both face and centre facades (**Figure 6**).

Figure 6.
Melia composite.

Figure 7.
Dalbergia sissoo.

4. *Dalbergia sissoo* (Indian rosewood, Shisham and local name Tahli) is a medium to large-sized deciduous tree. It is the state tree of Punjab. It can develop at an elevation up to 1500 m and mean yearly precipitation of 500–4500 mm. It fills well in a wide scope of soil types, from unadulterated sand and rock to rich alluvial soil of riverbanks. Be that as it may, development is delayed in inadequately circulated air through destinations, for example, those with weighty dirt soils. It is engendered however seed and cuttings. It is a decent coppice tree animal group.

Young branches and foliage structure a fantastic grub. The blossoms are the helpful wellspring of honey. Wood is utilized for excellent furnishings, cupboards, beautifying facades, marine and aeroplane grade compressed wood, fancy turnery, cutting, etching, apparatus handles and outdoor supplies. The revolution age for wood is around 30 years (**Figure 7**).

5. *Azadirachta indica* (Indian Lilac and local name Neem) is a quickly developing evergreen tree of the mahogany family. It can develop at an elevation up to 1000 m with a yearly precipitation of 400–1200 mm. It flourishes best on very much depleted profound and sandy soils. Blossoms show up from March to May and organic products for the most part mature during the period June and July. It is a decent coppice tree animal variety. This tree is really great for apiculture as neem honey orders premium cost in the market. Leaves are put in cabinets to forestall bugs eating the garments. Neem oil is utilized for getting ready beauty care products, for example, cleansers, neem cleansers, demulcents, and creams as well as toothpaste. Its stem is utilized as a toothbrush. Neem cake is broadly used to treat cash crops, especially sugarcane and vegetables. Neem wood is extremely weighty, sturdy, and medium intense utilized for making trucks, ships, farming executes, furniture cupboards, cut articles, and so on (**Figure 8**).

6. *Salix alba* (White willow and local name Willow) is a medium to the enormous estimated deciduous tree. It inclines toward soggy or wet weighty soil in a radiant position however can likewise develop well on ineffectively depleted or irregularly overflowed soils. The blossoms are delivered in catkins in late winter and

Figure 8.
Neem tree under banana cultivation.

Figure 9.
Salix alba.

organic products mature in midsummer. It is engendered through cuttings. The wood is extreme, solid, and light in weight yet has negligible protection from rot. Coppice shoots and branches from pollarded plants are utilized for making. Its lightweight wood is utilized for making cricket bats as it does not fragment without any problem (**Figure 9**).

10. Tree species of the Kandi region

The Kandi area is home to an immense scope of trees and bushes that assume a significant part in soil preservation as well as giving fuel, grain, and wood during a period of shortage. A portion of the significant tree types of Kandi area are talked about beneath:

1. *Grewia optiva* (Bhimal, local name Biul or Dhaman) is a moderate measured deciduous tree. It is conveyed in the Himalayas up to a height of 2000 m. It sheds its leaves in March–April and the new leaves seem a month after the fact. It blossoms in March and June and the age of the organic product is from October to December. It is a solid light demander and requires total light for its ideal development. It tends to be effectively spread through seeds or cuttings. Its completely adult leaves are an excellent feed for cows, particularly in winter. The lumber is utilized for shafts, shoulder posts, bunk outlines, oars, apparatuses, and hatchet handles. Its delicate branches are utilized for making bins. The branches are kept in water for not many days and their bark is utilized for rope making. It is

Figure 10.
Grewia optiva.

typically present on field bunds or on slants in the Kandi district and is a valued tree for individuals of the Kandi area (**Figure 10**).

2. *Dendrocalamus strictus* (local name Bamboo) has the capacity of delivering the most extreme biomass per unit region and time when contrasted with other timberland plants. In Punjab, it exists in the Kandi district, and the woods areas of the Talwara block are loaded with bamboo. A large portion of the types of bamboos require sandy topsoil to mud soil, very much depleted soil, and wet circumstances; however, *D. strictus* requires coarse-grained very much depleted, and dry areas. A couple of bamboos are strong; however, a large portion of them is empty. Bamboos bloom at long stretches; in this manner accessibility of seed is an issue. Engendering of bamboos by vegetative strategy is speedier; however, the estate so raised would blossom and will for the most part bite the dust when the parent bunches rose. The most widely recognized strategy for vegetative proliferation is a rhizome. In this technique, a year-old culm is cut at 30–60 cm stature with its rhizome and root foundation is recovered and the whole offset is planted in 45 cm^3 pits during the stormy season. Bamboos are utilized for an assortment of purposes viz., material, covering, development, fencing, food (seed and new shoots), grub, fuel, paper, and mash, crafted works, bin making, apparatus handles, stepping stools, windbreaks, and disintegration control, and so forth (**Figure 11**).

3. *Albizia lebbeck* (East Indian walnut, local name Siris) is a moderate to enormous estimated tree, with dull dim harsh bark. It fills in an assortment of soils and environments, the precipitation differing from 630 to 2550 mm and an elevation

Figure 11.
Dendrocalamus strictus.

Figure 12.
Albizia lebbeck.

up to 1500 m. In the Kandi locale, it exists on the field bunds as well as in the wood regions. It is a deciduous tree. Its leaves begin falling in October–November and the new leaves show up in March and April. In regular circumstances, siris produces plentiful seed crops every year. It very well may be raised effectively by direct seed planting or relocating. Seeds age from January to March; the best time for assortment is February. It has solid, flexible, and hardwood. The tree has an excellent grub. The heartwood is attractive, with light and dim streaks, which is utilized for furniture, the bureau works, inside enhancement, framing, and so forth. The tree is additionally appropriate for the end goal of building and agrarian execution. Its leaves, bark, blossoms, and cases are utilized therapeutically (**Figure 12**).

4. *Acacia catechu* (The cutch tree and local name Khair) is a little or medium estimated tree. It is available in dry areas as well as in sub-Himalayan parcels with precipitation up to 3800 mm. In the Kandi district, it is available in overflow in the non-arable and timberland regions claimed by individuals. The leaves are shed in February and the new leaves show up toward the end of April or during May. Blossoms likewise show up simultaneously as the new leaves. Khair is a solid light demander and ice strong. Direct planting of its seed is exceptionally effective and is the least demanding strategy for spread. It can likewise be spread by root or shoot cuttings. The seeds can be gathered by trimming little case-bearing

Figure 13.
Acacia catechu.

branches in December or early January. Khair is for the most part utilized for the development of cutch and Katha. It is additionally an excellent wood and utilized in house development, making sugarcane smashers, furrows, boats, and so on. Its heartwood is exceptionally sturdy. It is likewise utilized for fuel and charcoal making and its gum is of generally excellent quality (**Figure 13**).

5. *Acacia nilotica* (local name Babul or Kikar) is a tree of variable size. It is broadly planted or self-planted all through dry and hot areas of India. In the Kandi zone of the state, it happens as a confined tree or in gatherings of little fixes, in developed fields, town nibbling grounds, and waste terrains. Babul develops on an assortment of soils. The tree is much of the time found on alluvial topsoil in northern India. Blooming happens in June–July. The babul is a solid light demander and is ice delicate. The counterfeit recovery of babul presents no trouble. The decision between direct planting and planting is chosen by the goal of the manor and the site conditions. On the side of the road, woodland regions, rail line strips, and planting out of packs plants are inclined toward. The wood is principally utilized in construction work for posts, bars, and door jambs. It is one of the most loved lumbers for a wide range of farming executes, device handles,

Figure 14.
Acacia nilotica.

Figure 15.
Leucaena leucocephala.

and different purposes where hardness and durability are required. The bark is utilized for tanning and babul is the fundamental wellspring of gum arabic. Leaves and twigs are a decent wellspring of goat and sheep grub (**Figure 14**).

6. *Leucaena leucocephala* (Miracle tree, Ipil, White lead tree, and local name Subabul) is a little quickly developing tree. It performs well in a wide scope of precipitation from 650 to 3000 mm. It develops well on profound, all-around depleted, unbiased careous soils. Leucaena is neither lenient to inadequately depleted soils nor to ice. A quickly developing tree is liked for biomass creation. Nonetheless, it is considered a weed under unmanaged conditions as it structures thick bushes that group out any local vegetation of an area. This vegetable gives a fantastic wellspring of high-protein dairy cattle grub. It is accessible on field bunds in the Kandi district. Individuals of Kandi use a feed of subabul blended in with grub of biul and siris. It is utilized for an assortment of different purposes, for example, rear entryway trimming, disintegration control on bunds, kindling, fiber, and windbreaks (**Figure 15**).

11. Conclusion

Agroforestry has the double advantage of creating grain and wood from trees alongside crop creation from a solitary real estate parcel. It is helpful to develop teak, crease, sesame, and eucalyptus for wood, lumber for apparatuses, crease, bamboo and teak, bamboo for paper and mash, and karanj for biofuel. Manure trees are for land recovery, soil wellbeing, and food security; natural product trees for sustenance and pay; feed trees that further develop smallholder domesticated animals' creation; lumber and fuel wood trees for safe house and energy; therapeutic trees to battle infection, particularly where there is no drug store; and trees that produce gums, tars, or plastic items. Ranchers having enormous land holdings generally take on this course of action of tree planting. Significant tree species for block planting are poplar, eucalyptus, and deck. This vegetable gives a great wellspring of high-protein dairy cattle grain. Wood is utilized in building, furniture, paper, kindling, fuel, coal, toys, and so on. Alongside wood, timberlands give you honey, food, organic products, blossoms, creature feed, strands, bamboo, stick, therapeutic plants, sweet-smelling plants, tannins, normal colors, leaves, gum, excellent wood carvings, wood, kach, khirsal. The bark, roots, seeds, polish, silk, flavors, plant pesticides, and so forth are acquired from the backwoods.

Author details

Gyan Shri Kaushal* and Rajiv Umrao
Sam Higginbottom University of Agriculture, Technology and Sciences (SHUATS), Allahabad, Uttar Pradesh, India

*Address all correspondence to: gyanshri7@gmail.com

References

[1] Losada M, Santiago-Freijanes J, Rois-Díaz M, Moreno G, Herder MD, Aldrey-Vázquez J, et al. Agroforestry in Europe: A land management policy tool to combat climate change. Land Use Policy. 2018;**78**:603-613

[2] Csonka A, Bareith T, Gál VA, Fertő I. Spatial pattern of CAP measures promoting agroforestry in Hungary. AgBioforum. 2018;**21**(2):127-134

[3] Kaushal GS, Umrao R, Chaurasia S. Integration of Horticultural Crops in Agroforestry. Agroforestry Prospective, Strategies and Future Aspects. Taran Publication; 2021. pp. 67-75

[4] CEDERSSA. Bosques y Proyecto Nacional, Competitividad Forestal Presupuesto, Iniciativa Forestal. México: Palacio Legislativo de San Lázaro; 2011. pp. 21-28

[5] Rizvi SJH, Tahir M, Rizvi V, Kohli AA. Allelopathic Interactions in Agroforestry Systems. Taylor & Francis; 2010

[6] FAO. Greater efficiency of tree species, and the growth of agricultural crops. 2003. Available from: http://www.fao.org/3/XII/

[7] Garrity DP. Agroforestry and the achievement of the Millennium Development Goals. Agroforestry Systems. 2004;**61**:5-17

[8] World Bank. Sustaining Forest: A Development Strategy. Washington, DC: World Bank; 2004

[9] Bijalwan A, Dobriyal MJR. Agroforestry and horticulture: An employable and eco-friendly option. International Journal of Current Research in Biosciences and Plant Biology. 2015;**2**(8):81-86

[10] Gautam NC, Narayan LRA. Wastelands in India. Mathura, India: Pink Publishing House; 1988. p. 96

[11] Hossain MM. Forest Tree Species for Agroforestry, Agroforestry Prospective, Strategies and Future Aspects. Taran Publication; 2021. pp. 147-163

[12] Chavan S. National Institute of Inorganic Stress Management, Baramati, Dist., Pune. 2014

[13] Chakor BR. Agro-Forestry Planning an Overview. Agroforestry Prospective, Strategies and Future Aspects. Taran Publication; 2021. pp. 54-61

[14] Tyndall J, Colletti J. Mitigating swine odor with strategically designed shelterbelt systems: A review. Agroforestry Systems. 2007;**69**:45-65

[15] Cassman KG. Ecological intensification of cereal production systems: Yield potential, soil quality, and precision agriculture. Proceedings of the National Academy of Sciences of the United States of America. 1999;**96**:5952-5959

[16] Anderson SH, Udawatta RP, Seobi T, Garrett HE. Soil water content and infiltration in agroforestry buffer strips. Agroforestry Systems. 2009;**75**:5-16

[17] Lee KH, Isenhart TM, Schultz RC. Sediment and nutrient removal in an established multi-species riparian buffer. Journal of Soil and Water Conservation. 2003;**58**:1-8

[18] Allen S, Jose S, Nair PKR, Brecke BJ, Nkedi-Kizza P, Ramsey CL. Safety net role of tree roots: Experimental evidence from an alley cropping system.

Forest Ecology and Management. 2004;**192**:395-407

[19] Van Noordwijk M, Lawson G, Soumaré A, JJR G, Hairiah K. Root distribution of trees and crops: Competition and/or complementarity. In: Ong CK, Huxley P, editors. Tree–Crop Interactions: A Physiological Approach. Wallingford, UK: CAB International; 1996. pp. 319-364

[20] Nair VD, Nair PKR, Kalmbacher RS, Ezenwa IV. Reducing nutrient loss from farms through silvopastoral practices in coarse-textured soils of Florida, USA. Ecological Engineering. 2007;**29**:192-199

Chapter 19

The Implications of Conservation Agriculture in Forests Management against Soil Erosion and Degradation

Moses Z. Sithole, Azikiwe I. Agholor and Shalia M. Ndlovu

Abstract

Ecosystems play a huge role in support of human life, this is evident through their provision of food, fiber, water and fuel. However, these potentials are reduced through human activities, which comes with the lack of conservation of our forests. Deforestation is one of the major issues as far as sustainable development is concerned. Deforestation contributes towards soil erosion, particularly, in forests across the world. Soil erosion deprives human beings of the opportunity to enjoy the benefits of harvesting the forests' potential towards supporting human life, which includes the release of oxygen and the uptake of carbon dioxide. Thus, the concept of conservation agriculture becomes of paramount importance. Hence, this paper explored the implications of Conservation Agriculture in Forest management and evaluated policies in place to promote the adoption and use of conservation agriculture across the globe.

Keywords: soil degradation, soil erosion, conservation agriculture (CA), forest management, deforestation and sustainable agriculture

1. Introduction

The ecosystem plays a very paramount role in support of human life. This is manifested in the four forms established by the 2005 Millennial Assessment of Ecosystems (TMAE), namely: 1. Provision of habitat, food, fiber and water; 2. Regulation of climatic conditions, water and air quality; 3. Support system for the formation of soils, recycling of soil nutrients, as well as soil health and quality and preservation of biodiversity; and 4. Cultural support involves cultural beliefs, tourism and recreation as well as herbal benefits [1, 2]. It is moreover, emphasized by [2] that ecosystems are essential for human life, with which without, there is no life. It can further be argued that an ecosystem is an anchor to human life. Ecosystem refers to the community of living organisms, their diverse physical environment, and their interrelationships within a unit space [3]. It is that environment that supports the flow of energy. As per [4], ecosystems aid in carbon sequestration as well as

 IntechOpen

the preservation of both biotic and abiotic organisms. This, therefore, is evidence enough to argue that forests support human life and their management is a policy matter.

However, growing scientific concerns point to the fact that the support and services humanity receives from the world's ecosystems are on a rapid decline due to forest degradation [1, 2, 5]. Forest refers to a complex ecological system with interdepended, interrelated and interconnected elements of both the biotic and abiotic organisms largely dominated by trees [1]. Furthermore, this may not be limited to the only current dominance of trees, as geographers argue that forests exist in the past, the now and the future. However, 31% of the worldwide land is under threat of degradation and this has negative implications for human life [6]. Therefore, this threat calls for proper forest management so that future generations will benefit from the diverse services and support of the same forest and ecosystems we benefit from now. Moreover, the degradation of our global forests is a result of factors such as climate change, veld fires, pests and diseases, air pollution, land pollution, forest fragmentation, and soil erosion and degradation [1–3].

Forest management is the practical application of efforts in controlling the use and exploitation of our forests by means of policy, administration, social, economic, environmental and technical aspects [6]. It is furthermore, a critical process which determines the existence and preservation of our forests for continued and/or sustainable development across the world. Hence, the concept of Conservation Agriculture (CA), seeks to preserve and where possible improve the natural resources humanity has access to, for the production of food and fiber. Conservation Agriculture refers to a farming system that aid in the preservation of arable lands and natural resources while improving degraded lands [7]. This is possible when the three principles of CA are practiced across the Agriculture and Forestry sectors. The three pillars of CA are Minimal soil disturbance, permanent soil cover and crop diversification [7, 8]. CA has the potential to help achieve Sustainable Development Goals 13 and 15, namely: Climate Action and Life on Land, respectively. Therefore, this paper delineates the concept of CA in the context of forestry, highlights the CA practices for forest management, and presents the implications of CA in forest management. On the implications of CA in forest management, the chapter is mainly focused on the environmental aspects instead of a broader scope around social, economic and environmental aspects.

1.1 Delineation of conservation agriculture

Conservation Agriculture (CA) was first introduced in 1930, aiming at improving agricultural production and performance, profitability as well as sustainable farming and food security to fight poverty across the world [7]. Significantly, CA helps to further achieve Sustainable Development Goals 1, 13 and 15, that is Zero Hunger, Climate Action and Life on Land, respectively. Minimal soil disturbance, optimal soil cover and crop diversification are the three main pillars of CA [7, 8]. According to [8–11], CA is a resource-saving farming mechanism. Therefore, forestry cannot be an exception in the use of this beneficial farming system. This is to be done in line with the aim of forest management outlined in the introduction of this chapter, which is to ensure the preservation of forests and improve degraded forests across the world. **Figure 1** illustrates the three pillars of CA. The CA pillars are very practical, even in forest management. These come with enormous benefits such as reduced production costs, reduces soil compaction, improves soil health and structure, reduces soil

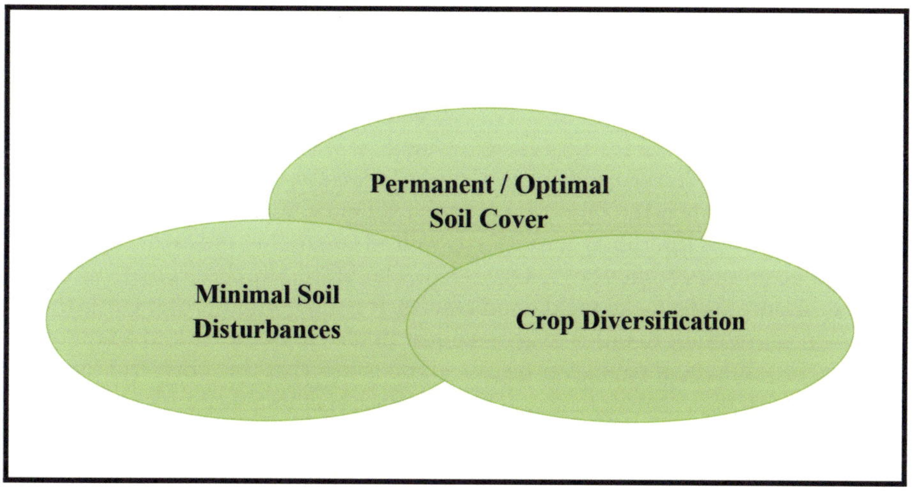

Figure 1.
The pillars of conservation agriculture.

erosion, minimized soil water evaporation, enhanced soil nutrient status, enhanced biological and microbial activities, improves soil water-holding capacity, improves Cation Exchange Capacity, improves the management of weeds, pests and diseases and leads to improved livelihoods and food security [9, 12–14].

Conservation Agriculture (CA) is an agricultural system with the power to conserve natural resources, reduction of greenhouse gas emissions as well as enhancement of soil health and fertility [10]. In the quest to conserve nature, the powerful, agricultural system (CA), uses its three pillars (already highlighted in sections above) as guiding principles to be practiced simultaneously. Conservation agriculture, through the application of the three pillars, prohibits the burning of the crop remains, while promoting the use of integrated pest management and the use of green manure which has the ability to produce residue soil cover. It further limits the movements of humans and machinery on agricultural soils which also leads to soil compaction [9, 12]. According to [15, 16] minimal soil disturbance refers to the tilling of the ground or turning of the soil with an aim of not disturbing the microbial life, without exposing the soil to harsh climatic conditions (especially too high temperatures) and erosion (either by wind or water).

Moreover, this practice involves the conservation tillage practices such as zero tillage and reduced tillage. Zero tillage and minimal tillage both involve direct seeding [17]. Minimal soil disturbance is the first pillar of conservation agriculture, and it helps to retain moisture in the soil, increases soil organic matter content, improves soil health and also helps to mitigate the effects of high climatic conditions [17–19]. On the other hand, permanent soil cover is mainly practiced through the organic mulch, green manure, retention of crop residue as well as plastic mulch. However, conservation agriculture promotes organic mulch as well as the retention of crop residue [7, 8, 20]. Soil cover protects the soil from harsh climatic conditions (such as the sun and rain). Furthermore, permanent soil cover helps to enhance the microbial activities in the soil, enhances soil fertility, and conserves soil moisture and nutrients. Permanent soil organic cover plays a huge role also in suppressing weeds and controlling pests and diseases [13, 19]. Moreover, it positively contributes to the improvement of soil's physical, chemical and biological properties.

Remarkably, an improvement in the soil's chemical, physical and biological properties improves the structure of the soil [18]. The more the soil structure is improved, the more soil fertility is improved. Therefore, permanent soil cover, significantly, benefit farmers in terms of soil water conservation, weeds, pests and diseases management, soil fertility enhancement, as well as the protection of the soil from climatically harsh conditions. Crop rotation assists in the improvement of crop nutrition, improves the farming system's resilience, manages pests, diseases and weeds, and improves microbial activities as well as enhancing the farm's agronomic output and economic efficiency [9, 13, 17–19]. Therefore, the three pillars of CA practiced simultaneously can yield good results. It is also evident, that CA is the vehicle farmers require, in order to achieve sustainable agriculture and adapt to the ever-changing climate. It can never be overemphasized that this powerful farming system is important for forest management for both sustainable development and nature conservation.

1.1.1 Delineation of conservation agriculture in the context of forestry

Agroforestry (AF) and Conservation Agriculture (CA) cannot be separated because they complement each other for the success of nature conservation and the Sustainable Development Goals (SDGs) 1, 13 and 15 as articulated in the introduction of this chapter. Agroforestry refers to a farming system that seeks to manage the land by means of combining agriculture, shrubs and trees [18]. Agroforestry also comes with some known benefits for both the environment and human life, such involves improved soil health, higher yields and chances of maximizing income for improved livelihoods [18, 21]. Forests are a support system for the ecological systems (ecosystems) and they do this through the continuous production and conservation of the soils, water and air [1, 21]. The conservation and enhancement of the soil and water in our forests through Forest Conservation Management Practices [FCMP] guarantees the prevention of land degradation, and desertification [21]. These further reduce the already existing risks of natural disasters such as floods, droughts and landslides. Conservation Agriculture is key to ensuring proper or conservative management of our forests across the world. Therefore CA in the context of forestry refers to a farming system which applies conservation principles in the management of forests to ensure the sustainable development and conservation of natural resources, simultaneously. These are made possible by the key principles of CA, namely: optimal ground or soil cover and minimal soil disturbance.

Ground cover is a strategy employed by forest managers, both in natural and planted forests to ensure reduced surface runoff and erosion in forests in order to achieve optimal forest performance [3, 5, 6]. On the other hand, minimal soil disturbance is the strategy to produce similar results as ground cover, however, it restricts ground tillers to turn the soils deep [7]. Ground cover, therefore, refers to conservation agricultural practice that involves allowing low-growing shrubs and/or planting low-growing plants, shrubs, grasses and wildflowers in order to prevent soil erosion [3–7]. While minimal soil disturbance, by definition refers to the farming practice that involves all ways of farming which avoids tilling the ground or soil [13]. This in the agricultural sector can be understood as direct seeding, which happens mostly, without disturbing the soil with the use of mechanical implements. Both these principles of Conservation Agriculture are essential to yield soil preservation, which then assures forest conservation. In the absence of soils, there can be no trees or forests. Hence, the CA practices aid to conserve and improve the soils in our forests.

1.2 The implications of conservation agriculture in forests management

The practice of Conservation Agriculture (CA) in forestry and agroforestry comes with a number of benefits for humanity. These involve but are not limited to financial benefits, that save money for the farmers by reducing production costs and leading to improved livelihoods. Though CA comes with three principles or pillars, as mentioned in subsequent sections above, its applicability in forestry and/or agroforestry

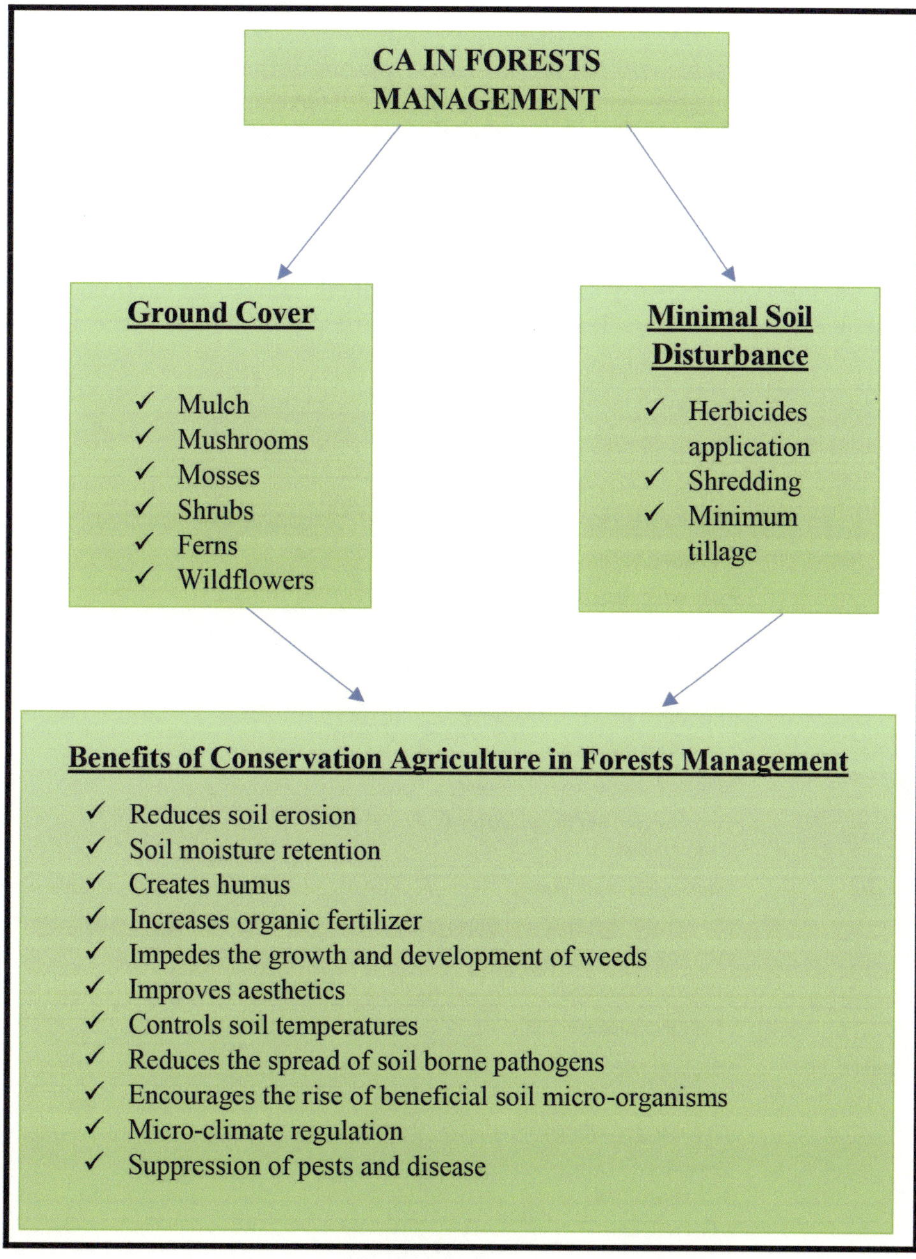

Figure 2.
Implications of conservation agriculture in Forest management (conceptual framework).

involves ground cover and minimal soil disturbance. **Figure 2** shows the CA practices applicable in forest management and their implications for forest management.

Figure 2 summarizes the implications of Conservation Agriculture (CA) in Forest Management and for the management of our soils to prevent soil degradation and erosion. These are achieved over a period of time and with persistence in practice. This means it is not possible to reap these benefits without having patiently and persistently practiced minimal soil disturbance and applied maximum or optimal ground cover for several years. To further expound on the benefits, **Table 1** is provided below.

Successful forest management results in sustainable development. Sustainable development is a complex process that requires numerous actors, including proper and scientific management of our forests. Forestry is one of the key players in economic emancipation, growth and development of both developing and third-world countries. **Table 1** shows the benefits and challenges of Conservation Agriculture (CA) in Forests Management. These benefits and challenges are classified into three groups: environmental, social and economic. The concept of sustainable development cannot be mentioned without touching on the three aspects of sustainability. This is because development that is not sustainable will result in the extinction of the human race. Therefore, the concept of CA in forest management comes with environmental, social and economic benefits. However, there are some challenges that farmers in the

Aspect	Benefits	Challenges
Environmental	• Storage of Carbon • Regulates micro-climate • Development and formation of aerosols • Improves soil health • Captures, stores and supplies water • Suppression of pests, diseases and weeds • Regulates atmospheric gaseous exchange • Promotes species diversification • Decreases pollution • Protects against natural disasters • Improves soil health and quality (soil conservation)	• Policy • Certification and regulations • Authorization of land-use
Social	• Sustainable development • Sustainable livelihood and employment creation • Poverty alleviation • Improves human health	• Insufficiency of technical skills • Lack of knowledge of sustainable (CA) forests management • Inadequate support
Economic	• Supplies wood for wood production output • Tourists attraction sites • Provides resources to produce pharmaceutical • Fuel supplies	• Policy • Certification and regulations

Table 1.
Benefits and challenges of CA in forests management.

forestry business are dealing with regarding CA. These are also categorized into three, namely: environmental, social and economic challenges.

The environmental challenges involve policy, certification and regulations, and Authorization of land use. While economic challenges involve Policy, Certification and regulations. Significantly, there the issue of policy comes in on both the environmental and economic challenges of the implementation of CA on forestry and agroforestry. And lastly, the social challenges associated with the implementation of CA in forestry and agroforestry. These involve the insufficiency of technical skills, the lack of knowledge on sustainable (CA) forests management as well as inadequate support from both government and farmers in the industry. The level of support in forest management has a significant influence on an agroforest's management and success. Moreover, these challenges can be addressed by a collective effort. That is the effort of the government, Non-Governmental Organizations (NGOs), the farmers, and society at large. Everyone has a role to play towards the management of our forests across the world.

1.3 Sustainable forest management and its underpinning pillars

To achieve forest sustainability, researchers have submitted the Sustainable Forest Management (SFM) concept, which comes with three pillars or principles [21–24]. This strategic approach to forest management came as a response to the global issues of deforestation and forest degradation. Forest degradation and deforestation have paused a challenge not only to hiking climatic conditions but to society at large by reducing the productivity of forests and the ecosystem services rendered by forests [21, 23]. SFM refers to a system of forest management through the application of the Sustainable Development (SD) pillars. The three pillars of SFM borrowed from the concept of SD are environmental protection and preservation, economic prosperity and socio-cultural acceptability principles [23]. SFM can be defined as a theory and/or system that seeks to ensure that forests worldwide provide the expected goods and services to meet the needs of both the current and future generations [23]. Furthermore, the SFM system contributes towards the Sustainable Development Goals 1, 13 and 15 as indicated in Section 2.1.1 of this chapter. The SFM framework has been put in place to prevent the massive forest losses experienced globally [23]. The economic loss which occurred as a result of the loss of forests globally has been a resulting decline in total forest area (land), which predominately leads to deforestation and degradation of forests.

The implementation of SFM strategies has been reported to have contributed towards food security, employment creation, economic emancipation, poverty alleviation, and preservation of forest ecosystems. This ensures that society continues to harvest the socio-cultural, economic and ecological benefits presented by well-managed forests worldwide. **Figure 3** shows the pillars of Sustainable Forest Management.

1.3.1 The pillars underpinning the theory of sustainable Forest management

See **Figure 3.**

1.3.2 The principles of sustainable Forest management

Table 2 summarizes the principles of the Sustainable Forest Management (SFM) system which is key to the conservation of our natural resources, food security and sustainable development. In a study by [22] in Sabah, Malaysia, it was discovered that

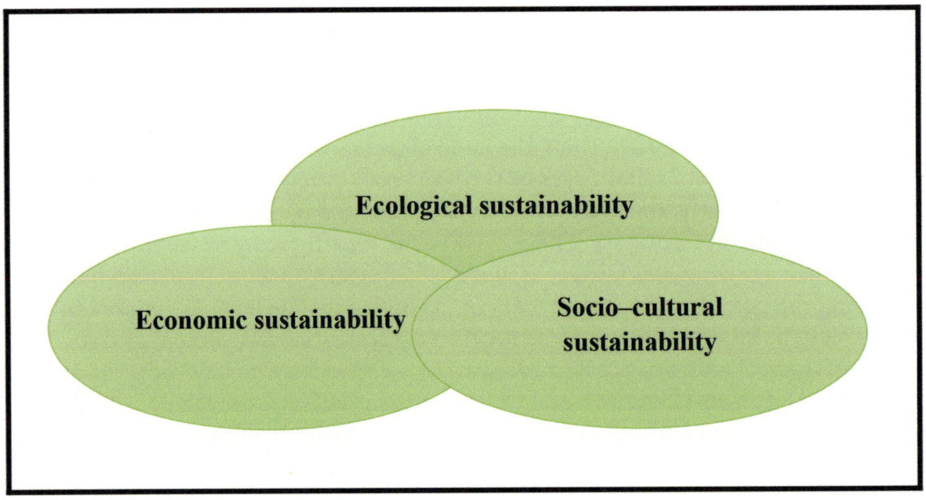

Figure 3.
Pillars of sustainable Forest management.

SFM principle	Pillar of CA in agroforestry	Pillar of SFM
Conservation of biodiversity	Permanent or optimal soil cover and minimal soil disturbance	Ecological sustainability
Maintenance of a healthy forest ecosystem and productivity	Permanent or optimal soil cover and minimal soil disturbance and where possible diversification of shrubs and trees in the forest	Ecological and economic sustainability
Maximizing societal benefits	Permanent or optimal soil cover and minimal soil disturbance	Socio-cultural sustainability
Soil and water conservation	Permanent or optimal soil cover and minimal soil disturbance	Ecological sustainability
Sustainable development of the society	Permanent or optimal soil cover and minimal soil disturbance	Economic and sociocultural sustainability
Forest contribution to world-wide ecological cycles	Permanent or optimal soil cover and minimal soil disturbance	Economical sustainability

Table 2.
Principles of sustainable forest management.

Sustainable Forest Management practices contribute immensely towards food security, with 51% of the participants responding “strongly agree” to the fact that SFM contributes towards food security. Furthermore, SFM was perceived to have a role towards increasing the streams of income generation for individuals and households, which is viewed to speed up economic growth and development [22, 23]. It is further perceived that successful forest management will continue to positively impact locals in terms of economic emancipation and poverty alleviation.

Table 2 shows the principles guiding the implementation of SFM globally. Contemporary literature submits that Sustainable Forest Management (SFM) strategies and practices involve reforestation, replanting forests after harvesting, treating tree diseases in early stages after diagnosis, a controlled burn of naturally revive forests, control of weeds and pests, as well as thinning out of some trees to allow adequate sunlight [22–25]. It is furthermore, evident from **Table 2**, that the use of SFM strategies can be improved and strengthened by the use and implementation of Conservation Agricultural practices. Following the guiding principles of SFM, implementing the SFM practices, and incorporating the pillars of Conservation Agriculture will ensure the achievement of the SFM objectives, which have been discussed in this chapter in terms of the benefits of SFM and CA. These include economic growth and development, environmental preservation and protection and sociocultural sustainability. Therefore, CA and SFM in forest management should be applied simultaneously to optimize the benefits of both strategies.

2. Conclusion and recommendations

There's a dualistic problem facing the world today, particularly, humankind. That is economic emancipation, growth and development and the degradation of the environment which is happening at a very high rate. This affects our forests which serve as a support system for human life. The forests support human life by providing oxygen, the abiotic and biotic life which is mostly dominated by trees as well as the control or management of the regional climatic conditions (temperatures). Forests play a huge role in producing goods that are essential for developing economies across the world. This is done through the provision of wood which is used to produce furniture and paper, fuel supplies, and supplies resources used in the production of pharmaceutical products. It also serves as a tourist attraction site and works a great deal in employment creation. Hence, the applicability of Conservation Agriculture (CA) in forest management is crucial. This chapter, therefore, aimed at evaluating the implications of CA in forest management and forestry. CA practiced in forestry promises reduced costs of production, improved livelihoods and poverty alleviation (**Table 1** and **Figure 2**). Therefore, it is concluded that CA is very important in forest management because it delivers the objectives of sustainable development. That is economic growth, social care and environmental protection. Therefore, it is recommended that CA be promoted by policy, and implementation of such policies be monitored. Though the applicability of CA in forestry depends on the specifications of each and every region, however, the society of farmers is encouraged to employ all the applicable principles of CA in their agroforest practices to ensure they benefit from this promising technology in the agricultural sector.

Acknowledgements

The authors would like to acknowledge the University of Mpumalanga for funding them in the publication of this book chapter.

Conflict of interest

The authors declare no interest.

Author details

Moses Z. Sithole*, Azikiwe I. Agholor and Shalia M. Ndlovu
University of Mpumalanga, Mbombela, South Africa

*Address all correspondence to: moses.sithole@ump.ac.za

References

[1] Smith P, Ashmore MR, Black HIJ, Burgess PJ, Evans CD, Quine TA, et al. The role of ecosystems and their management in regulating climate, and soil, water, and air quality. Journal of Applied Ecology. 2013;**50**:812-829

[2] Li Y, Wu Z, Xu X, Fan H, Tong X, Liu J. Forest disturbances and the attribution derived from yearly landsat time series over 1990-2020 in the Hengduan Mountains Regions of Southwest China. Journal of Forest Ecosystems. 2021;**8**(73):1-17

[3] Danley B, Wildmark C. Evaluating conceptual definitions of ecosystem services and their implications. Journal of Ecological Economics. 2016;**126**:132-138

[4] Angelstam P, Robert A, Marine E, Lars L, Mats N, Patru-Stupariu I, et al. Knowledge production and learning for sustainable Forest management on the ground: Pan-Earopean landscapes as a time machine. International Journal of Forest Research. 2011;**84**(5):581-596

[5] Szymanski C, Tabeni S, Alvarez JA, Campos CM. Diversity of plants mammals as indicators of the effects of land management types in woodland. Journal of Forest Ecosystem. 2021;**8**(74):1-15

[6] Food and Agriculture Organization. The State of Food Security and Nutrition in the World 2022. Rome, Italy: Food and Agriculture Organization; 2022

[7] FAO. What Is Conservation Agriculture? Rome, Italy: Food and Agriculture Organization (FAO); 2015. Available from: http://www.fao.org/ag/ca/1a.html [Accessed: April 25, 2022]

[8] Sithole MZ, Agholor IA. Assessing the adoption of conservation agriculture towards climate change adaptation: A case of Nkomazi, Mpumalanga Province. TIIKM Proceedings of the International Conference on Agriculture. 2021;**6**(1):68-80

[9] Muzangwa L, Mnkeni PNS, Chiduza C. Assessment of conservation agriculture practices by smallholder farmers in Eastern Cape Province of South Africa. Multidisciplinary Digital Publishing Institute. 2017;7(46):103390

[10] Thierfelder C, Mombeyarara T, Mango N, Rusinamhodzi L. Integration of conservation agriculture in smallholder farming systems of Southern Africa: Identification of key entry points. International Journal of Agriculture and. Sustainability. 2013;**11**(1):1-13

[11] Busari MA, Kukal SS, Kaur A, Bhatt R, Dulazi AA. Conservation tillage impacts on soil, crop and the environment. International Soil and Water Conservation Research. 2015;**3**(2):119-129

[12] Jug D, Jug I, Brozovic B, Vukadicovic V, Stipesevic B, Durdevic B. The role of conservation agriculture in mitigating and adaptation to climate change. Poljoprivreda. 2018;**24**(1):34-44

[13] FAO. Climate Smart Agriculture: Building Resilience to Climate Change. Berkeley, USA: FAO; 2018

[14] Joseph KX, Issahaku Z. Effects of soil conservation technologies in improving soil productivity in Northern Ghana. Journal of Soil Science and Environmental Management. 2015;**6**(6):158-167

[15] Kassam A, Friedrich T, Derpsch R, Kienzle J. Overview of the worldwide

spread of conservation agriculture. International Journal of Agricultural Sciences. 2015;**11**(2):14-27

[16] Jat ML, Jat HS, Jat RK, Tetarwal JP, Jat SL, Parihar CM, et al. Conservation agriculture-based sustainable intensification of cereal system for enhancing pulse production and attaining higher resource-use efficiency in India. Indian Journal of Agronomy. 2016;**6**(1):182-198

[17] Meena RP, Jha A. Conservation agriculture for climate change resilience: A microbiological perspective. 2017. Available from: http://www.researchgate.net/publication/321861539 [Accessed: May 15, 2022]

[18] Mar S, Nomura H, Takahashi Y, Ogata K, Yabe M. Impact of erratic rainfall from climate change on pulse production efficiency in lower Myanmar. Journal of Experimental Botany. 2018;**71**(1):443-450

[19] Rho H, Dotty SL, Kim SH. Endophytes alleviate the elevated CO_2 dependent decrease in photosynthesis in rice, particularly under nitrogen limitation. Journal of Environmental Biology. 2020;**71**:707-718

[20] Balyan S, Kumar M, Mutum RD, Raghuvanshi U, Agarwal P, Mathur S, et al. 'Identification of MIRNA-mediated drought responsive multi-tiered regulatory network in drought tolerant rice', Nagina 22. Scientific Reports. 2017;**7**:15446-15458

[21] Sims B, Friedrich T, Kassam A, Kienzle J. Agroforestry and conservation agriculture: Complementary practices for sustainable development. In: Proceedings of the II World Congress of Agroforestry, Nairobi, Kenya. 2009. Available from: http://www.lesspub.com/cgi-bin/site.pl?332&ceNews_newsID=6259

[22] Lintangah WJ, Atin V, Ibraham AL, Yahya H, Johnlee EB, Martin RA, et al. Sustainable Forest management contribution to food security: A stakeholders' perspective in Sabah, Malaysia. IOP Conference Series: Earth and Environmental Science. 2021;**2021**:1053012012. DOI: 10.1088/1755-1315/1053/1/012012

[23] FAO. Global Progress towards Sustainable Forest Management – Brightspots and Challenges. Rome, Italy. 2021. DOI: 10.1505/146554822835224856

[24] Lazdinis M, Angelstam P, Pulzl H. Towards sustainable forest management in the European Union through ploycentric forest governance and an integrated landscape approach. Journal of Landscape Ecology. 2019;**34**:1737-1749

[25] Neary DG. Forest soil disturbance: Implications of factors contributing to the wildland fire Nexus. Soil Science Society of America Journal. 2019;**83**:228-243

Chapter 20

Assessment of Land Degradation Factors

Tülay Tunçay and Oğuz Başkan

Abstract

Land degradation is a phenomenon that threatens food security and ecosystem balance observed on a global scale. At the beginning of the 20th century on a global scale, its importance was not yet understood due to low climate change, population growth, and industrialization pressure, but today, with the increasing effect of these factors, it has affected more than 25% of the world's terrestrial areas. Land use/cover change, destruction of forest areas, opening to agriculture, or conversion of forest areas to high economic plantations are the main factors of land degradation. Population growth and increasing demand for food, water, and energy are increasing pressure on natural resources, primarily agricultural and forest land. Due to its dynamic relationship with the climate change, land degradation creates more pessimistic results in arid and semi-arid areas that are more vulnerable and have a high population density. Despite the intergovernmental meetings, commissions, and decisions taken, land degradation continues on a global scale and the human-climate change dilemma creates uncertainties in achieving the targeted results.

Keywords: climate change, land degradation, ecosystem, soil, land use/cover

1. Introduction

Land degradation (LD) is the physical, chemical, and/or biological degradation of soil by natural or anthropogenic processes that result in the reduction or destruction of essential ecosystem functions. The leading causes of land degradation and, consequently, the main threats to its ecological functions are erosion, reduction of organic matter, loss of biodiversity, soil compaction, framing, point and diffuse pollution, and salinization [1].

LD is a widespread and global phenomenon that affects food security, ecosystem services, and human well-being [2]. Land degradation is a major threat to the food security of a rapidly growing global population, especially those living on limited land resources. Approximately 15% of the world's population is currently living as being deprived of food security [3, 4]. It seems inevitable that the problem of food security will be exacerbated by land degradation as the global population is projected to reach nine billion by 2050 [5].

While LD was approximately 36,108 ha worldwide in the early 1900s [6], today, more than 25% of the world's land area (37.25 million km^2) is affected by LD [7]. It is estimated that an area of about 5 to 10 million hectares is added to these areas each year [8]. In these areas, there is a decrease or degradation of soil quality and a decline in biological and economic productivity due to erosion and physical and chemical changes [7]. This negative change is observed mainly in arid and semi-arid areas [9].

About 40% of LD is observed in developing countries. Projections show that these areas will increase to 78% by 2100, covering 50% of population growth [10]. Climate change appears to be a significant risk to agriculture, biodiversity, and livelihoods in developing countries [11].

Population growth and increasing demand for food, water, and energy are increasing pressure on natural resources, primarily agricultural land. It is estimated that demand for food will increase by 50% and demand for water and energy by 40% by 2030 compared to current levels [12]. The increase in natural resource consumption and industrialization to meet growing population needs has led to conditions that threaten food security today due to climate change.

Although it is a global phenomenon, soil degradation (SD) is much more pronounced in drylands, where land is highly vulnerable to degradation due to drought and water scarcity than in non-drylands [13]. The percentage of global soils vulnerable to severe drought was more than doubled between the 1970s and the early 2000s [14]. It is estimated that drylands will increase by 10% by the end of this century [15]. The expansion of drylands combined with increasing aridity due to climate change has direct implications for desertification [16]. Vegetation change and LD have made these areas much more vulnerable in terms of agricultural production and sustainable management.

Drylands, home to 38% of the world's population (2.7 billion people), account for 44% of the world's cropland and 50% of its livestock [9], as well as about 30% of the global carbon [17]. Drylands are, therefore, central to sustaining the habitats, crops, and livestock that support most of the world's population [13].

Given the importance of soil carbon sequestration and storage, preventing, mitigating, and reversing land degradation can provide more than one-third of the most cost-effective greenhouse gas mitigation needed until 2030 to keep global warming below the targeted threshold of 2°C. The Paris agreement (2016) on climate change seeks to increase food and water security and help prevent conflict and migration (Intergovernmental Science and Policy Platform on Biodiversity and Ecosystems (IPBES)).

2. Assessing land degradation

After desertification was officially recognized as a severe problem at the United Nations Conference on Desertification (UNCD) [18], Action Plans to Combat Desertification (PACH) studies were conducted in 1979 and 1991. The GLASOD project of the Global Assessment of Soil Degradation of the Soil Degradation Database, initially led by ISRIC and funded by the United Nations Environment Program (UNEP) in the 1990s, produced the first world map of human-induced land degradation at a scale of 1:10,000,000. Subsequently, awareness of LD problems was raised by the map, which was based on the methodology developed by experts (200 soil scientists and 21 environmental experts from around the world) at the United Nations Conference on Environment and Development (UNCED) Summit in Rio de Janeiro in 1992. The assessment and monitoring of situations and processes of desertification at local, regional, or larger scales were carried out in the 1980s by the food and agriculture organization (FAO), and the United Nations Environment Program

(UNEP) developed the methodology for assessing and mapping desertification. In this context, many studies have been conducted worldwide to measure desertification, monitor the implementation of the desertification vonvention, and determine indicators for improvement.

Common to all studies is that LD is one of the world's most pressing environmental problems. The situation will inevitably deteriorate if precautionary measures are not taken, as it is related to releasing greenhouse gasses into the atmosphere. Globally, about 25% of all land is degraded, and more than 1.5 billion people live in these areas. It is estimated that 24 billion tons of fertile land are lost annually due to unsustainable agricultural and industrial practices to increase efficiency, and if these aberrations continue, 95% of the world's land could be degraded by 2050 [19].

It seems ordinary that LD is most prevalent in arid, semi-arid, and low-yield areas along with the pressure of population increase. Globally, half of the total land area of the European Union (4.18 million km^2) is degraded annually, with Africa and Asia being the most affected [19]. On the global scale, smallholder farmers living in arid and semi-arid areas are the most affected by LD. Because the process is slow and imperceptible, people in the region usually become aware of it too late, which increases the severity and negative consequences of LD. It is estimated that, by 2050, LD and climate change will lead to a decline in global crop yields by about 10%. Most of this will occur in India, China, and sub-Saharan Africa, where SD could halve crop production [16]. It is estimated that the world population's growing demand for agricultural products, including food, feed, fuel, and manufactured goods will increase by about 35% to 9.8 billion people by 2050 [20]. However, pressure on global land resources is also increasing as agricultural production systems become less resilient due to biodiversity loss and natural factors, such as climate change and extreme weather events. It is estimated that up to 700 million people will be displaced by 2050 due to issues related to scarce land resources. This number could reach 10 billion by the end of this century [19].

The components of desertification are the increase in population and food demand, increase in industrial raw materials, overconsumption, waste, and LD due to misuse. In addition to increasing food demand, food waste leads to increased consumption of natural resources. Globally, about 14% of food produced is lost between harvest and retail, while an estimated 17% of total global food production is wasted (11% in households, 5% in restaurants, and 2% in retail). Lost and wasted food accounts for 38% of total energy consumption in the global food system [20].

Because land degradation processes reduce carbon sequestration and increase greenhouse gases emissions (GHG), GHG reduction targets are unlikely to be met unless action is taken. Moreover, global food security targets will be missed if land degradation is not successfully addressed, as land degradation leads to productivity losses and thus reduced food supply [8, 21].

Human-induced accelerated erosion is directly influenced by land use/cover change (LU/LC). Overall, human activities have increased soil erosion to levels between 8% and 90% in many parts of the world. Globally, soil erosion caused by human activities has increased by about 60% during this period [22]. At the beginning of the twenty-first century, the global average value of potential erosion is 10.2 tons ha^{-1} per year and the global soil loss due to erosion processes ranges from 24 to 75 billion tons of fertile soil [8, 12, 14]. Furthermore, it is estimated that there is an economic loss of about $ 400 billion in crop production [23].

Stopping or controlling LD to protect natural resources and reclaim degraded land has become a global phenomenon to ensure food security, and solutions have been sought at the global level. To this end, in the sustainable development goals (SDGs) adopted by world leaders in 2015 (Goal 15.3), a consensus was reached on combating desertification, restoring degraded lands and soils, including lands affected by desertification, drought, and floods and achieving a balanced world of LD by 2030 [2]. Therefore, countries are expected to develop implementation strategies by setting their targets to reduce poverty, ensure food security, improve nutrition, and reduce land degradation over the following decades.

On the other hand, efforts to offset LD by rehabilitating degraded land alone may not yield sufficient results. LD, which develops slowly and takes time to show results, needs to be monitored spatially and temporally on a global scale because of its dynamic relationship with climate change.

3. Land use/cover changes

With rapid population growth and evolving technology, human pressure on natural resources is increasing, causing these resources to lose their regenerative power. Appropriate management systems must be developed to protect natural resources and ensure their sustainability by preventing SD. For sustainable land management, it is crucial to ensure the effective management of natural resources by considering the balance between protection and use. LD and loss of productivity are mainly due to human misuse of land.

Land use /Land cover Cchange (LU/LCC) is a phenomenon that directly affects the sustainable use of natural resources, such as climate change, biodiversity loss, LD, desertification, and carbon sink areas. The increasing demand for food and energy that has accompanied population growth has resulted in land being used beyond its capacity. Humans have altered about three-quarters of the world's land in the last millennium [24, 25]. LUC is estimated to have impacted nearly one-third (32%) of global land area over the past six decades (1960–2019), about four times greater than estimates from long-term land change assessments [26].

Humans cause 60% of land use change in the world's land areas, and 40% is caused by the effects of climate change [27]. The majority of human-caused land change appears as the opening of forest land for agriculture, human settlement, and industry. Land use change for more productive purposes increases global sustainability problems. Critical sustainability components such as biodiversity loss, habitat destruction, and degradation of carbon sinks are directly affected by land use/land cover change.

Unfortunately, despite this importance, there are still large spatial and temporal uncertainties related to LU/LCC [28, 29]. The vast majority of studies using satellite imagery, etc., show that forest areas have decreased and agricultural areas have increased [26]. Results obtained with multidimensional maps show that forest areas are decreasing worldwide [30]. Another study found that forest land increased in 2016 compared to 1982 [27], but it is unclear from which LU this land originated or whether it is a forest or industrial land. For example, 66,000 hectares of land in West Sumatra, Indonesia, were put into operation for logging and 33,000 hectares of land that were destroyed after the contract expired were given to private companies for palm plantations [31]. It appears that these uncertainties will continue with normalized difference vegetation index (NDVI) imagery using satellite technologies that are not supported by ground reality on a global scale. LU/LC algorithms that classify global land cover into different types may not be suitable for producing scientific approaches and realistic values [32]. Global and national LU/LC data produced using different approaches are inadequate or inappropriate for land use management and reporting [33, 34].

Common to all studies that address LU/LCC is that LU/LCC in arid and semi-arid areas is negatively evolving due to increasing population pressure. The basis of decision made on Climate Change is based on taking place in the scope of Paris agreement about land use-change [25, 35]. These areas most affected by LU/LCC are naturally arid and semi-arid. Due to their fragile nature, these areas are extremely sensitive to changes in LU and LC. Balancing LD in these areas under the control of meteorological conditions requires a very long-term process. On the other hand, changing LU/LC in these areas degrades carbon sinks and triggers climate change through greenhouse gasses released into the atmosphere.

Forest areas are affected the most by LU/LCC. In the last two decades (2000–2020), forest area has decreased by 1 million km^2, or 2.4% of the forest area in 2000. Restoration of forest areas or afforestation activities has increased globally only in the European continent by 1.7% (64,000 km^2). In other continents, restoration activities are not sufficient to prevent losses. In Africa and South America, this rate constitutes 30% of losses [32].

LU concepts and policies that lead to LD may make agroecological systems even more vulnerable to climate change [36]. Although replacing destroyed or degraded natural forest areas with plantations of high commercial value is considered beneficial in economic terms, these plantations that replace the natural ecosystem can lead to biodiversity loss [37] (e.g., the process of deforestation). The destruction of forested areas and their conversion into plantations with commercial value is one of the severe problems associated with LU/LCC today. Although the climate is significant for land productivity, the human factor is still the most important factor for climate change [2, 27, 38].

Short-term fluctuations in primary production make it extremely difficult to distinguish long-term fluctuations resulting from human-induced LD from the effects of periodic droughts [39–42]. Human influences are further masked by topography, soil types, vegetation types, and spatial variability in land use.

4. Climate

Climate change is recognized as one of the most critical factors contributing to LD, as defined by the UNCD. Changes in the surface energy budget resulting from land surface change have a major impact on the Earth's climate. Addressing climate as one of the factors leading to LD is more important than addressing the consequences of LD [43].

Precipitation and temperature are the main factors that determine the Earth's climate and thus the distribution and density of vegetation. The interaction of human activities on the distribution of vegetation through land management practices and precipitation events makes the land more vulnerable to degradation. This vulnerability may become a severe threat as climate change continues.

Looking at the relationship between LD and vegetation/vegetation in general, it is known that as vegetation declines, land degradation continues to increase through feedback from the land surface to the atmosphere. Therefore, the decrease in vegetation is thought to reduce evaporation and increase radiation reflected into the atmosphere (albedo), which decreases cloud formation, that is, precipitation. Precipitation is the most important climate factor in identifying areas at risk of LD and potential desertification [44].

More important than the increase in precipitation has been the increase in heavy precipitation [45]. Climate models conclude that heavy precipitation will increase in the 21st century [46]. Soil erosion is a consequence of varying precipitation amounts and intensities. On the other hand, the intensity of precipitation and its energy are more critical than the amount of precipitation. The studies found that every 1% increase in the total amount of precipitation without any change in the intensity of precipitation increases the erosion rate by 0.85%. It has been shown that rainfall and intensity change together, and erosion increases by 1.7% for each 1% increase in rainfall amount [47].

In recent years, how climate change may affect temperature and precipitation has been a matter of debate considerably. However, there is still no extensive research conducted on the impacts of climate change on the wind. Changing wind speed has a direct and indirect impact on wind energy, climate fluctuations, storm events, and social life (maritime, etc.) [48, 49]. Wind speed is also a significant factor in the revised wind erosion equation model, which is calculated by considering the soil properties, vegetation cover and height, land roughness, soil moisture content, and land use changes [50]. In addition, the change in wind speed has a significant impact on soil moisture, evaporation, and water resources in arid and semi-arid regions. The removal of the topsoil, which is rich in prolific mineral and organic matter, through the wind causes a decrease in agricultural production potential and, consequently, contributes to land degradation [51].

LD and associated climate change must be controlled to protect vegetation, biodiversity, and ecological balance in land areas. LD and climate change are evolving as two phenomena that interact due to their dynamic behavior. According to IPBES, the contribution of climate change to LD is 10% of human-induced greenhouse gas emissions and forest land destruction. After the industrial revolution, global carbon emissions are estimated at 270 ± 30 gigatons (Gt) from fossil fuel combustion and 136 ± 5 Gt from LU and tillage changes. Most importantly, greenhouse gas emissions stored in the soil are released into the atmosphere through SD, triggering climate change. In this way, 4.4 billion tons of CO_2 were released into the atmosphere between 2000 and 2009 [52, 53]. Although CO_2 emissions into the atmosphere are proportionally decreasing due to intergovernmental meetings and treaties, they continue to increase and reached 418.2 parts per million (ppm) in 2021. This rate is around 50% higher than when the industrial revolution began [54].

According to the European Commission report (EC), 2011–2020 were the warmest 10 years on record, and the global average temperature rose 1.1°C above pre-industrial levels in 2019. Human-caused global warming is currently increasing by 0.2°C per decade. Compared to pre-industrial times, an average annual increase of 2°C will lead to dangerous and potentially devastating changes on a global scale. The world could face unexpected and far worse consequences than predicted. To control and prevent this pessimistic picture, the international community has accepted the need to keep warming below 2°C and limit it to 1.5°C [55].

Climate change and related meteorological conditions combined with LD seem to be a paradox that needs to be resolved quickly. The warming of the world with the effects of increasing human-induced greenhouse gasses, such as LD and industrialization, has increased about twofold in decades after the 1980s, considering past years

(1880), reaching 0.18°C [56]. The change in precipitation regime and temperature increase with warming will affect all soil's physical, chemical, and biological properties. Increasing soil fragility, especially in arid and semi-arid areas, increases the possibility of fundamental problems, such as erosion, loss of organic matter, compaction and decrease in water retention, loss of plant nutrients, and decrease or extinction of biodiversity [57].

SD intensely impacts a large part of the European continent. 45% of the continent's soils contain little or very little organic matter [58]. Loss of natural soil functions due to drought, fire, and erosion leads to a significant increase in desertification risk in these areas [57]. Stabilizing LD and improving soil quality are linked to mitigating climate change [59].

The largest contributor to global warming is CO_2 released into the atmosphere by human activities. Atmospheric CO_2 concentrations in 2020 are 48% above pre-industrial (pre-1750) levels. Natural causes, such as solar radiation or changes in volcanic activity are estimated to have contributed less than plus/minus 0.1°C to the total warming between 1890 and 2010 [60].

Resulting in a temperature increase of 1 to3 °C in arid regions, the CO_2 concentrations are predicted to reach 700 ppm by 2050. This increase will increase the global potential evapotranspiration rate by 75–725 mm. Climate models predict that 50% of the world will experience regular droughts throughout the 21st century [59]. Higher temperatures, changing precipitation and regimes, and extreme weather will trigger LD, leading to water and wind erosion and further soil loss. Prolonged droughts could have devastating consequences for the 2.7 billion people living in semi-arid areas beyond meeting growing food needs.

Climate factors can lead to the degradation of clean water resources and soil quality. Considering that 10 countries are located in the region that is globally considered a risk zone with an elevation of only 10 m above sea level, the proportion of the population living in these areas varies from 38% (Gambia) to 88% (Bahama), and 634 million people live in these 10 countries [61], the dichotomy of LD and climate change is becoming more apparent. According to the February 2007 intergovernmental panel on climate change (IPCC) report [62], the sea level is projected to rise 59 cm by the end of the 21st century if fossil fuel consumption and economic growth continue unabated. If no further changes are made, a 38 cm rise in sea level is estimated to increase the number of people exposed to storm surges fivefold [63]. The potential sea level rise due to climate change may render clean water resources in these countries unusable due to salinization.

Temperature changes alter evapotranspiration, soil moisture, and infiltration. Because these potential changes alter surface runoff and infiltration rates, they can alter groundwater's quantity and storage capacity. Climate change projections predict that soil moisture content will decrease significantly during extremely dry periods in the summer, while increases in excessive precipitation during these periods will increase erosion.

A significant portion of the climate change observed over the past 50 years is due to human activities [45]. In particular, the increase in average temperatures is the most apparent human-induced meteorological change. On the other hand, meteorological components are interrelated in many ways. The change in average annual precipitation amount, time, intensity, and the length of dry periods are the most prominent examples of the multiple dynamic relationships of meteorological factors. According to the IPCC assessment, the rise in temperature and changes in other climate parameters during the twentieth century are the result of radiative forcing from human greenhouse gas emissions [64].

The role of temperature in land productivity trends in recent years is explained by the 0.6 C average increase in air temperature observed globally between 1960 and 2006 [65]. About 9% of the decline in land productivity is due to temperature alone [2, 66]. A study on desertification factors in China found that the value of net primary productivity (NPP) decreased by 35.2% due to temperature increases and precipitation decreases [67].

Climate change and human activities are the main factors affecting vegetation dynamics, especially in high mountain regions characterized by highly fragile ecosystems [68]. Studies examining the effects of factors, such as climate change, temperature change, precipitation change, and solar radiation change [68] found an increasing trend compared to long-term averages. Studies show that temperature increases, especially in winter, compared to long-term averages [69, 70].

Climate conditions were found to be more important than human factors in affecting vegetation [68]. In studying the effects of climate change and human activities on desertification, it was found that desertification increased by 55.8%, with 70.3% of this rate caused by humans and 21.7% caused by climate change. On the other hand, stabilization or recycling of LD was found to be 42.1% caused by humans and 48.4% caused by climate change [67].

Most studies on desertification have produced mixed results regarding climate change and human impacts causing desertification. In fact, some study results indicate that the leading cause of desertification is long-term overgrazing, deforestation, and conversion of grasslands to cropland, while human-induced factors are more effective [71–73]. Other studies explain climatic conditions, such as drought, strong wind, and water erosion, temperature fluctuations, and winter precipitation as the main causes of desertification [74, 75].

The IPBES report notes that LD is a major contributor to climate change, and that deforestation alone is responsible for about 10% of all anthropogenic GHG. Another major contributor to climate change is the release of carbon previously stored in the soil, and LD was responsible for annual CO_2 emissions of up to 4.4 billion tons between 2000 and 2009. LD is a significant contributor to climate change, and climate change is projected to be the primary cause of biodiversity loss. By 2050, LD and climate change will reduce crop yields by an average of 10% globally and up to 50% in certain regions.

5. Soil organic matter

Soil organic matter (SOM) is an essential component of the terrestrial ecosystem. Any change in its quantity and composition in the soil significantly impacts soil and air conditions. Terrestrial areas, which contain the most organic carbon after the oceans, are much more unstable and open to short-term changes compared to the ocean and atmospheric conditions. The carbon balance in terrestrial ecosystems can change significantly under human activities.

High temperatures and low precipitation in drylands generally result in low organic matter (OM) production and rapid oxidation. Low OM leads to poor aggregation and low aggregate stability, which means a high potential for wind and water erosion. Loss of natural soil functions due to drought, fire, and erosion leads to a significant increase in desertification risk in these areas. The risk of desertification is most likely to occur in areas where precipitation is decreasing, dry periods are increasing in the summer months, and mis-intensive LU is occurring. The increase in temperature negatively affects carbon accumulation in the soil, leading to a decrease in organic carbon and an increase in the amount of carbon in the atmosphere [76, 77].

Land management will continue to be the most important determinant of SOM content and susceptibility to erosion in the coming decades. However, changes in vegetation due to short-term weather conditions and short-term climate changes will significantly affect soil organic matter dynamics and erosion, especially in semi-arid regions.

Soil carbon stores have a major impact on global climate change, and LD due to natural conditions or human activities is one of the leading causes of changes in soil carbon storage [78]. In a study conducted in semi-arid steppe areas [79], it was found that a sudden change in soil moisture due to high inter-annual rainfall variability causes about 65–80% of the total carbon loss in soils with different vegetation [80].

About 1550 Gt of soil organic carbon and 950 Gt of soil inorganic carbon constitute the global carbon source (2500 gigatons, Gt). Soil carbon source is 3.3 times greater than atmospheric carbon (760 Gt). Soil organic carbon varies from 30 tons ha^{-1} at 1-meter soil depth in semi-arid climates to 800 tons ha^{-1} in organic soils in cold regions and plays a vital role in the global carbon cycle and balance [81, 82]. The amount of soil organic carbon (SOC) is in a dynamic balance between storage and loss [75]. Even small changes can significantly impact climate and ecosystem stability, as organic carbon plays a critical role in soil-atmosphere carbon exchange and plant growth and food production in SOC [83, 84]. It is an indicator of the importance of soils in reducing the effects of global warming by retaining carbon dioxide in the atmosphere.

SOC depletion is a specific form of degradation that causes a decrease in soil quality and fertility [85]. It has been reported that some croplands have lost half to two-thirds of their SOC pools following LU/LCC, with a cumulative carbon loss of 30–40 tons ha^{-1} [82]. Soil erosion results in the loss of a significant portion of SOC from the upper soil layer, where terrestrial ecosystems have more biological activity and organic matter. Regardless, lower concentrations of SOC reduce soil quality and productive capacity [84]. Therefore, understanding the spatial-temporal changes of SOC and the associated driving factors is crucial for assessing the feedback and maintaining ecosystem functions between the terrestrial carbon cycle and climate change [86, 87].

Human-induced desertification seems to be the main reason for the rapid release of SOC into the atmosphere. Due to the fragile ecosystem structures, especially in arid and semi-arid regions, unsustainable LU leads to increased carbon emissions released from the soil into the atmosphere [78]. As the SOC pool is depleted, 78 ± 12 Gt of carbon enters the atmosphere, which is about 1/3 of the acceleration of LD and erosion. The remaining 2/3 is mineralized. LD exacerbates CO_2 driven climate change by releasing CO_2 from cleared and dead vegetation and reducing the carbon storage potential of degraded land.

The slow decomposition of dead biomass (leaves, plant stems, and plant roots) in areas with low temperatures and adequate humidity leads to the accumulation of organic matter. Climatic conditions significantly impact the formation and storage of soil organic carbon. As temperature rise increases the decomposition of organic residues, it also increases carbon dioxide and methane gasses released from the soil.

In the process of LD, the decrease of soil organic carbon content and the decrease of vegetation may cause a more potent greenhouse effect due to greater warming of the surface soil [78, 88]. Today, the area affected by desertification worldwide is about 3.6 billion hm^2 [82].

Whether the soil C pool acts as a source or sink of atmospheric CO_2 is primarily controlled by changes in climate and soil water content (SWC) [89, 90]. The variability in precipitation associated with climate change leads to changes in SOM. While plant growth and C storage are enhanced by precipitation [91, 92], the opposite is true in areas with insufficient precipitation and high temperatures [93, 94]. On the other hand, an increase in soil temperature can cause a decrease in SOM despite increased precipitation [95, 96]. High air temperatures and adequate moisture can promote microbial activity in the soil, leading to faster decomposition of SOC [97]. For example, while an expected 3.3 ^{0}C increase in air temperature will result in a loss of 11–16% SOC in Europe, an average increase of 1°C in surface air temperature will result in a net loss of 5% in the worldwide SOC pool. Temperature increase alone leads to SOC losses in all soil moisture contents [80, 83, 96, 98].

6. Soil quality

Soil quality (SQ) stands out regarding the sustainable development of the global biosphere due to the fact that it is one of the most important terrestrial ecosystem functions. SQ is an indicator as a comprehensive reflection of the soil physically, chemically, and biologically. Revealing the dynamics of soil conditions, this sensitive indicator may change with the effect of different land uses and ecological restoration measurements [99–101]. Changes in land use have an impact on the physical and chemical properties of the soil, biological processes, and land productivity, ultimately leading to a change in SQ (**Figure 1**) [102, 103]. The impacts of the factors and their interaction on the soil quality were visualized in **Figure 1**.

In recent years, intensive efforts have been made on the concept of SQ and a reliable method to be used in measuring SQ. Some researchers define soil quality as the soil's capacity and suitability for use based on its functions. Capacity is defined as a common function of properties including climate, topography, vegetation, and parent material while suitability is associated with land use and management as a dynamic concept influenced by humans. However, it is known that there are some losses in the physical, chemical, and biological properties of the soil affected by land use and land management measures worldwide [104, 105]. In summary, SQ is how well the soil promotes and enhances plant and animal productivity and maintains or improves water and air quality. Maintaining and promoting SQ, ensuring ecosystem sustainability, and making rational management plans are fundamental requirements [106]. Accordingly, the welfare of a society largely depends on the productivity power of the land and the sustainable use of this power. In the opposite case when these soil conditions are not considered, it is inevitable that the land degradation will continue and it will lose its functionality within the ecosystem as a result of the disruption of its productivity and fertility parameters, as in other natural resources.

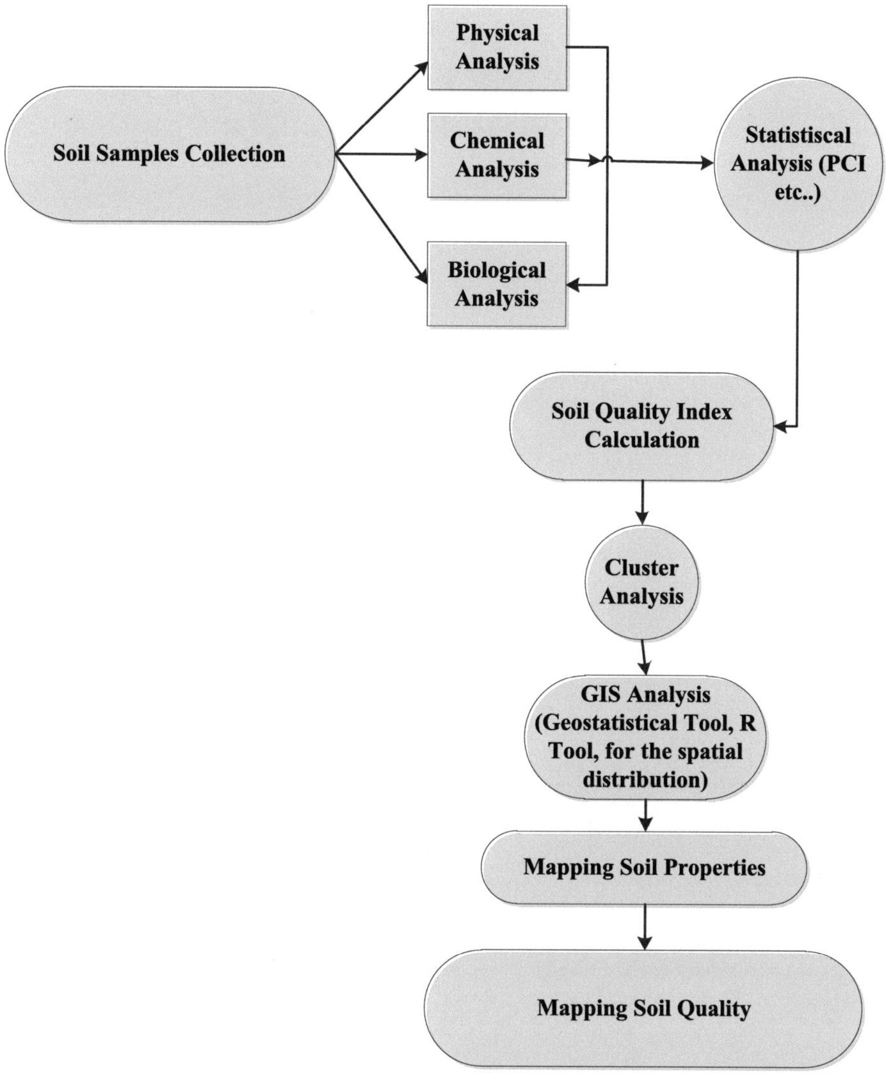

Figure 1.
Basic soil quality flow chart.

Author details

Tülay Tunçay[1] and Oğuz Başkan[2*]

1 Soil Fertilizer and Water Resources Central Research Institute, Ankara, Türkiye

2 Faculty of Agriculture, Department of Soil Science and Plant Nutrition, Siirt University, Siirt, Türkiye

*Address all correspondence to: ogbaskan@yahoo.com

References

[1] Montanarella L. Trends in land degradation in Europe. In: Sivakumar MVK, editor. Climate and Land Degradation: Environmental Science and Engineering. Berlin, Heidelberg: Springer; 2007. pp. 83-104. DOI: 10.1007/978-3-540-72438-4_5

[2] Montfort F, Bégué A, Leroux L, Blanc L, Gond V, Cambule AH, et al. From land productivity trends to land degradation assessment in Mozambique: Effects of climate, human activities and stakeholder definitions. Land Degradation and Development. 2021;**32**:49-65. DOI: 10.1002/ldr.3704

[3] FAO, IFAD, UNICEF, WFP, WHO. The State of Food Security and Nutrition in the World 2017: Building Resilience for Peace and Food Security. Rome: FAO; 2017

[4] FSIN. Global Report on Food Security 2018 1-195. 2018

[5] Montpellier P. Sustainable intensification: A new paradigm for African agriculture. Agriculture, Ecosystems and Environment. 2013;**116**:47-59. DOI: 10.1016/j.agee.2006.03.021

[6] Drenge HE, Chou NT. Global desertification dimensions and costs. In: Drenge HE, editor. Degradation and Restoration of Arid Lands. Lubbock: Texas Technical University; 1994. pp. 73-92

[7] ELD Initiative. The Economics of Land Degradation Initiative. The value of land: prosperous lands and positive rewards through sustainable land management [Internet]. 2015. Available from: https://www.eld-initiative.org/fileadmin/pdf/ELD-main-report_en_10_web_72dpi.pdf. [Accessed: July 10, 2022]

[8] Stavi I, Lal R. Achieving zero net land degradation: Challenges and opportunities. Journal of Arid Environments. 2015;**112**:44-51. DOI: 10.1016/j.jaridenv.2014.01.016

[9] Bestelmeyer BT, Okin GS, Duniway MC, Archer SR, Sayre NF, Williamson JC, et al. Desertification, land use, and the transformation of global drylands. Frontiers in Ecology and the Environment. 2015;**13**:28-36. DOI: 10.1890/140162

[10] Huang J, Yu H, Guan X, Wang G, Guo R. Accelerated dryland expansion under climate change. Nature Climate Change. 2015;**6**(2):166-171. DOI: 10.1038/nclimate2837

[11] IPCC. Climate Change 2014-Synthesis Report. Contribution of Working Groups I, II and III to the Fifth Assessment Report of the Intergovernmental Panel on Climate Change. 2014. 151 p. Pachauri RK, Meyer LA editors. IPCC, Geneva, Switzerland, p. 151

[12] Gnacadja L. Moving to zero-net rate of land degradation. Statement by executive secretary. UN convention to combat desertification, Rio de Janeiro [internet]. 2012. Available from: http://www.unccd.int/Lists/SiteDocumentLibrary/secretariat/2012/UNCCD%20ES%20Statement%20at%20PR%20in%20NY%20on%2026%20March%202012.Pdf. [Accessed: July 10, 2022]

[13] UNCCD. Zero net land degradation, a sustainable development goal for Rioþ20. 2012. Available from:

http://www.unccd.int/Lists/SiteDocumentLibrary/Rioþ20/UNCCD_PolicyBrief_ZeroNetLandDegradation.pdf. [Accessed: 10-07-22]

[14] Lee X, Goulden ML, Hollinger DY, Barr A, Black TA, Bohrer G, et al. Observed increase in local cooling effect of deforestation at higher latitudes. Nature. 2011;**479**:384-387. DOI: 10.1038/nature10588

[15] Feng W, Plante AF, Six J. Improving estimates of maximal organic carbon stabilization by fine soil particles. Biogeochemistry. 2013;**112**:81-93p. DOI: 10.1007/s10533-011-9679-7

[16] Reynolds JF, Stafford Smith DM, Lambin EF, Turner BL, Mortimore M, Batterbury SPJ, et al. Global desertification: Building a science for dryland development. Science. 2007;**2007**(316):847-851. DOI: 10.1126/science.1131634

[17] Hanan NP, Milne E, Aynekulu E, Yu Q, Anchang L. A role for drylands in a carbon neutral world? Frontiers in Environmental Science. 2021;**9**:786087. DOI: 10.3389/fenvs.2021.786087

[18] UNCOD. Round-up, plan of action and resolutions. In: United Nations Conference on Desertification. Nairobi, Kenya; 1977

[19] EC-JRC. Europen Commission – Joint Research Centre. World Atlas of Desertification [Internet]. 2018. Available from: https://wad.jrc.ec.europa.eu. [Accessed: July 10, 2022]

[20] UN. Stop Food Loss and Waste, for the people, for the planet [Internet]. 2022. Available from: https://www.un.org/en/observances/end-food-waste-day# [Accessed: July 10, 2022]

[21] Lal R, Safriel U, Boer B. Zero Net Land Degradation: A New Sustainable Development Goal for Rio+20. In: A Report Prepared for the Secretariat of the UNCCD. Bonn, Germany. 2012

[22] Yang D, Kanae S, Oki T, Koike T, Musiake K. Global potential soil erosion with reference to land use and climate changes. Hydrological Processes. 2003;**17**:2913-2928. DOI: 10.1002/hyp.1441

[23] FAO and ITPS. Food and Agriculture Organization of the United Nations and Intergovernmental Technical Panel on Soils. The Status of the World's Soil Resources (SWSR)- Main Report. 2015. Rome, Italy: Food and Agriculture Organization of the United Nations and Intergovernmental Technical Panel on Soils

[24] Luyssaert S, Jammet M, Stoy P, Estel S, Pongratz J, Ceschia E, et al. Land management and land-cover change have impacts of similar magnitude on surface temperature. Nature Climate Change. 2014;**4**:389-393. DOI: 10.1038/nclimate2196

[25] Arneth A, Shin YJ, Leadley P, Rondinini C, Bukvareva E, Kolb M, et al. Post-2020 biodiversity targets need to emrace climate change. Proceedings of the National Academy of Sciences, USA. 2020;**8-117**(49):30882-30891. DOI: 10.1073/pnas.2009584117

[26] Winkler K, Fuchs R, Rounsevell M, Herold M. Global land use changes are four times greater than previously estimated. Nature Communications. 2021;**12**:2501. DOI: 10.1038/s41467-021-22702-2

[27] Peng Song X, Hansen MC, Stehman SV, Potapov PV, Tyukavina A, Vermote EF, et al. Global land change from 1982 to 2016. Nature. 2018;**560**(7720):639-643. DOI: 10.1038/s41586-018-0411-9

[28] Bayer AD, Lindeskog M, Pugh TAM, Anthoni PM, Fuchs R, Arneth A. Uncertainties in the land-use flux resulting from land-use change reconstructions and gross land transitions. Earth System Dynamics. 2017;**8**:91-111. DOI: 10.5194/esd-8-91

[29] Prestele R, Arneth A, Bondeau A, de Noblet-Ducoudre N, Pugh TAM, Sitch S, et al. Current challenges of implementing anthropogenic land-use and land-cover change in models contributing to climate change assessments. Earth System Dynamics. 2017;**8**:369-386. DOI: 10.5194/esd-8-369-2017

[30] FAO. Global Forest Resources Assessment 2020: Main Report. Rome, Italy: FAO; 2020. p. 184. DOI: 10.4060/ca9825en

[31] Yonariza Y, Yurike Y. IOP Conference Series: Earth Environmental Sciences. 2020;**583**(1):012038. DOI: 10.1088/1755-1315/583/1/012038

[32] Potapov P, Hansen MC, Pickens A, Hernandez-Serna A, Tyukavina A, Turubanova S, et al. The global 2000-2020 land cover and land use change dataset derived from the Landsat archive: First results. Fronties in Remote Sensing. 2022;**3**:856903. DOI: 10.3389/frsen.2022.856903

[33] Fritz S, See L. Identifying and quantifying uncertainty and spatial disagreement in the comparison of global land cover for different applications. Global Change Bioloy. 2008;**14**:1057-1075. DOI: 10.1111/j.1365-2486.2007.01519.x

[34] Saah D, Tenneson K, Poortinga A, Nguyen Q, Chishtie F, Aung KS, et al. Primitives as building blocks for constructing land cover maps. International Journal of Applied Earth Observation Geoinformation. 2020;**85**:101979. DOI: 10.1016/j.jag.2019.101979

[35] Grassi G, House J, Dentener F, Federic S, den Elzen M, Penman J. The key role of forests in meeting climate targets requires science for credible mitigation. Nature Climate Change. 2017;**7**:220-226. DOI: 10.1038/nclimate3227

[36] Stringer LC, Dyer JC, Reed MS, et al. Adaptations to climate change, drought and desertification: Local insights to enhance policy in southern Africa. Environmental Science & Policy. 2009;**12**:748-765

[37] Bremer LL, Farley KA. Does plantation forestry restore biodiversity or create green deserts? A synthesis of the effects of land-use transitions on plant species richness. Biodiversity and Conservation. 2010;**19**(14):3893-3915. DOI: 10.1007/s10531-010-9936-4

[38] Chen C, Park T, Wang X, Piao S, Xu B, Chaturvedi RK, et al. China and India lead in greening of the world through land-use management. Nature Sustainability. 2019;**2**(2):122-129. DOI: 10.1038/s41893-019-0220-7

[39] Pickup G, Bastin GN, Chewings VH. Identifying trends in land degradation in non-equilibrium rangelands. Journal of Applied Ecology. 1998;**35**:365-377. DOI: 10.1046/j.1365-2664.1998.00319.x

[40] Dahlberg A. Interpretations of environmental change and diversity: A critical approach to indications of degradation—The case of Kalakamate, Northeast Botswana. Land Degradation and Development. 2000;**11**:549-562

[41] Dube OP, Pickup G. Effects of rainfall variability and communal and semi-commercial grazing on land cover in southern African rangelands. Climate

Research. 2001;**17**(2):195-208. DOI: 10.3354/cr017195

[42] Prince SD. Spatial and temporal scales of measurement of desertification. In: Stafford-Smith M, Reynolds JF, editors. Global Desertification: Do Humans Create Deserts? Berlin: Dahlem University Press; 2002. pp. 23-40

[43] Borrelli P, Robinson DA, Panagos P, Lugato E, Yang JE, Alewell C, et al. Land use and climate change impacts on global soil erosion by water (2015-2070). Proceedings of the National Academy of Sciences. 2020;**117**(36):21994-22001. DOI: 10.1073/pnas.2001403117

[44] Tabari H. Climate change impact on flood and extreme precipitation increases with water availability. Scientific Reports. 2020;**10**:13768. DOI: 10.1038/s41598-020-70816-2

[45] IPCC. The Scientific Basis, Contribution of Working Group I to the Third Assessment Report of the IPPC. United Kingdom and New York, NY, USA: Cambridge University Press; 2001. p. 881

[46] IPCC. Impacts, Adaptation, and Vulnerability, Contribution of Working Group the Third Assessment Report of the IPPC. United Kingdom and New York, NY, USA: Cambridge University Press; 2001. p. 881

[47] Pruski FF, Nearing MA. Climate-induced changes in erosion during the 21st century for eight U.S. locations. Water Resources Research. 2002;**2002**:38-12. DOI: 10.1029/2001WR000493

[48] Debernard JB, Roed LP. Future wind, wave and storm surge climate in the Northern Seas: A revisit. Tellus Series A-Dynamic Meteorology and Oceanography. 2008;**60**(3):427-438

[49] Borrelli P, Ballabio C, Panagos P, Montanarella L. Wind erosion suscetibility of Eurepean soils. Geoderma. 2014;**232-234**:471-478

[50] Fryrear DW, Bilbro JD, Saleh A, Schomberg H, Stout JE, Zobeck TM. RWEQ: Improved wind erosion technology. Journal of Soil and Water Conservation. 2000;**55**:183-189

[51] McLinnes KL, Erwin TA, Bathols JM. Global climate model projected changes in 10 m wind speed and direction due to anthropogenic climate change. Atmospheric Science Letters. 2011;**12**:325-333. DOI: 10.1002/asl.341

[52] IPBES. Summary for policymakers of the assessment report on land degradation and restoration of the Intergovernmental Science-Policy Platform on Biodiversity and Ecosystem Service, Bonn, Germany, 744 p. [Internet]. 2018. Available from: https://www.ipbes.net/system/tdf/spm_3bi_ldr_digital.pdf?file=1&type=node&id=28335. [Accessed: September 19, 2022]

[53] SEI. Stocholm Environment Institute. 2018. [Internet]. Available from: https://www.sei.org

[54] CO_2 Home page. CO_2 Earth. [Internet]. 2022. Available from: https://www.co2.earth/earths-co2-main-page [Accessed: July 10, 2022]

[55] PA. United Nations Framework Convention on Climate Change (UNFCCC). Climate change conference Paris 2015 [Internet]. 2015. Available from: http://www.un.org/sustainabledevelopment/cop21 [Accessed: May 01, 2022]

[56] NOAA. National Centers for Environmental Information, State of the Climate: Global Climate Report for Annual 2020 [Internet]. 2021. Available

from: https://www.ncdc.noaa.gov/sotc/global/202013. [Accessed: March 1, 2021]

[57] Başkan O. Climate Change and Soil. Istanbul, Türkiye. 2017

[58] EC. European Commission: Soil protection, The Story Behind the Strategy [Internet]. 2006. Available from: https://ec.europa.eu/environment/archives/soil/pdf/soillight.pdf. [Accessed: May 1, 2022]

[59] Webb PN, Marshall NA, Stringer LC, Reed MS, Adrian Chappell A, Herrick JE. Land degradation and climate change: Building climate resilience in agriculture. Frontiers in Ecology and the Environment. 2017;**15**(8):450-459. DOI: 10.1002/fee.1530

[60] IPCC. Special Report. Global warming of 1.5 0C. [Internet] 2022. Available from: https://www.ipcc.ch/sr15/chapter/chapter-1/

[61] Mcgranahan G, Balk D, Anderson B. The rising tide: Assessing the risks of climate change and human settlements in low elevation coastal zones. Environment and Urbanization. 2007;**19**:17-37. DOI: 10.1177/0956247807076960

[62] IPCC. Climate Change 2007: The physical science basis. In: Solomon S, Qin HD, Manning M, Chen Z, Marquis M, Averyt KB, Tignor M, Miller HL, editors. Contribution of Working Group I to the Fourth Assessment Report of the Intergovernmental Panel on Climate Change. Cambridge: Cambridge University Press; 2007. p. 996

[63] Nicholls RJ, Hoozemans FMJ, Marchand M. Increasing flood risk and wetland losses due to global sea-level rise: Regional and global analyses. Global Environmental Change. 1999;**9**:69-87. DOI: 10.1016/S0959-3780(99)00019-9

[64] Meadows ME, Hoffman MT. Land degradation and climate change in South Africa. The Geographical Journal. 2003;**169**(2):168-177. DOI: 10.1111/1475-4959.04982

[65] Winthrop M, Kajumba TC, Mcivor S. Mozambique Country Climate Risk Assessment Report, Ireland Irish aid, Resilience and Economic Inclusion Team. Ireland: Policy Unit; 2018. p. 44

[66] Burrell AL, Evans JP, Liu Y. The addition of temperature to the TSS-RESTREND methodology significantly improves the detection of dryland degradation. IEEE Journal of Selected Topics in Applied Earth Observations and Remote Sensing. 2019;**12**:1-7

[67] Zhou W, Gang C, Zhou F, Li J, Dong X, Zhao C. Quantitative assessment of the individual contribution of climate and human factors to desertification in Northwest China using net primary productivity as an indicator. Ecological Indicators. 2015;**48**:560-569. DOI: 10.1016/j.ecolind.2014.08.043

[68] Zhang Y, Zhang C, Wang Z, Chen Y, Gang C, An R, et al. Vegetation dynamics and its driving forces from climate change and human activities in the Three-River source region, China from 1982 to 2012. Science of the Total Environment. 2016;**563-564**:210-220. DOI: 10.1016/j.scitotenv.2016.03.223

[69] Xu W, Gu S, Zhao XQ, Xiao J, Tang Y, Fang J, et al. High positive correlation between soil temperature and NDVI from 1982 to 2006 in alpine meadow of the Three-River Source Region on the Qinghai–Tibetan Plateau. International Journal of Applied Earth Observation and Geoinformation. 2011;**13**:528-535. DOI: 10.1016/j.jag.2011.02.001

[70] Liang LQ, Li LJ, Liu CM, Cuo L. Climate change in the Tibetan plateau

three rivers source region: 1960-2009. International Journal of Climatology. 2013;**33**:2900-2916. DOI: 10.1002/joc.3642

[71] Wang T, Zhu ZD. Study on sandy desertification in China: 1. Definition of sandy desertification and its connotation. Journal of Desert Research. 2003;**23**(3):209-214

[72] Yang JP, Ding YJ, Chen RS. NDVI reflection of alpine vegetation changes in the source regions of the Yangtze and Yellow Rivers. Acta Geographica Sinica. 2005;**60**(3):467-478

[73] Wang XM, Chen FH, Dong ZB. The relative role of climatic and human factors in desertification in semiarid China. Global Environmental Change. 2006;**16**:48-57. DOI: 10.1016/j.gloenvcha.2005.06.006

[74] Hai CM, Ma L, Wang XM, Li L. Main factors analysis about soil desertification in typical section of interlock area of farming and pasturing: The case of Zhangbei county, Bashang area of Hebei province. Geographical Research. 2002;**21**(5):543-550

[75] Sun W, Li B. The relation between coupling among the principal components of desertification factors and desertification in rear hills of Bashang since 1950. Geographical Research. 2002;**21**(3):391-398

[76] Gorissen A, Tietema A, Joosten NN, Estiarte M, Penuelas J, Sowerby A, et al. Climate change effects carbon allocation to the soil in shrublands. Ecosystems. 2004;**7**:650-661. DOI: 10.1007/s10021-004-0218-4

[77] Wan Y, Lin E, Xiong W, Li Y, Guo L. Modeling the impact of climate change on soil organic carbon stock in upland soils in the 21st century in China. Agriculture, Ecosystems & Environment. 2011;**141**(1-2):23-31. DOI: 10.1016/j.agee.2011.02.004

[78] Mengmeng A, Sun Y, Yan B, Wei Y. A summary of the impact of land degradation on soil carbon sequestration. IOP conf. Series. Materials Science and Engineering. 2018;**394**:052028. DOI: 10.1088/1757-899X/394/5/052028

[79] Rey A, Oyonarte C, Morán-López T, Raimundo J, Pegoraro E. Changes in soil moisture predict soil carbon losses upon rewetting in a perennial semiarid steppe in SE Spain. Geoderma. 2017;**287**:135-146. DOI: 10.1016/j.geoderma.2016.06.05

[80] Zhao F, Wu Y, Hui J, Sivakumar B, Meng X, Liu S. Projected soil organic carbon loss in response to climate warming and soil water content in a loess watershed. Carbon Balance and Management. 2021;**16**:1-14. DOI: 10.1186/s13021-021-00187-2

[81] Piao S, Wang X, Wang K, Li X, Bastos A, Canadell JG, et al. Interannual variation of terrestrial carbon cycle: Issues and perspectives. Global Change Biology. 2019;**26**(1):300-318. DOI: 10.1111/gcb.14884

[82] Lal R. Soil carbon sequestration to mitigate climate change. Geoderma. 2004;**123**(1-2):1-22. DOI: 10.1016/j.geoderma.2004.01.032

[83] Crowther TW, Todd-Brown KE, Rowe CW, Wieder WR, Carey JC, Machmuller MB, et al. Quantifying global soil carbon losses in response to warming. Nature. 2016;**540**(7631):104-108. DOI: 10.1038/nature20150

[84] Wu Y, Zhao F, Liu S, Wang L, Qiu L, Alexandrov G, et al. Bioenergy production and environmental impacts. Geoscience Letters. 2018;**5**(1):1-9. DOI: 10.1186/s40562-018-0114-y

[85] Lal R, Safriel U, Boer B. Zero Net Land Degradation: A New Sustainable Development Goal for Rio+ 20. In: A Report Prepared for the Secretariat of the United Nations Convention to Combat Desertification. Bonn, Germany. 2012

[86] Zhao F, Wu Y, Yao Y, Sun K, Zhang X, Winowiecki LA, et al. Predicting the climate change impacts on water-carbon coupling cycles for a loess hillygully watershed. Journal of Hydrology. 2020;**581**:124388. DOI: 10.1016/j.jhydrol.2019.124388

[87] Ren W, Banger K, Tao B, Yang J, Huang Y, Tian H. Global pattern and change of cropland soil organic carbon during 1901-2010: Roles of climate, atmospheric chemistry, land use and management. Geography Sustainable. 2020;**1**(1):59-69. DOI: 10.1016/j.geosus.2020.03.001

[88] Schlesinger WH, Reynolds JF, Cuningham GL, Huenneke LF, Jarrell WM, Virginia RA, et al. Biological feedbacks in global desertification. Journal of Science. 1990;**247**:1043-1048. DOI: 10.1126/science.247.4946.1043

[89] Zhao F, Wu Y, Qiu L, Bellie S, Zhang F, Sun Y, et al. Spatiotemporal features of the hydro-biogeochemical cycles in a typical loess gully watershed. Ecological Indicators. 2018;**91**(3):542-554. DOI: 10.1016/j.ecolind.2018.04.027

[90] England JR, Paul KI, Cunningham SC, Madhavan DB, Baker TG, Read Z, et al. Previous land use and climate influence diferences in soil organic carbon following reforestation of agricultural land with mixed-species plantings. Agriculture, Ecosystems & Environment. 2016;**227**:61-72. DOI: 10.1016/j.agee.2016.04.026

[91] Quan Q, Tian D, Luo Y, Zhang F, Crowther TW, Zhu K, et al. Water scaling of ecosystem carbon cycle feedback to climate warming. Science Advances. 2019;**5**(8):1131. DOI: 10.1126/sciadv.aav1131

[92] Sharp ED, Sullivan PF, Steltzer H, Csank AZ, Welker JM. Complex carbon cycle responses to multi-level warming and supplemental summer rain in the high Arctic. Global Change Biology. 2013;**19**(6):1780-1792. DOI: 10.1111/gcbb.12149

[93] Kim J-S, Kug J-S, Jeong S-J, Huntzinger DN, Michalak AM, Schwalm CR, et al. Reduced north American terrestrial primary productivity linked to anomalous Arctic warming. Nature Geoscience. 2017;**10**(8):572-5766. DOI: 10.1038/NGEO2986

[94] Albert KR, Ro-Poulsen H, Mikkelsen TN, Michelsen A, Van Der Linden L, Beier C. Efects of elevated CO(2), warming and drought episodes on plant carbon uptake in a temperate heath ecosystem are controlled by soil water status. Plant, Cell & Environment. 2011;**34**(7):1207-1222. DOI: 10.1111/j.1365-3040.2011.02320.x

[95] Mahecha MD, Reichstein M, Carvalhais N, Lasslop G, Lange H, Seneviratne SI, et al. Global convergence in the temperature sensitivity of respiration at ecosystem level. Science. 2010;**13**(3):838-840

[96] Arora VK, Boer GJ, Friedlingstein P, Eby M, Jones CD, Christian JR, et al. Carbon-concentration and carbon-climate feedbacks in CMIP5 earth system models. Journal of Climate. 2013;**26**(15):5289-5314. DOI: 10.1175/JCLI-D-12-00494.1

[97] Lu M, Zhou X, Yang Q, Li H, Luo Y, Fang C, et al. Responses of ecosystem carbon cycle to experimental warming: A

meta-analysis. Ecology. 2013;**94**(3):726-738. DOI: 10.1890/12-0279.1

[98] Hakkenberg R, Churkina G, Rodeghiero M, Börner A, Steinhof A, Cescatti A. Temperature sensitivity of the turnover times of soil organic matter in forests. Ecological Applications. 2008;**18**(1):119-131. DOI: stable/40062115

[99] de Lima ACR, Hoogmoed W, Brussaard L. Soil quality assessment in rice production systems: Establishing a minimum data set. Journal of Environmental Quality. 2008;**37**:623-630. DOI: 10.2134/jeq2006.0280

[100] Guo LL, Sun ZG, Ouyang Z, Han DR, Li FD. A comparison of soil quality evaluation methods for Fluvisol along the lower Yellow River. Catena. 2017;**152**:135-143. DOI: 10.1016/j.catena.2017.01.015

[101] Han XS, Ma P, Guo YZ, Cai JJ, Wen SH. Effects of surface – Layer soil water – Stable aggregates under land use patterns. Journal of Arid Land Resources Environment. 2018;**32**:114-120. DOI: 10.13448/j.cnki.jalre.2018.057

[102] Bakhshandeh E, Hossieni M, Zeraatpisheh M, Francaviglia R. Land use change effects on soil quality and biological fertility: A case study in northern Iran. European Journal of Soil Biology. 2019;**95**:103119

[103] Levi N, Karnieli A, Paz-Kagan T. Using reflectance spectroscopy for detecting land-use effects on soil quality in drylands. Soil and Tillage Research. 2020;**199**:104571. DOI: 10.1016/j.still.2020.104571

[104] Nehrani SH, Askari MS, Saadat S, Delavar MA, Taheri M, Holden NM. Quantification of soil quality under semi-arid agriculture in the northwest of Iran. Ecological Indicators. 2020;**108**:105770. DOI: 10.1016/j.ecolind.2019.105770

[105] Jiang LJ, Jiapaer G, Bao A, Li Y, Guo H, Zheng G, et al. Assessing land degradation and quantifiying its drivers in Amudarya River delta. Ecological Indicators. 2019;**107**:105595. DOI: 10.1016/j.ecolind.2019.105595

[106] Delelegn YT, Purahong W, Blazevic A, Yitaferu B, Wubet T, Göransson H, et al. Changes in land use alter soil quality and aggregate stability in the highlands of Northern Ethiopia. Scientific Reports. 2017;**7**:13602. DOI: 10.1038/s41598-017-14128-y

Chapter 21

The Climate Change-Agriculture Nexus in Drylands of Ethiopia

Zenebe Mekonnen

Abstract

The objective of this chapter is to review the impacts of climate change on dryland agriculture and its possible solutions. Climate change poses significant challenges on dryland agriculture in Ethiopia. In turn, agriculture (malpractice) has contributed to climate change by emitting GHGs such as CO_2, CH_4 and N_2O. Globally, agriculture's contribution takes 14% of CO_2, 47% of CH_4 and 84% of N_2O. Agriculture contributes to 80% of total Ethiopia's GHGs emission: CH_4, N_2O and CO_2, respectively, contributed to 72, 15 and 14% to aggregated emission. To soothe the impacts of climate change, countries should act now differently together to stabilize the fractions of GHGs in the atmosphere at a level that would also stabilize the climate system. Adopting climate-compatible agricultural development strategies can enable to reduce agricultural GHGs emissions or sequestration enhanced while maintaining and even increasing food supply. It is understood that combating desertification, land degradation and mitigating the effects of drought are the basis for accelerated sustainable development, poverty reduction and ensuring food security in Ethiopia. Climate-smart dryland agriculture can maintain livestock and crop productivity, reduces GHGs emission, lessens the impact of climate change and reduces the trade-offs among agricultural development to fulfill food security, climate change and ecosystem degradation.

Keywords: abatement, food security, pastoralists, resilient, sustainable agriculture

1. Introduction

The Food and Agricultural Organization (FAO) stated that "the major challenge threatening the dryland communities is degradation of the natural resource base, which is leading to soil and vegetation loss, fertility decline, water stress, drying of water resources, lakes and rivers. This degradation is being exacerbated by increasing climate variability and change, with profound impacts on the livelihoods of dryland communities" [1]. Despite the fact that Ethiopia's contribution to global GHGs is about 0.04% [2], climate change poses significant challenges for agriculture in general and dryland agriculture in particular. In return, conventional agriculture in general and malpractice agriculture in particular have contributed to climate change by emitting greenhouse gases (GHGs) such as CO_2, CH_4 and N_2O. In this case, a paradigm shift at all levels is needed in such a way that agriculture should be at the core of sustainable development and poverty-reduction efforts as well as those related to lower-carbon and climate-resilient growth [2, 3].

According to the Intergovernmental Panel on Climate Change [4, 5], in Ethiopia, over the past five decades, the temperature has been increasing annually at a rate of 0.2°C. This has already led to a decline in agricultural production, and cereal production is expected to decline still further (12%) under moderate global warming [6]. Furthermore, it has led to a decline in biodiversity, a shortage of food and an increases in human and livestock health problems, as well as rural-urban migration and dependency on external support. Factors exacerbating the impact of climate change in Ethiopia are rapid population growth, land degradation, widespread poverty, dependency on rain fed agriculture, lack of awareness by policy and decision-makers about climate change and lack of appropriate policies and legislation [7, 8], National Meteorological Agency of Ethiopia [9]. More than 85% of the people in Ethiopia depend mainly on agriculture for their livelihoods. This will render them very vulnerable to climate variability and change. Consequently, a large number of people in Ethiopia are being affected chronically by drought and/or flooding, leading to deaths and loss of assets [10]. For instance in the period 1900–2019, there were 16 drought events that caused a total death of 402,367 people and a total affected population of 77,141,879 and resulted in total economic damage amounted to USD 1.5 billion [10]. This has obliged the country to make an appeal for international support. The problem is very serious in the arid and semi-arid areas, especially among the herders (**Table 1**) [12].

The livelihoods of pastoralists are highly dependent on natural resources and very sensitive to climate change, yet such events cannot be easily separated from other events such as land degradation and policy changes [12]. The study by Thomas *et al.* [13] showed that the agricultural development challenges related to climate change in Ethiopia's dryland agro-ecosystems are decline in crop yields and agricultural productivity, high variation in rainfall, water scarcity, drought and erosion, decline in livestock feed and the consequent decline in livestock productivity, prevalence and outbreak of pest and disease, increase in invasive plant and animal spices, loss of biodiversity and an increase in the vulnerability of pastoralist livelihoods.

Despite all those challenges for agricultural development in the dryland agro-climatic zones in Ethiopia, agriculture has remained conventional and traditional in such environments. Those conventional and traditional agricultural developments, combined with the impacts of climate change and variability, are not sustainable, retard climate change mitigation and adaptation initiatives, and exacerbate food insecurity. Therefore, the core objectives of this review were to assess the contribution of such conventional agricultural developments to GHGs emissions from global and Ethiopian perspectives; to give directions on how these unsustainable forms of agriculture could be transformed into sustainable developments by applying climate-smart technologies and proper resource management strategies.

2. Causes and challenges of climate change

Greenhouse gases allow the penetration of incoming solar radiation but absorb the outgoing long wave radiation from the earth's surface and re-radiate the absorbed

Dryland features	Descriptions
General characteristics	• The rainfall is low in amount, erratic and uneven in distribution and generally concentrated in a few heavy storms with high intensity, making droughts a common experience • Vegetation is consequently very sparse and generally degraded, leaving large areas of soil unprotected • Soils in many drylands have low organic matter content, are highly eroded and have low fertility • The high temperatures and strong winds result in high evapotranspiration rates, which further exacerbate the limited availability of moisture
Ecologies	• The ecology is fragile and the environment is unstable • Consist of a wide range of agro-ecologies including the arid, semi arid and dry sub-humid • The altitude ranges from −124 to 1500 m a.s.l. and the rainfall ranges from 200 to 700 mm annually, with a growing period of 90 to 180 days
Resource	• Consists of various types of plants, crops and domestic and wild animals • one of the main centers of biodiversity of sorghum, finger millet, field peas, chick-pea, cowpea, perennial cotton, safflower, castor bean, sesame and other crops • Center of livestock genetic diversity, for example the distinct breeds Borana, Jijiga cattle, the black headed Ogaden sheep, the Afar goat, the Somali goat and the camel resources
Population	• About a third of the populations in Ethiopia currently live in the dryland areas • The population of dryland areas is continually increasing, because people are moving from the highly degraded highlands • Populations in the drylands are exceeding the current carrying capacity and this land is becoming degraded
Farming systems	• The farming systems are mixed with highly integrated animal and crop production • Small landholding owned by households on which crops are produced and partially supports variable number of livestock • The holdings are small, marginal, unconsolidated and scattered, making it difficult for farmers to work on all their fields at the same time

Source: [1, 11] (Regional Learning and Advocacy Programme).

Table 1.
The characteristics of drylands in Ethiopia.

radiation back to the surface of the earth and by doing so they have caused global warming and climate change [4, 5].

The emission of GHGs from anthropogenic activities such as industrial processes, land use change and agriculture are the main causes of climate change. As indicated in **Figure 1**, agriculture's contribution to GHGs emissions is huge. It takes 14% of CO_2, 47% of CH_4 and 84% of N_2O to make up the global share of GHGs emissions [2, 14, 16–19]. These gases are the most persuasive GHGs that are emitted from unsustainable agricultural practices [20–22]. In Ethiopia, agriculture contributed 80% of total country's GHGs emission. Of this, CH_4, N_2O and CO_2 contributed 72%, 15% and 14% to aggregated emission respectively [23]. Agriculture includes cropland management; grazing land management/pasture improvement; management from agricultural

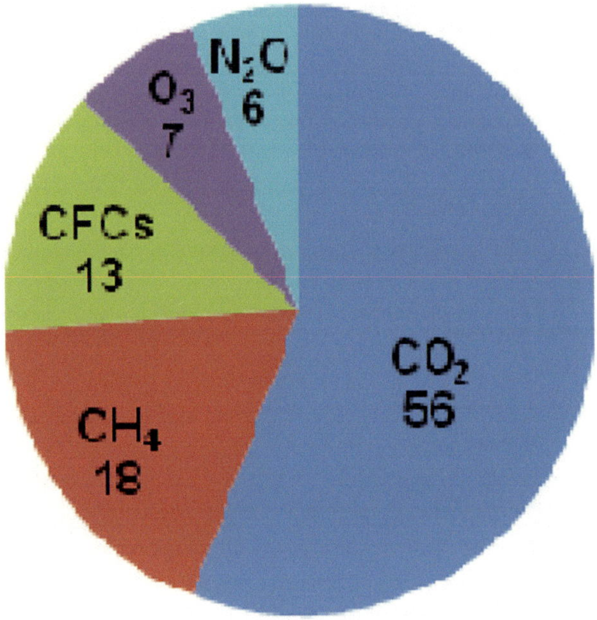

Figure 1.
Percentage global contribution of GHGs to climate change (Source: [14, 15]).

organic soils; restoration of degraded lands; livestock management; manure/bio-solid management; and bioenergy production [2, 4, 19]. These practices can result in GHGs emissions such as CH_4 from enteric fermentation and rice production, N_2O emissions from soils, N_2O and CH_4 from manure management and biomass burning, and CO_2 emissions and removals in agricultural soils. This in turn impacts agricultural developments by contributing to climate change.

To soothe the impacts of climate change, countries should act now, act together and act differently to stabilize the fractions of greenhouse gases in the atmosphere at a level that would also stabilize the climate system. This will give sufficient time to allow ecosystems to adapt naturally to climate change, to ensure that food production is not threatened and to enable economic development to proceed in a sustainable manner [24]. As was dealt in Kyoto Protocol, in order to promote sustainable agricultural development, countries should promote sustainable forms of agriculture in light of climate change [25]. Based on the results of the International Food Policy Research Institute [26], climate change was supposed to have reduced net crop revenue by −28% to −79% in Central Africa, by −7% to −32% in West Africa, by −12% to −17% in Southern Africa, by −11% to −12% in East Africa and by −4% to −7% in North Africa. In Ethiopia, the study by Deressa [26] showed that a unit increase in temperature during summer and winter would reduce net revenue by \$177.62 ha^{-1} and \$464.71 ha^{-1}, respectively. On the other hand, the marginal impact of increasing precipitation during spring would increase net revenue by \$225.09 ha^{-1}. How can agricultural GHGs emissions (**Table 2**) be reduced or sequestration enhanced while maintaining and even increasing food supply, particularly in dryland agriculture? As shown in **Figure 2**, this can be answered by adopting climate-compatible agricultural development strategies [29, 30].

Sub-sector in Agriculture	Main drivers	Emission in million tonnes of CO_2e		
		2010	2020	2030
Forestry	Deforestation Forest degradation	50	125	90
Livestock	Methane from enteric fermentation N_2O from manure left on pastures	65	146	125
Soil management	Crop production Fertilizer use Manure management	12	5.8	60

Source: [2, 27, 28].

Table 2.
Emission drivers in Ethiopia's agricultural sector.

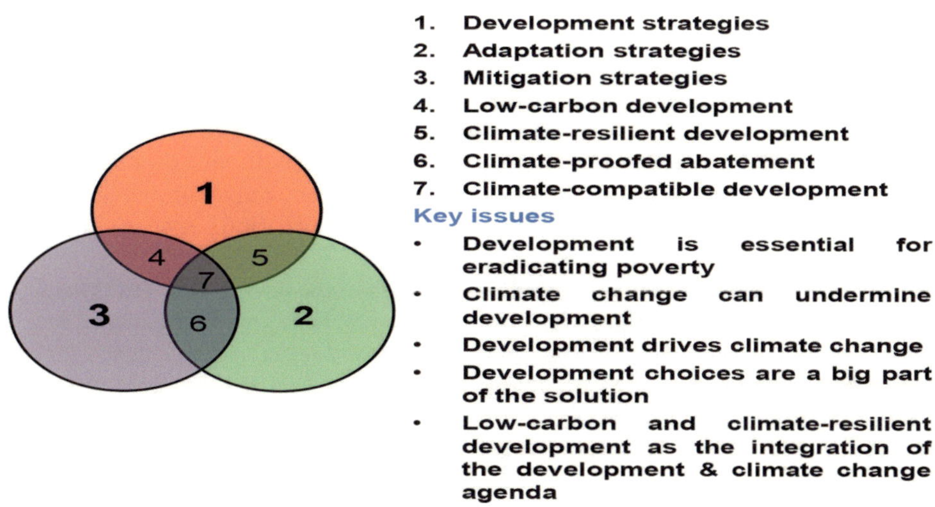

Figure 2.
Options of strategies and key issues in climate change-agricultural development nexus (Source: [29, 30]).

3. Climate-smart agriculture in drylands

Climate-smart agriculture can be defined as agriculture that sustainably increases productivity, resilience (adaptation), reduces/removes GHGs (mitigation), and enhances achievement of national food security and development goals [2, 31–33]. Making agriculture climate-smart is one of the means to tackle climate change and its impacts which is the focus of Sustainable Development Goals (SDGs) (Goal 13) and complements SDGs 1 and 2. Agricultural development in drylands is a victim of climate change impacts. It is anticipated that higher temperatures could reduce crop yields by 10–20% in Sub-Saharan Africa by 2050. In return, unsustainable agricultural development is one of the causes of climate change as it is responsible for 10–12% of anthropogenic GHGs emissions each year and much more (30%) if human beings take into account the clearance of forests to make way for crops and livestock [34, 35]. Agricultural development must be effective in terms of food production, reducing GHGs emissions and helping farmers adapt to climate change [36, 37]. To build the resilience of drylands, it is essential to make agricultural land management practices more sustainable; improving grassland management so as to enhance carbon sequestration; reforestation and restoration of dryland forests; improving the efficiency and productivity of livestock by rearing improved breeds and transforming high emitter livestock (*e.g.* cattle) to lower emitter ones (*e.g.* chickens); improving livestock feeds [2].

Climate change requires environmental conservation and global partnerships that are related to two of the Millennium Development Goals (MDGs): ensure environmental sustainability and develop a global partnership for development [38]. These have been strongly strengthened in the SDGs under goals 15 and 17 [39]. Parry [40] stated that climate change is a binary development issue. In the first case, unsustainable development, in the past and present, is the root cause of climate change. In the The second case, sustainable development is certainly a necessary, and probably sufficient condition for overcoming this challenge (**Figure 3**). Portfolios of mitigation and adaptation strategies to unsustainable development will not result in the right co-benefits. Rather sustainable transformations are important for the case in point [41, 42]. For instance, Denmark has reduced GHGs emissions by 28% in 1990–2009 because of a 31% reduction in N_2O emissions due to improved use of manure and a 40% reduction in the use of inorganic fertilizer in 1990–2000, with a further consensus to reduce GHGs emissions from agriculture by 50–70% without a decrease in food production [43]. Ethiopia has also planned to follow similar trends through its climate resilient green economy strategy. This creates a win-win situation between climate change and agricultural development [28, 44, 45].

Land degradation and human population growth in the drylands of Ethiopia, exacerbated by climate change such as severe droughts, have greatly impaired the country's economic and social development and its food security status. It is clear that combating desertification and land degradation, and mitigating the effects of drought are the basis for accelerated sustainable development, poverty reduction and insuring food security in Ethiopia. This requires the realization of strong partnership building and commitment at regional and international levels. Cognizant of this fact, the

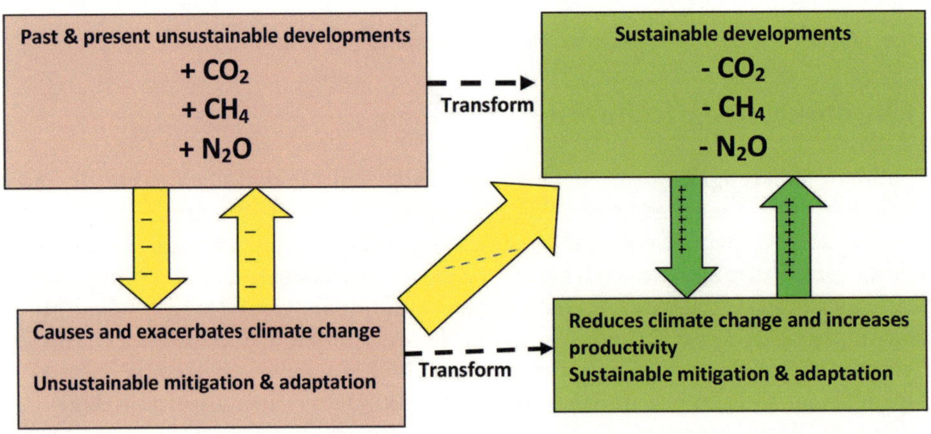

Figure 3.
The climate change and agricultural development relationships (negative signs before GHGs indicate emission reduction and the yellow arrows show negative impacts on each other & positive signs before GHGs indicate emission enhancement and the green arrows show win-win). The strategies that help to make such transformations are described in Sections 3.1 to 3.4 below.

Ethiopian Government was one of the pioneering governments to accept and endorse the Great Green Wall for the Sahel and Sahara Initiative (GGWSSI) and was ready for its implementation [46].

3.1 Grazing land management

Drylands are characterized by low and highly variable precipitation and warm temperatures. Livestock grazing is the predominant type of land use, providing a livelihood for a considerable number of people [47]. Optimal rangeland management depends on (i) the current state of the vegetation; (ii) the observed rainfall; and (iii) optimizing the stocking density and rate to reduce emission of GHGs, particularly methane. The stocking density refers to the number of livestock per hectare of rangeland while the stocking rate refers to the ratio of livestock to available forage on the pasture in a given year [48].

The livestock population of Ethiopia, which reached more than 160 million heads in 2011 and more than 224 million heads in 2020 [49, 50], is the largest in Africa and the 10th in the world. It constitutes a large component of the Ethiopian agricultural sector and is well integrated with the farming systems in general and provides the sole means of subsistence for the herders in the lowlands in particular. More than 50% of Ethiopia's land is utilized for grazing and browsing. Herders in the lowlands take the lion's share of this figure. Even if the world share of non-CO_2 emissions from the livestock sector of Ethiopia is the minimum as shown in **Figure 4** [28, 51], sector-wise Ethiopia's emission profile is dominated by emissions from agriculture contributing about 80% of the total. Whereas gas-wise it is dominated by CH_4 contributing 80% of the total CO_2 equivalent emissions in 1994 [52] and most of this contribution is from less productive livestock. Even in current times, cattle take more than 80% of the share of CH_4 emission in Ethiopia [53, 54]. Based on IPCC [55] guidelines, methane emissions from enteric fermentation are estimated using equation 1 for eight major livestock subcategories in Ethiopia (**Table 3**). The livestock subcategories are donkeys, camels, cattle, goats, mules, sheep, horses and poultry. Livestock population data for each subcategory is from CSA [49, 50]. The emission factors attributed to each livestock subcategory for enteric fermentation are all IPCC default values ascribed for Ethiopian conditions. The methane emissions resulting from equation 1 are then multiplied by 21, the global warming potential for methane at 100 years in the atmosphere, to yield the carbon dioxide equivalent in tonnes of CO_2e (**Table 3**). In order to optimize methane emissions while there is an increasing livestock population [50], there is a need to settle climate smart livestock production with proper rangeland management, improved feed and highly productive livestock breeds. If Ethiopia's livestock production is climate-smart and reduces emissions by 38%, the emission from the eight livestock subcategories (**Table 3**) is less 16,929,022 tCO_2e and 24,583,413 tCO_2e than conventional livestock production in 2011 and 2020 respectively. The Ethiopian Climate-Resilient Green Economy strategy states that, in agriculture, higher livestock productivity has the potential to reduce 45 x 10^6 tonnes of CO_2e emissions a year in 2030 [28]. Grazing lands are considered an important carbon sink-storing 10–30% of the global soil organic carbon. Improved grazing management on rangeland, such as species management, irrigation, rotational grazing, and fertilization, is expected to capture a significant amount of carbon. Studies indicated that there are potential soil carbon sequestration rates of 0.6 - 1.3 $tCO_2e\ ha^{-1}yr^{-1}$ from these improved managements [57].

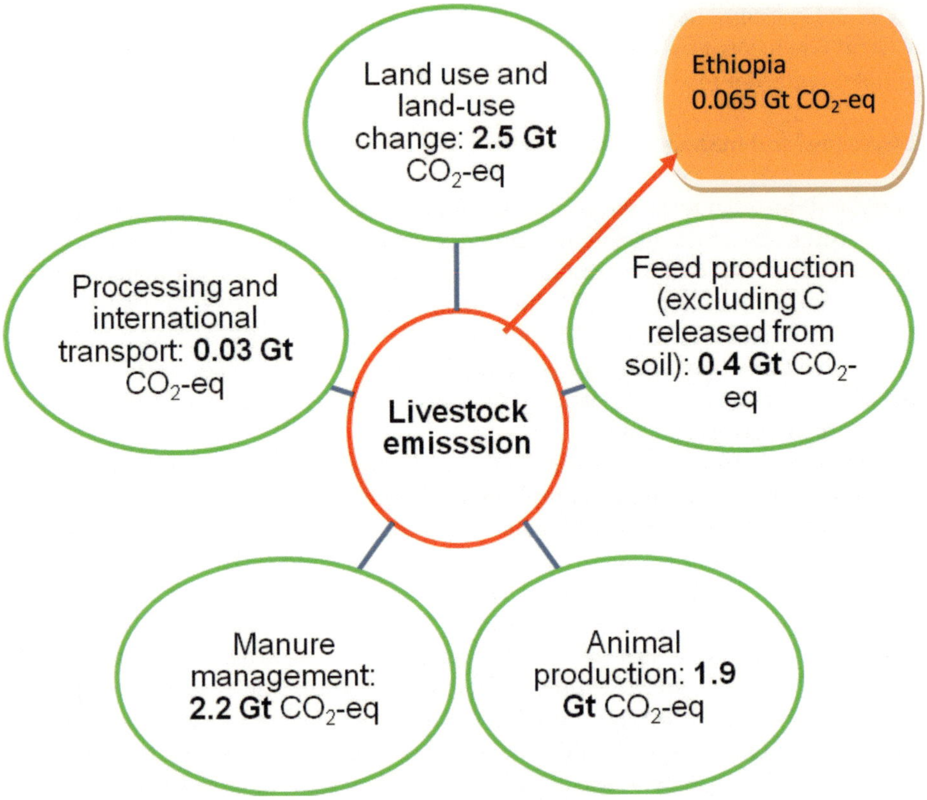

Figure 4.
Global non-CO_2 emission from the livestock sector (Ethiopia's contribution Ethiopia is 0.065 Gt CO_2-eq) (Source: [51]).

$$E = \frac{EFt * Nt}{1000} \tag{1}$$

Where: E is methane emissions from enteric fermentation, tCH_4/year; EFt is emission factor for the defined livestock population, kg CH_4/head/yr; Nt is the number of head of livestock species/category in the country, t is species/category of livestock Emission CO_2e /year (tonnes) = CH_4 emission/year (tonnes) * 21 (Global Warming Potential for CH_4).

3.2 Water management

Water and desertification are the most optimizing factors to foster economic, social and environmental development in the drylands and that the sustainable utilization of water resources is a priority at regional and national scales [58]. Climate change will have enormous effects on the hydrological cycles in drylands with less total rainfall, drier soils but with increased risks of floods from increased frequency and intensity of storm events [4]. There should be a need to enhance physical and economic water productivity. The former is defined as the ratio of the amount of agricultural output to the amount of water used and the latter is defined as the value derived per unit of water used [13].

Livestock categories	Default IPCC Emission factor (KgCH$_4$/head/yr) for Ethiopia [55, 56]		Number of livestock in Ethiopia [50]	Methane emission/year(Tonnes)			Emission CO$_2$e / year(tonnes)
	Enteric fermentation	Manure management		Enteric fermentation	Manure management	Total	
	a	B	c	d = ac/1000	e = bc/1000	f = d+e	g = 21*f
Donkeys	10	0.9	6,209,665	62,097	5,589	67,685	1,421,392
Camels	46	1.92	1,102,119	50,697	2,116	52,814	1,109,084
Cattle	31	1	53,382,194	1,654,848	53,382	1,708,230	35,872,834
Goats	5	0.17	22,786,946	113,935	3,874	117,809	2,473,979
Mules	10	0.9	385,374	3,854	347	4,201	88,212
Sheep	5	0.15	25,509,004	127,545	3,826	131,371	2,758,799
Horse	18	0.9	2,028,233	36,508	1,825	38,334	805,006
Poultry	0	0.02	49,286,932	0	986	986	20,701
Total			**160,690,467**	**2,049,484**	**71,945**	**2,121,429**	**44,550,007**
Number of livestock in Ethiopia [50]							
Donkeys	10	0.9	9,987,762	99,878	8,989	108,867	2,286,199
camels	46	1.92	7,702,493	354,315	14,789	369,103	7,751,173
cattle	31	1	65,354,090	2,025,977	65,354	2,091,331	43,917,948
Goats	5	0.17	50,501,672	252,508	8,585	261,094	5,482,967
Mules	10	0.9	357,603	3,576	322	3,898	81,855
Sheep	5	0.15	39,894,394	199,472	5,984	205,456	4,314,579
Horse	18	0.9	2,111,134	38,000	1,900	39,900	837,909

Livestock categories	Default IPCC Emission factor ($KgCH_4$/head/yr) for Ethiopia [55, 56]		Number of livestock in Ethiopia [50]	Methane emission/year(Tonnes)			Emission CO_2e / year(tonnes)
	Enteric fermentation	Manure management		Enteric fermentation	Manure management	Total	
	a	B	c	d = ac/1000	e = bc/1000	f = d+e	g = 21*f
Poultry	0	0.02	48,955,675	0	979	979	20,561
Total			**224,864,823**	**2,973,726**	**106,902**	**3,080,628**	**64,693,191**
Difference 2020–2011			**64,174,356**	**924,242**	**34,957**	**959,199**	**20,143,184**

Table 3.
Methane emission in Ethiopia's livestock sector.

The drylands of Ethiopia are characterized by scarce and unreliable rainfall. Due to this, within the context of dryland development, the Federal Constitution of Ethiopia in article 52(2d) provides legal provisions ("to administer land and other natural resources in accordance with Federal laws") which provide a basis for regional governments to take an active role in formulating and implementing appropriate policies and programmes for water development in dryland areas (**Figure 5**). Rainwater harvesting is a centuries old practice by the Ethiopian pastoralists and it has continued to be implemented in the current Government's efforts in soil and water conservation programmes to improve food security ([60], Tolossa *et al.* 2020).

Adoption of improved approaches and good practices to water development can strengthen the contribution of dry lands to national economies, and reduce their drain on resources by enhancing resilience and reducing the need for food and other cash interventions during emergencies brought on by climate extremes such as floods and droughts. Improving water development and management, particularly through ecosystem-based approaches, enhances the productivity and sustainability of soil, water and vegetation resources so as to make dryland agricultural development initiatives as sustainable as possible. This improves the resilience of both human communities and ecosystems to climate change in the drylands [59, 61, 62].

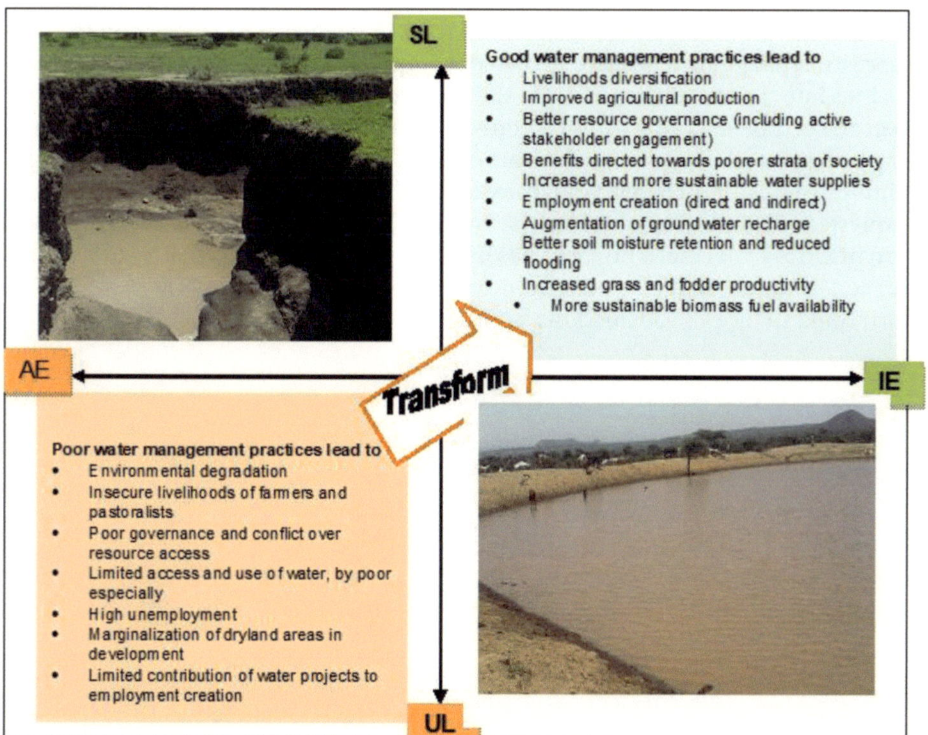

Figure 5.
*Impact matrix of water development in dry lands (Adapted from [59]) (UL is fragile and unsustainable livelihoods; **SL** is more secure livelihoods; **AE** is adverse environmental conditions; and **IE** is improved environmental conditions).*

3.3 Conservation agriculture

Conservation agriculture (CA) is a concept for resource-saving agricultural crop production that strives to achieve acceptable profits together with high and sustained production levels while concurrently conserving the environment. CA is based on enhancing natural biological processes above and below the ground. Interventions such as mechanical soil tillage are reduced to an absolute minimum, and the use of external inputs such as agrochemicals and nutrients of mineral or organic origin is applied at an optimum level and in a way and quantity that does not interfere with, or disrupt, the biological processes. CA is characterized by three principles (**Figure 6**) which are linked to each other, namely: continuous minimum mechanical soil disturbance; permanent organic soil cover; and diversified crop rotations in the case of annual crops or plant associations in the case of perennial crops [64–66].

Conventional tillage exposes the soil by deep cultivation and this in turn enhances CO_2 emissions from the soil. More than 97% of the world's food supply is produced on land that emits GHGs when intensively tilled, fertilized, and/or grazed by animals [67]. Conversion of 76% of the croplands in the USA, for example, to conservation tillage could sequester as much as 286–468 million metric tonnes (MMTs) CO_2e over 30 years showing that conservation agriculture could become a net sink for carbon [68] and play an important mitigation and adaptation role in climate change effects [69, 70].

A global estimate of carbon sequestration from the conversion of conventional tillage to conservation tillage will be as high as 4900 MMT CO_2e by 2020. Combining economics of fuel cost reductions and environmental benefits of conversion to conservation tillage are a positive first step for agriculture toward decreasing carbon emissions into the atmosphere [71]. In the same token, it was also calculated that, if 15% of the carbon in crop residues is converted to passive soil organic carbon (SOC), it may lead to a carbon sequestration rate of 200 MMT $CO_2e\ yr^{-1}$ when it is used with less intensive tillage. A change from conventional tillage to no-tillage has been found to sequester 4300–7100 kg of carbon $ha^{-1}yr^{-1}$ [72]. A traditional agricultural conservation practice in northern Ethiopia has been found to be effective for in-situ soil and water conservation, reducing runoff on average by 11% and soil loss by 36% [73]. This in turn could reduce GHG emissions from agricultural lands.

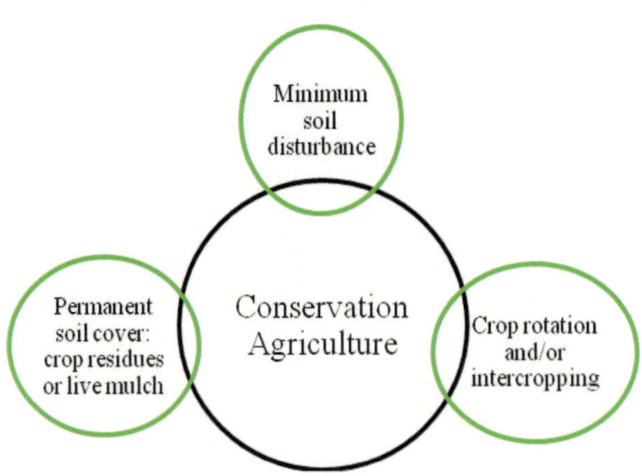

Figure 6.
The three pillars of conservation agriculture (Source: [63]).

Agriculture can contribute to the mitigation of climate change by adopting practices that promote the stashing of CO_2 as carbon in soil, crop biomass and trees, and by displacing the use of fossil fuels required for tillage, chemical manufacture, equipment manufacture, and grain handling operations [74–76]. In the Ethiopian case too, agricultural development as business as usual and contributing the largest share of Emission (**Figure 7**), without consideration of climate risks and opportunities, will lead to maladaptive practices weakening national resilience to climate change [78]. This is also emphasized with the Cancún Agreements that developing nations are, for the first time, officially encouraged to develop low-carbon development strategies.

3.4 Vegetation management

Plants are central in carbon, water and nitrogen cycles thereby necessitating the need for sustainable utilization of these resources with a view to contributing towards reducing the impact of climate change and variability. The ways in which these resources are used and managed, determine the future direction of climate change impacts in drylands [79]. Enhancing awareness on the importance of plant biodiversity and sustainable livelihoods in response to climate change and variability is vital in the fragile dryland ecosystem where there is direct dependence on natural resources for livelihood [80]. Adopting practices of adaptation and mitigation such as proper fire management, improved forest management, reforestation, reducing deforestation and forest degradation will enhance carbon sinks and help to minimize impacts of climate change. In addition to high temperature and changing rainfall patterns, the major threats affecting vegetation resources in drylands are the coping strategies put in place, such as firewood and charcoal sale, by community members during times of drought. These livelihood activities provide households with an alternative income source when livestock and crop production fail. But these activities become unsustainable as droughts become more frequent, leading to substantial deforestation and forest degradation. With expected future climate change and increasing drought risk (**Figure 8**), pressures on vegetation resources are likely to intensify, unless more sustainable alternative sources of fuel and income generating options are provided or put in place. Otherwise, the resulting deforestation and forest degradation will go on to diminish development efforts of local communities and make them vulnerable to climate change shocks [81, 82].

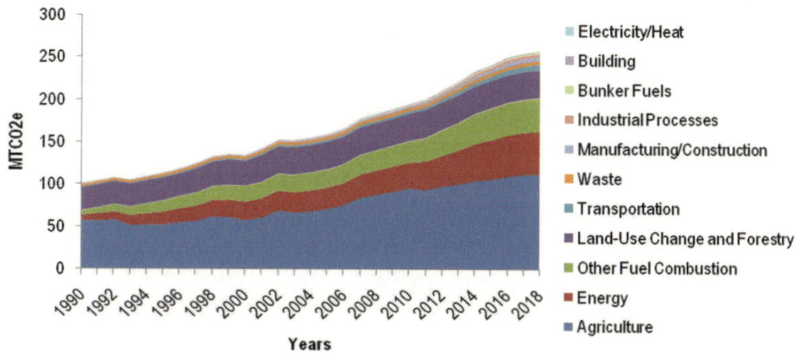

Figure 7.
All GHGs emission trend of Ethiopia by sector (Source: [77]).

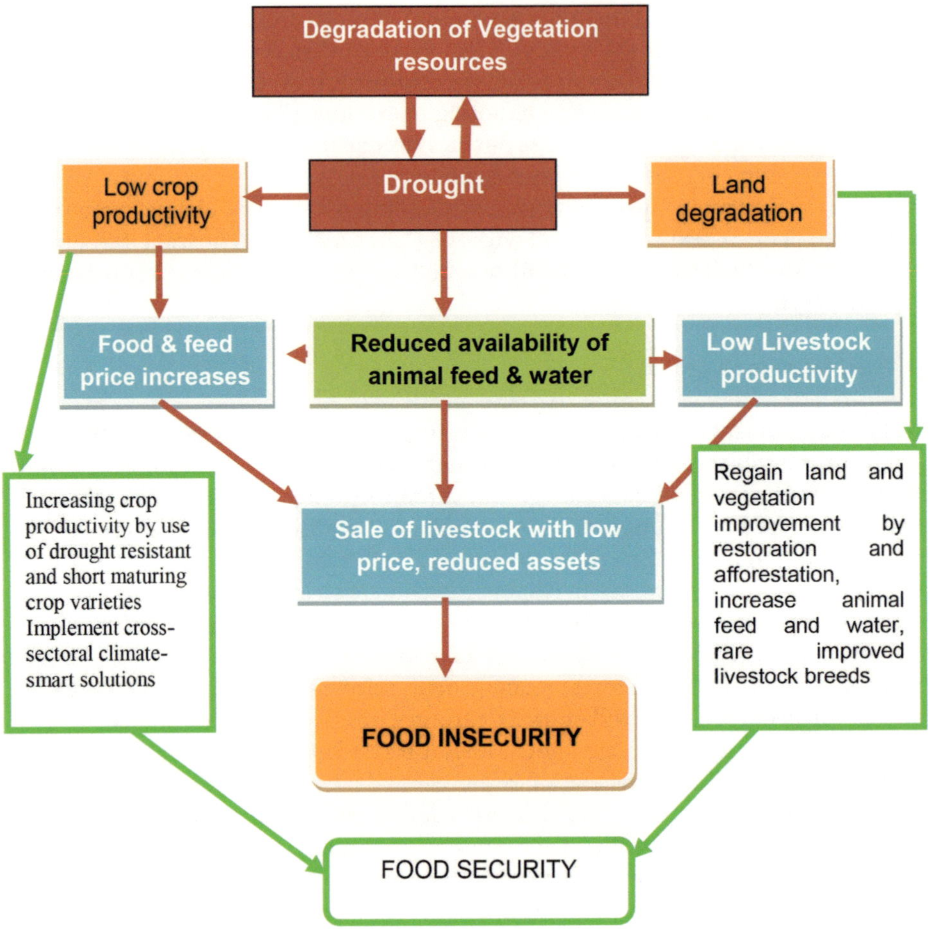

Figure 8.
Repercussions of vegetation degradation and drought in drylands of Ethiopia and how to reverse it by managing the resources and use of technology.

4. Conclusions

Climate change is a global concern whereby developing countries are the most affected by its impacts. Every ecosystem is affected by climate change impacts and in particular drylands are more vulnerable. Dryland agriculture in Ethiopia is more susceptible to the impacts of climate change as the system is already fragile, degraded and unstable with low, erratic and unevenly distributed rainfall patterns. To optimize the productivity of dryland agriculture and enhance food security for the growing population, the practices of agriculture should be climate compatible which encompasses sustainable development, adaptation and mitigation strategies. To this end GHG emissions are reduced or sequestration enhanced while maintaining and even increasing food supply to attain food security. Indeed, there is a need to reduce forest degradation and deforestation, improve rangeland management, improve livestock feeds and rare drought resistant breeds, use drought resistant and short maturing crop varieties, improve soil and water management (including water harvesting and conservation agriculture).

Achieving success in dryland agriculture by overcoming the challenges of climate change requires a comprehensive approach of technical, institutional and financial innovations, so that both adaptation and mitigation strategies are consistent with efforts to safeguard food security, maintain ecosystem services, provide carbon sequestration and reduce emissions. The dryland agriculture in Ethiopia needs reform to attain much greater harmony with the natural and human environment and follow the principles of green economy and making synergies with other sectors. At the end of the day, it is possible to create climate-smart dryland agriculture that maintains livestock and crop productivity as well as reduces GHGs emissions and lessens the impact of climate change. Therefore, productive and ecologically sustainable agriculture with strongly reduced GHGs emissions is fundamental so as to reduce trade-offs in dryland agricultural development to fulfil food security, mitigate climate change and improve ecosystem degradation.

Conflicts of interest

The author declares there are no conflicts of interest.

Author details

Zenebe Mekonnen
Ethiopian Environment and Forest Research Institute, Addis Ababa, Ethiopia

*Address all correspondence to: zenebemg2014@gmail.com

References

[1] FAO. Agricultural based Livelihood Systems in Drylands in the Context of Climate Change: Inventory of Adaptation Practices and Technologies of Ethiopia. FAO, Rome, Italy.pp.53. 2010a

[2] FDRE (Federal Democratic Republic of Ethiopia). Updated Nationally Determined Contribution, Addis Ababa, Ethiopia. 2021

[3] ADP (African Development Forum). Climate Change, Agriculture and Food Security. 7th African Development Forum, Acting on Climate Change for Sustainable Development in Africa, Issue paper #2, Addis Ababa, Ethiopia. 2010

[4] IPCC (Intergovernmental Panel on Climate Change). Climate change 2007: The Physical Science Basis. Summary for Policymakers. Contribution of Working Group I to the Fourth Assessment Report of the Intergovernmental Panel on Climate Change. IPPC secretariat, WMO, Geneva, Switzerland. 2007

[5] IPCC. In: Pachauri RK, Meyer LA, editors. Climate Change 2014: Synthesis Report. Contribution of Working Groups I, II and III to the Fifth Assessment Report of the Intergovernmental Panel on Climate Change. Geneva, Switzerland; 2014. 151 pp

[6] Ringer C. Food and water under global change: Developing adaptive capacity with a focus on rural Africa. Paper presented in the workshop "How can African agriculture adapt to climate change? 11-13 December 2008, Nazareth, Ethiopia. 2008

[7] Kassahun D. Impacts of climate change on Ethiopia: A review of the literature. In: Green Forum, editor. Climate change – a burning issue for Ethiopia: Proceedings of the 2nd Green Forum, Addis Ababa, 31 October–2 November 2007. 2008. pp. 9-35

[8] Sintayehu W. Climate change: Global and national response. In: Green Forum, editor. Climate change – a burning issue for Ethiopia: Proceedings of the 2nd Green Forum Conference held in Addis Ababa, 31 October–2 November 2007, Addis Ababa, Ethiopia. 2008. pp. 37-69

[9] NMA (National Meteorological Agency). Climate Change National Adaptation Program of Action (NAPA) of Ethiopia. Addis Ababa, Ethiopia. 2007

[10] WBG (The World Bank Group). Climate Risk Profile: Ethiopia. World Bank Publications, The World Bank Group, Washington, DC, USA. 2021

[11] REGLAP Secretariat. Key statistics on the drylands of Kenya, Uganda and Ethiopia. 2012

[12] Gebremichael Y, Kifle M. Local innovation in climate-change adaptation by Ethiopian pastoralists. PROLINNOVA–Ethiopia and Pastoralist Forum Ethiopia (PFE), Final report, Addis Ababa, Ethiopia. 2009

[13] Thomas RJ, de Pauw E, Qadir M, Amri A, Pala M, Yahyaoui A, et al. Increasing the resilience of dryland agro-ecosystems to climate change. SAT eJournal. 2007;**4**(1):1-37

[14] Smith P, Martino D, Cai Z, Gwary D, Janzen H, Kumar P, et al. Policy and technological constraints to implementation of greenhouse gas mitigation options in agriculture. Agriculture, Ecosystems and Environment. 2007;**118**:6-28

[15] FAO. The share of food systems in total greenhouse gas emissions. Global, regional and country trends, 1990-2019. In: FAOSTAT Analytical Brief Series No. 31. Rome, Italy; 2021

[16] Herzog T. World Greenhouse Gas Emissions in 2005. WRI Working Paper. World Resource Institute. 2005

[17] Lynch J, Cain M, Frame D, Pierrehumbert R. Agriculture's contribution to climate change and role in mitigation is distinct from predominantly fossil CO_2-emitting sectors. Frontiers in Sustainable Food Systems. 2021;**4**:518039

[18] Woldegiorgis A. Ethiopia: Greenhouse Gas Emissions and Sources. Ethiopian Energy Authority, Interim Report on Climate Change Country Studies. Addis Ababa, Ethiopia. 1995

[19] WRI CAIT (World Resources Institute Climate Analysis Indicators Tool). Emissions including Land-Use Change and Forestry. 2015

[20] GGWG (Greenhouse Gas Working Group). Agriculture's role in greenhouse gas emissions & capture. Greenhouse Gas Working Group Rep. ASA, CSSA, and SSSA, Madison, WI. 2010

[21] Majule AE, Rioux J, Mpanda M, Karttunen K. Review of climate change mitigation in agriculture in Tanzania. MICCA Programme, FAO. 2014

[22] Shehrawat PS, Muketshwar R. Greenhouse gases: Causes, losses and remedial measures for sustainable agriculture. Scientific Papers Series Management, Economic Engineering in Agriculture and Rural Development. 2015;**15**(2):363-370

[23] UNFCCC. Emissions Summary for Ethiopia. UNFCCC Secretariat. 1995

[24] UNFCCC (United Nations Framework Convention on Climate Change). New York, USA. 1992

[25] IFPRI (International Food Policy Research Institute). Micro-Level Practices to Adapt to Climate Change for African Small-scale Farmers. IFPRI Discussion Paper 00953. Addis Ababa, Ethiopia. 2010

[26] Deressa TT. Measuring the Economic Impact of Climate Change on Ethiopian Agriculture: Ricardian Approach. The World Bank Development Research Group, Sustainable Rural and Urban Development Team. Policy Research Working Paper 4342. Addis Ababa, Ethiopia. 2007

[27] DEA (Danish Energy Agency), the Organization for Economic Co-operation and Development (OECD) and the UNEP Risø Centre (URC). National Greenhouse Gas Emissions Baseline Scenarios: Learning from Experiences in Developing Countries. 2013

[28] FDRE (Federal Democratic Republic of Ethiopia). The path to sustainable development: Ethiopia's Climate-Resilient Green Economy Strategy. Addis Ababa, Ethiopia. 2011

[29] Metz B. Climateworks/ European Climate Foundation Low Carbon Growth Initiatives. CLEAN Expert Meeting, Amsterdam, The Netherlands. 2010

[30] Tompkins EL, Mensah A, King L, Long TK, Lawson ET, Hutton C, et al. An Investigation of the Evidence of Benefits from Climate Compatible Development. LS2 9JT, UK: Sustainability Research Institute (SRI), School of Earth and Environment, The University of Leeds, Leeds; 2013

[31] FAO. "Climate-Smart' Agriculture – Policies, Practices and Financing

for Food Security, Adaptation and Mitigation. FAO, Rome, Italy. 2010b

[32] Fisher S. Low carbon resilient development in the least developed countries. IIED Issue Paper. IIED, London, UK. 2013

[33] Niggli U, Fließbach A, Hepperly P, Scialabba N. Low Greenhouse Gas Agriculture: Mitigation and Adaptation Potential of Sustainable Farming Systems. Rome Italy: FAO; 2008

[34] Meridian Institute. "Agriculture and Climate Change Policy Brief: Main Issues for the UNFCCC and Beyond" Edited by Donna Lee; Adapted from "Agriculture and Climate Change: A Scoping Report" by Bruce Campbell, Wendy Mann, Ricardo Meléndez-Ortiz, Charlotte Streck, Timm Tennigkeit, and Sonja Vermeulen. 2011. Available at www.climate-agriculture.org

[35] Pye-Smith C. Promoting climate-smart agriculture in ACP countries. CTA Policy Brief No.9, Wageningen, The Netherlands. 2012

[36] Ickowicz A, Ancey V, Corniaux C, Duteurtre G, Poccard-Chappuis R, Touré I, et al. Crop–livestock production systems in the Sahel – increasing resilience for adaptation to climate change and preserving food security. In: Meybeck A, Lankoski J, Redfern S, Azzu N, Gitz V, editors. Building resilience for adaptation to climate change in the agriculture sector, Proceedings of a Joint FAO/OECD Workshop 23-24 April 2012, Rome, Italy. 2012

[37] Neufeldt H, Kristjanson P, Thorlakson T, Gassner A, Norton-Griffiths M, Place F, Langford K. ICRAF Policy Brief 12: Making Climate-Smart Agriculture Work for the Poor. World Agroforestry Centre (ICRAF): Nairobi, Kenya. 2011

[38] FAO. The state of food security in the world: Eradicating world hunger-key to achieving the Millennium Development Goals. Rome, Italy. 2005

[39] UNESC (United Nations Economic and Social Council). Report of the Inter-Agency and Expert Group on Sustainable Development Goal Indicators. 2016. E/CN.3/2016/2/Rev. 1

[40] Parry M. Climate change is a development issue, and only sustainable development can confront the challenge. Climate and Development. 2009;**1**:5-9

[41] Schmidt E, Tadesse F. The impact of sustainable land management on household crop production in the Blue Nile Basin, Ethiopia. Land Degradation and Development. 2019;**30**:777-787

[42] UNEP (United Nations Environment Programme). Emissions Gap Report 2019. Nairobi, Kenya: UNEP; 2019

[43] CTA (Technical Centre for Agricultural and Rural Cooperation). Climate-Smart Agriculture: Success Stories from Farming Communities Around the World. Wageningen, The Netherlands. 2013

[44] Bass S, Wang S, Ferede T, Fikreyesus D. Making Growth Green and Inclusive: The Case of Ethiopia. Paris: OECD Publishing; 2013

[45] Yirgu L, Nicol A, Srinivasan S. Warming to Change? Climate Policy and Agricultural Development in Ethiopia. Future Agricultures Consortium, Working Paper 071. 2013

[46] FDRE (Federal Democratic Republic of Ethiopia). National Strategy and Action Plan for the Implementation

of the Great Green Wall Initiative in Ethiopia. Addis Ababa, Ethiopia. 2012

[47] AU (African Union). Policy Framework for Pastoralism in Africa: Securing, Protecting and Improving the Lives, Livelihoods and Rights of Pastoralist Communities. Addis Ababa, Ethiopia: AU, Department of Rural Economy and Agriculture; 2013

[48] Quaas MF, Baumgärtner S. Optimal grazing management rules in semi-arid rangelands with uncertain rainfall. Natural Resource Modeling. 2012;**25**:364-387

[49] CSA (Central Statistical Agency of Ethiopia). Agricultural Sample Survey, Volume II, Report on Livestock and Livestock Characteristics (Private Peasant Holdings). Addis Ababa, Ethiopia. 2011

[50] CSA. Agricultural Sample Survey, Report on Livestock and Livestock Characteristics (Private Peasant Holdings), Volume II, Statistical Bulletin 587, Addis Ababa, Ethiopia. 2020

[51] Hristov AN, Oh J, Lee C, Meinen R, Montes F, Ott T, et al. Mitigation of Greenhouse Gas Emissions in Livestock Production – A Review of Technical Options for Non-CO_2 Emissions. Rome, Italy: FAO; 2013

[52] NMSA (National Meteorological Service Agency). Initial National Communication of Ethiopia to the United Nations Framework Convention on Climate Change (UNFCCC). Addis Ababa, Ethiopia. 2001

[53] Engdaw BD. Assessment of the trends of greenhouse gas emission in Ethiopia. Geography, Environment, Sustainability. 2020;**13**(2):135-146

[54] Menghistu HT, Abraha AZ, Mawcha GT, Tesfay G, Mersha TT, Redda YT. Greenhouse gas emission and mitigation potential from livestock production in the drylands of Northern Ethiopia. Carbon Management. 2021;**12**:289-306

[55] IPCC (Intergovernmental Panel on Climate Change). Guidelines for National Greenhouse Gas Inventories. Prepared by the National Greenhouse Gas Inventories Programme. 2006

[56] Brown S, Grais A, Ambagis S, Pearson T. Baseline GHG Emissions from the Agricultural Sector and Mitigation Potential in Countries of East and West Africa. CCAFS Working Paper no. 13. Copenhagen, Denmark: CGIAR Research Program on Climate Change, Agriculture and Food Security (CCAFS); 2012

[57] Eagle AJ, Olander LP. Greenhouse gas mitigation with agricultural land management activities in the United States—a side-by-side comparison of biophysical potential. Advance in Agronomy. 2012;**115**

[58] UNFCCC. Doha Declaration on Food Security. International Conference on Food Security in Drylands, 14-15 November 2012, Doha, Qatar. 2012

[59] Mtisi S, Nicol A. Good Practices in Water Development for Drylands. Glad Switzerland: IUCN; 2013

[60] Yosef BA, Asmamaw DK. Rainwater harvesting: An option for dry land agriculture in arid and semi-arid Ethiopia. International of Water Resources and Environmental Engineering. 2015;7(2):17-28

[61] Biggs EM, Bruce E, Boruff B, Duncan JM, Horsley C, Pauli N, et al. Sustainable development and the water–energy–food nexus: A perspective on livelihoods. Environmental Science & Policy. 2015;**54**:389-397

[62] CGIAR Research Program on Water, Land and Ecosystems (WLE). Healthy soils for productive and resilient agricultural landscapes. Colombo, Sri Lanka: International Water Management Institute (IWMI). CGIAR Research Program on Water, Land and Ecosystems (WLE). 12p. (WLE towards Sustainable Intensification: Insights and Solutions Brief 2). 2017

[63] IPCC (Intergovernmental Panel on Climate Change). Summary for Policymakers. In: Climate Change and Land: An IPCC Special Report on Climate Change, Desertification, Land Degradation, Sustainable Land Management, Food Security, and Greenhouse Gas Fluxes in Terrestrial Ecosystems. 2019. (https://www.ipcc.ch/site/assets/uploads/sites/4/2020/02/SPM_Updated-Jan20.pdf)

[64] FAO (Food and Agricultural Organization). The Economics of Conservation Agriculture. Rome, Italy. 2001

[65] FAO. What is Conservation Agriculture? In: Conservation Agriculture website of FAO. 2008, http://www.fao.org/ag/ca/1a.html

[66] Hobbs PR, Sayre K, Gupta R. The role of conservation agriculture in sustainable agriculture. Philosophical Transactions of the Royal Society. 2008;**363**:543-555

[67] Reicosky DC, Hatfield JL, Sass RL. Agricultural Contributions to Greenhouse Gas Emissions. In: Reddy KR, Hodges HF, editors. Climate Change and Global Crop Productivity. 2000. pp. 37-55

[68] Kern JS, Johnson MG. Conservation tillage impacts on national soil and atmospheric carbon levels. Soil Science Society of America Journal. 1993;**57**:200-210

[69] Karki TB, Gyawaly P. Conservation agriculture mitigates the effects of climate change. Journal of Nepal Agricultural Research Council. 2021;**7**:122-132

[70] O'Dell D, Eash NS, Hicks BB, Oetting JN, Sauer TJ, Lambert DM, et al. Conservation agriculture as a climate change mitigation strategy in Zimbabwe. International Journal of Agricultural Sustainability. 2020. DOI: 10.1080/14735903.2020.1750254

[71] Lal R. Residue management conservation tillage and soil restoration for mitigating greenhouse effect by CO_2 enrichment. Soil Tillage Research. 1997;**43**:81-107

[72] JMF J, Franzluebbers AJ, Weyers SL, Reicosky DC. Agricultural opportunities to mitigate greenhouse gas emissions. Environmental Pollution. 2007;**150**:107-124

[73] Nyssen J, Ovaerts B, Araya T, Cornelis WM, Bauer H, Haile M, et al. The use of the *marasha* ard plough for conservation agriculture in Northern Ethiopia. Agronomy and Sustainable Development. 2011;**31**:287-297

[74] Cole CV, Duxbury J, Freney J, Heinemeyer O, Minami K, Mosier A, et al. Global estimates of potential mitigation of greenhouse gas emissions by agriculture. Nutrient Cycling in Agroecosystems. 1997;**49**:221-228

[75] Fentie A, Beyene AD. Climate-smart agricultural practices and welfare of rural smallholders in Ethiopia: Does planting method matter? Land Use Policy. 2019;**85**:387-396

[76] Paustian K, Cole CV, Sauerbeck D, Sampson N. CO_2 mitigation by agriculture: An overview. Climate Change. 1998;**40**:135-162

[77] Ritchie H, Roser M. "CO_2 and Greenhouse Gas Emissions." 2020. *Published online at OurWorldInData.org.* Retrieved from: 'https://ourworldindata.org/co2-and-other-greenhouse-gas-emissions'

[78] Adem A, Bewket W. A Climate Change Country Assessment Report for Ethiopia. Addis Ababa, Ethiopia. 2011

[79] Omeny P, Oyieke H. Climate change and variability: The Definitions and Implications to Dryland Natural Resources. The 7th RPSUD Conference on "The Changing Climate: Management for Sustainable Utilization: Lessons from Global and African Experiences" held in Adam, Ethiopia July 16-18, 2008

[80] Amanuel W, Tesfaye M, Worku A, Seyoum G, Mekonnen Z. The role of dry land forests for climate change adaptation: The case of Liben Woreda, Southern Oromia, Ethiopia. Journal of Ecology and Environment. 2019;**43**(11)

[81] Mera GA. Drought and its impacts in Ethiopia. Weather and Climate Extremes. 2018;**22**:24-35

[82] Riché B, Hachileka E, Awuor CB, Hammill A. Climate related vulnerability and adaptive capacity in Ethiopia's Borena and Somali communities: Final Report. Save the Children UK, Addis Ababa, Ethiopia. 2009.

Chapter 22

Climate Change, Rural Livelihoods, and Human Well-Being: Experiences from Kenya

André J. Pelser and Rujeko Samanthia Chimukuche

Abstract

Over the next few decades, climate change is set to fuel the existing degradation of ecosystems across Africa, leading to dramatic consequences for poor rural populations that depend largely on agriculture and fishing for their livelihoods. This chapter draws on the findings of a study that explored how climate change affects the livelihoods and ultimately the well-being of farming and fishing households in a remote rural area in Kenya and discusses the coping strategies adopted by these communities. Understanding how climate change impacts people's livelihoods is important as a precursor to assist communities to adapt to and cope with the adverse effects of climate change. The results pointed to relatively wide utilization of traditional knowledge in coping strategies. Conversely, robust modern technologies for forecasting weather patterns remain under-utilized among the target population. The chapter concludes with recommendations to capitalize on and strengthen the existing coping strategies of the affected communities.

Keywords: climate change, Africa, Kenya, livelihoods, farmers, fishermen, coping strategies

1. Introduction

All across Africa, anthropogenic factors, such as rapid human population growth, urbanization, pollution, deforestation, and depletion of natural resources, have given rise to an unprecedented degradation of ecosystems (savanna woodlands, coral reefs, tropical forests, wetlands, etc.), which in turn has increased the vulnerability of these ecosystems to climate change. Two key sources of livelihood in Africa, namely farming and fisheries, are under severe pressure from global climate change. Projected trends in temperature and rainfall [1] are likely to exacerbate existing patterns of poverty, food insecurity, and forced migration in sub-Saharan Africa.

Although the African continent is responsible for only 3.7% of global greenhouse gas emissions, it bears a disproportionate component of the impacts of climate change [2, 3]. It is projected that, over the next few decades, continued global warming will fuel an increase in average temperatures and extreme heatwaves in African countries, with accompanying spikes in both the frequency and intensity of rainstorms across the continent [4, 5]. In recent times, the increase of surface temperature in Africa has happened at a faster rate than the average for the rest of the world [1]. Extreme temperature increases across Africa are attributed to human-induced actions that are driving climate change, with agriculture and water counting among the most vulnerable sectors [3, 5].

Africa is regarded as the most vulnerable continent to climate change impacts because of its low adaptive capacity and overdependence on natural resource-based livelihoods [3, 5, 6]. In Africa, climate-induced changes are likely to have dramatic effects on the livelihoods of poor rural communities in particular, on a continent already struggling to eradicate poverty as part of the United Nations' Sustainable Development Goals for 2030 [7]. Agriculture and fishing are usually the prominent livelihood activities in rural communities in Africa but are highly vulnerable to increased extreme weather-related events caused by climate change [5]. By 2080, agricultural productivity in sub-Saharan Africa is projected to drop from 21% to 9% because of climate change [8]. An increase of between 1.5°C and 2°C in global warming is projected to trigger severe economic and ecosystem impacts in the form of reduced food production, biodiversity loss, water shortages, loss of lives, and increased levels of human morbidity across the continent [5]. There is, however, a lack of extensive knowledge about how climate change affects farmers and fishermen at a finer scale, as most predictions in this regard are mainly based on global climate change models [1].

In the east African state of Kenya, poverty and food insecurity have been exacerbated by the changing climate of recent decades, to such an extent that climate change is now hampering efforts to achieve sustainable development in the country [9]. Identifying how climate change impacts people's livelihoods is, therefore, essential as a precursor to an understanding of how communities can adapt to and cope with the adverse effects of climate change. Using one of the remote rural areas of Kenya in the Suba district, with resource-poor households that rely on farming and fishing as sources of livelihood, this chapter explores how climate change is affecting these livelihoods and ultimately the well-being of such households. More specifically, the chapter seeks to address the following two questions:

I. How does climate change affect the livelihoods of farmers and fishermen and their families in the Suba district of Kenya?

II. What coping strategies are currently employed by the affected communities to mitigate the impact of climate change on their livelihoods?

2. Climate change impacts in Kenya and the Suba district of Kenya

Historically, Africa's contribution to carbon dioxide emissions has been minimal, but the continent has been affected by the adverse impacts of global warming more severely due to its combined pressures of high levels of poverty, weak health facilities, and limited capacity to adapt to the shifts in climate [5, 10]. Agriculture is the main source of livelihood for about 85% of Africa's population [2] and contributes significantly to the Gross Domestic Product of most African countries. Smallholder farmers in sub-Saharan Africa, however, have a low adaptive capacity to climate change because of several interlocking challenges, including high levels of poverty, poor access to credit for inputs, and poor infrastructure [11, 12]. Under climate change scenarios, crop yields are projected to fall even further in sub-Saharan Africa [5] where most of the people are already food insecure. Changes in rainfall patterns in sub-Saharan Africa, where agriculture is predominantly rainfed, are a serious threat to food security, nutrition, and general well-being of the people [13].

As for eastern Africa, the Sixth Assessment Report of the Intergovernmental Panel on Climate Change [1, 5] projects a shorter rainfall season, an increase in temperature, and a rise in malaria cases because of the rise in temperature. With a population of 54 million, Kenya is the economic and financial hub of the region with the country's economy mainly depending on tourism and rainfed agriculture [14]. Like the rest of eastern Africa, Kenya too is highly vulnerable to climate change impacts [15, 16] with an average annual temperature that is projected to increase by as much as 2.5°C between 2000 and 2050 [17]. Kenya is endowed with a diverse range of ecosystems that serve a number of pivotal functions in livelihood activities, such as agriculture, tourism, and fisheries. In Kenya, the evidence of climate change has been noted in soaring temperatures and irregular rainfall patterns that are characterized by intense downpours in the rainy season [18]. Analyses of climate data over the last 100 years suggest that Kenya has already been experiencing an increase in temperature of up to 0.8 degrees [19]. Rising temperatures have increased plant pests and diseases, thus affecting the quality of agricultural produce. For rural areas in Kenya, 75–80% of poor communities either directly or indirectly derive their sources of income from agriculture [2, 16]. Hence climate variability – even a small increase in drought frequencies and intensities – increases the vulnerability to food insecurity and water availability, especially in the arid and semi-arid regions of Kenya.

Agriculture in the Suba district of Kenya, as elsewhere in the country, has been drastically affected by intermittent droughts, soaring high temperatures, and delays in seasonal rainfall patterns that tend to destroy entire harvests [20, 21]. Climate change has altered rainfall patterns by delaying the onset of the March–September rain season, causing severe droughts and water stress [5]. The duration of the rainy season has shortened, but the intensity of the rains has increased [22]. Over-dependence on rainfed agriculture has made the communities more vulnerable to climate variability. Maize production, which is mainly produced under rainfed systems, has been reduced, leading to food insecurity. Livestock farming is also vulnerable to the changing climates as high temperatures decrease grazing fields and increase the prevalence of livestock pests and diseases [23, 24]. In addition, drought events directly result in a rise in food prices, leading to further food insecurity.

3. Theoretical framework: the sustainable livelihoods approach

The Sustainable Livelihoods Approach emphasizes natural resources as productive assets in sustaining rural livelihoods. A livelihood encompasses all the activities and capabilities that are required as a means of living and it is considered sustainable if it is able to withstand and recover from any adverse effects without eroding its natural resource base [25]. Climate change negatively impacts livelihood security and presents a livelihood disturbance, especially when adaptive mechanisms are limited. The Sustainable Livelihood Approach has proved useful to explain the adaption of rural households to the impacts of climate change, thus allowing for a more detailed look at livelihoods on a context-specific level [11]. This approach helps to develop intervention strategies that are, among others, people-centered, dynamic, responsive, and participatory and that happens in collaboration with public and private institutions [25].

There is wide consensus in the scientific community that anthropogenic climate change will affect ecosystem services, food production, and water resources – all of which are vital assets associated with human livelihoods [8, 26, 27]. The livelihoods framework states that people require several assets to achieve positive livelihood outcomes. These assets include human capital (education, health, etc.), physical capital (technology, infrastructure, etc.), social capital (networks, leadership, etc.), natural capital (water, land and produce, etc.), and financial capital (wages, savings, etc.) [25]. Communities that have access to more resources, or that possess highly diverse assets, are likely to have greater livelihood options and abilities to adapt to climate change effects. The increase in extreme weather events, however, has raised the level of vulnerability of marginalized communities through declining food security and disruption in livelihood activities that are essential for survival, including agricultural production and fishing [28].

4. Methodology

4.1 Description of the study area

The Suba district is located in the Nyanza province in southwestern Kenya along Lake Victoria (**Figure 1**). The district is one of the poorest in Kenya, comprising approximately

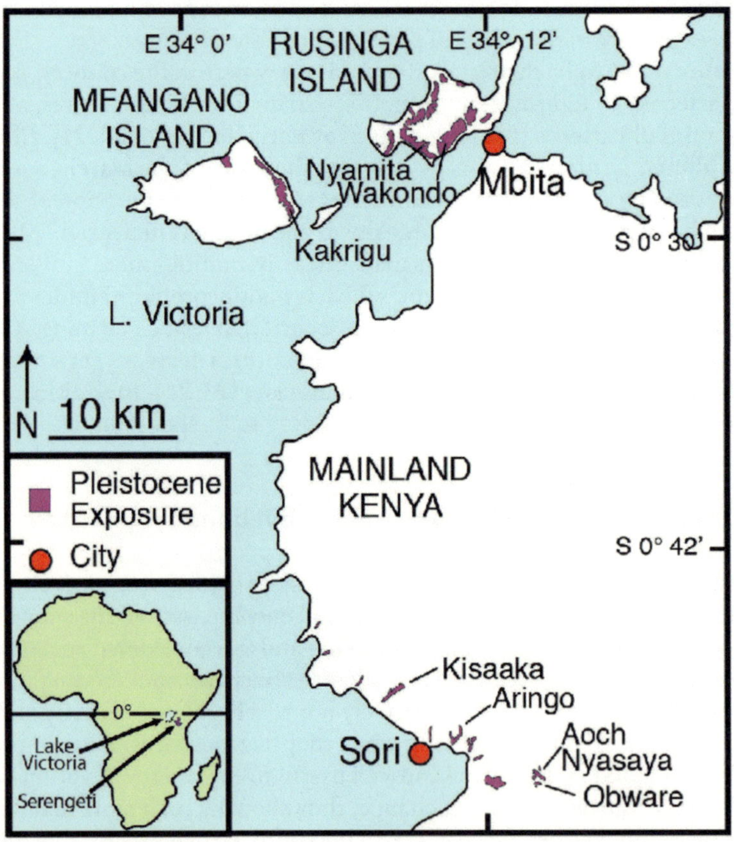

Figure 1.
Map of the study areas [29].

200,000 people who are densely populated in the Lake Victoria basin [30]. More than one-third (35.5%) of the population in Kenya lives below the poverty line of US$1.90 per day [31], but in the Suba district, the poverty incidence is as high as 52% [32].

Apart from mixed farming practices, some people in this district are highly dependent on the adjacent Lake Victoria for fishing at both subsistence and commercial scale. The Suba district is, therefore, mainly dominated by natural resource-based livelihoods comprising smallholder subsistence farming and fishing households. Unfortunately, the rapidly declining fish stock sizes and catches in Lake Victoria – the result of, among others, serious pollution problems over the past 20 years – are now threatening the food security and survival of more than one million Kenyans [16]. Major crops in the Suba district include maize, sorghum, cassava, and legumes, while bananas and sweet potatoes are grown widely as security crops that withstand drought periods and feed households in times of famine [33]. Livestock husbandry is also common in the form of households rearing cattle, goats, chickens, and donkeys.

The Suba district faces inadequate health provisions and high levels of poverty that have exacerbated the risk of climate-related diseases [32, 34]. Cholera, dysentery, and typhoid tend to increase as people are in constant contact with bacteria from inappropriate sewerage and waste disposal. The poor access to health care services is exacerbated by adverse climatic disasters that are posing health risks through malaria and cholera among households [35]. Communities in the Suba district are, therefore, now highly vulnerable to epidemics associated with extreme climate disasters.

Figure 1 shows the study areas comprising Mfangano Island, Rusinga Island, and Mbita point in the Suba district. All areas are located along the shores of Lake Victoria, giving inhabitants an opportunity for either fishing or farming livelihoods, depending on the landholdings and fishing rights, or availability of resources.

4.2 Research design and sampling

The research design was exploratory in nature and used a case study approach to investigate climate change impacts in the three selected villages of the Suba district. A quantitative research methodology was followed where the principal researcher [36] conducted face-to-face interviews using a semi-structured questionnaire as a measuring instrument.

A multi-stage sampling design was employed where, firstly, the three pre-selected villages of Rusinga Island, Mfangano Island, and Mbita point were purposefully sampled based on the already known prevalence of livelihood activities of farming and fishing in the area. Databases from the Kenya National Bureau of Statistics containing a list with the names and addresses of all 2640 households in the three villages served as the sampling frame. In each village a total of 30 households were systematically sampled, bringing the total sample size to 90 households. This relatively small sample size resulted in a confidence interval (error level) of 10.2% at a 95% confidence level, which means that caution should be taken when extrapolating the findings to the rest of the target population. In all cases, personal interviews were conducted with the heads of households allowing the latter to seek consensus responses from their families.

4.3 Measuring instrument, data collection, and data analysis

The Sustainable Livelihoods Framework was used to inform the construction of the questionnaire. More precisely, the questionnaire was divided into segments that comprised of—i) household demographics (assessment of the human capital), which included questions such as: *How many members are in your household? As a family, what do you practice as a primary source of livelihood?* ii) Livelihoods, assets or ownership of properties including livestock, production and diversity of produce/yield levels, accessibility to safe drinking water, current fishing or farming constraints, awareness or perceptions on climate change, observed and expected climate change phenomena, climate change impacts, and coping strategies. Some of the questions asked to respondents were: *What assets do you possess? What major crops and livestock do you produce? Have you experienced any changes in temperatures in the last few years? How do you cope with these impacts of climate change?*

The questionnaire was initially drafted in English and then translated into the dominant local languages of Kiswahili and Kijaluo. All selected respondents were informed beforehand about the research to gain oral consent for their participation in the study. The principal researcher also approached community leaders or gatekeepers of the communities and explained the study to them and obtained formal permission for entering the communities. Apart from the principal researcher, an additional three enumerators of Kenyan origin were employed to aid in data collection. Data quality checks were done close to the data sources as well as during data entry. Because of the statistical limitations of the sample size as explained earlier, data analysis was restricted to the use of descriptive statistics and no inferential tests were run on the data.

While a strength of this study was our ability to draw on data from three villages that were interlinked, a limitation was that the sample size was limited, and further research, preferably a longitudinal study, could be done to explore climate variability and the perceptions of farmers and fishermen at different intervals over a longer period of time, instead of a cross-sectional survey. The study results were based on the respondents' lived experiences and personal observations and as such might contain an element of subjectivity. Any bias in this study has nevertheless been minimized by asking respondents to make comparisons with historical trends benchmarked against previous farming or fishing seasons. Some of the responses, especially those pertaining to trends in temperature change and rainfall patterns, have also been compared with long-term official data and findings of other studies on climate trends in rainfall and temperature in Kenya.

5. Findings and discussion

5.1 Demographic profile of respondents

The survey results showed that the level of education among respondents (household heads) was relatively low, as almost 36% of the respondents had no formal education at all, while only 11% completed secondary education. Subsequently, it came as no surprise that only 41.8% of the respondents – dropping to as low as 23.3% in the case of Mfangano Island – had any awareness of climate change. The low level of formal education clearly impacted negatively on the perceptions and general knowledge of the respondents about climate change. More importantly, as pointed out in previous studies [1, 5, 10], such a lack of knowledge of climate change may affect the adaptation capabilities of the respondents through a lack of preparedness. This is supported by other studies that postulated the correlation between the level of literacy and the adaptive capacity of communities to climate change [21, 37, 38].

The most important asset for rural livelihoods in the study area was natural capital, namely access to land, livestock, and the fishing opportunities of Lake Victoria. The natural capital of the study area was supplemented by physical capital in the form of human-made necessities, such as shelter, boats, and fishing nets. Almost two in every three households (63.6%) had access to land and most owned that land privately. Approximately 33% of the respondents had farming as a source of livelihood, 41% solely relied on fishing, and 26% relied on both farming and fishing for survival. A lack of diversification into other income-generating opportunities made the community more vulnerable to droughts or floods that destroyed their crop production or reduced their fish catches – a situation that was no doubt aggravated by the low level of human capital. This further resulted in over-reliance on food aid during climate disaster periods.

5.2 Farmers' and fishermen's perceptions of climate change

Respondents in all three villages confirmed that they observed changes in climate over the five years preceding the interviews. The population might not be aware of the specific detail pertaining to climate change phenomena, but based on their traditional knowledge, they had noticed changes in climate patterns. Almost 85% of farmers and fishermen in the study area confirmed that they had experienced a significant increase in temperature, 43.3% experienced a decrease in rainfall, 83.3% reported changes in either the start or end of the rainy season, 30% reported an increase in floods, while 11.7% observed a decrease in the water of Lake Victoria.

These observations broadly correlate with official records of changing weather patterns in Kenya since the 1990s [6]. A further comparison of the respondents' observations with an analysis of climate changes in Kenya for the period 1977–2014 confirmed a decrease in rainfall as well as a rise in temperature for this period [39]. Moreover, from 1960 to 2014 the average temperature has increased in all 21 arid and semi-arid regions for which the trends were analyzed [39]. In the same study, most pastoralists in the arid and semi-arid regions of Kenya confirmed that they experienced much lower rainfall alongside a high frequency of unpredictable rains and rising temperatures during this period. These trends are confirmed by The World Bank Group [40] who also found that Kenya has experienced rising temperatures over large parts of the country since the early 1960s. Since 1960, the annual mean temperature increase has been put at 1°C, while the average rate per decade has been estimated at 0.21°C. Further confirmation of rising temperatures was found in another study that tested for variability and trends in temperature in Kenya, Ethiopia, and Tanzania for the period 1979–2010 [41]. The findings of that study pointed to consistent increases in extreme temperatures in these countries for the four decades under analysis – a trend that correlates with increases in global mean temperature. Most of the extreme temperatures that occurred since 2000 were also higher than the mean temperature for the long-term period 1979–2010 [41]. At the same, however, no statistically significant correlations were observed for trends in rainfall since 1960 [40]. A study of trends and variability in precipitation in East Africa (Kenya, Tanzania, and Ethiopia), too could not find any general pattern in rainfall for the period 1981–2016, as rainfall indices showed both increasing and decreasing trends in all three countries during the period under study [41].

Excessively heavy downpours caused floods and in some instances crop failure and hazardous conditions for fishermen in Lake Victoria's open waters. From these findings, it is evident that climate variability is increasing in the Suba district. Farmer and fishermen respondents in all the villages perceived these changes in climate patterns as natural phenomena, rather than seeking the causes in anthropogenic factors. This lack of an informed, holistic understanding of the interface between the causes of climate change and the imminent threats that it poses to their livelihoods – a function of the poor human capital in the Suba district – inevitably hampers the ability of the respondents to adjust in a more strategic and coherent way to their environment and its changing ecosystems.

5.3 Impacts of climate change on farmers and fishermen

Extreme climatic conditions, combined with other external factors, significantly contributed to low productivity for farmers and fishermen. The survey results indicated an increase in food insecurity resulting from reduced crop yields, low fish catches, reduced livestock, and a decrease in general food availability in the study area. Almost half of the respondents (49.4%) mentioned sensitivity – a change in crop yields in response to a change in the mean and variability of temperature, 73.7% observed increased livestock mortality and 80.6% reported reduced food availability. Famers stated that reduced crop yields resulted in high poverty levels as their sources of livelihood were now vulnerable to climate disasters.

Fishermen in all three villages indicated a slight decrease in fish catches, resulting in increased food insecurity. A substantial proportion of fishermen in all three villages (59.4%) further observed reduced species diversity in their catches. Several other factors that were fuelled by climate variability, such as the increased risk of fishing in Lake Victoria due to violent storms and floods (62.3% of respondents), had significantly reduced fish catches. As a result, one-third of all the fishermen respondents (33.1%) reported a decline in their households' annual net income.

5.4 Reduced human health

Adverse effects of climate change have resulted in an increase in communicable diseases that impacts human health. The target population mainly depended on unprotected water sources for domestic uses, thereby resulting in a surge of diarrheal diseases, cholera, and typhoid. Even though it was not directly investigated in the sampled population, such a lack of secure water sources may result in the surge of water-borne diseases in incidences of high rainfall and flooding.

A decrease in the availability of water for domestic purposes means that women had to travel longer distances to community boreholes – on average two to three kilometers per day – to fetch water. Few households had access to protected water sources and thereby resorted to unclean water sources. These findings are consistent with the prediction that water stress is estimated to increase due to climate change factors [42]. Such unprotected and unclean water was sourced from rivers, dams, ponds canals, and wells that are all liable to contamination by the disposal of both domestic and any other environmental waste. Water-borne diseases, such as cholera and dysentery, increase where there are no clean protected water sources [1, 5, 8, 43]. The respondents in the study area relied on using chlorine pills and boiling in some instances to treat their water for drinking. However, such methods were not always available to all as there were periods when communities struggle to access the chlorine pills.

5.5 Climate change adaptation and coping strategies of the respondents

Social capital in the form of informal networks within the three villages of Mbita point, Mfangano Island, and Rusinga Island were central to the everyday survival of the communities. Bonded ties and reciprocal relationships between friends and families enable the exchange of services and goods. The findings suggest that these informal institutions tend to be exclusive and are defined by kinship and neighborhoods. Such reciprocal links include the sharing of capital assets, information, cash loans, emotional support, food, and labor. For example, fishermen on Rusinga Island form groups for collective fishing using their physical capital in the form of cooperative assets, such as boats and nets. Maintaining these links with friends and families in the community offers households the opportunity to adapt or recover from climate disasters and helps them to buffer shocks.

Local communities also learned to adapt to the naturally changing environment by using their local and traditional knowledge to recognize and cope with these changes. It has been proven in several studies that indigenous knowledge of the environment helps local communities to respond actively to the challenges posed by climate variability [44–46]. The traditional climate prediction practices of using animal behavior patterns, moon characteristics, and tree phenology are still practiced in the study area. Even though the efficiency of some of the methods is debatable, these traditional practices have enabled the community to cope with climate variability to a certain extent. The use of indigenous knowledge and traditional warning systems to monitor weather changes is, however, being compromised as climate change is bringing more extreme and unpredictable weather patterns. Low mastery of this traditional knowledge by the younger community members further erodes its value in the community.

Respondents from fishing households tend to use traditional curing methods of preservation to enable their fish catches to last longer. This is because modern preservation techniques, such as ice, are expensive and inaccessible to them. Fishermen tend to smoke, oil fry, or salt-dry their fish for it to last longer in their food storage. These traditional methods of preservation can be effective in preserving fish catches for a certain period, thereby providing food and a source of income through market trade-in disaster periods [47].

Survey results also indicated that most respondents (71.3% of all households) relied heavily on food aid distributed by humanitarian aid agencies as a means of coping with droughts and floods. Monthly government handouts of maize, rice, and cooking oil also had buffered the population from hunger during droughts and floods. Furthermore, to address famine in the community, the Kenyan government had resorted to hand out cash for people who had been severely affected by droughts and floods [48].

Other coping mechanisms included borrowing money from other households, borrowing food on credit, and curtailing expenses on other things that were deemed as not basic, such as clothing and luxuries. For the farming households, the integrated crop-livestock system enabled the people to minimize the effects of droughts. In bad seasons, livestock was sometimes sold to cushion families against droughts and floods. Farmers also employed conservation farming practices, such as rainwater harvesting and crop diversification. Other, more resilient, crops such as sorghum, cowpeas, and groundnuts had been grown to buffer against climate shocks. Under periods of moderate drought stress, all these practices have helped to maintain crop productivity.

6. Conclusions and recommendations

The findings show that the Suba district has suffered significant changes in climate patterns in recent times. Some of the experiences and observations of the survey respondents are confirmed by analyses of long-term data for climate change in Kenya, such as those pointing at an incremental change in the mean temperature since the 1960s [39–41]. The survey results also suggested that many of the respondents did not attribute the climate change phenomena to anthropogenic activities, but rather to natural causes only, confirming that education levels do influence perceptions on climate change as suggested in other studies [21, 37, 38]. The research also showcased a relatively wide utilization of traditional knowledge in coping strategies. Indigenous knowledge in predicting seasonal weather and rainfall patterns, preserving grains for planting purposes, and various traditional farming support systems can be explored and adopted to lessen the impacts of climate change on their agricultural activities. Conversely, robust modern technologies for forecasting weather patterns remain under-utilized in this area. Merging local knowledge with modern science in Africa could, therefore, help in developing agronomical knowledge among farmers in climate change coping and adaptation.

The data confirmed that food availability in the Suba district had reduced due to climate change. Crop failure and low agricultural produce were greatly attributed to unreliable rainy seasons, droughts, and floods among other climate-induced factors, such as the increase of crop pests and diseases. Overall, low food production means low income, low self-sufficiency, and low levels of well-being in the household, thereby threatening human security. From the findings, it is further evidenced that climate change has proportionately caused more problems for the poor households in the Suba district due to their high levels of poverty, poor access to health care facilities, and their limited capacity to adapt to climate change. Low levels of human and physical capital, as put forward by the sustainable livelihoods approach, therefore, have a detrimental impact on households' resilience and ability to cope with the changes brought by climate change.

The formation of network groups by the three villages created viable opportunities for intervention strategies that supported livelihood stability and climate change adaptation in a community participatory approach. Development practitioners in the three villages should, therefore, be careful not to undermine the traditional safety nets present, such as informal coping networks and traditional leadership when introducing or considering new community-based projects. Maintaining these networks of social capital has brought an element of flexibility to livelihood coping strategies during droughts and floods.

Institutional support from both government and donor aid agencies to mitigate the impact of climate change on the target population mostly came in the form of emergency support. This support was only a reaction, and not anticipatory, making the rural population continually dependent in a way that may not be sustainable in the long term. Climate services that should be considered by institutions include meteorological forecasts at finer (local) scales to help farmers adjust their planting and marketing strategies in line with reliable forecasts. Fishermen may also benefit from such meteorological forecasts, where they can avoid fishing in erratic thunderstorms. In addition, such reliable meteorological services will help disaster managers to be better prepared. The introduction of schemes to increase borehole ratio per radius in communities should also be considered. This will help communities to avoid the long-distance travel to access clean and safe water, which is often more than 3 km away, except where communities benefit from nearby institutions, such as schools. Sustainable technologies, such as rainwater harvesting, can also be introduced to store rainwater in tanks for the domestic and agricultural needs of the communities.

The introduction of incentives and compensation for environmental services can support coping strategies implemented by the communities in the study area. Communities might help in reforestation by planting trees and grasses in designated areas to maintain abundant grazing lands for livestock. Such initiatives can be combined with other mitigation efforts, such as rehabilitation of degraded landscapes and general landscape management. This will, however, require substantial funding from local and international institutions, together with dedicated leadership to manage the program sustainably.

Lastly, exploring social and predictive adaptive behaviors among vulnerable communities and identifying them through formative research can be helpful in designing successful programs that support or contribute to climate change adaptation.

Acknowledgements

The authors would like to express their sincere gratitude to the following agencies and people who, in one way or another, assisted with this project: The Kenyan National Bureau of Statistics, the Ministry of Fisheries of the Suba district in Kenya, the communities, gatekeepers (chiefs and elders) and local government councils of Mbita, Mfangano Island, and Rusinga Island for their willingness to cooperate in the study.

Author details

André J. Pelser[1*] and Rujeko Samanthia Chimukuche[2]

1 Department of Sociology, University of the Free State, Bloemfontein, South Africa

2 Africa Health Research Institute, Durban, South Africa

*Address all correspondence to: pelseraj@ufs.ac.za

References

[1] Intergovernmental Panel on Climate Change. Climate Change 2021: The Physical Science Basis - Summary for Policymakers [Internet]. 2021. Available from https://www.ipcc.ch/report/ar6/wg1/downloads/report/IPCC_AR6_WGI_SPM_final.pdf. [Accessed: March 2, 2022]

[2] Kogo BK, Kumar L, Koech R. Climate change and variability in Kenya: A review of impacts on agriculture and food security. Environment, Development and Sustainability. 2021;**23**:23-43. DOI: 10.1007/s10668-020-00589-1

[3] Nyiwul L. Climate change adaptation and inequality in Africa: Case of water, energy and food insecurity. Journal of Cleaner Production. 2021;**278**:123393. DOI: 10.1016/j.jclepro.2020.123393

[4] Ongoma V. Insights for African countries from the latest climate change projections. The Conversation. 2021. Available from: https://theconversation.com/insights-for-african-countries-from-the-latest-climate-change-projections-165944. [Accessed: March 08, 2022]

[5] Intergovernmental Panel on Climate Change. Climate Change 2022: Impacts, Adaptation and Vulnerability [Internet]. 2022. Available from: https://www.ipcc.ch/report/ar6/wg2/. [Accessed: March 8, 2022]

[6] Rahut DB, Aryal JP, Marenya P. Understanding climate-risk coping strategies among farm households: Evidence from five countries in eastern and southern Africa. The Science of the Total Environment. 2021;**769**:145236. DOI: 10.1016/j.scitotenv.2021.145236

[7] United Nations Department of Economic and Social Affairs. Sustainable Development: The 17 goals [Internet]. 2015. Available from: https://sdgs.un.org/goals. [Accessed: March 7, 2022]

[8] Intergovernmental Panel on Climate Change. Climate change 2007: Impacts, Adaptation and Vulnerability [Internet]. 2007. Available from: https://www.ipcc.ch/site/assets/uploads/2018/03/ar4_wg2_full_report.pdf. [Accessed: March 3, 2022]

[9] Oluoko-Odingo A. Determinants of poverty: Lessons from Kenya. Geochemical Journal. 2009;**74**(4):311-331. DOI: 10.1007/s10708-008-9238-5

[10] Collier P, Conway G, Venables T. Climate change and Africa. Oxford Review of Economic Policy. 2008;**24**(2):337-353. Available from: https://citeseerx.ist.psu.edu/viewdoc/download?doi=10.1.1.523.8782&rep=rep1&type=pdf

[11] Mubaya CP. Farmer Strategies towards Climate Variability and Change in Zimbabwe and Zambia (Thesis). Bloemfontein: University of the Free State; 2010

[12] Ringler C. Climate change and hunger: Africa's smallholder farmers struggle to adapt. EuroChoices. 2010;**9**(3):16-21. DOI: 10.1111/j.1746-692X.2010.00175.x

[13] Ramin BM, McMichael AJ. Climate change and health in sub-Saharan Africa: A case-based perspective. EcoHealth. 2009;**6**(1):52-57. DOI: 10.1007/s10393-009-0222-4

[14] Climatelinks. Kenya: Climate change fact sheet [Internet]. 2021. Available from: https://www.climatelinks.org/sites/default/files/asset/document/2021-09/USAID_Kenya_Climate_Change_Country_Fact_Sheet.pdf. [Accessed: March 14, 2022]

[15] Herrero M, Ringler C, Van de Steeg J, Thornton P, Zhu T, Bryan E, Omolo A, Koo J, Notenbaert A. Climate Variability and Climate Change: Impacts on Kenyan Agriculture [Internet]. Nairobi: International Livestock Research Institute; 2010. Available from: https://cgspace.cgiar.org/bitstream/handle/10568/3840/climateVariability.pdf. [Accessed: March 27, 2022]

[16] Ogello EO, Munguti JM. Aquaculture: A promising solution for food insecurity, poverty and malnutrition in Kenya. African Journal of Food Agriculture Nutrition and Development. 2016;**16**(4):11331-11350. DOI: 10.18697/ajfand.76.15900

[17] Bhalla N. As Climate Change Threatens Kenyan Tea, Millions of Workers Seen at Risk [Internet]. Nairobi: The Thomson Reuters Foundation; 2021. Available from: https://www.reuters.com/article/us-climate-change-kenya-tea-idUSKBN2CR1Q6. [Accessed: March 10, 2022]

[18] Government of Kenya. National Climate Change Response Strategy: Executive Brief [Internet]. 2010. Available from: https://cdkn.org/sites/default/files/files/National_Climate_Change_Response_Strategy_Executive_Brief.pdf. [Accessed: March 20, 2022]

[19] Lim B, Spanger-Siegfried E, Burton I, Malone E, Huq S. Adaptation Policy Frameworks for Climate Change: Developing Strategies, Policies and Measures. Cambridge: United Nations Development Programme; 2004. p. 263. Available from: https://www.preventionweb.net/files/7995_APF.pdf

[20] Wenndt AJ. An Analysis of Climate Change and its Impacts on Small-Holder Push Pull Farmers in Western Kenya [Internet]. Mbita Point: World Food Prize Foundation; 2011. Available from: http://www.push-pull.net/Anthony.pdf. [Accessed: March 10, 2022]

[21] Mburu BK, Kung'u JB, Muriuki JN. Climate change adaptation strategies by small-scale farmers in Yatta District, Kenya. African Journal of Environmental Science and Technology. 2015;**9**(9):712-722. DOI: 10.5897/AJEST2015.1926

[22] Huho JM, Ngaira JKW, Ogindo HO, Masayi N. The changing rainfall patterns and the associated impacts on subsistence agriculture in Laikipia East District, Kenya. Journal of Geography and Regional Planning. 2012;**5**(7):198-206. DOI: 10.5897/JGRP12.018

[23] Nkonki-Mandleni B, Anim FDK. Climate change and adaptation of small-scale cattle and sheep farmers. African Journal of Agricultural Research. 2011;7(17):2639-2646. DOI: 10.5897/AJAR10.747

[24] Kasulo V, Chikagwa- Malunga S, Chagunda MGG, Roberts DJ. The perceived impact of climate change and variability on smallholder dairy production in northern Malawi. African Journal of Agricultural Research. 2012;7(34):4830-4837. DOI: 10.5897/AJAR12.259

[25] Serrat O. The Sustainable Livelihoods Approach. In: Knowledge Solutions. Singapore: Springer; 2017. p. 1140. DOI: 10.1007/978-981-10-0983-9

[26] Ribot J. Vulnerability does not fall from the sky: Toward multiscale, pro-poor climate policy. In: Mearns R, Norton A, editors. Social Dimensions of Climate Change: Equity and Vulnerability in a Warming World. Washington: The World Bank; 2010. p. 348. DOI: 10.1596/978-0-8213-7887-8

[27] Basar MA. Climate Change, Loss of Livelihood and the Absence of

Sustainable Livelihood Approach: A Case Study of Shymnagar, Bangladesh [Thesis]. Lund: Lund University; 2007

[28] International Institute for Sustainable Development (IISD). Livelihoods and climate change, 2003 [Internet]. 2003; Winnipeg: IISD. Available from: https://www.iisd.org/system/files/publications/natres_livelihoods_cc.pdf. [Accessed: March 10, 2002]

[29] Creative Commons. Location of Karungu and Mfangano and Rusinga Islands in western Kenya along the eastern shore of Lake Victoria [Internet]. 2022. Available from: https://www.researchgate.net/figure/Location-of-Karungu-and-Mfangano-and-Rusinga-Islands-in-western-Kenya-along-the-eastern_fig 1_321192315. [Accessed: March 17, 2022]. Licence Available from: https://creativecommons.org/licenses/by/4.0/legalcode

[30] Kaneko S, Mushinzimana EM, Karama M. Demographic surveillance system (DSS) in Suba District, Kenya. Tropical Medicine and Health. 2007;**35**(2):37-40. Available from: https://core.ac.uk/download/pdf/58750722.pdf

[31] The Borgen Project. The impact of Covid-19 on poverty in Kenya [Internet]. 2021. Available from: https://borgenproject.org/tag/poverty-in-kenya/ [Accessed: March 31, 2022]

[32] Kenya National Bureau of Statistics. Socio Economic data of Kanya, 2011 [Internet]. 2015. Available from: https://kenya.opendataforafrica.org/SEDK2015/socio-economic-data-of-kenya-2011?region=1000490-suba. [Accessed: March 31, 2022]

[33] Obiero KO, Raburu PO, Okeyo-Owuor JB, Raburu EA. Community perceptions on the impact of the recession of Lake Victoria water on Nyando wetlands. Science Research Essays. 2012;**7**(16):1647-1166. DOI: 10.5897/SRE11.324

[34] Karanja DMS. Health, diseases and nutrition in the Lake Victoria Basin. In: Environment for Development: An Ecosystems Assessment of Lake Victoria Basin. Nairobi: United Nations Environment Programme; 2006. Available from: http://hdl.handle.net/1834/7366

[35] Opere A, Ogallo LA. Natural disasters in Lake Victoria Basin (Kenya): Causes and impacts on environment and livelihoods. In: Environment for Development: An Ecosystems Assessment of Lake Victoria Basin. Nairobi: United Nations Environment Programme; 2006. Available from: http://hdl.handle.net/1834/7361

[36] Chimukuche RS. Challenges of Climate Change for Rural Community Livelihoods and Development: The Case of Farmers and Fishermen in Suba District of Kenya (Thesis). Bloemfontein: University of the Free State; 2014

[37] Conchrane L, Costolanski P. Climate change vulnerability and adaptability in an urban context: A case study of Addis Ababa, Ethiopia. International Journal of Sociology and Anthropology. 2013;**5**(6):192-204. DOI: 10.5897/IJSA2013.0459

[38] Mirza MMQ. Climate change and extreme weather events: Can developing countries adapt? Climate Policy. 2003;**3**(3):233-248. DOI: 10.1016/S1469-3062(03)00052-4

[39] Kenya Markets Trust. Contextualising Pathways to Resilience in Kenya's ASALs under the Big Four Agenda [Internet]. 2019. Available from: https://www.kenyamarkets.org/wp-content/

uploads/2019/10/Contextualising-Pathways-to-Resilience-in-Kenyas-ASALs-under-the-Big-Four-Agenda.pdf. [Accessed: April 13, 2022]

[40] The World Bank Group. Climate Change Knowledge Portal: Kenya [Internet]. Washington: The World Bank; 2021. Available from: https://climateknowledgeportal.worldbank.org/country/kenya/climate-data-historical. [Accessed: April 13, 2022]

[41] Gebrechorkos SH, Hülsmann S, Bernhofer C. Changes in temperature and precipitation extremes in Ethiopia, Kenya, and Tanzania. International Journal of Climatology. 2018;**39**(1): 18-30. DOI: 10.1002/joc.5777

[42] Meadows ME. Global change and southern Africa. Geographical Research. 2006;**44**(2):135-141. DOI: 10.1111/J.1745-5871.2006.00375.X

[43] Cairncross S, Valdmanis V. Water supply, sanitation and hygiene promotion. In: Jamison DT, Breman JG, Measham AR, Alleyne G, Claeson M, Evans DB, Jha P, Mills A, Musgrove P, editors. Disease Control Priorities in Developing Countries. 2nd ed. Washington: The World Bank; 2006. pp. 771-792. Available from: https://core.ac.uk/download/pdf/13105977.pdf

[44] Egeru A. The role of indigenous knowledge in climate change adaptation: A case study of the Teso sub-region, eastern Uganda. Indian Journal of Traditional Knowledge. 2012;**11**(2):217-224. Available from: https://www.researchgate.net/publication/280546104_Role_of_Indigenous_Knowledge_in_Climate_Change_Adaptation_A_case_study_of_the_Teso_Sub-Region_Eastern_Uganda

[45] Guthiga P, Newsham A. Meteorologists meeting rainmakers: Indigenous knowledge and climate policy process. IDS Bulletin. 2011;**42**(3):104-109. DOI: 10.1111/idsb.2011.42

[46] Osbahr HC, Twyman C, Adger WN, Thomas DSG. Evaluating successful livelihood adaptation to climate variability and change in southern Africa. Ecology and Society. 2010;**15**(2):20-27. DOI: 10.5751/ES-03388-150227

[47] Ofulla AVO, Onyuka JHO, Wagai S, Anyona D, Dida GO, Gichuki J. Comparison of different techniques for processing and preserving fish *Rastrineobola agentea* from Lake Victoria. World Academy of Science, Engineering and Technology. 2011;**60**:1643-1647. Available from: https://www.researchgate.net/profile/Gabriel-Dida/publication/225273972_Comparison_of_Different_Techniques_for_Processing_and_Preserving_fish_Rastrineobola_argentea_from_Lake_Victoria_Kenya/links/0c960515a62bfc9b99000000/Comparison-of-Different-Techniques-for-Processing-and-Preserving-fish-Rastrineobola-argentea-from-Lake-Victoria-Kenya.pdf

[48] Migiro K. Kenya Sees More Cash Handouts Ending Hunger for 3 Million [Internet]. Nairobi: The Thomson Reuters Foundation; 2013. Available from: https://news.trust.org/item/20130724135202-t47ib/. [Accessed: March 25, 2022]

Chapter 23

Impact of Biofertilizers on Plant Growth, Physiological and Quality Traits of Lettuce (*Lactuca sativa* L. var. *Longifolia*) Grown under Salinity Stress

Hayriye Yildiz Dasgan and Tugce Temtek

Abstract

This study aims to reveal the responses of biofertilizers to the detrimental effects of salt stress on lettuce cultivation. Presidential variety lettuce seeds belonging to Syngenta company were used as plant material. Microalgae *Chlorella vulgaris*, beneficial bacteria, and mycorrhizal fungi are used to reduce salt damage in lettuce plants grown under salt stress. The experiment was carried out on eight different applications; (1) control, (2) salt (50 to 75 mM NaCl), (3) micro microalgae, (4) microalgae + salt, (5) bacteria (6) bacteria + salt, (7) mycorrhiza, (8) mycorrhiza + salt. The biofertilizers decreased the salt's detrimental effects and increased the lettuce weight. Compared to salty conditions, microalgae + salt, mycorrhiza + salt, and bacteria + salt applications increased lettuce weight by 19.2, 21.3, and 20.08%, respectively. Biofertilizers increased pH, EC, total soluble solids, titratable acid, and total dry matter in lettuce leaves under salt stress. Biofertilizers had a stress-reducing effect under salinity and increased leaf osmotic potential, leaf water relative content, and leaf stomatal conductance. Microalgae *Chlorella vulgaris*, mycorrhiza, and beneficial bacteria are recommended as stress relievers when growing lettuce in saline agricultural soils or with saline irrigation water.

Keywords: microalgae, mycorrhiza, beneficial bacteria, stomatal conductance, lettuce weight

1. Introduction

When plants are exposed to adverse environmental conditions, such as nutrient deficiency, lack of water, low or high temperature, ultraviolet radiation, salinity, insufficient oxygen, heavy metal toxicity, diseases, and pests, their growth is adversely affected. This condition is called stress. Stress can last for a long time or be temporary for a short time. Agricultural productivity is decreasing due to the detrimental impacts of climate change. Therefore, in order to extend sustainable agriculture and to increase crop products for food in the world, it seems necessary to use the appropriate solutions to decline the negative effects of stresses on agricultural plants [1]. Salinity is one of the most important abiotic stress factors that adversely affects growth and development in plants, limiting yield and quality. The salinity-affected area is expected to reach about 50% of total agricultural land by 2050. Salinity stress generates various detrimental effects on plants' morphological, physiological, biochemical, molecular, and agronomic characteristics and decreases productivity. Reduced plant growth under salinity stress is due to decreased nutrients, hormonal imbalance, generation of reactive oxygen species (ROS), ionic toxicity, and osmotic stress [2].

In recent years, improvements in beneficial microorganisms have raised the tendency to use biofertilizers as valuable tools in sustainable agriculture. Biofertilizers have various benefits for plant growth. They regulate the soil texture and activate the soil biologically. It has been reported that many biofertilizers suppress plant pathogens and protect the plant against soil-borne diseases, so they are known as environmentally friendly. In terms of agricultural sustainability, biofertilizers do not harm the ecological system and do not contain harmful substances, they are proportionally cheaper when compared to commercial chemical fertilizers. Biofertilizers stimulate plant growth and produce phytohormones, thus increasing the yield and quality of the plant. In the fight against salinity, biofertilizer applications are widely preferred all over the world because they significantly increase salt tolerance [3].

One of the most effective alternatives among biofertilizer applications is mycorrhiza. Mycorrhizal fungi, which have the ability to establish a symbiotic relationship with plant roots, take carbohydrates that they cannot synthesize from the plant itself and contribute to the ability of plants to take in more water and nutrients by expanding their root domain thanks to their hyphae [4, 5]. It has been reported that the positive effect of mycorrhiza is not only to increase the intake of water and nutrients but also to increase the tolerance of plants to abiotic and biotic stress conditions [4, 6, 7]; mycorrhiza and beneficial bacteria have taken their place in the biofertilizer industry in recent years. The effectiveness of these fertilizers has positive effects on the nutrition of the plants by increasing the solubility of nutrients in the root area, with benefits, such as lowering the pH in the root zone, secretion of chelators, production of special ion carrier proteins [8–12]. While the solubility and availability of nutrients, such as phosphate, Fe, Zn, and Mn, increase by decreasing the pH in the root zone, some bacteria also fix the nitrogen to the soil from the air. It is reported that PGPR (plant growh promoting rhizobacter) bacteria that promote plant growth produce hormones, fix nitrogen in the air, and dissolve phosphate [13].

Chlorella vulgaris, one of the microalgae species with the highest biotechnological applicability, has been widely commercialized and is used as a food supplement for humans and as a feed additive for animals. These algae, a member of *Chlorophyta*, is seen as an alternative protein source due to their high protein content of 42–58% and has been cultivated for various purposes by many countries [14]. Instead of chemical fertilizers, which are generally used as a nitrogen source in agriculture, the use of *C. vulgaris* with high protein content will be a cheaper and environmentally friendly application. However, studies on the use of microalgae as biofertilizers, both in the world and in our country, are limited.

The origin of lettuce is accepted as Anatolia, Caucasus, and Turkestan regions. Some researchers stated that different forms of salad and lettuce are found in central Europe and southern Europe and the Canary Islands, some African countries, Mesopotamia, Kashmir, Nepal, and even Siberia [12]. Its Latin name "*Lactuca sativa* L. var. *longifolia*" was used in this study. It is also called Romain lettuce or Cos lettuce.

This lettuce is a species whose leaves are longer than wide, the leaves overlapping each other and often forming a loose and oval core.

The aim of this study is to determine the effect of microalgae, bacteria, and mycorrhiza biofertilizers on plant growth, yield, and plant nutrient content of lettuce grown under salt stress. It also revealed the effects of using less chemical fertilizers in lettuce cultivation, thus saving fertilizer and protecting the environment, as well as the yield and quality of the plant.

2. Material and method

2.1 Plant material and growing conditions and experimental design

The present study was conducted in a glasshouse between the spring and the summer of 2019 at the University of Cukurova, Adana, Turkiye (36°59'N, 35°18'E, 20 m above sea level). Lettuce seeds (*Lactuca sativa* L. var. *longifolia* "Presidental") were provided by Syngenta seed company. Seed sowing was done on 28 October 2019 and seedlings were planted on 13 December 2019 (**Figure 1**). The lettuce plant was grown in a two-liter capacity pot filled with cocopeat substrate and irrigated nutrient solution. Lettuce plants were grown with eight treatments and three replications. Randomized blocks experimental design, with 10 plants in each replication (30 plants per treatment) was used. The treatments are as follows:

1. Control

2. Salt

3. Microalgae

4. Microalgae + Salt

5. Bacteria

6. Bacteria + Salt

7. Mycorrhiza

8. Mycorrhiza + Salt

The pH and EC of the nutrient solution during the trial were measured daily. The pH was kept between 5.5 and 6.0, and the EC was fixed at 1.5 and 2.7 μS cm^{-1}. Two stock solutions for nutrition were used: stock A (Potassium nitrate, calcium nitrate, ammonium nitrate, and Fe-EDDHA) and stock B (potassium sulfate, mono-potassium sulfate, magnesium sulfate, microelements, zinc sulfate, boric acid, manganese sulfate, and ammonium molybdate). The nutrient solution consists of nitrogen (N) 212 ppm, phosphorus (P) 30 ppm, potassium (K) 305 ppm, calcium (Ca) 205 ppm, magnesium (Mg) 60 ppm, Iron (Fe) 3.0 ppm, manganese (Mn) 0.4, Boron (B) 0.40 ppm, Zinc (Zn) 0.50 ppm, copper (Cu) 0.05 ppm, and molybdenum (Mo) 0.07 ppm. The first salt application of 50 mM NaCl was made on February 28, 2020. The salt concentration was increased to 75 mM NaCl on 20 March 2020. Lettuce plants were harvested on April 6, 2020.

Figure 1.
Lettuce seedlings were transferred to pots containing cocopeat.

2.2 Biofertilizers

"*Chlorella vulgaris*" strain of microalgae was used in 2×10^7 mL^{-1} concentration. The inoculation was diluted 40 times before its use [16]. For 1 L nutrient solution, 25 mL of algae was added from the 2×10^7 mL^{-1} concentration. Three bacterial species (*Bacillus subtilis*, *Bacillus megaterium*, and *Pseudomonas fluorescens*) were obtained from NGB Company (Next Generation Biotechnology) with a trading name "Rhizofill." A mixture of 50 mL of the colony (1×10^9 mL^{-1}) was added every 10 days in a 50 L nutrient solution. The commercial mycorrhiza was obtained from ERS Company (Bioglobal). The mixture was composed of *Glomus intraradices*, *Glomus aggregatum*, *Glomus mosseae*, *Glomus clarum*, *Glomus monosporus*, *Glomus deserticola*, *Glomus brasilianum*, *Glomus etunicatum*, and *Gigaspora margarita* (1×10^4 w: w). During seed planting, 1000 spores' $plants^{-1}$ were used.

2.3 Leaf stomatal conductance

During the experiments, by using Delta T Devices brand AP4 model portable porometer, the gas passing from stomas in mature leaves was recorded [17].

3. Leaf osmotic potential

To determine the osmolality (c), 1 g of fresh weight from fully expanded leaves was homogenized in a mortar and mixed with distilled water to reach a final volume of 20 mL. After extraction using a millipore filter, the sap was utilized to determine the osmolality using a freezing point osmometer (Gonotec Osmomat 030, Germany). The osmotic potential was determined using the following formula according to the Van't Hoff equation [18]:

$$\psi s\left(MPa\right) = -c\left(mos\ mol\ kg^{-1}\right) \times 2.58 \times 10^{-3}.$$

3.1 Leaf relative water content

Fresh, turgor, and dry weights of the leaf samples were determined. The relative water content of the leaves was calculated with the following formula:

$$FW - DW\,/\,TW - DWx100$$

3.2 Lettuce weight, leaf area, and its number at harvest

The yield of lettuce is expressed as g plant^{-1}. At the same time, the number of leaves per plant was recorded. Afterward, the leaf area was determined by leaf area meter (Li-3100, LICOR, Lincoln, NE, USA) and indicated as cm^2 plant^{-1}.

3.3 Evaluation of dry matter and total soluble solids

Dry weight (DW) was obtained in a forced-air oven at 70°C until constant weight. Dry matter (DM) was measured by weighting fresh (FW) and dried lettuce material and expressed in percentage (DM = 100 × DW/FW). Besides, total soluble solution (TSS) was measured with a digital refractometer and was expressed in percentages.

3.4 Measurement of EC, pH of lettuce leaf

The electrical conductivity (EC) and lettuce leaf pH were determined.

3.5 Statistical analysis

Data were exposed to ANOVA test using SAS-JUMP/7. In addition, Fisher's LSD test was used to compare the averages at a 5% significance level.

4. Results and discussion

It was determined that lettuce weight was statistically affected by salt and biofertilizers. The lowest lettuce weight was obtained from salt (229 g) and then control (235 g) treatments (**Figures 2** and **3**). The heaviest lettuce was obtained from mycorrhiza (369 g) followed by microalgae (346 g). The biofertilizers decreased the salt effect and increased the lettuce weight. Compared to salty conditions, microalgae + salt, mycorrhiza + salt, and bacteria + salt applications increased lettuce weight by 19.2, 21.3, and 20.08%, respectively. Biofertilizers provided an increase in lettuce weight compared to control. Under without salt conditions, microalgae, bacteria, and mycorrhiza applications increased by 47.2, 30.6, and 57.2%, respectively (**Figure 3**). The fungal colonization, PGPR, and microalgae biofertilizers may stimulate the rate

Figure 2.
Images of lettuce plants with the following applications: control (a), bacteria + salt (b), mycorrhiza + salt (c), microalgae + salt (d) 16 days before harvest (20 March 2020).

of photosynthesis. Mycorrhiza may benefit plants by stimulating growth-regulating substances, increasing photosynthesis, improving osmotic adjustment under drought and salinity stresses, and increasing resistance to pests [19]. As cytokinin hormone

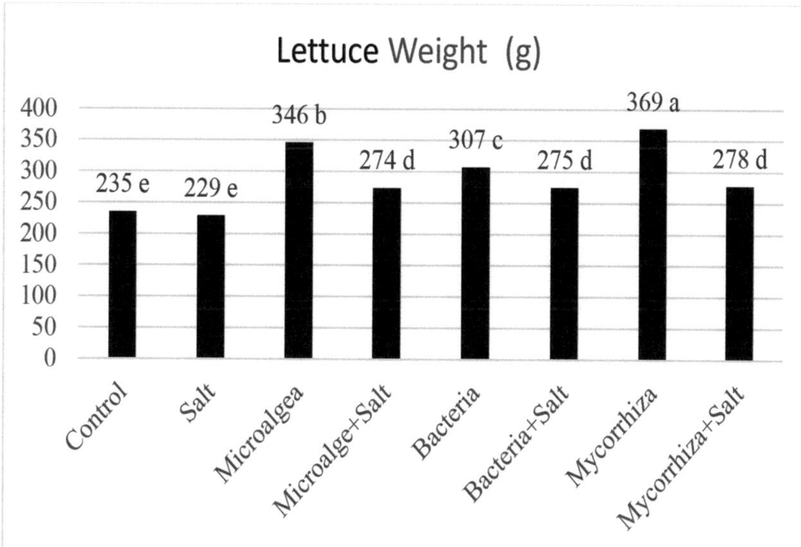

Figure 3.
Effects of the biofertilizers on lettuce weight (g plant^{-1}) under salt stress.

Treatments	Leaf number per plant	Dry matter (%)
Control	37.00 de	6.97 c
Salt	36.00 e	7.95 b
Microalgae	45.33 a	7.88 c
Microalge + salt	39.33 cd	8.16 b
Bacteria	42.00 bc	7.66 bc
Bacteria + salt	38.33 de	8.59 ab
Mycorrhiza	44.00 ab	5.19 d
Mycorrhiza + salt	37.00 de	9.42 a
P	0.0001	0.0001
$LSD_{0.05}$	2.768	0.926

LSD; minimum significant difference, mean followed by the same letter in each column are not significantly different according to LSD test (probability level of 0.05).

Table 1.
The effect of biofertilizers on lettuce leaf number and dry matter under salt stress.

production is a relatively common trait of PGPR and mycorrhizal fungi [20], cytokinin production may ameliorate salt stress. The cytokinins can enhance stomatal opening and photosynthesis. Stimulation of shoot biomass of lettuce plants grown in saline by the cytokinin-producing microorganisms implies considerable root-to-shoot cytokinin signaling [3].

Salt stress had a decreasing effect on the number of leaves in the lettuce plant. Compared to salt stress microalgae, bacteria and mycorrhiza applications increased the number of leaves by 9.3, 6.4, and 2.8%, respectively (**Table 1**). Under without salt conditions, the increasing effects of biofertilizers on leaf number were 26, 17, and 22% in microalgae, bacteria, and mycorrhiza, respectively.

The differences between the applications were found to be statistically significant for leaf dry matter. Compared to saline conditions, an increase in biofertilizer + salt applications has been achieved. Microalgae + salt, bacteria + salt, and mycorrhiza + salt increased the dry matter by 2.64, 8.05, and 18.4%. The higher plant dry matter accumulation with biofertilizers under salinity could be related to a higher source activity due to higher stomatal conductance and photosynthesis [21]. Beneficial microorganisms increase the production of cytokinins and they can enhance stomatal opening under salinity stress. Compared to control conditions, microalgae and bacteria applications increased dry matter production. Algal biofertilizer increased by 13.05% and bacterial biofertilizer increased by 9.8%. On the contrary, dry matter in lettuce leaves decreased in the mycorrhiza application may be due to faster growth.

The "L" represents brightness from the color parameters measured using a Hunter colorimeter (**Table 2**). The brightness values of lettuce plant increased in biofertilizer + salt applications compared to salty conditions. Increases of 23.01% were achieved in microalgae + salt application, 1.89% in bacteria + salt application, and 27.80% in mycorrhiza + salt application. When comparing control conditions and biofertilizer applications, increases in biofertilizer applications were determined. An increase of 58.53% was achieved in algae application, 49.43% in bacteria application, and 63.86% in the mycorrhizal application. Compared to saline conditions, an increase of 6.93% in microalgae + salt application, 7.40% in bacteria + salt application, and 9.82% in mycorrhiza + salt application was determined in "a" value. Kardüz et al. [22] found a significant effect of mycorrhiza application on the "a" value, which shows the green color of the lettuce leaves. Compared to the control, the "b" color value increased by 11.46, 1.83, and 8.85% in algae, bacteria, and mycorrhiza applications, respectively. According to saline conditions, an increase of 14.88, 23.75, and 11.34% was determined in microalgae + salt, bacteria + salt, and mycorrhiza + salt applications, respectively.

The highest EC value in lettuce leaves was 19.45 in bacteria + salt application and the lowest value was 8.32 in the microalgae application (**Figure 4**). Compared to saline conditions, 2.27 and 4.37% decreases in EC values were determined in microalgae + salt and mycorrhiza + salt applications, respectively. Compared to the control, decreases of 36.63, 33.96, and 12.03% were determined in the EC values of microalgae, bacteria, and mycorrhiza applications, respectively.

Lettuce leaves had the highest pH value of 5.95 in mycorrhiza and the lowest value of 5.81 in bacteria + salt application. Compared to the control, increases of 1.19 and 1.70% were recorded in the algae and mycorrhiza treatments, respectively. Compared to saline conditions, 1.35 and 0.68% decreases were recorded in microalgae + salt and mycorrhiza + salt applications, respectively (**Figure 4**). The salt stress decreased the TSS value. Lettuce leaves have the highest and lowest TSS of 5.60 and 2.73%, respectively, in control and mycorrhiza treatments.

TSS increases were recorded in biofertilizer + salt applications compared to a single biofertilizer application. Under salt stress, bacteria + salt was the biofertilizer application that showed the highest TSS value of 4.95% (**Figure 4**).

The highest acidity value in lettuce leaves was determined as 3.06% in mycorrhiza + salt application and 1.91% in microalgae application. Compared to saline conditions, 16.98, 40.56, and 44.33% acidity increases were recorded in microalgae + salt, bacteria + salt, and mycorrhiza + salt applications, respectively. Compared to the control, acidity decreases of 24.50, 47.82, and 2.37% were determined, respectively, in microalgae, bacteria, and mycorrhiza biofertilizers (**Figure 4**). TSS, titratable acidity,

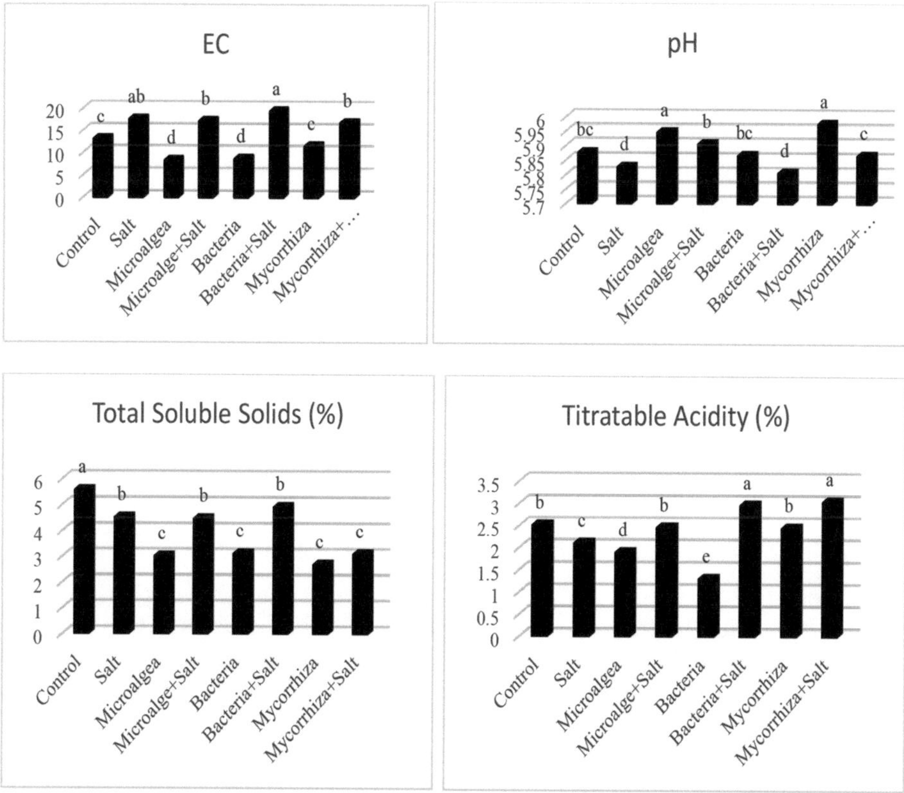

Figure 4.
The effect of biofertilizers on pH, EC, total soluble solids, and acidity of lettuce grown under salt stress.

Treatments	L	A	B
Control	32.93 e	−11.19 ab	39.87 de
Salt	37.94 d	−12.83 c	37.63 e
Microalgae	52.20 ab	−10.59 a	44.44 ab
Microalgae + salt	46.67 c	−11.94 b	43.23bc
Bacteria	49.21 a-c	−11.16 ab	40.60 cd
Bacteria + salt	38.66 d	−11.88 b	46.57 a
Mycorrhiza	53.96 a	−11.91 b	43.40 bc
Mycorrhiza + salt	48.49 bc	−11.57 b	41.90 b-d
P	0.0001	0.0021	0.0004
$LSD_{0.05}$	4.796	0.828	2.965

LSD; minimum significant difference, mean followed by the same letter in each column are not significantly different according to LSD test (probability level of 0.05)

Table 2.
The effect of biofertilizers on lettuce leaf color "L," "a," and "b" values under salt stress.

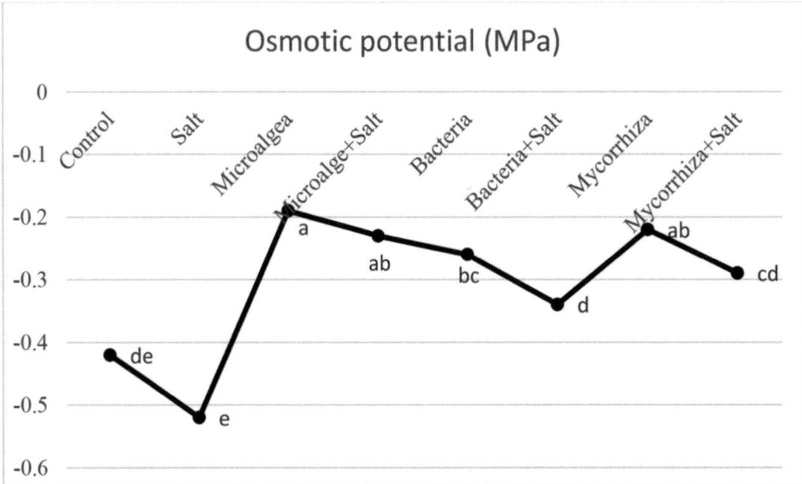

Figure 5.
Effects of biofertilizers on osmotic potential of lettuce leaf under salt stress and control conditions.

ascorbic acid, and dry matter were reported to be higher in the fruits of tomato plants inoculated with mycorrhiza than in those that were not inoculated [23].

The osmotic potential was found to be low in saline conditions (−0.52). Biofertilizer + salt combinations reduced this effect and increased the osmotic potential compared to saline conditions. The highest osmotic potential was obtained with −0.19 MPa in microalgae + salt application, followed by mycorrhiza with −0.22 Mpa and microalgae + salt applications with −0.23 Mpa. Biofertilizers had a stress-reducing effect under salinity and increased osmotic potential (**Figure 5**). Root colonization by AMFs can induce the production of the major groups of organic solutes and induce the accumulation of specific osmolytes, such as proline, soluble sugars, and amino acids [3].

Relative water content in the lettuce leaf was determined as the lowest in the salt application (68.5%) and the highest (85.6%) in mycorrhizal plants. Biofertilizers increased the relative water content and reduced stress in lettuce leaves under salt stress (**Figure 6**). Mycorrhizal plants generally show higher stomatal conductance and transpiration rates than non-mycorrhizal plants [24], even under salinity stress [25], this has been associated with improved leaf water status.

The lowest stomatal conductivity was determined in salt stress (115 mmolm^{-2}s^{-1}) and the highest (310 mmol mmolm^{-2}s^{-1}) in mycorrhizal biofertilizer alone. Biofertilizers increased the stomatal conductivity of lettuce leaves under salt stress and control conditions (**Figure 7**). Compared to saline conditions, salt + biofertilizer applications had a positive increasing effect on stomatal conductivity; microalgae + salt, bacteria + salt, and mycorrhiza + salt applications increased 63.8, 59.2, and 50.0%, respectively. Higher stomatal conductance and higher photosynthesis were reported under NaCl stress in pepper plants inoculated with PCPG [21]. Yao et al. [26] reported that PGPR prevented salinity-induced ABA accumulation in cotton seedlings. The ABA may mediate stomatal and photosynthetic responses to salinity stress [27], and the effects of plant-microorganism interactions on ABA status may enhance the growth of salinized plants.

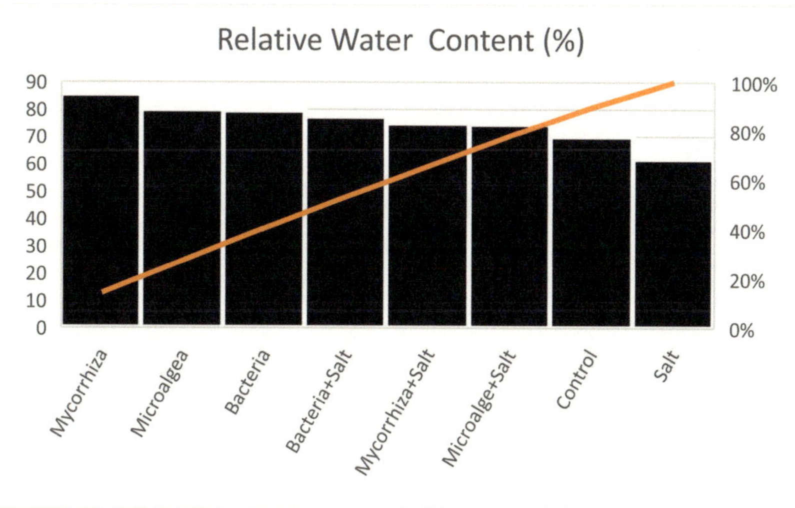

Figure 6.
Effects of biofertilizers on relative water content of lettuce leaf under salt stress and control conditions.

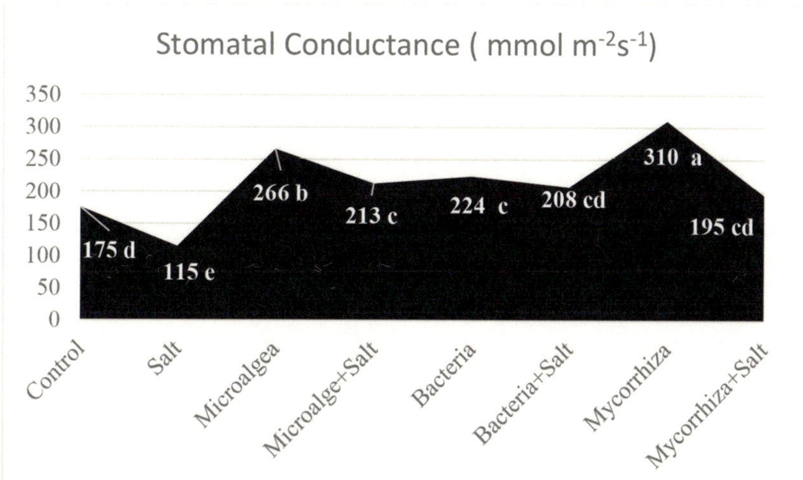

Figure 7.
Effects of biofertilizers on stomatal conductance of lettuce leaf under salt stress and control conditions.

5. Conclusion

The biofertilizers decreased the salt's detrimental effects and increased the lettuce weight. Compared to salty conditions, microalgae + salt, mycorrhiza + salt, and bacteria + salt applications increased lettuce weight by 19.2, 21.3, and 20.08%, respectively. Biofertilizers increased pH, EC, total soluble solids, titratable acid, and total dry matter in lettuce leaves under salt stress. Biofertilizers had a stress-reducing effect under salinity and increased leaf osmotic potential, leaf water relative content, and leaf stomatal conductance. Microalgae *Chlorella vulgaris*, mycorrhiza, and beneficial bacteria are recommended as stress relievers when growing lettuce in saline agricultural soils or with saline irrigation water.

Conflicts of interest

The authors declare that they have no conflicts of interest.

Thanks

The authors thank us for using the greenhouse and laboratory facilities of the Department of Horticulture, Faculty of Agriculture, Çukurova University.

Author details

Hayriye Yildiz Dasgan* and Tugce Temtek
Agricultural Faculty, Cukurova University, Department of Horticulture, Adana, Turkiye

*Address all correspondence to: dasgan@cu.edu.tr

References

[1] Etesami H, Noori F. Soil salinity as a challenge for sustainable agriculture and bacterial-mediated alleviation of salinity stress in crop plants. In: Kumar M, Etesami H, Kumar V, editors. Saline Soil-based Agriculture by Halotolerant Microorganisms. Singapore: Springer; 2019. DOI: 10.1007/978-981-13-8335-9_1

[2] Kumar A, Singh S, Gaurav AK, Srivastava S and Verma JP. Plant growth-promoting bacteria: Biological tools for the mitigation of salinity stress in plants. Frontiers in Microbiology. Frontiers Media S.A. 2020;**11**:1-15. DOI: 10.3389/fmicb.2020.01216

[3] Dodd IC, Pérez-Alfocea F. Microbial amelioration of crop salinity stress. Journal of Experimental Botany. 2012;**63**(9):3415-3428. DOI: 10.1093/jxb/ers033

[4] Altuntas O, Abak K, Dasgan HY. Serada biber yetiştiriciliğinde arbusküler mikorizal fungus kullanımının bitki gelişimi ve verime etkileri [The Effects of using arbuscular mycorrhizal fungi on the plant development and yield in pepper growth in greenhouses]. Selcuk Tarim Bilimleri Dergisi. 2015;**2**(2):144-151

[5] Ortas I, Kaya Z, Cakmak I. Influence of VA-mycorrhiza inoculation on growth of maize and green pepper plants in phosphorus and zincdeficient soils. In: Horst WJ et al., editors. Plant Nutrition-Food Security and Sustainability of Agro-ecosystems. Dordrecht: Kluwer Academic Publishers; 2001. pp. 632-633

[6] Carvalho LM, Correia PM, Martins-Louçao AM. Arbuscular mycorrhizal fungal propogules in a salt marsh. Mycorrhiza. 2004;**14**:165-170

[7] Ruiz-Lozano JM. Arbuscular mycorrhizal symbiosis and alleviation of osmotic stress. New perspectives for molecular studies. Mycorrhiza. 2003;**13**:307-317

[8] George E, Häussler K, Kothari SK, Li XL, Marschner H. Contribution of mycorrhizal hyphae to nutrient and water uptake of plants. In: Read DJ, Lewis DH, Fitter AH, Alexander IJ, editors. Mycorrhizas in Ecosystems. UK: CAB International Cambridge; 1992. pp. 42-48

[9] Marschner H. Zinc uptake from soils. In: Robson AD, editor. Zinc in Soils and Plants. Developments in Plant and Soil Sciences. Vol. 55. Dordrecht: Springer; 1993

[10] Dasgan HY, Kusvuran S, Ortas I. Responses of soilless grown tomato plants to arbuscular mycorrhizal fungal (Glomus fasciculatum) colonization in re-cycling and open systems. African Journal of Biotechnology. 2008;7(20): 3606-3613

[11] Dasgan HY, Aydoner G, Akyol M. Use of some microorganisms as bio-fertilizers in soilless grown squash for saving chemical nutrients. Acta Horticulture. 2012;**927**:155-162

[12] Dasgan HY, Cetinturk T, Altuntas Ö, The effects of biofertilisers on soilless organic grown greenhouse tomato. Acta Horticulture. 2017;**1164**:555-561

[13] Cakmakci R, Erat M, Erdogan UG, Donmez MF. The influence of PGPR on growth parameters, antioxidant and pentose phosphate oxidative cycle enzymes in wheat and spinach plants. Journal of Plant Nutrition and Soil Science. 2007;**170**:288-295

[14] Safi C, Zebib B, Merah O, Pontalier PY, Vaca-Garcia C.

Morphology, composition, production, processing and applications of Chlorella vulgaris: A review. Renewable and Sustainable Energy Reviews. 2014;**35**:265-278

[15] Esiyok D. Winter and Summer Vegetable Cultivation. Palme Publisher; 2012:404 s. Bornova/İzmir

[16] Ergun O, Dasgan HY, Isik O. Effects of microalgae Chlorella vulgaris on hydroponically grown lettuce. Acta Horticulturae. 2020;**1273**:169-176. DOI: 10.17660/ActaHortic.2020.1273.23

[17] Akhoundnejad Y, Dasgan HY. Physiological performance of some high temperature tolerant tomato genotypes. International Journal of Scientific and Technological Research. 2018;**4**(7):57-74

[18] Altuntas O, Dasgan HY, Akhoundnejad Y, Kutsal IK. Does silicon increase the tolerance of a sensitive pepper genotype to salt stress? Acta Scientiarum Polonorum, Hortorum Cultus. 2020;**19**(2):87-96. DOI: 10.24326/asphc.2020.2.9

[19] Al-Karaki GN. Nursery inoculation of tomato with arbuscular mycorrhizal fungi and subsequent performance under irrigation with saline water. Scientia Horticulturae. 2006;**109**:1-7

[20] Dodd IC, Zinovkina NY, Safronova VI, Belimov AA. Rhizobacterial mediation of plant hormone status. Annals of Applied Biology. 2010;**157**:361-379

[21] del Amor F, Cuadra-Crespo P. Plant growth-promoting bacteria as a tool to improve salinity tolerance in sweet pepper. Functional Plant Biology. 2012;**39**:82-90

[22] Kardüz Y, Tuzel Y, Oztekin GB. Mycorrhiza application in salad-lettuce cultivation in capillary system. Ege Univesity Journal. 2015;**52**(2):15159. DOI: 10.20289/euzfd.15563

[23] Kowalska I, Konieczny A, Gąstoł M. Effect of mycorrhiza and the phosphorus content in a nutrient solution on the yield and nutritional status of lettuce grown on various substrates. Journal of Elementology. 2015;**20**(3):631-642. DOI: 10.5601/jelem.2014.19.4.789

[24] Ruiz-Lozano JM, Alguacil MM, Bárzana G, Vernieri P, Aroca R. Exogenous ABA accentuates the differences in root hydraulic properties between mycorrhizal and non mycorrhizal maize plants through regulation of PIP aquaporins. Plant Molecular Biology. 2009;**70**:565-579

[25] Sheng M, Tang M, Chen H, Yang B, Zhang F, Huang Y. Influence of arbuscular mycorrhiza on photosynthesis and water status of maize leaves under salt stress. Mycorrhiza. 2008;**18**:287-296

[26] Yao LX, Wu ZS, Zheng YY, Kaleem I, Li C. Growth promotion and protection against salt stress by Pseudomonas putida Rs-198 on cotton. European Journal of Soil Biology. 2010;**46**:49-54

[27] Dodd IC. Hormonal interactions and stomatal responses. Journal of Plant Growth Regulation. 2013;**22**:32-46

Chapter 24

Potential Allelopathic Effect of Species of the Asteraceae Family and Its Use in Agriculture

Ana Daniela Lopes, Maria Graciela Iecher Faria Nunes, João Paulo Francisco and Eveline Henrique dos Santos

Abstract

Some species are capable of producing substances that affect seed germination, stimulating, or retarding this process, and can also suppress the development of other plants, acting as an antagonistic plant. This can occur naturally, through the release of exudates, or through the action of essential oil, extracts obtained from different parts of the plant, or plant residues with potential allelopathic action. The aim of this chapter is to present the main plant genera of the Asteraceae family with potential phytotoxic or allelopathic activity, with a suppressive effect on the growth of herbicide-tolerant weeds. The genus defined were *Acmella*, *Artemisia*, and *Bidens*, highlighting the form of use—plant extract, essential oil, or plant residues. The Asteraceae family is considered a repository of species to be explored for allelopathy with several associated secondary metabolites such as terpenes, saponins, alkaloids, alkamides, cinnamic acid derivatives, and flavonoids. In addition to these, for the genus *Bidens*, the presence of the acetylenic compound phenylheptatriine (PHT) is considered an important allelochemical with potent allelopathic action. The presence of this compound is associated with the cytotoxic activity of representatives of this genus, which can be a source of prospecting for new molecules to be used as bioherbicides.

Keywords: allelopathy, allelochemicals, bioherbicides, metabolites, weeds

1. Introduction

The term allelopathy has been referenced and conceptualized in different ways and from different perspectives. It can be understood as the ability of plants to interfere with other organisms in the environment [1]; as a process where chemical compounds are released into the environment by an organism and, once released into the environment, interact and can influence the growth and development of biological systems, including inhibition or stimulation effects [2, 3]. Older definitions, such as that of Hans Molisch in 1937, consider allelopathy as the direct or indirect result of the transfer of chemical substances from one plant to another [4]. Thus, in 1996, the International Allelopathy Society (IAS) defined allelopathy as the science that studies any process, essentially involving secondary metabolites produced by plants, algae, bacteria, and fungi, that influence the growth and development of agriculture and biological systems, including positive (stimulation) or negative (inhibitory) effects [5, 6]. It can be seen that, despite the different definitions, these refer, in short, to the central role of secondary metabolites in allelopathy [1], which are involved in defining the characteristics of natural ecosystems and agroecosystems [7].

The compounds identified with potential allelopathic activity are known as allelochemicals and, since their discovery, research has been carried out with the objective of isolating and identifying the substances responsible for this phenomenon and grouping them [8]. Allelochemicals can range from simple hydrocarbons to complex compounds of high molecular weight and can be classified into 10 categories according to their structures and properties: (1) water-soluble organic acids, straight chain alcohols, aliphatic aldehydes, and ketones; (2) simple lactones; (3) long-chain fatty acids and polyacetylenes; (4) quinones (benzoquinone, anthraquinone, and quinone complex); (5) phenolics; (6) cinnamic acid and its derivatives; (7) coumarins; (8) flavonoids; (9) tannins; and (10) steroids and terpenoids (sesquiterpene lactones, diterpenes, and triterpenoids) [9].

The performance of bioassays makes it possible to identify the phytotoxicity of different species, explained by the delay in seed germination, inhibition of plant growth, or any adverse effect caused by specific substances (phytotoxins) or growth conditions [10]. Unlike herbicides, allelochemicals act at low, but constant concentrations, over a long period of time. This fact, associated with the increase in global demand for organic products in the last two decades [11], makes it urgent to invest in studies on the use of allelochemicals as natural pesticides, in order to promote more sustainable agriculture, minimize the effects of pesticides on the environment and human health [12]. Among the benefits associated with the use of allelopathic compounds for the development of new agrochemicals, the fact that most of them are biodegradable and less polluting than traditional pesticides is highlighted due to their shorter half-lives [13].

Allelochemicals can be found in different parts of the plant, including flowers, leaves, stems, roots, or fruits of different species [2, 14]. Secondary metabolites present in medicinal and weed plants have been reported as potent growth inhibitory agents, indicating that such plants act as a depot for allelopathic compounds [15]. Many of these species are representatives of the Asteraceae family, which is composed of approximately 1000 genera, comprising more than 25,000 species of flowering plants [16], representing the largest family among flowering plants in the world, with distribution on all continents, except for Antarctica [17, 18]. This family includes food crops such as lettuce (*Lactuca sativa* L.), endive (*Cichorium endivia* L.), edible safflower seeds (*Carthamus tinctorius* L.), and sunflower (*Helianthus annuus* L.), species used in oil production (Enclyclopaedia Britannica 2015), medicinal species like *Achillea millefolium* L. [19], *Vernonia* spp. [20], and *Matricaria chamomilla* L. [21]; and species used in the bioremoval of pollutants in urban areas, such as metals and xenobiotics (*Solidago*, *Tanacetum*, and *Rudbeckia*) [22, 23]. Several secondary metabolites are present in the Asteraceae family such as terpenes [24], including sesquiterpene lactones [25, 26], saponins [27], alkaloids [28], alkamides [29], cinnamic acid derivatives, and flavonoids [30].

In this sense, this review intends to highlight, as a target for the search for natural alternatives in crop protection, genera of the Asteraceae family that grow spontaneously in different environments [31], some of which are capable of influencing the development of other species by allelopathy. Therefore, three genera of plants belonging to the Asteraceae family with recognized allelopathic activity were selected:

Acmella, *Artemisia*, and *Bidens*. *Artemisia* and *Bidens* along with *Ambrosia*, *Bellis*, *Helianthus*, and *Tagetes* are the main genera of the Asteraceae family with allelopathic or phytotoxic activity. Furthermore, the phytotoxic potential of *Acmella oleracea* was recently described, for the first time, against the weeds *Calopogonium mucunoides* and *Ipomoea purpurea* [8], confirming allelopathy of its extract.

A. oleracea (L.) R. K. Jansen (synonymies *Spilanthes oleracea* L., *S. oleracea* Jacq., and *Spilanthes Acmella* auct. Non (L.) Murr.) popularly known as "jambu" is a plant native to the regions of Asia and South America (especially in the northern region of Brazil), where it is widely used in regional cuisine [32–34]. It is commonly used in folk medicine with proven healing, antispasmodic, anti-inflammatory, antimalarial activity, in the treatment of rheumatism, as a tonic [35, 36], antioxidant, antinociceptive, anti-inflammatory, diuretic, and anesthetic [37]. It also has larvicidal [38], insecticidal [39], acaricide [40], and anthelmintic [41] effects. Acidic amino acids, triterpenes, stigmasterol, and alkaloids predominate in the phytochemical profile of *A. oleracea*, however, the biological activity seems to be related to the abundant presence of N-alkylamides, especially spilanthol [37, 42]. Studies on allelopathic activity and the metabolites involved, however, are still poorly explored, except for the work of [43] who evaluated the inhibition effect, phytotoxicity, and metabolites present in the aqueous methanolic extract of *A. oleracea* (L.) R. K. Jansen on the growth of *Lolium multiflorum* Lam. and *Echinochloa crus-galli* (L.) P. Beauv.

The genus *Artemisia* has more than 350 species and is considered a promising source of biologically active compounds with the potential to provide new herbicides and growth regulators. The different species that make up the genus have phytotoxic compounds for monocots, dicots, photosynthetic bacteria, and endomycorrhizal fungi [44], especially *Artemisia annua* L., which center of origin is Asia [45] but after domestication it began to be cultivated in different countries such as Austria, Brazil, Spain, the United States, France, Poland, and Romania [46]. A. *annua* stands out in the Asteraceae family both for the variety of natural products characterized (almost 600 in total, including around 50 amorphane and cadinane sesquiterpenes), and by the highly oxygenated nature of secondary metabolites of the terpenoid class [47]. There are several studies that report the allelopathic activity of artemisinin and its synthetic derivatives, which can act as inhibitors and promoters of complex signals in response to biotic and abiotic factors [48–52], from the aqueous extract, essential oil, or biomass. Plant of its own species (**Figure 1**).

Bidens pilosa L. is an annual plant native to tropical America and widely distributed in tropical and subtropical regions of the world. The genus has about 280 species and is widespread in both cultivated and uncultivated areas, being considered one of the most harmful weeds in agriculture, promoting crop losses in more than 40 countries [53]. *B. pilosa* shows rapid growth exhibiting allelopathic effect on various cultures [53–57]. It is also used as a medicinal plant, cover plant, and source of nectar for bees. Its roots, leaves, and seeds have antibacterial, antidysenteric, anti-inflammatory, antimalarial, antiseptic, anticancer, antipyretic, hepatoprotective, hypotensive, hypoglycemic, diuretic, and antidiabetic activity [58, 59].

Xuan and Khanh [60] systematized a literature review based on 218 literary sources reported over 40 years highlighting chemical constituents, nutraceuticals, ethnomedical, biological, and pharmacological uses, and the effects and toxicity of *B. pilosa*. In this survey, the authors reported that the main compounds (301 compounds) belong to the group of polyacetylenes, polyacetylene glycosides, flavonoids, flavone glycosides, aurones, chalcones, okanine glycosides, phenolic acids, terpenes, pheophytins, fatty acids, and phytosterols, the which were identified and isolated

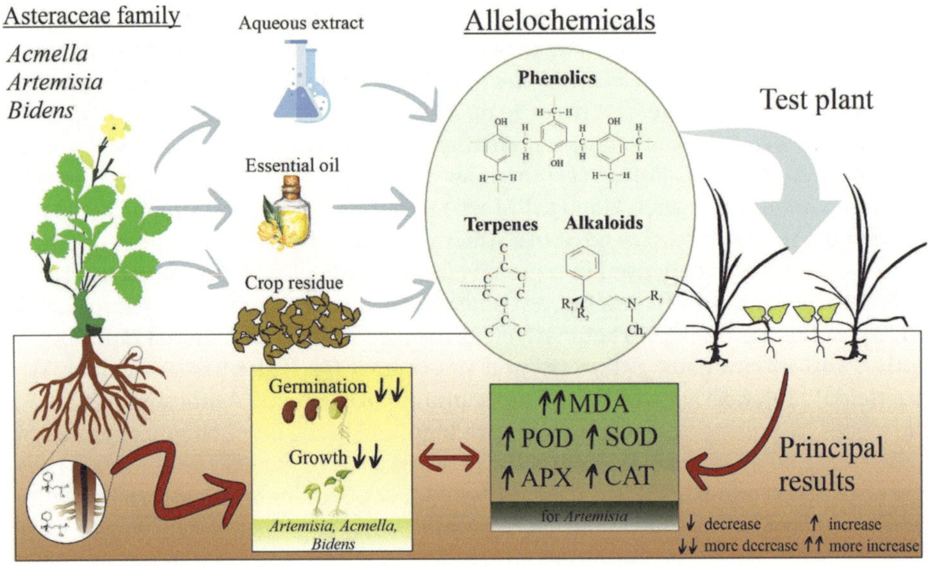

Figure 1.
Graphic summary referring to the main genera of plants of the Asteraceae family used how potential allelopathic in agriculture.

from different parts of this plant and considered as bioactive compounds potentially responsible for the pharmacological action, biological and allelopathic properties of the species, which will be described in more detail below.

It is important to highlight, however, that most of the articles that suggest the allelopathic effect of the crude extract of a plant species do so through bioassays, useful tools for previous studies on the allelopathic potential of a species or compound, however, not suitable for sufficient to relate the results obtained in vitro, in the laboratory, with field conditions [3, 61, 62]. For this reason, results from bioassays developed under natural conditions were included in this review [63, 64]. Under these conditions, the biosynthesis of allelochemicals and their release can be influenced by temperature, luminosity, humidity, interaction with soil biota, and nutrient availability [3, 65, 66]; in addition to the fact that plants are evaluated at the initial stage of development, considered the most sensitive stage to allelochemical activity [67, 68].

2. Genus in Asteraceae family with potential allelophatic

2.1 Genus: Acmella

2.1.1 Extract

Among the species belonging to the genus *Acmella*, *Acmella oleracea* (L.) R.K. Jansen stands out, popularly known as "jambú" [32, 69]. The allelopathic activity of this plant is still poorly explored. Kato-Noguchi [43] evaluated the activity of methanolic extracts of *A. oleracea*, from the whole dried plant, at concentrations of 3, 10, 30, 100, and 300 mg/ml. The biological activity was tested against *Lepidum sativum*, *Lactuca sativa*, *Lolium multiflorum*, and *Echinochloa crus-galli*, for this, 10 pre-germinated seeds were used, using Tween 20 as control. After 48 hours, the length of the roots and shoots of the seedlings were measured and the concentrations necessary to inhibit 50% of the growth (IC50) of the roots and shoots were determined. The *A. oleracea* extract showed dose-dependent action. At a concentration of 30 mg/ml the extract inhibited root growth in 4.7, 0, 21.3, and 57.1%, and for the shoot in 7.8, 0, 43.4, and 98.9% for *Lepidium sativum*, *L. sativa*, *L. multiflorum*, and *E. crus-galli*, respectively, when compared to the control.

When the IC50 values in the roots were analyzed, they were 5.9, 0.4, 25.9, and 7.8 mg/ml and in the shoot 4.3, 0.8, 126.0, and 15.1 mg/ml, for *L. sativum*, *L. sativa*, *L. multiflorum*, and *E. crus-galli*, respectively. The extracts were purified and two compounds were isolated, identified as (E,E)-2,4-undecadien-8,10-diinoic acid isobutylamide and nona-(2Z)-en-6,8-diinoic-2-phenylethylamide. The first compound inhibited the growth of roots and shoots of *L. sativum* and root of *E. crus-galli* at concentrations greater than 0.3 mM. The second compound inhibited the growth of *L. sativum* shoots and *E. crus-galli* roots at concentrations greater than 1 mM. Thus, these two compounds may be associated with the allelopathic activity of *A. oleracea* [43].

Araújo [8] analyzed the allelopathic activity of the hydroalcoholic extract, hexane, dichloromethane, and ethyl acetate fraction from the dry leaf of *A. oleracea*. The phytotoxic activity was tested on *L. sativa* using concentrations between 0.1 and 1 mg/ml of extracts and fractions. *Calopogonium mucunoides* was also used as a test species for the fractions at a concentration of 0.2 mg/ml. Ten seeds of *L. sativa* and five seeds of *C. mucunoides* were sown in Petri dishes containing a filter paper. Water and DMSO were used as negative controls and menadione as positive. Germination was recorded after 24 hours by root protrusion and root length was measured after 5 days. For *L. sativa*, both the hydroalcoholic extract and the fractions affected its germination, and the dichloromethane and ethyl acetate fractions were more harmful. At the concentration of 1 mg/ml, the inhibition was 76 and 60% of root growth for the dichloromethane and ethyl acetate fractions, respectively. The IC50 for the dichloromethane fraction was 0.48 mg/ml, which is similar to the results of Kato-Noguchi [43]. All fractions inhibited the germination of *C. mucunoides*, around 60%, at the concentration studied, and compared to lettuce seeds, the latter seems to be more resistant to *A. oleracea* allelochemicals. The dichloromethane fraction presented as major compounds: p-methoxy-cinnamic and 3,4-dimethoxy-cinnamic acids, 3,4-dimethoxy- and 3,4,5-trimethoxybenzoic acids, and palmitic acid. The presence of fatty acids and phenolic acids are related to the allelopathic activity presented by this fraction. The IC50 values found were 1.13, 0.94, 0.36, 0.37, and 0.19 mg/ml for hydroalcoholic extract, hexane, dichloromethane, ethyl acetate, and aqueous residue, respectively.

2.1.2 Vegetable residue

Suwitchayanon, Kunasakdakul and Kato-Noguchi [70] evaluated the allelopathic activity of 14 plants, including *Acmella* oleracea. *L. sativa* was used as a test plant. *A. oleracea* plant material was dried and ground and added to the Petri dish together with agar and finally, a new layer of agar was added. As a control, only agar was used. Each plate received five *L. sativa* seeds and incubated for 48 hours. All the plants studied showed a greater inhibitory effect on root development than on the hypocotyl. *A. oleracea* at a concentration of 50 mg dry plant weight inhibited radicle appearance in 71.8% and hypocotyl development in 47.4%. There is a need for further studies using this technique with different concentrations of the plant drug.

2.2 Genus: *artemisia*

2.2.1 Extract

Many researches have evaluated the allelopathic effects of aqueous extracts, or even volatile organic compounds, released by different parts of *Artemisia* plants [71–73]. Water stress conditions [74] and drought conditions [75] can intensify the allelopathic effect of the aqueous extract of these plants. Seeking answers to these statements, [71] developed an experiment of laboratory simulations of stress in relation to artemisinin production, obtained by the aqueous extract of *A. trifida* in environments with abiotic stress conditions. In view of the reported results, Guo [71] organized their experiment through tests of growth and development of rye subjected to four treatments, one characterized by the application of Hoagland's nutrient solution, a second with artemisinin solution at a concentration of 20 ml/l and, the others characterized by the combinations of the previous treatments, being one with cultivation of rye under the addition of extract of *A. trifida* at 5% (mass/volume) and Hoagland solution and, finally, cultivation with 20 mg/l of artemisinin solution together with *A trifida* extract. In Guo's research [71], abiotic stress was characterized by the submission of rye plants to freeze/thaw cycles and, throughout development, biomass accumulation, photosynthetic parameters, relative water content, malondialdehyde acid (MDA) and activity of superoxide dismutase, catalase, peroxidase, and ascorbate peroxidase enzymes. By applying the methodology and collecting the results, the authors concluded that the allelopathic effect of the *A. trifida* extract can be intensified by the abiotic stress studied, promoting inhibition of growth, biomass, photosynthesis, in addition to triggering MDA, and osmotic regulatory substances on rye seedlings.

Pastures are the food base for the herds that drive the world economy, but most of them are degraded by either anthropogenic actions or agricultural activities. Leaving this comfort zone for a widely accepted response of grassland degradation, Wang [72] went further and asked himself: is the allelopathic effect of *A. frigida* enough to explain, in part, the degradation of grasslands in northern China? And they added: does the allelopathic effect of *A. frigida* intensify as pasture degradation progresses? To answer the hypotheses raised in the study, the authors collected samples of *A. frigida*, randomly distributed in the experimental plots, from pastures with different grazing intensities for 180 days, considering such intensities as low, medium, and high. As a control, an area without animal grazing was kept next to the other areas. From the samples collected, Wang [72] cut 3 mm sections of the aerial part of *A. frigida* plants and obtained the aqueous extract by adding the plant tissues in an Erlenmeyer flask and placing them in an incubator at 20°C for 48 hours. After the time, the material was filtered through a sieve with a diameter of 0.5 mm. Concentrations of 0.25 g/ml (0.25 g of *A. frigida* aerial part per ml of distilled water) of the aqueous extract from each grazing intensity plot were kept at 4°C for a period of 10 days.

The allelopathic effect was evaluated in two different ways by Wang [72]. At first, the authors verified the effect of seed germination of *Melilotus officinalis*, *M. sativa*, and *A. splendens* (plants found in grazing areas) when submitted to aqueous extracts of *A. frigida*. In a second moment, the authors cultivated *A. frigida* plants in pots for 2 months to verify the allelopathic potential on their own species. In this experiment, the authors used activated carbon mixed with soil to absorb the allelochemicals released by mugwort plants. As a first result, the authors verified that 1,8-cineole and β-terpineol may be the main components contained in *A. frigida* induced by the grazing gradient. It is also important to report that the authors concluded that as grazing is intensified, the allelopathic effect of *A. frigida* increases significantly, since the concentration of secondary metabolites increases in the chemical composition of the extract. This variation in the concentration of chemical components verified by the authors in the aqueous extracts significantly inhibited seed germination and seedling growth of all grass species tested in the experiment.

Zhou [76] studied the phytotoxic effect of the aqueous extract of *Artemisia ordosica* leaves against two soil microalgae (*Chlorella vulgaris* and *Nostoc* sp.). The algae cells were grown in pyrex flasks with the addition of 0, 1, 5, 10, or 25 g/l of *A. ordosica* extract, with a light:dark cycle of 12 hours:12 hours. The cultivation without the addition of the extract served as a control. The concentration of chlorophyll a after 48 hours of incubation was also evaluated. The aqueous extract of *A. ordosica* showed 23 main compounds including alcohols (D-pinitol, diethylene glycol, 2,3-butanediol, inositol, glycerol, and tetriite), organic acids (glycolic acid, palmitic acid, octadecanoic acid, acid, and pentaric acid), phenolic derivatives (3,4-dihydroxyphenylglycol, 3-vinylcatechol, 1-(5-hydroxy-2-methoxyphenyl)-1,2-ethanediol, and 2-(2-hydroxyethyl)phenol), and sugars (inositol, D-glucopyranose, sucrose, and L-mannitol). The growth rates of *C. vulgaris* increased significantly at the concentration of 1 g/l and the chlorophyll a content increased with the time of culture under 5 g/l of the extract of *A. ordosica*. The highest concentrations (10 and 25 g/l) inhibited the growth of *C. vulgaris* and the concentration of chlorophyll a decreased. The growth rates of *Nostoc* sp. were not affected in the presence of 1 g/l of extract, and the other treatment concentrations inhibited the microalgae growth.

Luo [77] evaluated the allelopathic activity of aqueous extracts of leaves, stems, and roots of *Artemisia halodendron* against *Agriophyllum squarrosum*, *Setaria viridis*, *Artemisia scoparia*, *Lespedeza davurica*, *Chenopodium acuminatum*, and *Corispermum macrocarpum*. The extracts were used at concentrations of 10, 20, 40, 60, 80, and 100 g/l, and distilled water was used as a control. The analyzes were performed in Petri dishes adding the extracts and 30 seeds of the test plants. Afterward, the plates were incubated for 21 days with a photoperiod of 14 hours with light and 10 hours in the dark. At the end of the experiment, the seeds that showed growth of both the radicle and the bud were analyzed. The germination rate of seeds varied significantly between the six species under treatment and between the concentrations evaluated. The present study suggests a significant negative allelopathic effect of *A. halodendron* in other species. The extracts significantly reduced the germination of *A. squarrosum* seeds, and this suppression increased with increasing concentration, with the leaf extract being more effective when compared to the stems and roots. For *Centrosema macrocarpum* the extracts did not differ in their effects on seed germination.

Li [78] analyzed the biological activity of *Artemisia argyi* against *Brassica pekinensis*, *L. sativa*, and *Oryza sativa*. The extracts were prepared with dried leaves of *A. argyi* having water, 50% alcohol, and 100% alcohol as solvents. The concentrations of the extracts used were 50, 100, and 150 mg/ml. Twenty seeds of the tested species were added to Petri dishes with two layers of filter paper and then treated with extracts prepared from *A. argyi* at the different study concentrations. Ultrapure water was used as a control. The plates were grown at 25°C, 85% humidity, and a controlled cycle of 12 hours light/12 hours dark. The number of germinated seeds was counted from the second day after treatment and the counting lasted 1 week. The root length, stem length, and biomass of each treatment were also analyzed.

The three types of extracts were analyzed by UPLC-Q-TOF-MS, the aqueous extract presented as main components: caffeic acid, schaftoside, 4-caffeoylquinic acid, 5-caffeoylquinic acid, 3,5-dicapheoylquinic acid; 50% ethanol extract:

4,5-dicapheoylquinic acid, 3-caffeoylquinic acid, schaftoside, rutin, kaempferol 3-rutinoside, 3,4-dicapheoylquinic acid, 3,5-caffeoylquinic acid, 3-caffeoyl acid, 1-p-coumaroylquinic acid, 1,3,4-tri-caffeoylquinic acid, and eupatiline, and for the 100% ethanol extract the main metabolites were eupatiline, jaceosidine, and casticin. There was a concentration-dependent increase in inhibition of the extracts used. The aqueous extract exerted a significant inhibitory effect on germination and biomass production of the evaluated plants. The 50% ethanol extract showed inhibition on the germination and biomass index of *B. pekinensis* and *L. sativa*, but moderate inhibitory effects on *O. sativa*. The ethanol extract (100%) showed only inhibitory effects on the germination of *B. pekinensis* and *L. sativa* and no effect on *O. sativa*. However, it exerted inhibitory effects on the biomass of the three plants at high concentrations. Thus, it is possible to affirm that aqueous extract of *A. argyi* exhibited the strongest allelopathic effect. For example, *B. pekinensis* can be significantly inhibited by a low concentration of extract (50 mg/ml), showing the order of inhibition efficiency of: germination index > germination speed index > root length > germination rate > stem length > biomass. In summary, according to the comprehensive allelopathy index of the indicators studied, the order in which they were sensitive to the aqueous extract was *B. pekinensis* > *L. sativa* > *O. sativa*. *O. sativa* was selected as a test plant for RNA isolation and sequencing, transcriptome data and RT-qPCR verification showed that suppression of chlorophyll synthesis and photosynthesis was one of the main mechanisms of the inhibitory effect of *A. argyi* on test plants.

2.2.2 Essential oil

Researches with aqueous extract of *Artemisia* are broader and gain prominence in research, however, the essential oil of this plant can also present allelochemicals. Reports in the literature confirm that *Artemisia* essential oils can exert high phytotoxicity against weeds and other undesirable plants in a crop [79, 80] because they present terpenoids, especially monoterpenes and sesquiterpenes, which are the main components of the essential oil and are often responsible for its plant inhibitory activity [81].

Within this line of research, Yang [82] investigated the chemical composition of allelochemicals present in the essential oil of *Artemisia ordosica* and the effects that this could have on growth rates, photosynthetic activity, and oxidative damage in plant species present in biological soil crusts. In the work of Yang [83], the experimental units were composed of soil collected, superficially, on which a total of 0, 1, 3, 5, or 10 ml of *A. ordosica* essential oil were applied. These volumes were dissolved in 50 ml of dimethylsulfoxide. The essential oil in this study was obtained by hydrodistilling, for 3 hours, 100 g of air-dried leaves of *A. ordosica*, using the steam distillation method as described in [84]. The authors evaluated possible alterations in the plants present in the biological crust of the soil by means of analyses, carried out in triplicate, of chlorophyll a fluorescence, presence of reactive oxygen, in addition to the activities of the enzymes peroxidase, superoxide dismutase (SOD), and malondialdehyde acid (MDA). As highlighted results observed by the authors, mention is made of the 37 chemical components identified in the essential oil of *A. ordosica*, with emphasis on terpenoids, alcohols, esters, and acetones. The high concentrations of allelochemical compounds present in the essential oil resulted in the inhibition of photosynthetic activity (Fv/Fm), decrease in photosynthetic parameters ET0/ABS, ET0/TR0, and RC/CS0 due to the decrease in electron flow and photon absorption and reduction of the PSII reaction center. There were also increases in POD, SOD activities, and the presence of MDA. The results indicated that the essential oil of

A. ordosica inhibited the growth and negatively affected the development of plants present in the biological crust of the soil surface.

The allelopathic potential of *Artemisia* essential oil was tested on seed germination of nine weeds and two wheat varieties by Benarab [85]. In this study, the essential oil was obtained by hydrodistillation of the aerial part of the plants. After characterization by gas chromatography coupled with mass spectrometry, the essential oil was tested on seed germination and seedling growth of weeds and wheat. The concentrations used were 0.2, 0.4, 0.6, and 0.8 μl/ml, with 1 ml of these added in Petri dishes used for germination tests. The authors found 36 chemical compounds in the essential oil of *Artemisia herba-alba*, especially camphor (28.58%), cis-thujone (22.03%), eucalyptol (11.65%), and trans-thujone (7.03%). They found that essential oil extracts have a significant effect on weed germination inhibition, with the best results having been observed at a concentration of 0.2 μl/ml. These positive results for the allelopathic potential of the essential oil evaluated, suggest that, according to Benarab [85], Benvenuti [79], Önen [80], the same can be tested as a bio-herbicide.

With objectives similar to those of the works mentioned above, Benvenuti [79] collected 20 species of Asteraceae, during the spring–summer period, in several places in the Tuscany region. From each of the species, the yield and quality of the essential oils obtained were evaluated, as well as the verification of the inhibition of germination and growth of weeds. The essential oils were obtained by hydrodistilling the dried and ground flower buds in a Clevenger apparatus for 2 hours and their yield was calculated considering the values of the dry biomass of the flower buds collected in an area of 1 m^2. As some species did not show sufficient oil yields, the authors used only 10 species of *Asteraceae* in this experimental design, which consisted of Petri dishes lined with filter paper (Whatman no. 1, Whatman, Maidstone, UK) moistened with 7 cm^3 of distilled water, on which 50 seeds of each weed species were placed. Essential oils in concentrations of 10 and 100 μl/l were added to the plates. A second experiment was tested by the authors. In this, the essential oils were sprayed at concentrations of 10, 100, and 1000 mg/l, in weeds at two different phenological stages, when they had expanded cotyledons and when the third true leaf appeared. After analyzing the results, Benvenuti [79] found total inhibition of germination of the weed *A. retroflexus* at a concentration of 10 μl/l, and as the main result of experiment two, a reduction in weed fresh weight (about 20–30% after 10 days) and chlorophyll content (destroyed after the same period of time) was observed after application of the essential oil of *Artemisia annua*, thus confirming the total and rapid effectiveness of these essential oils and, according to the authors of the work, the essential oil of *A. annua* can be used as a natural herbicide.

2.2.3 Vegetable residue

Li [78] also evaluated the allelopathic activity of *Artemisia argyi* leaf powder against *B. pekinensis*, *L. sativa*, *O. sativa*, *Portulaca oleracea*, *Oxalis corniculata*, and *S. viridis*. The experiment was carried out in pots containing sand soil and *A. argyi* leaf powder for soil preparation, *A. argyi* powder was mixed with sand soil in a ratio of 100:0, 100:2, 100:4, or 100:8. Fifty seeds of *B. pekinensis*, *L. sativa*, *O. sativa*, *P. oleracea*, *O. corniculata*, and *S. viridis* were sown independently in each pot. Germination rate and plant height were measured on the ninth day for *B. pekinensis*, *L. sativa*, and *O. sativa* and on the thirteenth day for *P. oleracea*, *O. corniculata*, and *S. viridis*. When the soil:powder ratio of *A. argyi* leaves was 100:2, the germination rate of *B. pekinensis* and *L. sativa* was inhibited, as was the height of plants of *B. pekinensis*, *L. sativa*, *O. sativa*, and *P. oleracea*. The rates analyzed were dependent on the proportion of powder, the higher the concentration, the greater the inhibition. In the ratio of 100:8, the germination inhibition rates of *B. pekinensis*, *L. sativa*, *O. sativa*, *P. oleracea*, *O. corniculata*, and *S. viridis* were 71.82, 93.20, 31, 75, 65.47, 63.60, and 60.78%, respectively. These authors also analyzed the activity of *A. argyi* leaf powder in a field experiment. *Chrysanthemum morifolium* seedlings were transplanted to beds divided into plots. The *A. argyi* powder was uniformly applied to the plots at concentrations of 0, 0.1, and 0.2 kg/m^2. After 30 days, the biomass of each treatment group, growth, and yield of *C. morifolium* was evaluated. *A. argyi* leaf powder inhibited the germination and growth of the test plants. Only one species grew in the group treated with 0.2 kg/m^2. After the harvest of *C. morifolium*, there were no significant differences in the number of flowers and in the weight of the flowers between the groups with the addition of *A. argyi* leaf powder and the control group. Therefore, *A. argyi* leaf powder did not inhibit the growth of *C. morifolium* in the field, it only exerted an inhibitory effect on weed seeds. Thus, the powder from the leaves of *A. argyi* can be used as a herbicide.

2.3 Genus: *bidens*

2.3.1 Extract

B. pilosa is a weed widely distributed in subtropical and tropical regions with phytotoxic action already described for different species. The allelopathic activity of *B. pilosa* was evaluated by Deba [86] from the aqueous extract of the acid fraction (100, 200, and 500 ppm) of its leaves, stems, and roots. The species used as indicator plants were *Raphanus sativus* and *E. crus-galli*. For this, 10 seeds of each species were placed in Petri dishes lined with filter paper soaked with 8 ml of each solution of *B. pilosa* extract. As a control treatment, distilled water was used. After 7 days at room temperature (25–28°C) the percentage of germination and the length of the shoot and roots of *R. sativus* and *E. crus-galli*.

The extracts of *B. pilosa* showed a strong reduction in the growth of the indicator plants, however, the inhibition of germination did not exceed 20% in all treatments compared to the control. At all concentrations tested, the inhibitions promoted by stem and root extracts against *E. crus-galli* and *R. sativus* ranged from 70 to 90%. The herbicidal activity of *B. pilosa* was proportional to the applied doses, with almost complete inhibition of the growth of hypocotyls and radicles of *E. crus-galli* and *R. sativus* at a concentration of 500 ppm.

Chemical analysis of *B. pilosa* extracts by GC–MS detected 15 compounds including pyrocatechin, salicylic acid, p-vinylguaiacol, dimethoxyphenol, eugenol, 4-ethyl-1,2-benzenediol, iso-vanillin, 2-hydroxy-6-methylbenzaldehyde, vanillin, vanillic acid, p-hydroxybenzoic acid, protocatechuic acid, p-coumaric acid, ferulic acid, and caffeic acid. All of them were found in root extracts, except vanillin and iso-vanillin. Demethoxyphenol, eugenol, iso-vanillin, and vanillic acid, in turn, were not found in the stems of *B. pilosa*; and, in the leaves, salicylic acid, dimethoxyphenol, and vanillic acid were not observed. In all parts of the evaluated plant, the phenolics pyrocatechin, p-vinylguaiacol, 4-ethyl-1,2-benzenediol, 2-hydroxy-6 methylbenzaldehyde, p-hydroxybenzoic acid, protocatechuic acid, p-coumaric acid, and ferulic acid were verified, and caffeic acid, present in much greater amounts than the other phenolics (117.4, 298.7, and 350.3 μg/g in leaves, stems, and roots, respectively). In roots, ferulic acid was higher than pyrocatechin; however, it was lower in leaves and stems. In short, the total amount of these phenolic acids in the root.

Khanh [87] reported the allelopathic effect of leaves, stems, and roots of the acidic ethyl acetate fraction of *B. pilosa* extract on the same species tested (*E. crus-galli* and *R. sativus*) by Deba [86]. Root and stem extracts showed inhibitory effects of 70–90% on the emergence of *E. crus-galli* and *R. sativus*, for concentrations from 100 to 500 ppm. The potential allelopathic effect of *B. pilosa* extract on crops and weeds has already been described for other crops, with reports of inhibition of germination and growth of seedlings of soybean, mung bean, rice, corn, radish, cucumber, lettuce, sorghum, peanuts, and vines [87–91].

Hsu and Kao [54] evaluated the allelopathic effect of the aqueous extract of leaves, stems, and roots of *Bidens pilosa var. radiata* on the germination and growth of seedlings of the same species and of the sympatric species, *Bidens bipinnata* and *Ageratum conyzoides*. The plants of *B. pilosa* var. *radiata* were collected in an abandoned area of a farm in southern Taiwan, and then separated into leaves, stems, and roots, which were used to prepare the aqueous extract. The effect of the extract on seedling growth was verified by measuring the radicle and hypocotyl length of the germinated seeds of the tested model plants. The control treatment consisted of the use of distilled water. In this treatment, 99% of *B. pilosa* var. *radiata* germinated after 8 days of incubation. The final germination percentage of *B. pilosa* var. *radiata* was significantly reduced by the application of root extracts (39%), however, it was not affected by the stem and leaf extracts.

For the species *B. bipinnata*, after incubation in distilled water for 17 days, 97% of seed germination was verified. In relation to the control, the final percentage of germination reduced by 36% and 15% for treatments with extracts of roots and leaves, respectively. The stem extract also reduced the percentage of seed germination, however, without significant effect. Differently from what was observed for *B. pilosa* var. *radiata* and *B. bipinnata*, the germination percentage of *A. conyzoides* seeds was not affected by any of the extracts was higher than in the leaves and stem.

Stem and leaf extracts reduced the overall growth of new tissues in *B. pilosa* var. *radiata*, but contributed to the elongation of hypocotyls and radicle. Both the stem and leaf extracts significantly stimulated hypocotyl elongation and inhibited radicle elongation. On the other hand, the root extract of *B. pilosa* did not affect the elongation of hypocotyls or radicles. The seedlings of *B. bipinnata* did not have the global growth affected by any type of extract, however, the hypocotyl elongation was stimulated by the leaf extract. In contrast, the growth of hypocotyls and roots of *A. conyzoides* was significantly stimulated by the three types of extracts of *B. pilosa* var. *radiata* evaluated. According to [54] such differential responses reveal that the aqueous extract of *B. pilosa* var. *radiata* was more harmful for seed germination and seedling development of *B. pilosa* and *B. bipinnata* than for *A. conyzoides*. This pattern indicates a high capacity of *B. pilosa* to compete with similar taxa, given the results found when the species is sympatric, as is the case of *B. bipinnata*.

Mao [55] described the allelopathic effect of the aqueous extract of *B. pilosa*, which, at concentrations of up to 20 mg/ml, had some facilitating effect on the growth of *Trifolium repens* and *Medicago sativa* pasture buds, however, at high concentrations (100 mg/ml or more) it showed considerable inhibitory effect on seed germination and seedling growth, noting, therefore, that the allelopathic inhibitory effects generally increase with increasing concentrations.

Lima [56] evaluated the effect of the ethanolic extract (0.5% w/v) of the aerial part of the species *B. pilosa* and *B. alba* on the growth of seedlings of *L. sativa* L. For this, a sheet of Whatman number 6 paper was placed in a 15 cm Petri dish and soaked with a 0.5% solution of extract, resulting in a final concentration of 5 mg of dry extract. Ethanol was used as a control treatment. After drying in an oven at

40°C for a period of 12 hours, and subsequent evaporation of the solvent, the filter papers corresponding to each species were placed in Gerbox boxes, on which 3 ml of distilled water were added and 25 seeds were distributed in each box. The boxes were kept under ideal conditions for germination (17°C, with minimum limits of 16.8°C and maximum of 17.2°C) and after 7 days, the radicle and hypocotyl length were measured and compared to the control treatment. Both species inhibited radicle and hypocotyl growth. The extract of *B. pilosa* promoted an inhibition of 47.29% on radicle growth and 60.63% on hypocotyl growth (60.63%), while *B. alba* inhibited radicle growth by 95.94% and hypocotyl in 56.50%.

The allelopathic potential described for *B. pilosa* is associated with the fact that this species presents a variety of secondary metabolites, among them, phenolic compounds, saponins, flavonoids, flavones glycosides, polyacetylenes, terpenes chalcone glycosides, phenylpropanoids glycosides, terpenoids [92–94], with emphasis on the group of polyacetylenes, phenolics, and terpenoids. Polyacetylenes are derived from natural hydrocarbons characterized by one or more acetylene groups in their structures and are produced predominantly in the roots of plants of the Asteraceae family [92].

Phytochemical studies on *B. pilosa* have reported the presence of alkaloids, flavonoids, triterpenes, acyl chalcones, polyacetylenes, 1-phenyl-1,3,5-heptatriino [95], phenolic compounds such as quercitin 3-O-rutinoside, phenolic acids such as chlorogenic acid and 3,4-di-O-caffeoylquinic, 3,5-di-O-caffeoylquinic, and 4,5-di-O-caffeoylquinic acids [96]. In this work, the authors attribute the allelopathic activity presented by the ethanolic extract of *B. pilosa* to the presence of polyacetylene and phenolic compounds. Some acetylenic compounds are toxic to a variety of organisms and harmful to plants [97], and the acetylenic compound phenylheptatriine (PHT) is considered an important allelochemical of *B. pilosa* [98], with potent allelopathic action. This is the most studied compound and its activity was first verified in leaves of *B. pilosa* [92]. Deba [86] and Priestap and Bennett [100] also identified and isolated PHT from *B. pilosa* oils using GC, GC–MS, and HPLC techniques, confirming that this is the main constituent in all parts of *B. pilosa*. Among the identified compounds, PHT showed the highest concentration (30–48%) in *B. pilosa*, followed by 1-phenylhept-5-ene-1,3-diyno (0.2–37.1%) and 7-phenylhept-2-ene-4,6-diinyl acetate (1.3–22.5%) [100] which may be involved in phytotoxic action. PHT was also verified in leaves of *B. alba* with a varying concentration in nature and in response to the photoenvironment [97]. Other polyacetylenes such as [cisdehydromatricariaester (cis-DMK), lachnophylum ester (LE), matricaria ester (ME), dehydromatricaria lactone, α-tertienyl, and thiophene polyacetylenes] isolated from higher plants are reported to have allelopathic action [98, 100].

Phenolics represent secondary metabolites with an allelopathic effect associated with the inhibition of seed germination and the establishment of plants in plant communities. The allelopathic activity of simple phenols (benzoic and cinnamic acid derivatives, flavonoids, and tannins) is well documented in the literature. GC–MS and HPLC analyzes indicated the presence of 15 phenolic compounds (salicylic acid, vanillin, phydroxybenzoic acid, caffeic acid, p-coumaric acid, and ferulic acid) in the leaves, stem, and root of *B. pilosa* [101], the main allelochemicals in nature according to Blum [102]. Deba [101] verified for all parts of the plant that the caffeic acid content was higher (117.4, 298.7, and 350.3 μg/g in leaves, stems, and roots) than pyrocatechin (18.5, 32.9, and 29.6 μg/g) and ferulic acid. Caffeic acid is one of the main allelochemicals present in the genus *Leonurus* [103] with proven allelopathic activity. Compounds such as p-hydrobenzoic acid, vanillic acid, ferulic acid, p-coumaric acid, and syringic acid are also considered the main allelochemicals in nature and can be found in both invasive weeds and species [102, 104, 105].

Substances such as phenolic acids, polyphenols, and flavonoids have allelochemical characteristics and can act as photosynthesis inhibitor herbicides, altering electron transport, and phosphorylation in photosystems. Furthermore, phenolic acids induce an increase in the activity of oxidative enzymes, having as a final consequence the modification of membrane permeability and the formation of lignin that contributes to the reduction of root elongation [106]. The results described by Lima [56] indicate that the two species analyzed can be used in the search for new herbicide molecules with less impact on agroecosystems and human health.

2.3.2 Essential oil

Studies point to the potential use of *B. pilosa* essential oil as an allelopathic, because it is in these secondary metabolites that terpenoids are present, whose natural functions are variable, including signal molecules, allelochemicals, phytoalexins, visual pheromones, pigments, photoprotective agents, member constituents, and reproductive hormones [4]. Terpenoids, including volatile terpenes, are the main components of essential oils, causing an allelopathic interaction between plants that have volatile allelochemicals [107].

Deba [101] and Priestap and Bennett [99] evaluated all the plant structures and the EOs (leaves, stems, roots, flowers, and oils), they reported that the oils contain from 60 to 114.6% of the total components detected. Most of the identified compounds (terpenes, thiophenes, and polyacetylene constituents) are referred to as allelochemicals, such as β-caryophyllene (0.10–7.50%) found in all parts of *B. pilosa*, limonene (0.2–2.12%), 4-terpineol (0.14–0.41%), β-linalool (0.09–0.43%), β-pinene (0.07–0.39%), α-pinene (0.2–5.97%), linalool (0.1–0.14%), sabinene (0.2–0.6%), and eugenol isobutyrate (0.5% at the root).

The studies by Cantonwine and Downum [97] and Zeng and Luo [108] compared the chemical composition of the essential oil of *B. pilosa* obtained from plants in Japan and Argentina and found that the main total components of β-copanene (11.2%), germacrene D (39.5%), 1-phenylhept-5-ene-1,3-diyno (27.0%), α-humulene (3.3%), and 1-phenylhepta-1,3,5-triine (78.9%) were not found in essential oil from Japan. On the contrary, the main components of *B. pilosa* from Japan, β-Bourbonene (2.09%), megastigmatrienone (7.39%), and diphenylenemethane (3.71%) were not detected in plants from Argentina. Very similar compounds of essential oils were found in plants from both areas, however, the percentage composition of essential oils (%) of plants from Japan was lower than that of plants from Argentina. According to the authors, it should be taken into account that the phytotoxic components of *B. pilosa* increase in drought conditions and the PHT compound vary significantly with geographic and seasonal factors.

The results found agree with those found by Ni [109] and Chen [110] when confirming the variations in the chemical components of *Mikania micrantha*, which is justified by the fact that the production and release of secondary substances by plants are highly influenced by the environment. Such a metabolic response suggests that plants, as a defense strategy, promote the release or increase of their secondary metabolites in order to adapt to the environment and geographic area in which they are inserted.

2.3.3 Vegetable residue

Hsueh et al. [57] confirmed the allelopathic effect of *B. pilosa* L. var. radiata Sch. Beep. on *Cyperus rotundus* L. To verify such activity, the authors evaluated, in a greenhouse, the allelopathic effect of residues of *B. pilosa* var. *radiata* on *C. rotundus*

(1); interspecific competition between *B. pilosa* var. *radiata* and *C. rotundus* (2); the sprouting of tubercles of *C. rotundus*, in the field, in an area with or without removal of plant residues of *B. pilosa* var. radiata (3); and in addition to the effects of the use of *B. pilosa* var. radiata as a cover crop on the reproduction of *C. rotundus* (4).

The experiment 1 was carried out in pots under semi-natural conditions (greenhouse). The soil used was collected in an area of vegetable cultivation severely infested by *C. rotundus*, which was mixed with different proportions of residues of *B. pilosa* var. radiata, 0, 1.4, 2.8, and 4.2 g/pot, equivalent to 0, 0.1, 0.2, and 0.3 kg/m^2, respectively. Pre-sprouted tubers of *C. rotundus* were planted at densities of 3, 6, and 9 pre-sprouted tubers, corresponding to low, medium, and high density treatments, respectively. After 4 weeks, the plants were harvested, divided into shoots and roots and dried for 48 hours at 80°C. Inhibition of *C. rotundus* seedling growth was observed in the proportion of 0.1 kg/m in the density of 3 plants/pot. The dry mass of shoot, root, and total of *C. rotundus* were 72, 51, and 61% lower than the control (without *B. pilosa* residue-treatment 0 kg/m^2), respectively.

The results of this experiment indicated that the phytotoxicity of *B. pilosa* residues to shoot and root (including the number of tubers) and seedling growth of *C. rotundus* was density dependent. The residues exhibited greater phytotoxicity for the seedlings at the lowest density (3 plants/pot) and the growth reduction decreased with increasing density, providing evidence for the existence of density-dependent phytotoxicity in the residue of *B. pilosa*. Weidenhammer [111] were the first to provide experimental evidence that the density-dependent phytotoxicity of allelochemicals can be used to distinguish allelopathy from intraspecific competition (or other microbial activities). Allelochemicals can cause a greater reduction in growth in the target plant at low density than at high density due to the dilution of the phytotoxicity of allelochemicals. Studies have indicated that leachate or allelopathic plant residues can inhibit the growth of *C. rotundus*, as verified by Babu [112] when testing the leachate from fresh leaves of *E. globulus* which significantly reduced the dry mass of *C. rotundus* root. [112]. Residues of *Helianthus annuus* L., *Sorghum bicolor* (L.) Moench and *Brassica campestris* L. reduced plant density and dry mass of *C. rotundus* [113, 114]. Due to the complexity of the ecological environment, it becomes difficult to distinguish allelopathy from leachate or waste from competition (including intraspecies or plant-microbial interactions) [115, 116].

The influence of *B. pilosa* residue on the shoot/root ratio of *C. rotundus* also showed a density-dependent phytotoxic effect, since the proportions decreased with increasing plant density for all investigated *B. pilosa* proportions. Ratio values greater than 1 indicate that root growth was more suppressed by *B. pilosa* residue than shoot growth, and the default biomass allocation was altered. According to the optimal partition theory (OPT), plants tend to partition more biomass in the root than in the shoot when there is a deficiency of the main nutrients and, thus, result in a lower proportion of the shoot/root ratio [117]. Williams [118] showed that *C. rotundus* allocated more biomass to the root instead of prioritizing high-density growth. The shoot/root ratio was inverse to the tendency for the application of *B. pilosa* residues in each density treatment, presuming that the nutrient absorption capacity of the *C. rotundus* root was impaired due to the deleterious effects of *B. pilosa* allelochemicals, such as phenolic acids, which resulted in an increase in shoot/root ratio [119, 120]. Also, according to Duke [121] plant growth can be stimulated at high density due to the phenomenon of hormesis (stimulation in subtoxic concentration), observed in the treatment with 9 plants/pot.

The relationship between soil characteristics and allelochemicals can affect the retention, transport, and transformation processes of allelochemicals in soil [122].

Batish [15] from experiments in pots found that although the content of soil organic matter, available nitrogen, pH, and EC were altered by the addition of allelochemical residues, the phytotoxicity of the residues of *Chenopodium murale* L. had negative effects on chickpeas and peas. In the field, Iqbal and Cheema [123] reported an improvement in cotton yield and a reduction in the population of *C. rotundus* when established in the area, intercropping allelopathic plants with cotton. However, few studies have investigated the simultaneous effect of interspecific competition and allelopathy, considering that in natural or semi-natural conditions the distinction of these processes is difficult.

For the interspecific competition tests, the modified protocol of Snaydon [124]. A plastic pot (13.5 cm in diameter and 13 cm in depth) was similarly subdivided by a plastic plate that served as an underground partition. The plastic plate was sealed with neutral silicone gel on the sides and bottom of the vessel. Another 30 × 30 cm plastic plate was fixed vertically to the upper edge of the pot, acting as a partition above the ground. This arrangement, with the arrangement of *B. pilosa* var. radiata and *C. rotundus* were designed to compare different competitions: NO competition, SHOOT competition, ROOT competition, and FULL competition. For each treatment, 1.2 kg of soil was mixed thoroughly with (AC treatment) or without (N treatment) fine powdered activated carbon (pure grade) in a 50:1 ratio, added to the two subdivisions of each pot. All pots were sprayed daily with water (50 ml per pot) and no fertilizer was added during the experiment.

From the relative arrangement of the subterranean partition associated with the disposition of *B. pilosa* and *C. rotundus*, four competition modes were obtained: (1) both the shoot and root of the two species were separated (NO competition), (2) only the root was separated (SHOOT competition), (3) only the shoot was separated (ROOT competition), and (4) neither the shoot. The findings of this experiment showed that under SHOOT competition, the growth of *B. pilosa* and *C. rotundus* was not suppressed with the addition of activated charcoal (AC treatment) compared to the AC treatment under NO competition. However, shoot and root growth of *C. rotundus* was reduced in the N treatment, presuming that this reduction under SHOOT competition may be a result of phytotoxicity in the leachate from the shoot of *B. pilosa*. In the ROOT and FULL competitions, the growth of *C. rotundus* increased, while *B. pilosa* decreased in the AC treatment compared to N. *C. rotundus* tends to express predominant root competition with coexisting plants, which was confirmed by Tuor and Froud-Williams [125] who demonstrated that this species can significantly reduce the dry mass of the aerial part and the height of corn and soybean when in conditions of complete competition in relation to situations without competition. Horowitz [126] indicated that citrus seedling growth was significantly inhibited by *C. rotundus* despite being fertilized with nitrogen. According to the author, the phytotoxic substances produced by *C. rotundus* may partially contribute to the competition with citrus. Although previous studies demonstrated that *C. rotundus* could compete with coexisting plants through allelopathy, the allelopathic effects of *C. rotundus* on *B. pilosa* were not observed, since the growth of *C. rotundus* was suppressed by *B. pilosa* when activated carbon was not applied.

C. rotundus shoot/root ratio responses to the allelopathic effects of *B. pilosa* roots did not increase under the influence of *B. pilosa* residues, as observed in experiment 1. This can be explained in two ways. The first one establishes that the allelochemicals in the residues and the exudates of the root can be different. Deba [101] found that the main phenolic compounds in the leaves, stems, and roots of *B. pilosa* may be similar in composition but different in content. And the second considers that allelopathy combined with root competition promoted a different inhibition mechanism in this experiment.

El-Rokiek [127] pointed out that phenolics (ferulic acid, caffeic acid, for example) present in mango leaves can inhibit seedling growth and tuber sprouting of *C. rotundus*. Fifteen phenolic compounds isolated from *B. pilosa* have also been reported to possess the allelochemicals of phenolic compounds (e.g., caffeic, ferulic, p-coumaric, p-hydroxybenzoic, salicylic acid, among others) [60, 128]. Such phenolics have been reported to cause deleterious damage to plant roots. In this regard, Einhellig [129] in his studies indicated that salicylic acid seems to cause damage to the membrane structure and permeability of root cells, while caffeic acid decreases the levels of nitrogen, phosphorus, potassium, iron and molybdenum in cowpea. Ferulic acid was responsible for inhibiting nitrogen uptake in the roots of corn seedlings [119], while transcinnamic, ferulic, and p-coumaric acids reduced net nitrogen uptake and plasma membrane H^+-ATPase activity [130]. Such activity is possibly associated, as presented elsewhere, with PHT, a supposed allelochemical of polyacetylene, which supposedly releases phytotoxic radicals acting as an inhibitory mechanism. The presence of PHT has already been verified in leaves of *B. pilosa*, with reports of suppression of the growth of seedlings of *A. syriaca*, *C. album*, *P. pratense*, and *T. pratense* with LC50 of 0.66, 0.83, 2.88, and 1.43 ppm, respectively [99]. The inhibitory effect was also correlated with nutrient deficiency conditions [122], in which plants with allelopathic effect exhibited greater inhibitory effects on neighboring plants, and contributed to stimulate the allelochemical exudation of *B. pilosa* by competing for nutrients with *C. rotundus* (under ROOT and FULL competitions).

In experiment 3, tuber sprouting of *C. rotundus* in the field was evaluated. In order to prevent the interference of invasive *C. rotundus* plants, this was carried out in a three-year strip and vegetation (24 m long and 1.5 m wide) containing residues of *B. pilosa* var. radiata. In this range, eight 30 cm × 30 cm blocks were randomly defined. In half of them, the shoots and litter were removed in order to keep the soil surface uncovered (VN), and in the other, they were kept (VS). On adjacent land, four 1 m × 1 m plots were randomly selected in which weeds, including *C. rotundus* tubers, were removed by hand weeding. In the center of each plot (30 cm × 30 cm) an opaque plastic sheet was placed. This treatment was conducted in order to compare tuber sprouting with the treatments previously described. Twenty-five tubers (0.5–1.0 cm in diameter) of *C. rotundus* were planted in each block.

The highest percentage of tuber sprouting, sprouts per block, and dry mass per sprout occurred in the treatment with opaque plastic sheet cover, followed by the area with tubercles and the presence of *B. pilosa* vegetation, and the area without sprouts and litter. The percentage of tuber sprouting (52%), average sprouts per block (18 sprouts per block) and dry mass per sprout (5.26 mg) reduced when *C. rotundus* tubers were planted in areas containing *B. pilosa*. The clean treatment, without the buds and litter of *B. pilosa*, indicated that even in the absence of these, the allelochemicals remained in the soil and continued to inhibit the sprouting of *C. rotundus* tubers.

B. pilosa exhibited a strong phytotoxic effect on *C. rotundus* tuber reproduction. The percentage of tuber sprouting, average number of sprouts per block, and dry mass per sprout were lower in treatments with and without removal of sprouts and litter than in treatments with opaque plastic. Scavo [122] highlights that in natural environments, the concentration, movement, and persistence of allelochemicals have the power to influence the phytotoxicity of potential donor plants on the target species. Water-soluble allelochemicals such as phenolics have been reported to have a short residence time in soil due to rapid leaching and degradation [128, 129].

The founds of Hsueh et al. [57] indicated that the use of land without cover, such as the VN treatment, blocked the entry of allelochemicals, such as phenolics and PHT, into the soil. It is estimated that the phytotoxic effect of *B. pilosa* remaining in the soil is degraded within a brief period, when the release of allelochemicals from the plant surface and litter to the evaluated block ceases. Off-block replenishment of *B. pilosa* allelochemicals was also limited due to low rates of chemical diffusion in the soil, sorption of soil particles and organic matter, and microbial degradation [131].

Pre-cutting of *C. rotundus* plants before the beginning of the experiment predicted a break in the apical dominance of the tubers [132, 133], however, there was a stimulus for the sprouting of dormant tubers. Additionally, the fact of belonging to the C4 photosynthetic pathway plant group [134] the *C. rotundus* shoots grew faster than the *B. pilosa* seedlings. Plant residues from sorghum stalks (15 ton/ha) incorporated into the soil had a lower inhibitory effect on the density of *C. rotundus* than surface-applied residue at the same concentration as Mahmood and Cheema [135]. Later, Khaliq [113] tested the combination of sorghum, sunflower and brassica residues (7.5 ton/ha each) reporting a reduction in *C. rotundus* plant density by 87% compared to the control. Boz [135] reported a decrease in the density of *P. oleracea* L., *Amaranthus retroflexus* L., and *Echinochloa colonum* L. when using wheat and rye straw, however, they did not show an inhibitory effect on *C. rotundus*.

The effects of vegetation and coverage with *B. pilosa* var. radiate residues were also evaluated for the reproduction of *C. rotundus*. For this, a field infested by *C. rotundus* was divided into 24 plots (3 m long and 1 m wide), each one separated by a 0.8 m wide ditch. Soil samples were collected for each plot in order to investigate the numbers and dry mass of *C. rotundus* tubers in the top 15 cm of soil. All plots were mowed and the seeds of *B. pilosa* (8 g/polt) were randomly sown in half of the plots. In another 12 plots, *B. pilosa* was not sowed. At 69 days after sowing, all above-ground parts of the plant were cut, weighed, and kept in the plot, and tuber counts and dry mass were weighed. Half of the *B. pilosa* and *C. rotundus* residue plots were covered with an opaque plastic sheet while the other half was not covered.

In the studies of Hsueh et al. [57], although the tubers proliferated before the cover crops were cut, the density of *C. rotundus* plants was significantly reduced in the treatment with *B. pilosa* residue, and decreased with increasing dry mass of the residue. The density of *C. rotundus* plants was higher in both species of cover crops in the opaque plastic treatment. No negative effect has been reported associated with the increase in temperature promoted by the presence of opaque plastic [136], on the reproduction of tubers, provided that this does not exceed a temperature of 45°C for a period of more than 7 hours per day [137]. Despite this, according to Campbell [98] and Stevens [138] the phytotoxicity of polyacetylenes such as PHT, due to the characteristics of photosensitization and high activity, can be reduced in the dark and at temperatures above 30°C. Thus, it is possible that the allelochemicals of *B. pilosa* are degraded more quickly in treatments with opaque plastic covering, both for *B. pilosa* and for *C. rotundus* due to the increase in temperature under the leaf.

According to Xuan [139] there is a strong correlation between the ability to compete and invasive capacity of a culture or native species with its allelopathic potential. Emphasis is given to *B. pilosa*, with recognized phytotoxic effect on its sympatric species, such as *B. bipinnata* L. and *Pteris multifida* Poir., in ecosystems [54, 140]. *B. pilosa* and *B. pilosa* var. *radiata* have been investigated in Southeast Asia for their ability to manage weeds. Hong [141] evaluated 10 allelopathic species for weed control in Vietnam. In this study, after the application of 2 t/ha of *B. pilosa* in the area and a reduction of 84.9% and 81.8% was verified for the density and dry mass of plants, respectively. The application of *B. pilosa* biomass also showed greater weed suppression than manual weeding and herbicide treatments stimulating the height of rice plants in relation to the control (without plant residue) and contributing with tiller numbers and panicle formation of rice grains (11.8–16.9%). in relation to herbicide treatment (5.6–7.8%). Furthermore, the introduction of *B. pilosa* plant material exceeded rice yield by more than 20% when compared to herbicide treatment and was similar to hand weeding. The data obtained allow us to conclude that the application of *B. pilosa* biomass in the rice field significantly suppresses the infestation and increases the rice yield. The weed reduction may have occurred due to the release of phytotoxins (allelochemicals) released from the decomposing biomass of *B. pilosa* in the soil. The allelopathic activity of *B. pilosa* on rice growth and yield was stimulated by the presence of allelochemical compounds and nutritional effects of *B. pilosa*, which generally contain nutrients (P, K, Ca, and Mg), which improved growth and rice productivity [141].

Krumsri [142] also examined the phytotoxic effects of *B. pilosa* residue on *E. crus-galli* with an increase in this when applied in the fresh form, in relation to the dry residue. In the same study, the authors found that both soil cover and incorporation with *B. pilosa* residues reduced the density of *E. crus-galli* when using plants harvested at 60 days of growth. Poonpaiboonpipat and Poolkum [143] reinforced the promising results of the *B. pilosa* var. radiata use in paddy field (4 ton/ha) in inhibiting weed growth (86.73%) and increasing productivity by 81.03%. Khanh [87] evaluated the effect of root exudates of *B. pilosa* during its initial growth stage (20-day exudates in agar culture) which promoted 70% inhibition on germination, root length and shoots of *Leucaena leucocephala*, and 50% inhibition in *E. crus-galli*, *M. sativa*, and *O. sativa*. *B. pilosa* was not inhibited. The results of these studies support the potential use of residues of *B. pilosa* and *B. pilosa* var. *radiata*, as well as the extract and essential oil, as natural herbicides in weed management in agroecosystems.

3. Conclusions

The potential of the Asteraceae family as a repository of species with phytotoxic or allelopathic activity is evident. Although the number of studies that associate allelopathy with metabolites present in the crude extract, essential oil, or in plant biomass residues is increasing and presents promising results, studies are still needed to prove the interaction of these compounds in the soil. According to Zhang [144] among the four pathways by which allelochemicals are released into the environment, plant residues exert the most negative effect. Still, the importance of allelopathy in nature requires further investigation, as allelopathic effects were smaller in studies where soil microorganisms were likely to be abundant and when the study duration was longer. For this, two points should be considered in future studies to confirm allelopathy. The first one is the identification of one or more phytotoxins produced by the plant under investigation, or the identification of a compound(s) that can be converted into a phytotoxin in the soil, after its release. The second refers to the quantitative determination of the compound(s), that is, if they are found in sufficient quantity (in time and/or space) in the soil in which the plant grows or grew, and if in these natural conditions can affect other species. This stage is especially difficult and challenging because, in the natural environment, plants are exposed to different interactions, effects of antagonism and synergism with other compounds in the soil, growth stages and physiological factors, state of the recipient plant, soil microbiota (especially with rhizosphere microorganisms), soil moisture, temperature, among others. Zhang [144] after a systematic review found that the effects of allelopathic were weaker between closely related species (or individuals of the same species) than between distant species, suggesting that allelopathy would favor the coexistence of closely related species, the opposite of the predicted effect in competition for resources. These notes further reinforce the need to investigate allelopathy and competition for resources together in order to explain the results of species coexistence.

Acknowledgements

The authors thank Universidade Paranaense and Coordenação de Aperfeiçoamento de Pessoal de Nível Superior – Brazil (CAPES) for the financial support and the fellowship.

Conflict of interest

The authors declare no conflict of interest.

Author details

Ana Daniela Lopes[1*], Maria Graciela Iecher Faria Nunes[1], João Paulo Francisco[2] and Eveline Henrique dos Santos[1]

1 Laboratory of Biotechnology of Plant Products and Microorganisms, Postgraduate Program in Biotechnology Applied to Agriculture, Paranaense University, Umuarama, Brazil

2 Department of Agronomic Sciences, Postgraduate Program in Agricultural Sciences, State University of Maringa, Umuarama, Brazil

*Address all correspondence to: anadanielalopes@prof.unipar.br

References

[1] Qian H, Xu X, Chen W, Jiang H, Jin Y, Liu W, et al. Allelochemical stress causes oxidative damage and inhibition of photosynthesis in *Chlorella vulgaris*. Chemosphere. 2009;**75**(3):368-375. DOI: 10.1016/j.chemosphere.2008.12.040

[2] Rice EL. Allelopathy. 2nd ed. New York, NY: Academic Press; 1984. p. 368. DOI: 10.1016/C2009-0-03024-4

[3] Reigosa M, Gomes AS, Ferreira AG, Borghetti F. Allelopathic research in Brazil. Acta Botânica Brasílica. 2013;**27**:629-646. DOI: 10.1590/S0102-33062013000400001

[4] Mizutani J. Selected allelochemicals. CRC Critical Reviews in Plant Sciences. 1999;**18**(5):653-671. DOI: 10.1080/07352689991309432

[5] Macias FA, Gallindo JCG, Molinillo JMG. Plant biocommunicators: Application of allelopathic studies. In: Luijendijk TJC, editors. 2000 Years of Natural Products Research – Past, Present and Future. 2000. pp. 137-161. DOI: 10.1271/nogeikagaku1924.73.1201

[6] De' Albuquerque MB, RCD S, Ima LM. Allelopathy, an alternative tool to improve cropping systems. Agronomy for Sustainable Development. 2011;**31**(379):395. DOI: 10.1051/agro/2010031

[7] Inderjit I, Duke SO. Ecophysiological aspects of allelopathy. Planta. 2003;**217**:529-539. DOI: 10.1007/s00425-003-1054-z

[8] Araújo CA, Morgado CS, Gomes AKC, Gomes ACC, et al. Asteraceae family: A review of its allelopathic potential and the case of *Acmella oleracea* and *Sphagneticola trilobata*. Rodriguésia. 2021;**72**:e01622020. DOI: 10.1590/2175-7860202172137

[9] Soltys D, Krasuska U, Bogatek R, Gniazdowska A. Allelochemicals as bioherbicides – present and perspectives. In: Price AJ, Kelton JA, editors. Herbicides – Current Research and Case Studies in Use. London: IntechOpen; 2013. p. 664. DOI: 10.5772/56743

[10] Blok C, Baumgarten A, Baas R, Wever G, et al. Analytical methods used with soilless substrates. In: Raviv M, Lieth JH, Bar-Tal A, editors. Soilless Culture Theory and Practice. 2nd ed. Academic Press Ed: London; 2019. pp. 509-564. DOI: 10.1016/B978-0-444-63696-6.00011-6

[11] Isik D, Kaya E, Ngouajio M, Mennan H. Summer cover crops for weed management and yield improvement in organic lettuce (*Lactuca sativa*) production. Phytoparasitica. 2009;**37**:193-203. DOI: 10.1007/s12600-009-0021-z

[12] Kropff MJ, Walter H. EWRS and the challenges for weed research at the start of a new millennium. Weed Research. 2000;**1**(7):10. DOI: 10.1046/j.1365-3180.2000.00166.x

[13] Daizy R, Batish DR, Singh HP, Kohli RK, et al. Plant flavonoids: Biological molecules for useful explotation. Australian Journal of Plant Physiology. 1995;**22**(87):99. DOI: 10.1071/PP9950087

[14] Gusman GS, Vieira LR, Vestena S. Allelopathy of pharmaceutically important plant species for cultivated species. Biotemas. 2012;**4**:37-48. DOI: 10.5007/2175-7925.2012v25n4p37

[15] Batish DR, Lavanya K, Singh HP, Kohli RK. Phenolic allelochemicals

released by *Chenopodium murale* affect the growth, nodulation and macromolecule content in chickpea and pea. Plant Growth Regulation. 2007;**51**(2):119-128. DOI: 10.1007/s10725-006-9153-z

[16] Bessada SMF, Barreira JCM, Oliveira MBPP. Asteraceae species with most prominent bioactivity and their potential applications: A review. Industrial Crops and Products. 2015;**76**:604-615. DOI: 10.1016/j.indcrop.2015.07.073

[17] Nakajima NJ, Semir J. Asteraceae from Serra da Canastra National Park, Minas Gerais, Brasil. Brazilian Journal of Botany. 2001;**24**(4):471-478. DOI: 10.1590/ S0100-84042001000400013

[18] Roque N, Bautista HP. Asteraceae: Characterization and Floral Morphology. Salvador: EDUFBA; 2008. p. 75. Available from: http://repositorio.ufba.br/ri/handle/ri/25086

[19] Baretta IP, Felizardo RA, Bimbato VF, MGJ dos S, et al. Anxiolytic-like effects of acute and chronic treatment with *Achillea millefolium* L. extract. Journal of Ethnopharmacology. 2012;**140**(1):46-54. DOI: 10.1016/j.jep.2011.11.047

[20] Toyang NJ, Verpoorte R. A review of the medicinal potentials of plants of the genus *Vernonia* (Asteraceae). Journal of Ethnopharmacology. 2013;**146**(3):681-723. DOI: 10.1016/j.jep.2013.01.040

[21] Singh O, Khanam Z, Misra N, Srivastava MK. Chamomile (*Matricaria chamomilla* L.): An overview. Pharmacognosy Reviews. 2011;**5**(9):82-95. DOI: 10.4103/0973-7847.79103

[22] Gawronski SW, Gawronska H. Plant taxonomy for phytoremediation. In: Marmiroli N, Samotokin B, Marmiroli M, editors. Advanced Science and Technology for Biological Decontamination of Sites Affected by Chemical and Radiological Nuclear Agents. Dordrecht: Springer Netherlands; 2007. pp. 79-88. DOI: 10.1007/978-1-4020-5520-1

[23] Nikolic M, Stevovic S. Family Asteraceae as a sustainable planning toll in phytoremediation and its relevance in urban areas. Urban Forestry & Urban Greening. 2015;**14**(4):782-789. DOI: 10.1016/j.ufug.2015.08.002

[24] Fathi E, Majdi M, Dastan D, Maroufi A. The spatio-temporal expression of some genes involved in the biosynthetic pathways of terpenes/ phenylpropanoids in yarrow (*Achillea millefolium*). Plant Physiology and Biochemistry. 2019;**142**:43-52. DOI: 10.1016/j.plaphy.2019.06.036

[25] Macías FA, Santana A, Yamahata A, Varela RM, et al. Facile preparation of bioactive seco-guaianolides and guaianolides from *Artemisia gorgonum* and evaluation of their phytotoxicity. Journal of Natural Products. 2012;**75**:1967-1973. DOI: 10.1021/ np300639b

[26] Silva, BP. Allelopathic potential of *Cosmos sulphureus* Cav. [thesis]. Jaboticabal: Universidade Estadual Paulista; 2017

[27] Stavropoulou MI, Angelis A, Aligiannis N, Kalpoutzakis E, et al. Phytotoxic triterpene saponins from *Bellis longifolia*, an endemic plant of Crete. Phytochemistry. 2007;**144**:71-77. DOI: 10.1016/j.phytochem.2017.08.019

[28] Castells E, Mulder PPJ, Pérez-Trujillo M. Diversity of pyrrolizidine alkaloids in native and invasive *Senecio pterophorus* (Asteraceae): Implications for toxicity. Phytochemistry. 2014;**108**:137-146. DOI: 10.1016/j.phytochem.2014.09.006

[29] Skaf J, Hamarsheh O, Berninger M, Balasubramanian S, et al. Improving antitrypanosomal activity of alkamides isolated from *Achillea fragrantissima*. Fitoterapia. 2018;**125**:191-198. DOI: 10.1016/j.fitote.2017.11.001

[30] Ccana-Ccapatinta GV, Ferreira PL, Groppo M, Costa FB. Caffeic acid ester derivatives and flavonoids of genus *Arnaldoa* (Asteraceae, Barnadesioideae). Biochemical Systematics and Ecology. 2019;**86**:103911. DOI: 10.1016/j.bse.2019.103911

[31] Bandyopadhyay S, Mitra S, Mukherjee SK. Traditional uses of some weeds of Asteraceae b the ethnic communities of Koch Bihar District, West Bengal. International Journal of Pharmacology. 2014;**4**:31-34. DOI: 10.3923/ijp.2014.275.281

[32] Chung KF, Kono Y, Wang CM, Peng CI. Notes on *Acmella* (Asteraceae:Heliantheae) in Taiwan. Botanical Studies. 2008;**49**:73-82. DOI: 10.1186/1999-3110-54-10

[33] Aguiar JPL, Yuyama LKO, Souza FCA, Pessoa A. Iron bioavailability from jambu (*Spilanthes oleracea* L.): Study in mice. RPAS. 2014;**5**:19-24. DOI: 10.5123/S2176-62232014000100002

[34] Dias AMA, Santos P, Seabra IJ, Junior RNC, et al. Spilanthol from *Spilanthes acmella* flowers, leaves and stems obtained by selective supercritical carbon dioxide extraction. Journal of Supercritical Fluids. 2012;**61**:62-70. DOI: 10.1016/j.supflu.2011.09.020

[35] Prachayasittikul S, Suphapong S, Worachartcheewa A, Lawung R, et al. Bioactive metabolites from *Spilanthes Acmella* Murr. Molecules. 2009;**14**:850-867. DOI: 10.3390/molecules14020850

[36] Stein R, Berger M, Cecco BS, Mallmann LP, et al. Chymase inhibition: A key factor in the anti-inflammatory activity of ethanolic extracts and spilanthol isolated from *Acmella oleracea*. Journal of Ethnopharmacology. 2020;**270**:116610. DOI: 10.1016/j.jep.2020.113610

[37] Savic S, Petrovic S, Savic S, Cekic N. Identification and photostability of N-alkylamides from *Acmella oleraceae* extracts. Journal of Pharmaceutical and Biomedical Analysis. 2020;**195**:113819. DOI: 10.1016/j.jpba.2020.113819

[38] Simas NK, Dellamora ECL, Schripsema J, Lage CLS, et al. Acetylenic 2-phenylethylamides and new isobutylamides from *Acmella oleracea* (L.) R. K. Jansen, a Brazilian spice with larvicidal activity on Aedes aegypti. Phytochemistry Letters. 2013;**6**:67-72. DOI: 10.1016/j.phytol.2012.10.016

[39] Benelli G, Pavela R, Drenaggi E, Maggi F. Insecticidal efficacy of the essential oil of jambú (*Acmella oleracea* (L.) R.K. Jansen) cultivated in central Italy against filariasis mosquito vectors, houseflies and moth pests. Journal of Ethnopharmacology. 2019;**30**:272-279. DOI: 10.1016/j.jep.2018.08.030

[40] Marchesini P, Barbosa AF, Sanches MNG, Nascimento RM, et al. Acaricidal activity of *Acmella oleracea* (Asteraceae) extract against *Rhipicephalus microplus*: What is the influence of spilanthol? Veterinary Parasitology. 2020;**283**:109170. DOI: 10.1016/j.vetpar.2020.109170

[41] Singh M, Roy B, Tandon V, Chaturvedi R. Extracts of dedifferentiated cultures of *Spilanthes Acmella* Murr. possess antioxidant and anthelmintic properties and hold promise as an alternative source of herbal medicine. Plant Biosystems - An International Journal Dealing with all

Aspects of Plant Biology. 2014;**148**:259-267. DOI: 10.1080/11263504.2013.766278

[42] Nascimento LES, Arriola NDA, da Silva LAL, Faqueti LG, et al. Phytochemical profile of different anatomical parts of jambu (*Acmella oleracea* (L.) R.K. Jansen): A comparison between hydroponic and conventional cultivation using PCA and cluster analysis. Food Chemistry. 2020;**332**:127393. DOI: 10.1016/j.foodchem.2020.127393

[43] Kato-Noguchi H, Suwitchayanon P, Boonmee S, Iwasaki A, et al. Plant growth inhibitory activity of the extracts of *Acmella oleracea* and its growth inhibitory substances. Natural Product Communications. 2019;**14**:1-5. DOI: 10.1177/1934578X19858805

[44] Ferreira JFS, Janick J. Allelopathic plants XVI. *Artemisia* species. Allelopathy Journal. 2004;**14**:167-176. DOI: 10.26651/2020-49-1-12502

[45] Feng X, Cao S, Qiu F, Zhang B. Traditional application and modern pharmacological research of *Artemisia annua* L. Pharmacology & Therapeutics. 2020;**216**:107650. DOI: 10.1016/j.pharmthera.2020.107650

[46] Klayman DL. *Artemisia annua*. In: Kinghorn AD, Balandrin MF, editors. Human Medicinal Agents from Plants, ACS Symposium Series. Washington, DC: American Chemical Society; 1993. pp. 242-255. DOI: 10.1021/bk-1993-0534.ch017

[47] Brown GD. The biosynthesis of artemisinin (Qinghaosu) and the phytochemistry of *Artemisia annua* L. (Qinghao). Molecules. 2010;**15**(11):7603-7698. DOI: 10.3390/molecules15117603

[48] Hierro JL, Callaway RM. The ecological importance of allelopathy. Annual Review of Ecology, Evolution, and Systematics. 2021;**52**(1):25-45. DOI: 10.1146/annurev-ecolsys-051120-030619

[49] Liu H, Huang J, Yuan L. In vitro effect of artemisinin on microbial biomasses, enzyme activities and composition of bacterial community. Applied Soil Ecology. 2008;**124**:1-6. DOI: 10.1016/j.apsoil.2017.10.022

[50] Morvillo CM, de la Fuente EB, Gil A, Martinez-Ghersa MA, et al. Competitive and allelopathic interference between soybean crop and *annua*l wormwood (*Artemisia annua* L.) under field conditions. European Journal of Agronomy. 2011;**34**(4):211-221. DOI: 10.1016/j.eja.2011.01.004

[51] Ni LX, Wang N, Liu XY, Yue FF, et al. Algaecide, toxic response of aquatic organisms to guide application of artemisinin sustained-release granule. Water Science and Engineering. 2020;**13**(2):106-115. DOI: 10.1016/j.wse.2020.06.007

[52] Yan ZQ, Wang DD, Ding L, Cui HY, et al. Mechanism of artemisinin phytotoxicity action: Induction of reactive oxygen species and cell death in lettuce seedlings. Plant Physiology and Biochemistry. 2015;**88**:53-59. DOI: 10.1016/j.plaphy.2015.01.010

[53] Holm LG, Plucknett DL, Pancho JV, Herberger JP. The World's Worst Weeds Distribution and Biology. Honolulu: The University Press of Hawaii; 1991. p. 609p

[54] Hsu HM, Kao WY. Contrasting effects of aqueous tissue extracts from an invasive plant, *Bidens pilosa* L. var. radiata, on the performance of its sympatric plant species. Taiwania. 2009;**54**(3):255-260. DOI: 10.6165/tai.2009.54(3).255

[55] Mao DJ, Xie JF, Quan GM, Zhang J. Effects of *Bidens pilosa* aqueous extracts

on germination and seedling growth of two pastures. Pharmaceutical Biology. 2010;**1996**:34-87

[56] Lima CP, Cunico MM, Miguel OG, Miguel MD. Efeito dos extratos de duas plantas medicinais do gênero *Bidens* sobre o crescimento de plântulas de *Lactuca sativa* L. Revista de Ciências Farmacêuticas Básica e Aplicada. 2011;**32**(1):83-87. DOI: 10.4322/2179-443X.0758

[57] Hsueh M-T, Fan C, Chang W-L. Allelopathic effects of *Bidens pilosa* L. var. *radiata* Sch. Bip. on the tuber sprouting and seedling growth of *Cyperus rotundus* L. Plants. 2020;**9**(6):742. DOI: 10.3390/plants9060742

[58] Tsao R, Eto M. Light-activated plant growth inhibitory activity of cis-dehydromatricariaester, rose bengal and fluoren-9-one on lettuce (*Lactuca sativa* L.). Chemosphere. 1996;**32**:1307-1317. DOI: 10.1016/0045-6535(96)00042-2

[59] Towers GHN, Arnason CK, Wat CK, Lambert JD. Controlling weeds using a naturally occurring polyacetylenes. Canadian Patent CA: 1172460; 1984

[60] Xuan TD, Khanh TD. Chemistry and pharmacology of *Bidens pilosa*: An overview. Journal of Pharmaceutical Investigation. 2016;**46**:91-132. DOI: 10.1007/s40005-016-0231-6

[61] Inderjit I, Weston LA. Are laboratory biossays for allelopathy suitable for prediction on field responses? Journal of Chemical Ecology. 2000;**26**:2111-2118. DOI: 10.1023/A:1005516431969

[62] Lankau R. Soil microbial communities alter allelopathic competition between *Alliaria petiolata* and a native species. Biological Invasions. 2010;**12**:2059-2068. DOI: 10.1007/s10530-009-9608-z

[63] Callaway RM, Aschehoug ET. Invasive plants versus their new and old neighbors: A mechanism for exotic invasion. Science. 2000;**290**(5491):521-523. DOI: 10.1126/science.290.5491.521

[64] Cummings JA, Parker IM, Gilbert GS. Allelopathy: A tool for weed management in forest restoration. Plant Ecology. 2012;**203**:1975-1989. DOI: 10.1007/s11258-012-0154-x

[65] Hadacek F. Secondary metabolites as plants traits: Current assessment and future perspectives critical. CRC Critical Reviews in Plant Sciences. 2002;**21**:273-322. DOI: 10.1080/0735-260291044269

[66] Macías FA, Molinillo JMG, Varela RM, Galindo JGC. Allelopathy – a natural alternative for weed control. Pest Management Science. 2007;**63**:327-348. DOI: 10.1002/ps.1342

[67] P C. Seeds: The ecology of regeneration in plant communities. Seed Science Research. 2003;**13**:247-248. DOI: 10.1079/9780851994321.0000

[68] Adkins SW, Ashmore S, Navie SC. Seeds: Biology, Development and Ecology. Cambridge: CABI Digital Library; 2007. p. 496. DOI: 10.1079/9781845931971.0000

[69] Hind N, Biggs N. Plate 460. *Acmella oleracea* compositae. Curtis's Botanical Magazine. 2003;**20**(1):31-39

[70] Suwitchayanon P, Kunasakdakul K, Kato-Noguchi H. Screening the allelopathic activity of 14 medicinal plants from northern Thailand. Environmental Control in Biology. 2017;**55**(3):143-145. DOI: 10.2525/ecb.54.143

[71] Guo J, Bao G, Yang Y, Xi J, et al. Impact of repeated freeze-thaw cycles environment on the allelopathic effect to

Secale cereale L. seedlings. Chemosphere. 2022;**308**(P3):136476. DOI: 10.1016/j.chemosphere.2022.136476

[72] Wang Q, Zhang H, Yang Q, Wang T, et al. The impact of grazing intensity on the allelopathic effect of *Artemisia frigida* in a temperate grassland in northeastern China. Flora. 2022;**288**:152005. DOI: 10.1016/j.flora.2022.152005

[73] Zhao-Jiang Z, Ru-Min Z, Pei-Jun G, Guo-Sheng W, et al. Allelopathic effects of *Artemisia frigida* Willd. on growth of pasture grasses in Inner Mongolia, China. Biochemical Systematics and Ecology. 2011;**39**:377-383. DOI: 10.1016/j.bse.2011.05.010

[74] Wang RL, Xia WN, Liu SW, Qin Z, et al. Effects of water stress on the growth and allelopathic potential of invasive plant *Mikania micrantha* H.B.K. Allelopathy Journal. 2016;**39**(2):143-152. DOI: 10.26651/2017-40-2-1071

[75] Wu R, Wu B, Cheng H, Wang S, et al. Drought enhanced the allelopathy of goldenrod on the seed germination and seedling growth performance of lettuce. Polish Journal of Environmental Studies. 2020;**30**(1):423-432. DOI: 10.15244/pjoes/122691

[76] Zhou X, Zhang Y, An X, De Philippis R, et al. Identification of aqueous extracts from *Artemisia ordosica* and their allelopathic effects on desert soil algae. Chemoecology. 2019;**29**(2):61-71. DOI: 10.1007/s00049-018-00276-8

[77] Luo Yongqing DZ, Yan Z, Zhao X, et al. *Artemisia halodendron* litters have strong negative allelopathic effects on earlier successional plants in a semi-arid sandy dune region in China. Frontiers in Plant Science. 2020;**11**:961. DOI: 10.3389/fpls.2020.00961

[78] Li J, Chen L, Chen Q, Miao Y, et al. Allelopathic effect of *Artemisia argyi* on the germination and growth of various weeds. Scientific Reports. 2021;**11**(1):1-15. DOI: 10.1038/s41598-021-83752-6

[79] Benvenuti S, Cioni PL, Flamini G, Pardossi AJWR. Weeds for weed control: Asteraceae essential oils as natural herbicides. Weed Research. 2017;**57**(5):342-353. DOI: 10.1111/wre.12266

[80] Önen H, Ozer Z, Telci I. Bioherbicidal effects of some plant essential oils on different weed species. Journal of Plant Diseases and Protection. 2002;**18**:597-605. DOI: 20.500.12881/7781

[81] Weston LA, Duke SO. Weed and crop allelopathy. CRC Critical Reviews in Plant Sciences. 2003;**22**(3-4):367-389. DOI: 10.1080/713610861

[82] Yang X, Deng S, De Philippis R, Chen L, et al. Chemical composition of volatile oil from *Artemisia ordosica* and its allelopathic effects on desert soil microalgae, *Palmellococcus miniatus*. Plant Physiology and Biochemistry. 2012;**51**:153-158. DOI: 10.1016/j.plaphy.2011.10.019

[83] Yang X, Huang B, Cui J, Chen X, et al. Mechanism of the allelopathic effect of macroalgae *Gracilaria bailiniae* on *Nitzschia closterium*. Ecotoxicology and Environmental Safety. 2022;**241**:1-12. DOI: 10.1016/j.ecoenv.2022.113767

[84] Kong C, Huang SS, Hu F. Allelopathy of *Ageratum conyzoides* V. biological activities of the volatile oil from ageratum on fungi, insects and plants and its chemical constituents. Acta Ecologica Sinica. 2001;**21**:584-587. DOI: 10.1023/A:1016229616845

[85] Benarab H, Fenni M, Louadj Y, Boukhabti H. Allelopathic activity of essential oil extracts from *Artemisia herba-alba* Asso. on seed and seedling germination of weed and wheat crops. Acta Scientiarum Naturalium Universitatis Sunyatseni. 2020;7(1):86-97. DOI: 10.2478/asn-2020-0009

[86] Deba F, Xuan TD, Yasuda M, Tawata S. Chemical composition and antioxidant, antibacterial and antifungal activities of the essential oils from *Bidens pilosa* L. var. *radiata*. Food Control. 2007;**19**:346-352. DOI: 10.1016/j.foodcont.2007.04.011

[87] Khanh TD, Cong LC, Xuan TD, Uezato Y, et al. Allelopathic plants: 20. Hairy Beggarticks (*Bidens pilosa* L.). Allelopathy Journal. 2009;**24**:243-254

[88] Ben HZ, Lin SP, Rong JW, Lin HX, et al. Allelopathic potential of native plants on invasive plant *Mikania micrantha* H.B.K in South China. Allelopathy Journal. 2008;**22**:189-195

[89] Bhatt BP, Tomar JMS, Misra LK. Allelopathic effects of weeds on germination and growth of legumes and cereal crops of North Eastern Himalayas. Allelopathy Journal. 2001;**8**:225-231

[90] Singh R, Hazarika UK. Allelopathic effects of *Galinsoga parviflora* Car. and *Bidens pilosa* L. on germination and seedling growth of soybean and groundnut. Allelopathy Journal. 1996;**3**:89-92

[91] Stevens GA Jr, Tang CS. Inhibition of seedling growth of crop species by re-circulating root exudates of *Bidens pilosa* L. Journal of Chemical Ecology. 1995;**11**:1411-1425

[92] Bohlman F, Burkhardt T, Zdero C. Naturally Occurring Acetylenes. London: Academic Press; 1973. p. 547. DOI: 10.1086/407964

[93] Brandao MGL, Krenttlia U, Soares LSR, Nery CGC, et al. Antimalarial activity of extracts and fractions from *Bidens pilosa* and other *Bidens* species (Asteraceae) correlated with the presence of acetylene and flavonoid compounds. Journal of Ethnopharmacology. 1997;**57**:131-138. DOI: 10.1016/s0378-8741(97)00060-3

[94] Subhuti D. *Bidens*: A popular remedy escapes notice of Western practitioners. 2007 Available from: http://www.itmonline.org/arts/*Bidens*.htm

[95] Abajo C, Boffill MA, J del C, Méndez MA, et al. In vitro study of the antioxidant and immunomodulatory activity of aqueous infusion of *Bidens pilosa*. Journal of Ethnopharmacology. 2004;**93**(1-2):319-323. DOI: 10.1016/j.jep.2004.03.050

[96] Chiang Y-M, Chuang D-Y, Wang S-Y, Kuo Y-H, et al. Metabolite profiling and chemopreventive bioactivity of plant extracts from *Bidens pilosa*. Journal of Ethnopharmacology. 2004;**95**(1-3):409-419. DOI: 10.1016/j.jep.2004.08.010

[97] Cantonwine EG, Downum KR. Phenylheptatriyne variation in *Bidens alba* var. radiata leaves. Journal of Chemical Ecology. 2001;**27**(2):313-326. DOI: 10.1023/A:1005680422159

[98] Campbell G, Lambert JDH, Arnason T, Towers GHN. Allelopathic properties of alphaterthienyl and phenylheptatriyne, naturally occurring compounds from species of Asteraceae. Journal of Chemical Ecology. 1982;**8**:961-972. DOI: 10.1007/BF00987662

[99] Priestap HA, Bennett BC. Investigation of the essential oils of *Bidens pilosa* var. *minor*, *Bidens alba* and *Flaveria linearis*. Journal of Essential Oil Research. 2008;**20**:396-402. DOI: 10.1080/10412905.2008.9700039

[100] Quintana N, Weir TL, Du J, Broeckling CD, et al. Phytotoxic polyacetylens from roots of Russian Knapweed (*Acroptilon repens* (L.) DC.). Phytochemistry. 2008;**69**:2572-2578. DOI: 10.1016/j.phytochem.2008.07.015

[101] Deba F, Xuan TD, Yasuda M, Tawata S. Herbicidal and fungicidal activities and identification of potential phytotoxins from *Bidens pilosa* L var. *radiata* Scherff. Weed Biology and Management. 2007;7:77-83. DOI: 10.1111/j.1445-6664.2007.00239.x

[102] Blum U, Wentworth TR, Klein K, Worsham AD, et al. Phenolic acid content of soils from wheat-no till, wheat-conventional till, and conventional till soybean cropping systems. Journal of Chemical Ecology. 1991;**17**:1045-1068. DOI: 10.1007/BF01402933

[103] Mandal S. Allelopathic activity of root exudates from *Leonurus sibiricus* L. (Raktodrone). Weed Biology and Management. 2001;**1**:170-175. DOI: 10.1046/j.1445-6664.2001.00027.x

[104] Khanh TD, Elzaawely AA, Chung IM, Ahn JK, et al. Role of allelochemicals for weed management in rice. Allelopathy Journal. 2007;**19**:85-96. DOI: 10.26651/2018-44-1-1149

[105] Khanh TD, Chung IM, Tawata S, Xuan TD. Allelopathy for weed management in sustainable agriculture. CAB Reviews Perspectives in Agriculture Veterinary Science Nutrition and Natural Resources. 2007;**34**:1-17. DOI: 10.1079/PAVSNNR20072034

[106] Carmo FMS, Borges EEL, Takaki M. Alelopatia de extratos aquosos de canela-sassafrás (*Ocotea odorifera* (Vell.) Rohwer). Acta Botânica Brasílica. 2007;**21**(3):697-705. DOI: 10.1590/S0102-33062007000300016

[107] Dudai N, Poljakoff-Mayber A, Mayer M, Putievsky E, et al. Essential oils as allelochemicals and their potential use as bio-herbicides. Journal of Chemical Ecology. 1999;**25**:1079-1089

[108] Zeng RS, Luo SM. Relationship between allelopathic effects of *Bidens pilosa* aqueous extracts and rainfall. Journal of South China Agricultural University. 1995;**16**:69-72

[109] Ni GY, Li FL, Chen BM, Song LY, et al. Allelopathic plants 21: *Mikania micrantha* H.B.K. Allelopathy Journal. 2007;**19**:287-296. DOI: 10.26651/2019-46-1-1202

[110] Chen B, Ni G, Ren W, Feng S. Effects of aqueous extracts of *Mikania micrantha* on litter decomposition of native plants in South China. Allelopathy Journal. 2007;**20**:307-331. DOI: 10.26651/2022-57-2-1408

[111] Weidenhamer JD, Hartnett DC, Romeo JT. Density-dependent phytotoxicity: Distinguishing resource competition and allelopathic interference in plants. Journal of Applied Ecology. 1989;**26**(2):613. DOI: 10.2307/2404086

[112] Babu RC, Kandasamy OS. Allelopathic effect of *Eucalyptus globulus* Labill. on *Cyperus rotundus* L. and *Cynodon dactylon* L. Pers. Journal of Agronomy and Crop Science. 1997;**179**(2):123-126. DOI: 10.1111/j.1439-037X.1997.tb00507.x

[113] Khaliq A, Matloob A, Irshad MS, Tanveer A, et al. Organic weed management in maize (*Zea mays* L.) through integration of allelopathic crop residues. Pakistan Journal of Weed Science Research. 2010;**16**(4):409-420

[114] Matloob A, Khaliq A, Farooq M, Cheema ZA. Quantification of allelopathic potential of different crop residues for the purple nutsedge suppression. Pakistan Journal of Weed Science Research. 2010;**16**(1):1-12

[115] Weidenhamer JD. Distinguishing allelopathy from resource competition: The role of density. In: Zeng RS, Mallik AU, Luo SM, editors. Allelopathy. Dordrecht: Kluwer Academic

Publishers; 2006. pp. 85-103. DOI: 10.1007/1-4020-4280-9_4

[116] Fuerst EP, Putnam AR. Separating the competitive and allelopathic components of interference. Journal of Chemical Ecology. 1983;**9**(8):937-944. DOI: 10.1007/BF00982203

[117] Kobe RK, Iyer M, Walters MB. Optimal partitioning theory revisited: Nonstructural carbohydrates dominate root mass responses to nitrogen. Ecology. 2010;**91**(1):166-179. DOI: 10.1890/09-0027.1

[118] Williams RD, Quimby PC, Frick KE. Intraspecific competition of purple nutsedge (*Cyperus rotundus*) under greenhouse conditions. Weed Science. 1977;**25**:477-481. DOI: 10.1017/s0043174500033944

[119] Bergmark CL, Jackson WA, Volk RJ, Blum U. Differential inhibition by ferulic acid of nitrate and ammonium uptake in *Zea mays* L. Plant Physiology. 1992;**98**(2):639-645. DOI: 10.1104/pp.98.2.639

[120] Schenk HJ. Root competition: Beyond resource depletion. Journal of Ecology. 2006;**94**(4):725-739. DOI: 10.1111/j.1365-2745.2006.01124.x

[121] Duke SO. Proving allelopathy in crop–weed interactions. Weed Science. 2015;**63**:121-132. DOI: 10.1614/WS-D-13-00130.1

[122] Scavo A, Abbate C, Mauromicale G. Plant allelochemicals: Agronomic, nutritional and ecological relevance in the soil system. Plant and Soil. 2019;**442**(1-2):23-48. DOI: 10.1007/s11104-019-04190-y

[123] Iqbal J, Cheema ZA, An M. Intercropping of field crops in cotton for the management of purple nutsedge (*Cyperus rotundus* L.). Plant and Soil. 2007;**300**(1-2):163-171. DOI: 10.1007/s11104-007-9400-8

[124] Snaydon RW. An analysis of competition between plants of *Trifolium repens* L. populations collected from contrasting soils. Journal of Applied Ecology. 1971;**8**:687-697. DOI: 10.2307/2402677

[125] Tuor FA, Froud-Williams RJ. Influence of nitrogen on competition between purple nutsedge, maize and soybean. International Journal of Pest Management. 2002;**48**(1):73-79. DOI: 10.1080/09670870110094378

[126] M H. Competitive Effects of three perennial weeds, *Cynodon dactylon* (L.) Pers., *Cyperus rotundus* L. and *Sorghum halepense* (L.) Pers., on young citrus. Journal of Horticultural Sciences. 1973;**48**(2):135-147. DOI: 10.1080/00221589.1973.11514513

[127] El-Rokiek KG, El-Masry RR, Messiha NK, Ahmed SA. The allelopathic effect of mango leaves on the growth and propagative capacity of purple nutsedge (*Cyperus rotundus* L.). Journal of American Science. 2010;**6**(9):151-159. DOI: 10.7537/marsjas060910.16

[128] Blum U. Effects of microbial utilization of phenolic acids and their phenolic acid breakdown products on allelopathic interactions. Journal of Chemical Ecology. 1998;**24**:685-708. DOI: 10.1023/A:1022394203540

[129] Einhellig FA. Mechanism of action of allelochemicals in allelopathy. In: Inderjit I, Dakshini KMM, Einhellig FA, editors. Allelopathy, ACS Symposium Series. American Chemical Society. Washington, DC; 1994. pp. 96-116. DOI: 10.1021/bk-1995-0582.ch007

[130] Abenavoli MR, Lupini A, Oliva S, Sorgonà A. Allelochemical effects on net

nitrate uptake and plasma membrane H^+-ATPase activity in maize seedlings. Biologia Plantarum. 2010;**54**(1):149-153. DOI: 10.1007/s10535-010-0024-0

[131] Schmidt S, Ley RE. Microbial competition and soil structure limit the expression of allelochemicals in nature. In: Inderjit, Dakshini KMM, Chester LF, editors. Principles and Practices in Plant Ecology: Allelochemical Interactions. Boca Raton, FL: CRC Press; 1994. pp. 339-351

[132] Bangarwa SK, Norsworthy JK, Jha P, Malik M. Purple nutsedge (*Cyperus rotundus*) management in an organic production system. Weed Science. 2008;**56**:606-613. DOI: 10.1614/WS-07-187.1

[133] Peerzada AM. Biology, agricultural impact, and management of *Cyperus rotundus* L.: The world's most tenacious weed. Acta Physiologiae Plantarum. 2017;**39**:270. DOI: 10.1007/s11738-017-2574-7

[134] Mahmood ARIF, Cheema ZA. Influence of sorghum mulch on purple nutsedge (*Cyperus rotundus* L.). International Journal of Agriculture and Biology. 2004;**6**(1):86-88

[135] Boz O. Allelopathic effects of wheat and rye straw on some weeds and crops. Asian Journal of Plant Sciences. 2003;**2**:772-778. DOI: 10.3923/ajps.2003.772.778[136]

[136] Patterson DT. Suppression of purple nutsedge (*Cyperus rotundus*) with polyethylene film mulch. Weed Technology. 1998;**12**:275-280. DOI: 10.1017/S0890037X00043815

[137] Wang G, McGiffen ME Jr, Ogbuchiekwe EJ. Crop rotation and effects on *Cyperus rotundus* and *C. esculentus* population dynamics in southern California vegetable production. Weed Research. 2008;**48**:420-428. DOI: 10.1111/j.1365-3180.2008.00649.x

[138] Stevens KL. Polyacetylenes as allelochemicals. In: Putnam AR, Tang CS, editors. The Science of Allelopathy. John Wiley & Sons; 1986. pp. 219-228

[139] Xuan TD, Anh LH, Khang DT, Tuyen PT, et al. Weed allelochemicals and possibility for pest management. International Letters of Natural Sciences. Netherlands. 2016;**56**:25-39. DOI: 10.18052/www.scipress.com/ILNS.56.25

[140] Zhang K, Shen Y, Fang YM, Liu Y. Changes in gametophyte physiology of *Pteris multifida* induced by the leaf leachate treatment of the invasive *Bidens pilosa*. Environmental Science and Pollution Research. 2016;**23**:3578-3585. DOI: 10.1007/s11356-015-5589-x

[141] Hong NH, Xuan TD, Eiji T, Khanh TD. Paddy weed control by higher plants from Southeast Asia. Crop Protection. 2004;**23**:255-261. DOI: 10.1016/j.cropro.2003.08.008

[142] Krumsri R, Suwunnamek U, Homhaul W, Laosinwattana C, et al. Allelopathic effects of *Bidens pilosa* var. radiata and its preliminary utilization to control weeds in rice. International Journal of Agricultural Technology. 2015;**11**:1875-1886

[143] Poonpaiboonpipat T, Poolkum S. Utilization of *Bidens pilosa* var. *radiata* (Sch. Bip.) Sherff integrated integrated with water irrigation for paddy weed control and rice yield production. Weed Biology and Management. 2019;**19**:31-38. DOI: 10.1111/wbm.12173

[144] Zhang Z, Liu Y, Yuan L, Weber E, et al. Effect of allelopathy on plant performance: A meta-analysis. Ecology Letters. 2021;**24**(2):348-362. DOI: 10.1111/ele.13627

Chapter 25

Studies on the Short-Term Effects of the Cease of Pesticides Use on Vineyard Microbiome

Simona Ghiță, Mihaela Hnatiuc, Aurora Ranca, Victoria Artem and Mădălina-Andreea Ciocan

Abstract

In this chapter, an overview of the impact of phytosanitary treatments on the vineyard microbiome is provided, together with the results of the research we conducted. The studied plant material consisted of grapevine from the cultivars Sauvignon blanc and Cabernet Sauvignon, cultivated within the plantation of the Research Station for Viticulture and Enology from Murfatlar, Romania. For each cultivar, a treated plot and an untreated plot were established. For each of those, the phyllosphere microbiota was quantified using the epifluorescence microscopy method, followed by automated image analysis using CellC software. At the same time, the soil fungal diversity was evaluated in three stages during the year 2021, using microscopic morphological criteria. The results give useful information regarding the phytosanitary state of the studied plant, as well as the short-term effects produced by the ceasing of pesticide application on the grapevine microbiota.

Keywords: grapevine, phyllosphere microbiota, soil fungi, biodiversity, vineyard microbiome

1. Introduction

It is widely known that plants, like all the other living organisms, are colonized by a multitude of microorganisms [1]. Microbial communities are generally described from the composition (abundance and diversity of the populations that establish the community) and function (behavior and metabolic activity) points of view [2]. The extremely complex functions of microbial communities are not yet significantly understood, even though numerous studies were conducted on the composition of microbial communities [3]. Being modulated by abiotic and biotic factors, predation, competition, and cooperation interactions take place between the members of microbial communities. The environmental effects generated by the microbial activity have the ability of altering the aforementioned interactions furthermore [4].

Microorganisms that have a close relationship with the host plant, regardless of the environmental variables, form the core microbiome [5]. This core microbiome comprises keystone microorganisms that possess genes that can improve the fitness of the holobiont, which were selectively chosen from an evolutionary point of view [6]. Bacterial strains belonging to the families *Pseudomonadaceae, Hyphomicrobiaceae,* and *Micrococcaceae* have been found in relation with the grapevine in any environmental condition, being part of its core microbiome [7].

On the other side of the spectrum, microorganisms found in a lower abundance, which are deeply influenced by the geographical location and habitat characteristics, represent the satellite taxa [8]. Despite their low abundance, satellite microorganisms have critical roles, such as protecting the plant against pathogens through the emission of volatile compounds with antifungal properties [9].

Although not long ago, the focus of plant-microorganisms interactions has been pathogenicity [10]; recently, the ability of some plant-associated microorganisms to directly or indirectly improve plant fitness and performance has been described, as well as the potential of using microbes as a replacement for some synthetic phytosanitary products [1]. A better grasp of the concept of the holobiont can lead to a better future of viticulture, as biocontrol, biofertilization, and biostimulation are realistic and achievable options to reduce the impact of biotic and abiotic stressors, as well as the use of chemical pesticides and fertilizers [11].

The aim of this study is to illustrate the impact of ceasing pesticide use for one year in vineyards on phyllosphere microbiota and soil fungi.

2. Vineyard microbiota

The term "microbiota" refers to the ensemble of microorganisms that exist in a defined environment [12]. In the vineyard ecosystem, endogenous factors, such as the plant age and cultivar, or exogenous factors, such as the cultivation system, geographic location of the plantation, farming techniques, seasonality, human intervention, soil characteristics, and surrounding plants, among others, can influence the presence of grapevine associated microorganisms to a certain extent [13]. It has been pointed out that the effects produced by the composition of the vineyard microbial communities may affect the phytosanitary status of the grapevine, and therefore wine quality, in a significant way [14]. The unique combination of bacteria, fungi, and other microscopic organisms that are found in association with the grapevine and the vineyard soil has even been termed "microbial terroir," as it was found in the recent years to imprint distinctive traits on wine [15]. Zoochory, hydrochory, anemochory, and anthropochory are the known microorganisms dispersal methods that transfer microbiota from the surrounding environment to the grapevine or from one grapevine to another [16]. Perennial plant structures, such as canes, spurs, and bark harbor a greater, more stable microbial diversity, being at the same time one of the sources for the microbiota of the ephemeral structures – leaves, flowers, and berries [17].

2.1 Grapevine as a holobiont: The phyllosphere

Like many other eukaryotes, the grapevine is colonized by a multitude of microorganisms that play a certain role in its growth and survival. First used in 1991 [18], the term "holobiont" evolved to describe a host and the microbial community associated with it [19]. The holobiont concept states that, as it is the case for the animal kingdom, the plant's health state is deeply influenced by the composition of its microbiota [16]. Depending on their role in relation to the plant, microbial species may be beneficial, pathogenic, or neutral [4].

Microorganisms may be found on the surface of grapevine organs, composing the epiphytic microbiota [20], or they may reside inside the plant tissues, making up the endophytic microbiota [21]. Some microbial species can be found both outside the plant structures and inside their tissues [16], meaning that microorganisms find gateways in piercing wounds caused by insects, stomata, or intercellular junctions, among others [22].

The sum of aerial plant organ surfaces represents the phyllosphere [23]. Depending on the plant organ they populate, phyllosphere microorganisms can be a part of the microbiota of the following plant compartments: leaves – phylloplane; flowers – anthosphere; fruits – carposphere; and stems – caulosphere. The phyllosphere is an open system colonized by complex microbial communities, even though the habitat can be considered hostile, as it is exposed to temperature oscillations, UV radiation, and plant-secreted antimicrobial compounds, as well as low water and nutrient accessibility [24]. The phyllosphere is dominated by the phylloplane, represented by the photosynthetically active foliar surface [23]. Considering the fact that the grapevine is a woody perennial plant that sheds leaves each autumn, the phylloplane is an ephemeral environment [24].

From a nutrient perspective, the foliar ecosystem is oligotrophic, due mainly to the presence of the hydrophobic cuticle that prevents plant metabolites from leaching and reduces water evaporation [25]. However, the presence of stomata, hydathodes, veins, and trichomes can assure a better nutrient supply for microorganisms [24]. Due to the distribution of such structures at the foliar level, the abaxial and adaxial sides of the leaf are colonized by different microorganisms [26]. In most cases, bacterial and fungal cells are found forming aggregates, held together by extracellular polymeric substances that can prevent desiccation [24]. Some phyllosphere inhabitants have also been found to protect their host plants through the substances they produce, that act like pesticides, stimulators, or fertilizers [27].

The other more intensely studied ephemeral grapevine organ is the fruit, the interest shifting often in this case from the health state of the plant to the impact on wine-making [16].

2.2 Vineyard soil microbiota

The soil is an everchanging complex environment, dominated by microbial activity [28]. As they have an important role in the cycle of nutrients and the decomposition of organic matter, microorganisms are a decisive factor in determining soil quality. A wide range of organisms coexists in the soil, including bacteria, fungi, archaea, viruses, oomycetes, protists, and arthropods, which are involved in complex trophic networks [29]. Bacteria and fungi are the dominant taxonomic groups, accounting for approximately 90% of the microbiota found in soil samples [30]. Bacteria are the most abundant soil microorganisms and are the first to react and reproduce when their optimal conditions are met [31]. However, in spite of having longer generation times, fungi are more efficient at decomposing organic substrates and have more stable populations [32, 33]. Although some species are phytopathogenic, there are numerous fungal species capable of antagonizing plant pathogens, stimulating vegetative growth, and decomposing plant residues [28].

The composition of soil microbial communities differs significantly both from a quantitative and qualitative point of view, being deeply influenced by the presence of nutrients, water, applied substances, and farming techniques, to name a few [34]. Most of the times, the exogenous factors play a significant role, but it has been pointed out that plant genotypes also possess the tools to intervene in the selection of root-associated microorganisms [35]. Soil is often regarded as a reservoir for the microbiota of the leaves, flowers, and grapes, as more similarities have been described between each of those aerial plant compartments and the soil than between each other [16]. Understanding the interaction between the plant and soil microbiota is essential in order to have a better grasp of the way farming practices affect the soil habitat [36].

Depending on the soil's relation to the plant, several compartments are distinguished: the endorhizosphere, the rhizosphere, and the bulk soil [37]. The rhizosphere is the most intensely studied soil region in relation to the plant, represented by the soil located in the immediate proximity to the plant root system. Microorganisms found in bulk soil are mostly inactive in comparison with those found in the rhizosphere [36] because the latter is characterized by a high nutrient content due to the release of rhizodeposits. These secretions contain sugars, amino acids, organic acids, flavonoids, and terpenoids [11, 38], which trigger a chemotactic response for some microorganisms [39]. The composition of rhizosphere microbial communities fluctuates in accordance with the root exudate patterns specific to the plant's vegetative cycle and health status [40, 41]. However, it has been pointed out that the high nutrient content makes the microbial diversity poorer in the rhizosphere in comparison with the bulk soil [36], as carbon inhibits the growth of some microorganisms when it is found in such quantities [6].

2.3 The impact of pesticide use on vineyard microbiota

Like any other crop, grapevine is susceptible to diseases, which are most often controlled by using chemical pesticides. Fungal diseases pose the biggest threat, making it necessary to use fungicides constantly [16]. Due to the fact that grapevine is one of the crops that require very frequent applications of phytosanitary products, the number of pesticide treatments and the maximum allowed quantity per year has been regulated by the European Commission [42]. In conventionally treated vineyards, chemical pesticides are used, raising the incidence of problems regarding pesticide resistance and the presence of residual pesticides [43]. In organically treated vineyards, copper-based fungicides are viewed as the most important treatments for the most commonly occurring diseases, although recently copper was added to the list for substitution candidates [42].

It is a known fact that phytosanitary treatments present the unwanted potential of affecting the structure and function of the microbiota, as their spectrum is too broad to include only the target microorganisms [27]. The composition of the soil microbiota is sensitive to the action of the chemical treatments. As it was pointed out in a study comparing conventional, organic, and biodynamic cultivation systems, the greatest microbial diversity and richness were found in the soil where grapevine was grown organically [16]. As soil is one of the main reservoirs for the phyllosphere microbiota, the composition of its microbial communities has a critical role in determining the microbial communities found in relation with other plant compartments [44]. Environmental factors, pathogens, and the plant itself are elements that have a well-established role in the manifestation of diseases. The other factor that intervenes in disease development is thought to be the composition of the microbiota [45], although the mechanisms that can successfully manipulate the microbial communities in order to inhibit the occurrence of diseases is not particularly well understood [43].

An essential aim of organic viticulture is reducing the use of pesticides without affecting the yield and production of grapes. Biological fungicides based on microorganisms have been recently developed and present an advantage, as they may be applied at any given time without worrying about the residual presence of pesticides on grapes [46]. Reducing the input of synthetically obtained pesticides can also be achieved by certain farming practices, such as altering the plant microclimate in order to avoid the optimal conditions for the development of pathogens, reducing the overwintering inoculum, or applying treatments only when alerted by devices that use mathematical models and monitor environmental conditions [43].

3. Materials and methods

3.1 Experimental site

Samples were collected from the vineyards of the Research Station for Viticulture and Enology from Murfatlar (RSVEM), Romania. The biological material consisted of grapevines of the Cabernet Sauvignon and Sauvignon blanc cultivars, grafted on *Vitis berlandieri* × *V. riparia* Selection Oppenheim 4 rootstock. The plantations are situated at 50 and 27 m altitude, respectively, and both have a North––South row orientation. The GPS coordinates are 44 10′48.84″ N 28 25′29.18″ E for the Cabernet Sauvignon plantation and 44 10′30.94″ N 28 25′16.70″ for the Sauvignon blanc plantation. Vines were planted in 2011 and 2008, respectively, at 2,2/1,1 m distances and are trained on a double-Guyot training system. The management of the soil, under and between the rows, is done by keeping the soil bare. The soil is of the calcic Chernozem type, with a medium texture and a humus percentage of 2.3%.

For each cultivar, a treated and an untreated plots were established. For the untreated plots, no treatments were applied in the year 2021 in order to observe the short-term effects of ceasing pesticide use on grapevine microbiota. For the conventionally treated plots, the usual treatment scheme has been applied, which involved 8 treatments during the studied year: the first treatment was applied during the dormancy period, consisting of calcium polysulfide; the second during BBCH 53, with products that have cymoxanil, mancozeb, copper oxychloride, and sulfur as active ingredients; the third during BBCH 60, with oxathiapiprolin, folpet, fenhexamid, proquinazid, and alpha cypermethrin; the fourth during BBCH 69, with oxathiopiprolin, folpet, proquinazid, fludioxonil, and cyprodinil; the fifth during BBCH 73, with fosetyl Aluminum, folpet, myclobutanil, and emamectin benzoate; the sixth during BBCH 77, with dimethomorph, mancozeb, metrafenone, boscalid, and hexythiazox; the seventh during BBCH 81, with dithianon, dimethomorph, sulfur, and emamectin benzoate; and the eighth, during BBCH 85, with copper hydroxide, sulfur, and fenhexamid.

IoT sensors were installed in the experimental plots, which were used to monitor, among others, leaf moisture. The PHYTOS 31 leaf wetness sensor measures the dielectric constant on the upper surface of the device, the value being proportional to the present water amount.

At harvest, the average production per vine for each of the 4 studied variants was calculated, in order to assess the impact of pesticide use cease on grape production.

3.2 Phyllosphere microbiota visualization and quantification

For the study of phyllosphere microbiota, sampling was carried out in 2021 during the phenophase BBCH 79 (when most of the bunches were compacted). The samples consisted of 6 leaves taken from one grapevine per variant, on which the sensors were placed, from the base, middle, and top of the canopy. The samples were processed immediately in the microbiology laboratory of Constanta Maritime University.

Considering the fact that the phyllosphere is an oligotrophic system, in which the distribution of nutrients is heterogeneous, squares of approximately 1 cm^2 were randomly chosen for each part of the leaf, which were cut with a sterile scalpel. In order to observe the cut sections under the epifluorescence microscope (N-400FL type with blue filter), they were subjected to an adhesive tape gluing process. The adhesive tape was stained with specific fluorochromes and then placed on a microscopic slide. By applying this technique, it is possible to recover the cuticle from leaves, trapping the microorganisms between the tape and the cuticle. Thus, a very high recovery of cells is permitted, while preserving spatial information.

Using this method, a total number of cells/analyzed surface is obtained, at the same time observing the physiological state of the microorganisms, using specific fluorochromes: SYBR Green (SYBR Green I nucleic acid gel stain. 10,000× in DMSO, Sigma Aldrich 5 ml) and Propidium Iodide (≥94% HPLC, Sigma Aldrich 10 mg). The staining solution is prepared in a ratio of 1:1, and applied for 8–10 minutes, according to [47–49].

The efficiency of fluorescent compounds for evaluating the integrity of cell membranes is determined by selectivity, brightness, excitation, and maximum emission. The final SYBR Green concentration is 10 µl/ml and 10 µg/ml for Propidium Iodide. For each sample, 20 microscope fields were quantified. To visualize the bacteria and fungi on the phyllosphere, the blue filter with a wavelength of 450–480 nm was used, specifically for the chosen fluorochromes. Images were taken with a digital camera and further used for automatic processing, using the "CellC" cell counting software, according to [50].

3.3 Soil fungi identification

For the identification of soil fungi, sampling was carried out in three stages during the year 2021, according to the BBCH phenophases of the grapevine: the first stage was BBCH phenophase 11 (appearance of the first leaf), the second was BBCH phenophase 79 (when several bunches were compacted), and the third was BBCH phenophase 97 (end of leaf fall). From each plot, the soil was collected from the horizon 0–10 cm, from the base of the grapevine trunks, analyzing a total of 12 samples.

Soil samples were processed in the RSVEM microbiology laboratory. The applied technique involves the cultivation of fungi on solid culture media, using the method of serial decimal dilutions. A volume of 0.1 ml of each dilution was spread on the surface of the Rose Bengal CAF Agar (RBCA) medium in triplicate. The plates were incubated at 25° Celsius, being checked initially after 72 h, then daily to observe the growth of the colonies. To avoid redundant isolation of the strains, for each morphotype with specific traits, the colonies present on the 3 plates were counted for the optimal dilution, after which they were isolated on potato dextrose agar (PDA). The modified slide culture method [51] was applied in order to allow a more efficient observation of the fungal structures under the microscope. The fungal strains were identified based on morphological criteria to the genus level, according to [52, 53].

Although this method offers information on the main fungal taxa present in soil, it is important to mention that a more detailed research would have been possible with the aid of molecular identification techniques.

3.4 Data analyses

For the identified soil fungi, the frequency for each genus was calculated according to the formula Di = (Ni/N) ×100, where Di = the frequency of genus i; Ni = UFC number for gender i; and N = total number of CFUs. According to this formula, the genus frequency can be grouped into several classes: <0.5% = rare, ≥0.5 < 1.5% = occasional, ≥1.5 < 3.0% = common, and ≥3.0% = abundant [54]. The ANOVA test was applied to determine whether there were statistically significant differences between the number of CFUs for the sampling phenophases, and the t test was used for treatment types and grapevine cultivars, taking into account a significance level of 5%. For the fungal populations, Sørensen's similarity index and Shannon diversity index were calculated.

4. Results and discussion

4.1 Phyllosphere microbiota

Microscopy analyses revealed that bacteria are prevalent in epidermal cell grooves, around trichomes and the stomatal opening, and less prevalent on the elevated surface of epidermal cells. Bacteria are the most abundant microbial group in the phyllosphere, followed by fungi. The measured density of bacteria is from 10^3 to 10^7 cells per square centimeter of leaf tissue (**Figure 1**), while the density of fungal structures is ranging from 10^2 to 10^4 cells per square centimeter of leaf tissue (**Figure 2**).

As it can be seen, the untreated plots show a significantly higher number of microorganisms per square centimeter, at least an order of magnitude greater in comparison with the treated ones, for both cultivars and for all canopy compartments. Microorganisms are placed in higher density on the abaxial side of the leaf, respectively, on the leaves from the base of the canopy. The difference in microbial density between the two vine varieties can also be attributed to the mesoclimate (hill vs. valley), as the plots where Cabernet Sauvignon is cultivated are located at a higher altitude and micro currents can form, that can reduce the humidity conditions favorable to the microbiota development.

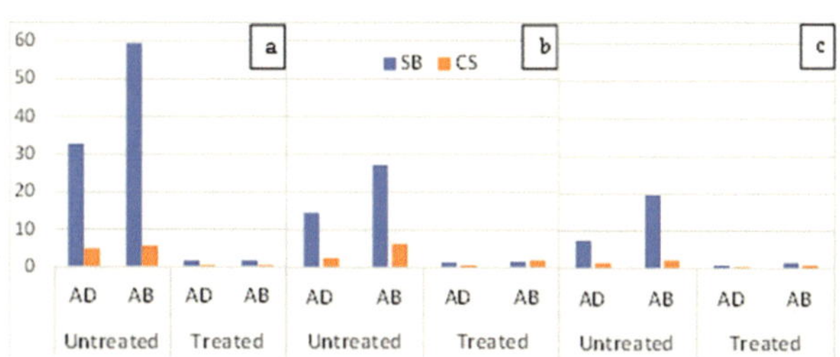

Figure 1.
Average number of bacterial cells ×10^5 on leaves from the base (a), middle (b), and top (c) of the canopy, from the adaxial (AD) and abaxial (AB) sides, for sauvignon blanc (SB) and cabernet sauvignon (CS).

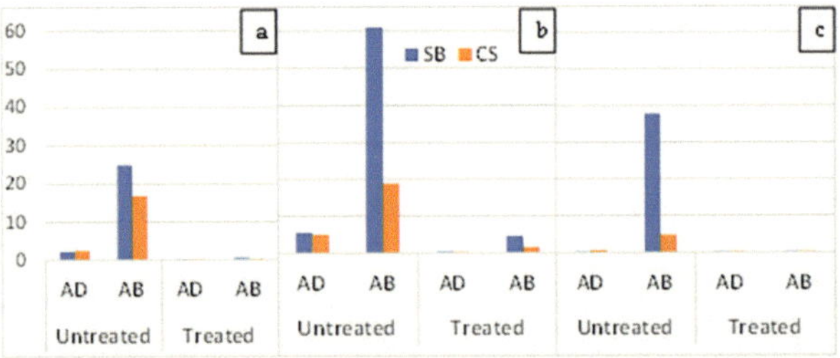

Figure 2.
Average number of fungal structures ×10⁴ on leaves from the base (a); ×10³ on leaves from the middle (b), and top (c) of the canopy, from the adaxial (AD) and abaxial (AB) sides, for sauvignon blanc (SB) and cabernet sauvignon (CS).

The presence of the analyzed microbiota on the abaxial surface of the leaf is probably due to stomata, which represent a natural entry pathway for endophytic microorganisms. The laminar layer may also play a significant role, as moisture emitted by stomata can be retained at this level, reducing the water stress of epiphytic microorganisms. Leaf wetness was analyzed in both plots using IoT sensors (**Figure 3**). In the studied period, the highest leaf wetness values were observed for the leaves from the untreated Sauvignon blanc plots, which also harbored the greatest number of microorganisms.

Epiphytic microbiota has a first contact with the leaf cuticle, which may contain a higher or lower amount of wax that may prevent bacterial colonization [55]. Bacterial aggregates can lead to the formation of biofilms on the leaf surface, which represent a form of adaptation that offers protection against desiccation. Phyllosphere-colonizing bacteria can alter the environment in order to modify the plant's immune system, reflected in differential host responses. Clearly, these bacteria are very dense ($10^7/cm^2$) and contribute to many processes in the behavior of the individual plants. The results of the analysis done directly on the leaf surface show that biofilms may be tens of micrometers thick and could form extensive networks that cannot be quantified (**Figure 4**). Biofilms contain multiple microbial species and could create physical barriers on the leaf surface and establish chemical gradients, promoting metabolic exchange. The biofilm could protect the microbial community under adverse conditions and confer them a survival and colonization selective advantage. Extracellular polymeric substances are usually produced, having the role of maintaining the foliar surface hydrated and concentrating detoxifying enzymes at the same time [24].

Even though the results illustrate the fact that pesticide use influences phylloplane microbiota in a negative way, it is important to mention the impact of ceasing

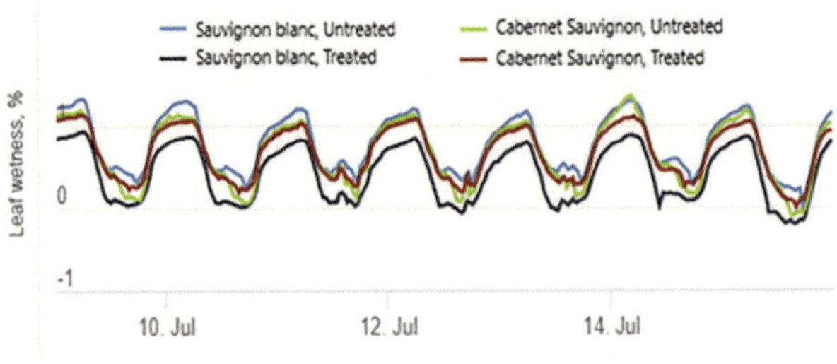

Figure 3.
Leaf wetness in the studied plots.

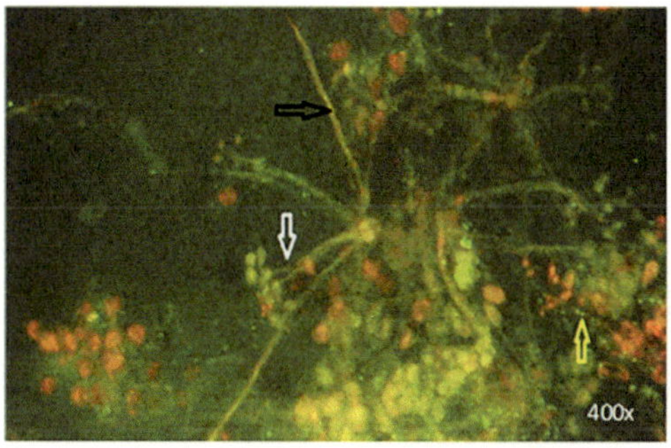

Figure 4.
Fluorescence micrograph of the microorganisms colonizing a grapevine leaf. Yellow arrow – Bacteria present on plant veins; black arrow – fungal hypha; and white arrow – fungal structures.

pesticide use on grape production, as the untreated grapevines were affected by diseases; for the Sauvignon blanc cultivar, there was a 50.6% decrease in grape production for the untreated variant when compared with the treated one, while for the Cabernet Sauvignon cultivar, the untreated variant had on average 33.6% lower production in comparison with the treated variant. Thus, the variant that harbored the highest number of microorganisms per square centimeter also showed the lowest grape production.

4.2 Soil fungi

A total of 123 strains were isolated, 44 for the BBCH 11 phenophase, 29 for the BBCH 79 phenophase, and 50 for the BBCH 97 phenophase. In terms of frequency (**Figure 5**), out of the 12 genera identified, the following were classified as abundant: *Penicillium* (37.87%), *Aspergillus* (26.08%), *Fusarium* (10.77%), *Paecilomyces* (5, 00%), *Cladosporium* (4.26%), and *Botrytis* (3.00%). Due to the absence of sporulation, for 8,51% of the strains, an indentification based on morphological criteria was not possible.

None of the isolated fungal strains presented sexual structures, only the anamorphic stage being observed. As a taxonomic classification, all genera belong to

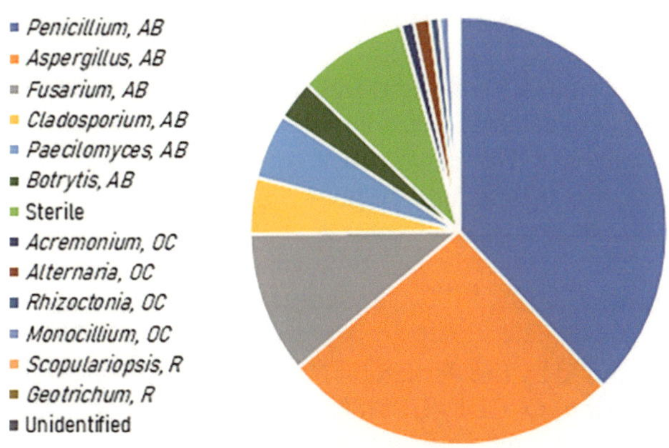

Figure 5.
Frequency of the isolated fungal strains. AB – Abundant; OC – Occasional; R – rare.

the phylum Ascomycota, except the genus *Rhizoctonia*, which belongs to the phylum Basidiomycota. The most predominant class is Eurotiomycetes, represented by 3 genera (68.96%), followed by class Sordariomycetes, represented by 4 genera (12.77%).

A very recent study pointed out that most of the isolated genera, such as *Acremonium*, *Alternaria*, *Aspergillus*, *Fusarium*, and *Penicillium* are very common in vineyard soil [56]. The genera *Penicillium*, *Aspergillus*, *Cladosporium*, *Acremonium*, *Alternaria*, *Botrytis,* and *Scopulariopsis* have also been reported as endophytes of the grapevine [57].

From a statistical point of view, the differences between the treated and the untreated experimental plots in terms of the diversity of isolated genera are not significant ($P = 0.55$, $F < F$ crit).

No statistically significant differences were reported with respect to the studied phenophases or the grapevine cultivars compared to the types of identified fungi. The calculated Shannon index had a higher value for the untreated plots (2.253), in comparison with the treated plots (2.139), whereas the calculated value for Sørensen's similarity index was 73.68%.

Fungal communities found in agricultural soils are influenced by factors, such as soil type, available nutrients, edaphic properties, plant communities, and agrotechnical practices, as well as climatic conditions [58]. The importance of the latter has been highlighted in a study that showed that climatic factors were probably the leading element that caused a variation in fungal communities from 1 year to another [59]. Water stress is a factor known to impact the composition of soil fungal communities [60].

A great number of soil micromycetes are active where readily assimilable elements are found, thus making the soil a "world of asexual microfungi" [61]. Fungi are generally involved in the decomposition of organic matter, the cycling of nutrients, soil aggregates formation, and the mobilization of minerals, among others [62]. Moreover, fungi are extremely adaptable, as they are able to react to detrimental conditions by modifying their form [63].

5. Conclusions

Concerning phylloplane microbiota, the differences between the treated and untreated plots were obvious, with the untreated leaves showing considerably greater numbers of microorganisms for both of the studied cultivars. Thus, the effects of ceasing pesticide use can be readily seen on ephemeral plant structures, such as the leaves.

However, when comparing soil fungi from a quantitative point of view, no significant differences can be seen after only 1 year between the treated and the untreated plots, statistically speaking. This can be due to the fact that pesticides can still persist in the soil residually, affecting microbial populations.

Acknowledgements

This work was funded by the Romanian National Authority for Scientific Research and Innovation, CCCDI - UEFISCDI, for the COFUND-ICT-AGRI-FOOD-MERIAVINO-1, project number 203, within PNCDI III.

Author details

Simona Ghiță[1], Mihaela Hnatiuc[1], Aurora Ranca[2], Victoria Artem[2] and Mădălina-Andreea Ciocan[2*]

1 Constanta Maritime University, Romania

2 Research Station for Viticulture and Enology Murfatlar, Romania

*Address all correspondence to: ciocanm@statiuneamurfatlar.ro

References

[1] Jacoby R et al. The role of soil microorganisms in plant mineral nutrition—Current knowledge and future directions. Frontiers in Plant Science. 2017;**8**:1. Article: 1617

[2] Little AE et al. Rules of engagement: Interspecies interactions that regulate microbial communities. Annual Review of Microbiology. 2008;**62**:375-401

[3] Rainey PB, Quistad SD. Toward a dynamical understanding of microbial communities. Philosophical Transactions of the Royal Society B: Biological Sciences. 2020;**375**(1798):20190248

[4] Manriquez B, Muller D, Prigent-Combaret C. Experimental evolution in plant-microbe systems: A tool for deciphering the functioning and evolution of plant-associated microbial communities. Frontiers in Microbiology. 2021;**12**:2. Article: 619122

[5] Toju H et al. Core microbiomes for sustainable agroecosystems. Nature Plants. 2018;**4**(5):247-257

[6] Lemanceau P et al. Let the Core microbiota Be functional. Trends in Plant Science. 2017;**22**(7):583-595

[7] Zarraonaindia I et al. The soil microbiome influences grapevine-associated microbiota. MBio. 2015;**6**(2):e02527-e02514

[8] Hanski I. Dynamics of regional distribution: The Core and satellite species hypothesis. Oikos. 1982;**38**(2):210-221

[9] Compant S et al. A review on the plant microbiome: Ecology, functions, and emerging trends in microbial application. Journal of Advanced Research. 2019;**19**:29-37

[10] Teixeira PJP et al. Beyond pathogens: Microbiota interactions with the plant immune system. Current Opinion in Microbiology. 2019;**49**:7-17

[11] Berlanas C et al. The fungal and bacterial rhizosphere microbiome associated with grapevine rootstock genotypes in mature and young vineyards. Frontiers in Microbiology. 2019;**10**:12-13. Article: 1142

[12] Marchesi JR, Ravel J. The vocabulary of microbiome research: A proposal. Microbiome. 2015;**3**(1):31

[13] Griggs RG et al. Sources and assembly of microbial communities in vineyards as a functional component of winegrowing. Frontiers in Microbiology. 2021;**12**:2. Article: 673810

[14] Liu D et al. From the vineyard to the winery: How microbial ecology drives regional distinctiveness of wine. Frontiers in Microbiology. 2019;**10**:2. Article: 2679

[15] Bokulich NA et al. Associations among wine grape microbiome, metabolome, and fermentation behavior suggest microbial contribution to regional wine characteristics. MBio. 2016;**7**(3):e00631-e00616

[16] Bettenfeld P et al. The microbiota of the grapevine holobiont: A key component of plant health. Journal of Advanced Research. 2021 [In press]

[17] Vitulo N et al. Bark and grape microbiome of Vitis vinifera: Influence of geographic patterns and agronomic management on bacterial diversity. Frontiers in Microbiology. 2019;**9**:2. Article: 3203

[18] Margulis L, Fester R. Symbiosis as a Source of Evolutionary Innovation:

Speciation and Morphogenesis. Cambridge, MA: MIT Press; 1991

[19] Simon J-C et al. Host-microbiota interactions: From holobiont theory to analysis. Microbiome. 2019;**7**(1):5

[20] Vandenkoornhuyse P et al. The importance of the microbiome of the plant holobiont. New Phytologist. 2015;**206**(4):1196-1206

[21] Doty SL. Functional importance of the plant endophytic microbiome: Implications for agriculture, forestry, and bioenergy. In: Doty SL, editor. Functional Importance of the Plant Microbiome: Implications for Agriculture, Forestry and Bioenergy. Cham: Springer International Publishing; 2017. pp. 1-5

[22] Mercado-Blanco J. Life of Microbes Inside the Plant. Cham: Springer; 2015. pp. 25-32

[23] Goswami S, Goel N, Majumdar RS. Phylloplane microbes impact host physiology: A review. Journal of Plant Protection Research. 2021;**61**(3):213-221

[24] Vorholt JA. Microbial life in the phyllosphere. Nature Reviews Microbiology. 2012;**10**(12):828-840

[25] Schlechter RO, Miebach M, Remus-Emsermann MNP. Driving factors of epiphytic bacterial communities: A review. Journal of Advanced Research. 2019;**19**:57-65

[26] Vacher C et al. The Phyllosphere: Microbial jungle at the plant-climate Interface. Annual Review of Ecology Evolution and Systematics. 2016;**47**:5

[27] Perazzolli M et al. Resilience of the natural phyllosphere microbiota of the grapevine to chemical and biological pesticides. Applied and Environmental Microbiology. 2014;**80**(12):3585-3596

[28] Gomes NCM et al. Dynamics of fungal communities in bulk and maize rhizosphere soil in the tropics. Applied and Environmental Microbiology. 2003;**69**(7):3758-3766

[29] Li J et al. Rhizosphere microbiome: The emerging barrier in plant-pathogen interactions. Frontiers in Microbiology. 2021;**12**:2. Article: 772420

[30] Xu H, Yu M, Cheng X. Abundant fungal and rare bacterial taxa jointly reveal soil nutrient cycling and multifunctionality in uneven-aged mixed plantations. Ecological Indicators. 2021;**129**:107932

[31] Deng J et al. Variations in soil bacterial community diversity and structures among different revegetation types in the Baishilazi nature reserve. Frontiers in Microbiology. 2018;**9**:2. Article: 2874

[32] Buerkert A et al. Chapter 18 - nutrient and carbon fluxes in terrestrial agro-ecosystems. In: Marschner P, editor. Marschner's Mineral Nutrition of Higher Plants. Third ed. San Diego: Academic Press; 2012. pp. 473-482

[33] Setati ME et al. The vineyard yeast microbiome, a mixed model microbial map. PLoS One. 2012;**7**(12):e52609

[34] Helgason BL, Walley FL, Germida JJ. Long-term no-till management affects microbial biomass but not community composition in Canadian prairie agroecosytems. Soil Biology and Biochemistry. 2010;**42**(12):2192-2202

[35] Zhou J et al. Wine terroir and the soil Bacteria: An amplicon sequencing–based assessment of the Barossa Valley and its sub-regions. Frontiers in Microbiology. 2021;**11**:2. Article: 597944

[36] Essel E et al. Bacterial and fungal diversity in rhizosphere and bulk soil

under different long-term tillage and cereal/legume rotation. Soil and Tillage Research. 2019;**194**:104302

[37] Martínez-Diz MDP et al. Soil-plant compartments affect fungal microbiome diversity and composition in grapevine. Fungal Ecology. 2019;**41**:234-244

[38] Bakker P et al. The rhizosphere revisited: Root microbiomics. Frontiers in Plant Science. 2013;**4**:1. Article: 165

[39] Kent AD, Triplett EW. Microbial communities and their interactions in soil and rhizosphere ecosystems. Annual Review of Microbiology. 2002;**56**:211-236

[40] Dunfield KE, Germida JJ. Seasonal changes in the rhizosphere microbial communities associated with field-grown genetically modified canola (*Brassica napus*). Applied and Environmental Microbiology. 2003;**69**(12):7310-7318

[41] Bulgarelli D et al. Revealing structure and assembly cues for Arabidopsis root-inhabiting bacterial microbiota. Nature. 2012;**488**(7409):91-95

[42] Rantsiou K et al. Impact of chemical and alternative fungicides applied to grapevine cv Nebbiolo on microbial ecology and chemical-physical grape characteristics at harvest. Frontiers in Plant Science. 2020;**11**:2. Article: 700

[43] Pertot I et al. A critical review of plant protection tools for reducing pesticide use on grapevine and new perspectives for the implementation of IPM in viticulture. Crop Protection. 2017;**97**:70-84

[44] Coller E et al. Microbiome of vineyard soils is shaped by geography and management. Microbiome. 2019;**7**(1):140

[45] Lamichhane JR, Venturi V. Synergisms between microbial pathogens in plant disease complexes: A growing trend. Frontiers in Plant Science. 2015: 1-2. Article: 385

[46] Escribano-Viana R et al. Impact of chemical and biological fungicides applied to grapevine on grape biofilm, must, and wine microbial diversity. Frontiers in Microbiology. 2018;**9**:59-59

[47] Luna GM, Manini E, Danovaro R. Large fraction of dead and inactive bacteria in coastal marine sediments: Comparison of protocols for determination and ecological significance. Applied and Environmental Microbiology. 2002;**68**(7):3509-3513

[48] Lunau M et al. An improved method for counting bacteria from sediments and turbid environments by epifluorescence microscopy. Environmental Microbiology. 2005;**7**(7):961-968

[49] Manini E, Danovaro R. Synoptic determination of living/dead and active/dormant bacterial fractions in marine sediments. FEMS Microbiology Ecology. 2006;**55**(3):416-423

[50] Selinummi J et al. Software for quantification of labeled bacteria from digital microscope images by automated image analysis. BioTechniques. 2005;**39**(6):859-863

[51] Rosana Y et al. Modified slide culture method for faster and easier identification of dermatophytes. Microbiology Indonesia. 2014;**8**(3):7

[52] Watanabe T. Pictorial Atlas of Soil and Seed Fungi: Morphologies of Cultured Fungi and Key to Species. Boca Raton: CRC Press; 2002

[53] Barnett H, Hunter B. Illustrated Genera of Imperfect fungi. St. Paul, MN: APS Press; 1998. p. 218

[54] Schnittler M, Stevenson S. Myxomycete biodiversity in four different Forest types in Costa Rica. Mycologia. 2000;**92**:630

[55] Marcell LM, Beattie GA. Effect of leaf surface waxes on leaf colonization by Pantoea agglomerans and Clavibacter michiganensis. Molecular Plant-Microbe Interactions. 2002;**15**(12):1236-1244

[56] Yan H et al. Diversity of soil fungi in the vineyards of Changli region in China. Canadian Journal of Microbiology. 2022;**68**(5):341-352

[57] Dissanayake A et al. Direct comparison of culture-dependent and culture-independent molecular approaches reveal the diversity of fungal endophytic communities in stems of grapevine (Vitis vinifera). Fungal Diversity. 2018;**90**:4

[58] Magdoff F, Van Es H. Building Soils for Better Crops: Ecological Management for Healthy Soils. Washington DC: SARE Outreach; 2021

[59] Chou MY et al. Vineyard under-vine floor management alters soil microbial composition, while the fruit microbiome shows no corresponding shifts. Scientific Reports. 2018;**8**(1):11039

[60] Frac M et al. Fungal biodiversity and their role in soil health. Frontiers in Microbiology. 2018;**9**:707

[61] Oliveira LG et al. Diversity of filamentous fungi isolated from the soil in the semiarid area, Pernambuco, Brazil. Journal of Arid Environments. 2013;**95**:49-54

[62] Yuvaraj M, Ramasamy M. Role of Fungi in Agriculture. Chapter in Biostimulants in Plant Science. Rijeka: IntechOpen; 2020. p. 12

[63] Sun J-M et al. Analysis of the genetic structure of Sclerotinia sclerotiorum (lib.) de Bary populations from different regions and host plants by random amplified polymorphic DNA markers. Journal of Integrative Plant Biology. 2005;**47**(4):385-395

Chapter 26

The Evaluation of the Macrophyte Species in the Accumulation of Selected Elements from the Varkenslaagte Drainage Line in the West Wits, Johannesburg South Africa

Tinyiko Salome Mthombeni

Abstract

Although mining has over the centuries improved the livelihoods and economies of many countries, the results have not spared the environment's luxurious legacy. Acid mine drainage contaminated sites with heavy metals that affect negatively and positively the macrophytes plants that grow on those sites. Accumulated elements by macrophytes planted on artificial wetlands portray the relative bioconcentration and translocation factors. Various elements were measured in the sediment, water, and macrophytes from the sampled sites and the results indicate that concentrations accumulated by plants play a significant role in biological and chemical processes in soil-water-plant relations. When comparing the drinking water quality standards by international organizations that were used as a guideline for the comparisons of elements concentration levels of elements found in water, Iron (Fe), Nickel (Ni), Manganese (Mn), and Copper (Cu) were found to be above the international water quality standards for drinking water and their average concentrations were 2230, 282, 5950, and 14,080 µg/l respectively. The sequence of elements accumulation by the macrophytes differed per plant and each of the three macrophytes plants was a hyperaccumulator of a certain element.

Keywords: macrophytes, acid mine drainage, phytoremediation, artificial wetlands, elements

1. Introduction

Although South Africa was the biggest producer of gold globally, the industry has been experiencing several drawbacks such as mine closure of older mines and shafts, declining mineral production, exhaustion of gold reserves, global low gold prices, the high energy requirement for deep-level mining, high wage demands and social unrests as well as the generation of acid mine drainage from the mines and tailings storage facilities [1]. The cessation of the large mining operations has detrimental effects as access to gold reserves are far underground, and mining operations resorted to dewatering activities to keep the groundwater level away from the mining operations [2]. Cessation of mining further resulted in flooding of the voids, a substantial cause of groundwater and surface water contamination by acidic water [3]. The acidic Sulfur rich wastewater or effluent from mining and industrial environments has greater consequences for Acid Mine Drainage in both actively operating and abandoned mines [4]. During the active mining process, the extraction of the gold-bearing conglomerate layer is crushed and gold become extracted [5]. Once the gold is extracted, the crushed rock is deposited on heaps known as slimes or tailings dumps, and generally, the gold-bearing conglomerates contain approximately 3% pyrite which gets deposited in slimes and tailings dumps. AMD is defined as a natural process (more correctly termed "acid rock drainage", or ARD) that occurs when sulfur-containing minerals become exposed to water and oxygen, in the presence of bacteria known as the *Acidithiobacillus* and *Ferrooxidans* [6]. Sulfides in pyrite rock (Fool's gold) then react with oxygen and water and leading to the production of sulfuric acid.

The sulfuric acid percolates through the slimes dam and dissolves some of the heavy metals. The resultant acidic, net acidic, and saline plume enters the surrounding soils, and eventually enters the groundwater and surface water bodies. AMD is a slow process characterized by low pH and high salinity levels with higher concentrations of sulfate, iron, aluminum, manganese, and the possibility of radionuclides [7]. The dark, reddish-brown and low pH water (often lower than 2.5) is difficult to rectify. The most important salts and heavy metals associated with AMD pose serious contamination threats to the environment and human health [8]. Metals such as Fe, Mn, Al, and other heavy metals as well as metalloids such as arsenic [9]. Heavy metals such as mercury and metalloid such as arsenic can become toxic and pose additional risks to the environment even when they are introduced to the water system in minute amounts [10].

Factors such as pH, redox potential, and soil types have a greater impact on the uptake of the element in the sediments. Soil types have a significant effect on the uptake of metals. In addition, clay soils, in particular, have higher sorption capacity which in turn reduces the availability of metals. Clay soils with high organic content enhance the conditions that favor successive precipitation of sulfides, and this can reduce the available elements at the lower depths. In addition, soils with less organic matter content tend to release elements from the sediments and improve metals uptake, a requirement of plant growth. Salinity levels also improve the rates of metal uptake especially Cr, Ni, and Zn [11]. According to [12], variations in water pH affect the ability of elements to be soluble. Furthermore, this also impacts the deposition ability of metals in the sediments as well as in the water column. The concentrations of Zn, Mn, and Ni in sediment have a direct bearing on the increased uptake of elements by plants [13].

Many aquatic macrophytes are classified as heavy metal accumulators and they are known to accumulate metals to various degrees and store them in below-ground tissues (rhizomes, roots) or above-ground tissues (leaves, stems). In some cases, aquatic macrophytes have been found to absorb higher concentrations of metals than are found in the water [14]. Plant species differ in their ability and tolerance to metal uptake and accumulation. Some plant species can accumulate high concentrations of a single metal and translocate it to the roots, rhizomes, stems, and/or leaves, while others can accumulate more than one element in different parts. Another category of

Site name	Site characteristics	GPS Location
Site 1	The first artificial wetland situated at the foothill of the mine dump. It received AMD water seeping from the tailings dam. The pH of this site was lower than one on many occasions when recorded. The soil was dominated by surface salt crusting with some visible salt crumbs attached to the base of the plant stem. Gravel was the dominating soil structure with visible red oxides on the soil surface. The plants at this site portrayed stressed conditions with inhibited growth and some dying off before the winter season. The flowing water into this site was through the canal but most of the inflows were through seepage from the tailings dam. As the sites and canal are situated on the foot of the mining dump, it was evident that A M D leaches from the mine dump to the canals and sites, and forms a lot of surface precipitation and salt crusting as shown in **Figures 2** and **3**.	S 26°25 48′ 96″ E 27° 22 16′ 46″
Site 2	The characteristics of this site were almost similar to those of the first wetland. The plants were also shorter with observable signs of stress from the elements at these sites as shown in **Figures 4** and **5**.	S26° 25 49 61″ E27° 25 15′ 92″
Site 3	This was the fourth wetland along the canal. The improvement in pH, temperature, and electrical conductivity was measured. Salt crusting and crumbling at this site were less reduced when compared to the first two sites. The growth of the macrophytes plants was also improving.	S 26° 25 5 l′ 37″ E 27° 22 14′ 51″
Site 4	The wetland consisted of tall green macrophytes plants with some growth of algae species observed. The pH at this site was in an improved range of 5–7. Salt crumbling and crusting were no longer visible at this site. The soil structure consisted of dark clay soils without gravel on the top layers.	S 26° 25 53′ 86″ E 27° 22 13′ 24″
Site 5	This was the biggest wetland of them all, it consisted of tall green macrophytes plants with deep waters. The pH condition at this site was 7. Some development of another aquatic biodiversity such as tadpoles was observed.	S 26° 25 54′ 78″ E 27° 22 12′ 78″

Table 1.
Descriptive summary of the study sites.

plants is known as the excluders and can tolerate metal-rich environments by reducing the number of elements translocated from the below-ground parts to the above-ground parts. For plants to survive, they must adapt to the chemical and physical characteristics of the soil, water, atmosphere, and climate. Plants that grow and survive in metal-contaminated environments (metallophytes) have the distinct characteristic of tolerance. Plants also may be categorized as metal excluders, indicators, accumulators, or even hyperaccumulators [15]. Hyperaccumulators can translocate metals to their above-ground organs, and thereby extract and accumulate quantities that exceed any other species in the same environment (**Table 1**) [25].

1.1 Study methods

The study area is found in the Gauteng Province, Southern Johannesburg, Under the Westonaria Municipality, **Figure 1**. The area is also called the West Wits, it is located in Carletonville, and it is situated along the 15 kilometers (km) Varkenslaagte drainage line, also known as the old canal of the West Wits Operations (AngloGold Ashanti). The area is divided by a rocky ridge called Gatsrand. It covers approximately 3785.5 ha and approximately 38.58% which equates to 14,604 ha under deep mining (Mponeng, Tautona, and Savuka mine). The rest of the area is occupied by mining-related infrastructures such as tailing dams, mining plants, shafts, related operations, and residential wells as excavations. The Northern part of the mining area has been

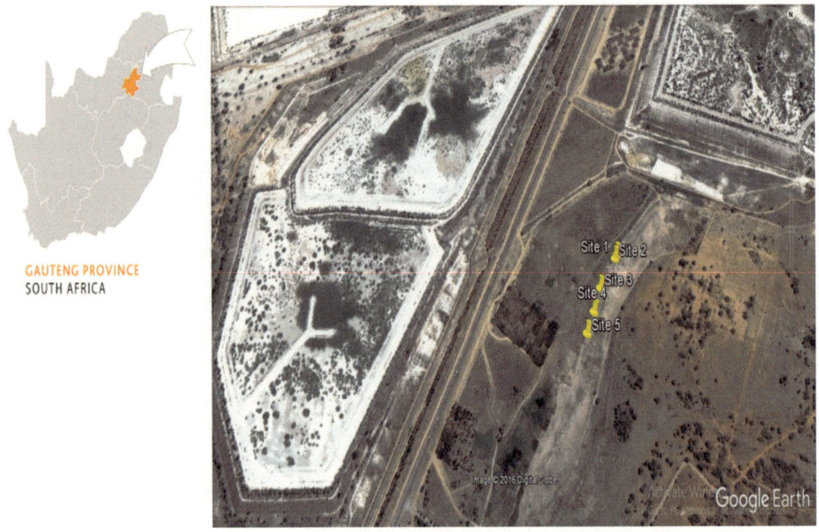

Figure 1.
Map of South Africa showing the location of Western Deep Levels in Carletonville.

Botanical name	Common name	Process of metal accumulation and types of metals accumulated	Reference (s)
Achillea millefolium	Yarrow	Accumulation of Pb, Cd, and Cu	[16]
Azolla pinnata	Water velvet	Biosorption and bioaccumulation of metals of Cu, Pb, Cr, Cd, and Zn	[17]
Bacopa monnieri	Water hyssop	Accumulation of Al, Asm Cd, Cr, Cu, Fe, Hg, Mn, Ni, Pb, and Zn	[18]
Hydrilla verticillita	Hydrilla	Hyperaccumulation of Cr and Cd	[19]
Myriophyllum aquaticum	Parrot feather	Translocation and degradation of metals	[20]
Phragmites australis	Common reed	Reed bed treatment systems for accumulation of Zn, Cu, Se, Pb, and Cd.	[21]
Brassica juncea	Indian mustard	Hyperaccumulation of As, Cd, Mo, and Cr.	[22]
Allium schoenoprasum	Chives	Hyperaccumulation of Pb, Zn, Ni, Cu, Co, and Cd.	[23]
Thlaspi caerulescens	Alpine pennycress	Hyperaccumulation of Zn, Cu, Pb, and Cd	[24]

Table 2.
Some herbaceous plants used in mine sited for phytoremediation.

converted into agricultural land and mining activities while only a small portion of the area is still in its natural state as shown in **Figure 1**.

The study site consisted of 17 artificial wetlands and only five selected wetlands were studied these were artificial wetlands 1, 2, 4, 5, and 7, and regarded as site 1, site 2, site 3,

Figure 2.
Some P. communis *species with AMD water flowing between the plants.*

Figure 3.
Metal precipitation on the soil crusts with some metals sticking to the basal part of the S. corymbosus *and* P. communis *species.*

site 4, and 5 in this study. The summary of the investigated sites is presented in **Table 2**. The selected five sites out of the seventeen, were due to clear observations made in terms of growth and development of the macrophytes and other physical characteristics (algae growth and tadpoles occurring) on these sites when compared to the rest of the other sites. Accessibility was also another factor considered when conducting sampling at those sites. The effectiveness of rehabilitation was much clearer on these sites. The wetlands were grown with macrophyte species of *Typha Capensis*, *Schoenoplectus Corymbosus*, and *Phragmites Communis*.

Figure 4.
Some metal precipitates at the edge of the artificial wetland with some precipitates attached to the lower parts and leaves of the P. communis *spp that fall off often become covered by the salts.*

Figure 5.
Some AMD and metal precipitation of the floodplain of the right side adjacent to the canal.

1.2 In situ physico-chemical measurements, water sampling, and analysis

In situ water measurements were done at each site. At each constructed wetland, the Physico-chemical measurements were done in a sequence of inflow, midflow, and outflow, and at each site, a total of three samples were collected, which amounts to the number of water samples 15, when collected from all the five sites. The parameters measured include pH, Conductivity, Dissolved Oxygen, and Redox Potential. Water measurements were done using ThermoScientific Orion Star A329 portable pH/ISE/

Variable	Descriptive Statistics of Raw Water Data (n = 15)							
	Mean	p-value	Median	Min.	Max.	Range	Variance	Std. Dev
Temperature (°C)	23.63	0.999	23.40	20.20	27.40	7.20	5.01	2.2429
pH	6.15	1.000	6.06	5.69	6.70	1.01	0.0934	0.3056
EC (mS/cm)	6.925	1.045	4.54	1.49	41.40	39.91	99.163	9.958
Redox (mV)	99.91	2.1725	94.40	54.40	194.30	139.9	1517.20	38.95
DO (%)	14.63	1.5675	6.11	0.0235	0.1436	0.7234	570.62	23.88

Table 3.
Summary of descriptive statistics of summer in situ *water measurements.*

Conductivity/RDO/DO Meter. The grab method was used to collect water from the study sites. Water samples collected in 500 ml water bottles were kept in a fridge on-site at a temperature of 5°C. Later in the lab, the water inside the bottles was filtered using 0.45 μm Millipore. The water samples were analyzed directly using ICP-OES after filtration. The concentration levels of these elements in water were compared to the international standards for drinking water and presented in **Table 3**.

1.3 Sediment sampling, preparation, and analysis

Sediment cores were extracted from the selected five sites mentioned above. PVC (Polyvinylchloride) pipe was inserted into the ground using a hammer and a plank. The sediment cores were pulled out using a long steel bar inserted into two holes made on the top sides of the pipe. The pulled-out cores were wrapped in refuse bags to prevent the sediments from falling out of the pipes. The pipes were taken to the laboratory and stored in the cold storage room at a temperature of 5°C before preparation and analysis. The pipe cores with sediment were cut vertically into two halves. Sediments inside the pipe were separated into three distinctive parts of 0–2, 2–10, and 10–30 cm, respectively using a plastic ruler.

A subsample of 20 g was extracted and preserved inside a plastic bag and freeze-dried for 3 days. The freeze-dried samples were milled using a mortar and pestle. The samples were further prepared and put inside the Spectroscout Geo + XRF Fluorescence small containers with a thin film inserted in the middle of the containers. The thin film served as the base of the container, and it was also polished before it was put into the Spectroscout Geo + XRF Analyzer Pro for analysis. The sediments were analyzed for the presence and concentration of Fe, S, Mg, Cr, Mn, Co, Ni, Cu, Zn, Mo, and Pb using the X-Ray Fluorescence.

1.4 Plants sampling, preparation, and analysis

Samples of Typha capensis, Schoenoplectus corymbosus, and Phragmites australis species were collected from the five sampling sites. In each sampled site, the plants' species were also sampled by using the sequence of inflow of the site, on the midflow, and also on the outflow of the constructed wetlands. Three samples of each of the plants were collected per site, and a total number of 45 samples were collected from all five sites in one season. The macrophyte species were uprooted using a spade. The plants were taken out of the sites and washed with local and rinsed with distilled water, for quantitative removal of soil and other foreign particles.

Variable	Plant species	Translocation factor (Autumn)		Bioconcentration factor (Autumn)			Translocation factor (Summer)		Bioconcentration factor (Summer)		
		Leaves	**Rhizomes**	**Leaves**	**Rhizomes**	**Roots**	**Leaves**	**Rhizome s**	**Leave s**	**Rhizome s**	**Root s**
Zinc	*P. communis*	0.35	0.32	26.64	24.66	76.14	0.3	0.41	59.28	82.8	200.8
	S. corymbosus	0.97	0.72	40.1	29.86	41.36	0.48	0.78	65.58	106.27	135.6
	T. capensis	0.49	0.81	22.91	38.02	46.75					
Nickel	*P. communis*	0.29	0.25	027.8	3.88	17.76	0.1	0.27	3.96	10.5	39.3
	S. corymbosus	0.35	0.45	3.95	5.05	11.35	0.22	0.81	4.27	15.9	19.6
	T. capensis	0.24	0.17	3.86	2.79	16.21					
Iron	*P. communis*	0.21	0.25	2.09	2.48	9.88	0.08	0.3	1.13	4.35	14.44
	S. corymbosus	0.13	0.31	1.19	2.8	9.03	0.12	0.79	1.29	8.55	10.79
	T. capensis	0.08	0.19	0.71	1.75	1.98					
Manganese	*P. communis*	1.02	0.23	33.29	7.62	32.57	0.02	0.13	6.23	32.98	253.5
	S. corymbosus	2.13	0.52	38.22	17.95	34.45	0.42	0.77	45.33	84.45	109
	T. capensis	4.16	0.69	151.1	25.19	36.3					
Copper	*P. communis*	0.26	0.28	8.1	8.44	30.65	0.14	0.26	12.86	24.28	92.4
	S. corymbosus	0.35	0.64	8.02	14.63	22.86	0.24	0.76	11.72	36.34	48.14
	T. capensis	0.36	0.54	7.1	10.68	19.94					
Magnesium	*P. communis*	1.26	0.29	74.53	17.1	58.93	1.21	0.41	151.05	51.25	125.5
	S. corymbosus	2.26	1.18	196.83	102.88	87.12	2.16	1.08	278.52	0.05	129.2
	T. capensis	1.55	0.66	250.78	106.05	161.5					
Cobalt	*P. communis*	0.17	0.18	5.67	5.76	32.51	0.06	0.18	7.05	33.4	55.14
	S. corymbosus	0.28	0.33	6.8	7.96	23.91	0.13	0.61	5.66	16.63	92.3
	T. capensis	0.15	0.1	8.35	5.88	57.18					

DOI: http://dx.doi.org/10.5772/intechopen.105708

Variable	Plant species	Translocation factor (Autumn)		Bioconcentration factor (Autumn)			Translocation factor (Summer)		Bioconcentration factor (Summer)		
		Leaves	Rhizomes	Leaves	Rhizomes	Roots	Leaves	Rhizome s	Leave s	Rhizome s	Root s
Sulfur	*P. communis*	1.29	0.34	135.54	46.62	175.3	0.95	0.53	274.72	223.1	166.6
	S. corymbosus	1.68	1	216.11	128.84	128.9	1.65	1.34	228.58	128.08	0.01
	T. capensis	0.78	0.27	145.35	50.71	185.3					
Phosphorus	*P. communis*	1.8	0.01	45.53	0.27	81.89	1.32	0.52	0.06	59.93	0.06
	S. corymbosus	1.37	3.49	105.8	269.32	77.22	1.67	2.42	0.11	231.14	95.55
	T. capensis	1.93	0.89	131.58	60.95	105.8					
Chromium	*P. communis*	0.44	0.32	4.42	3.27	10.09	0.16	0.35	2	4.23	12.18
	S. corymbosus	0.09	0.42	0.94	4.22	10.01	0.15	1.43	1.29	12.44	12.44
	T. capensis	0.07	0.15	0.53	1.21	7.9					
Molybdenum	*P. communis*	0	0	1.94	0.06	0	0	0	1.09	0.24	0
	S. corymbosus	0	0	2.36	0	0	0	0	2.67	1.03	0
	T. capensis	0	0	1.03	0.73	0					
Lead (Pb)	*P. communis*	0.37	0.2	3.98	2.21	10.64	0.16	0.33	2.73	5.77	17.55
	S. corymbosus	0.03	0.04	2.35	3.16	7.93	0.32	0.96	2.96	8.84	9.23
	T. capensis	0.16	0.16	11.98	1.96	1.78					

Table 4.
Translocation and bioconcentration factors for autumn and summer seasons.

The plants were taken to the laboratory where they were thoroughly washed and rinsed with tap and distilled water. They were then separated into leaves, rhizomes, and roots. The stems were not included in the study. The plant leaves, roots, and rhizomes were also chopped into pieces of +/− two cm and homogenized and a subsample of +/− 20 g was extracted from all the samples and freeze-dried for two three days to ensure that the plants are much dry for easy grinding. The homogenized plants' samples were milled using a Fritsch Pulverisette 6 Mill into pulverized powder before metals analysis.

1.5 Statistical analysis

Statistical analysis was done using the Statistica Analysis Package Version 10 computer package. Differences amongst the means were determined by analysis of variance. The Pearson correlation coefficients (r) were used to express the associations of quantitative variables. The basic descriptive statistics were performed to determine the Mean, P-value, Median, Minimum, Maximum, and standard deviation to indicate the relationship between the levels at which the elements were absorbed by the water and sediments as well as the level at which they were carried by the water and the relationship was again determined at $p \leq 0.05$. The sample size ($n = 15$) makes statistical comparisons easier as it is not limited by the sample size.

2. Results

2.1 Zinc (Zn)

The results for the Physico-chemical parameters are presented in **Table 4**. The average concentration for Zn accumulated by the plants in the autumn and summer seasons were in a range of 124.83 and 230.47 µg/g respectively, as shown in **Figure 6**. The result of the higher accumulation of Zn in summer could be that the new shoots

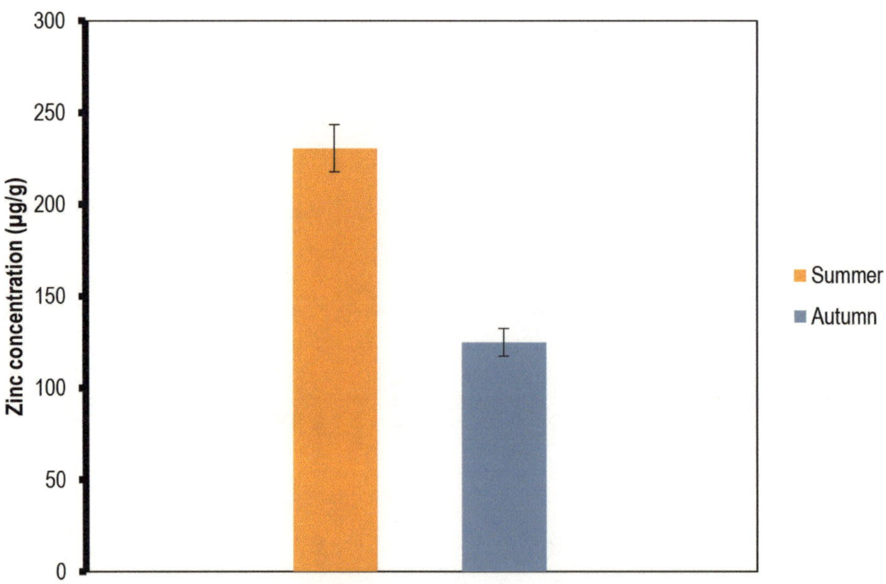

Figure 6.
Variations in the average Zinc concentration accumulated by plant in summer and autumn.

are sprouting and as a result, more Zn is taken up by the plants as compared to autumn when the plants shed their leaves.

2.2 Nickel (Ni)

When plant samples were analyzed for Ni content, the higher average concentration for Ni was measured in summer. The average concentration of Ni accumulated by the plants during the autumn and summer seasons were 36.36 and 49.55 µg/g, respectively as shown in **Figure 7**. Ni accumulation in the plants in the two seasons follows that of Zn.

2.3 Iron (Fe)

During the investigation period, it was observed that Fe was accumulated by the plants as FeO. A higher amount of Fe in the plants was accumulated in the summer season. The average concentration for Fe accumulated by the plants during the summer and autumn seasons was 25657.92 and 24807.34 µg/g, as shown in **Figure 8**. Unlike Zn, Fe accumulation in the two seasons did not differ much.

2.4 Manganese (Mn)

Plants accumulated Mn in the form of MnO. Mn was highly accumulated by the plants during the summer season and this also follows the trends of Zn and Ni. The average concentration for Mn accumulated by the plants were 5984.96 and 3950.78 µg/g in the summer and autumn seasons respectively as shown in **Figure 9**.

2.5 Copper (Cu)

In all the sampled plants, Cu was more highly accumulated by the plants in summer than in autumn in the same way as Zn, Ni, and Mn. The average

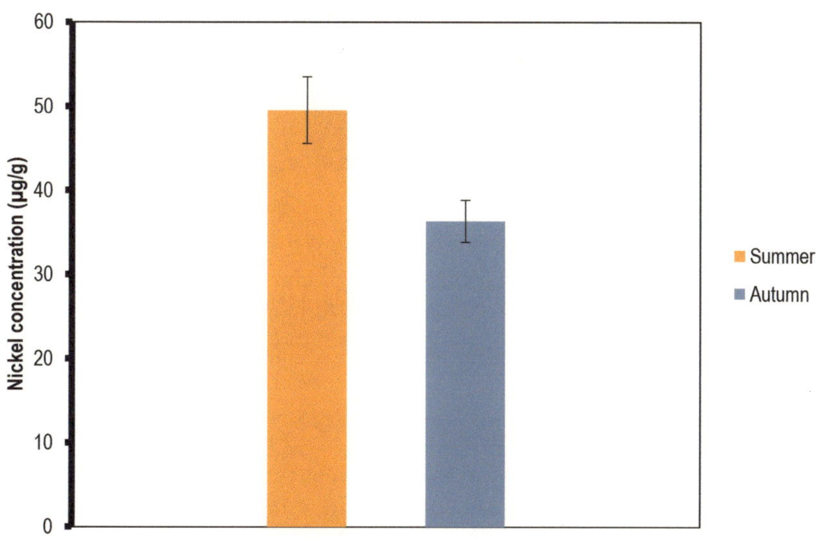

Figure 7.
Variation in the average Nickel concentration accumulated by plant in summer and autumn.

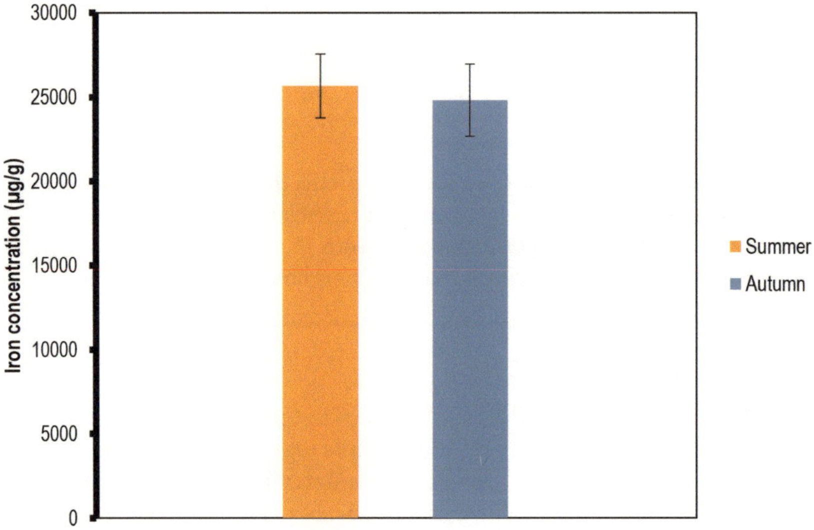

Figure 8.
Variation in the average Iron concentration accumulated by plant in summer and autumn.

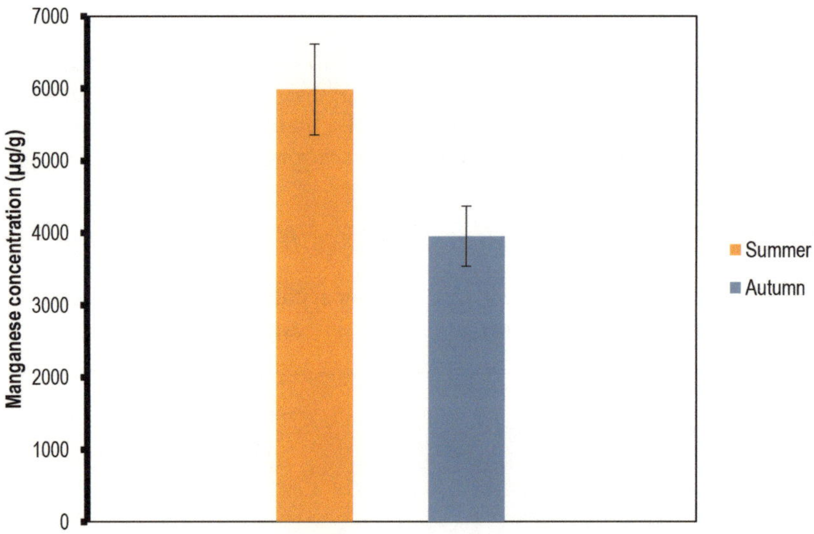

Figure 9.
Variation in the average Manganese concentration accumulated by plant in summer and autumn.

concentration of Cu accumulated in the plants sampled during the summer and autumn months was 78.30 and 45.63 µg/g as shown in **Figure 10**.

2.6 Magnesium (mg)

During the autumn and summer seasons, the average concentration for Mg accumulated in the plants was 18561.31 and 14976.21 µg/g respectively, as shown in **Figure 11**. A higher concentration of magnesium was accumulated by the plant during autumn. When the plants were measured for Mg, seasonal differences in the accumulation of Mg were not very pronounced.

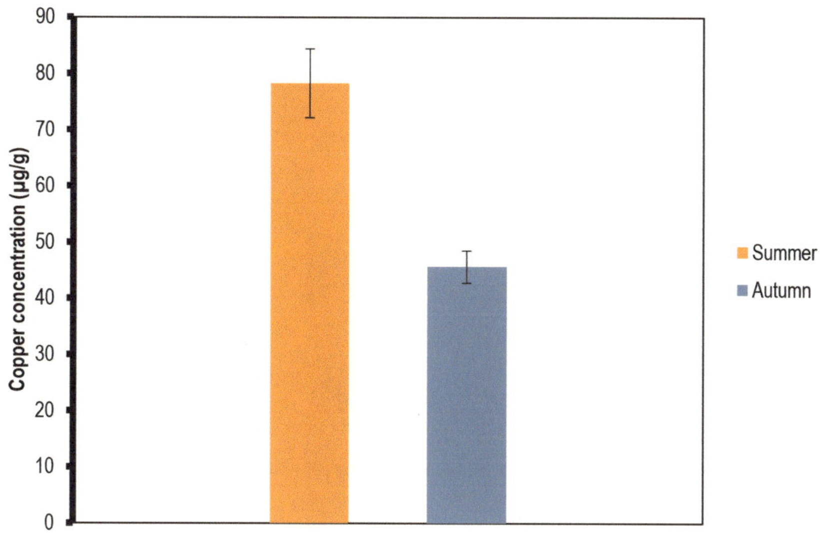

Figure 10.
Variation in the average Copper concentration accumulated by plant in summer and autumn.

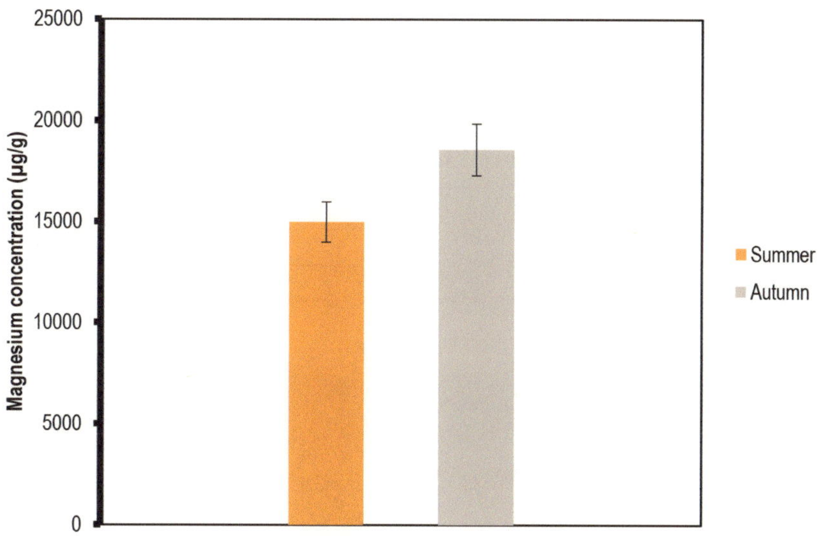

Figure 11.
Variation in the average Magnesium concentration accumulated by plant in summer and autumn.

2.7 Cobalt (Co)

During the sampling periods, harvested plants were measured for the presence of Co. After the plant has been measured for the metals accumulated by the plant, it was observed that the plant harvested in summer has accumulated and stored in higher amounts of Co than in autumn, and this also follows the trends like that of Zn, Ni, Cu, and Mg. The average concentration for Co measured in summer was 58.13 µg/g, whilst the average concentration of Co in autumn was 43.24 µg/g as shown in **Figure 12**.

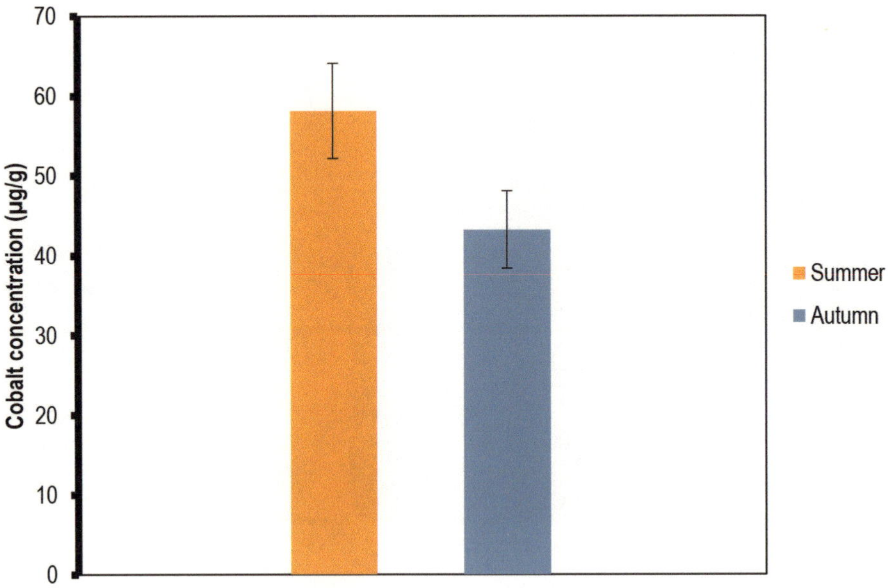

Figure 12.
Variation in the average Cobalt concentration accumulated by plant in summer and autumn.

2.8 Sulfur (S)

In summer, the plants have accumulated higher concentrations of S. The average concentration of S in the plants was 22826.03 µg/g in summer whilst in autumn it was 22749.46 µg/g as presented in **Figure 13**. The uptake of S is also not very pronounced and the average values, do not differ much between the plants measured from the sites as shown in **Figure 13**.

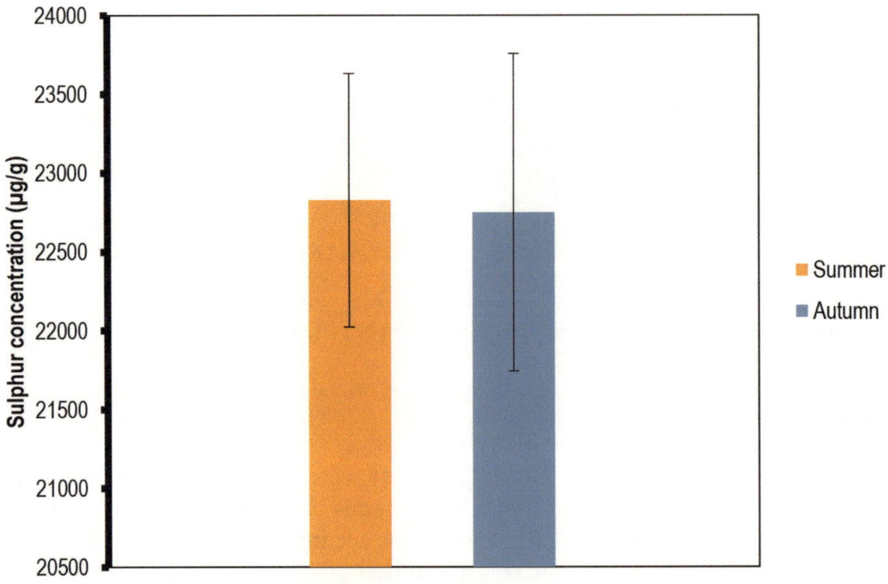

Figure 13.
Variation in the average S concentration accumulated by plant in summer and autumn.

2.9 Phosphorous (P)

Phosphorous was highly accumulated by the plants in autumn. The average concentration for P accumulated by the plants in autumn was 3759.61 μg/g whilst in summer it was 3487.44 μg/g as presented in **Figure 14**.

2.10 Chromium (Cr)

Higher amounts of Cr were measured in the plants in autumn. The average concentration for Cr in autumn was 68.76 μg/g and in summer it was 64.82 μg/g as shown in **Figure 15**.

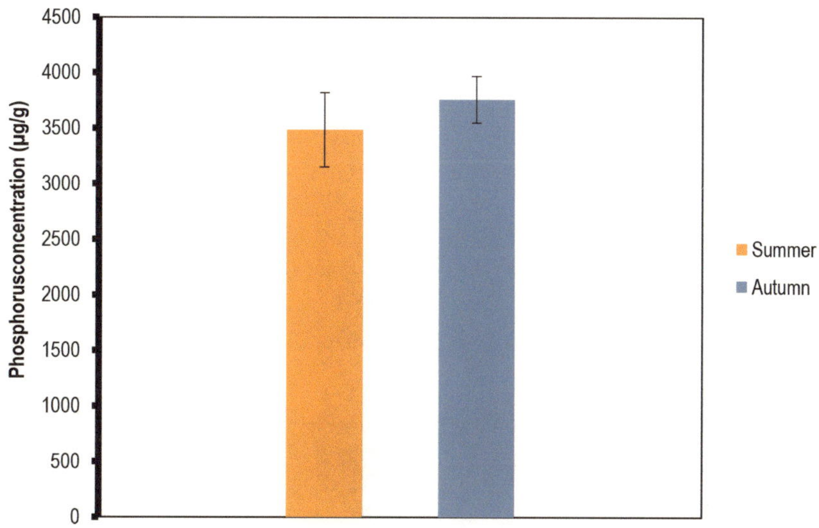

Figure 14.
Variation in the average P concentration accumulated by plant in summer and autumn.

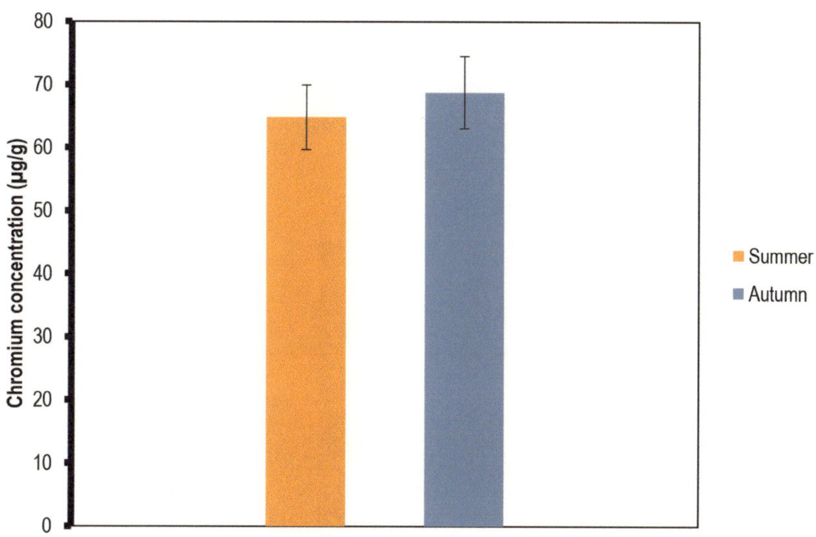

Figure 15.
Variation in the average Cr concentration accumulated by plant in summer and autumn.

2.11 Molybdenum (Mo)

In the plants sampled measured for Mo in autumn and summer, it was found that Mo was highly accumulated by the plants in autumn. The average concentration for Mo measured in autumn was 0.09 μg/g and in summer was 0.07 μg/g, as portrayed in **Figure 16**.

2.12 Lead (Pb)

The average concentration for Pb in the summer and autumn months was 12.082 and 12.084 μg/g respectively as shown in **Figure 17**. Higher concentrations for Pb were measured in the plants in summer.

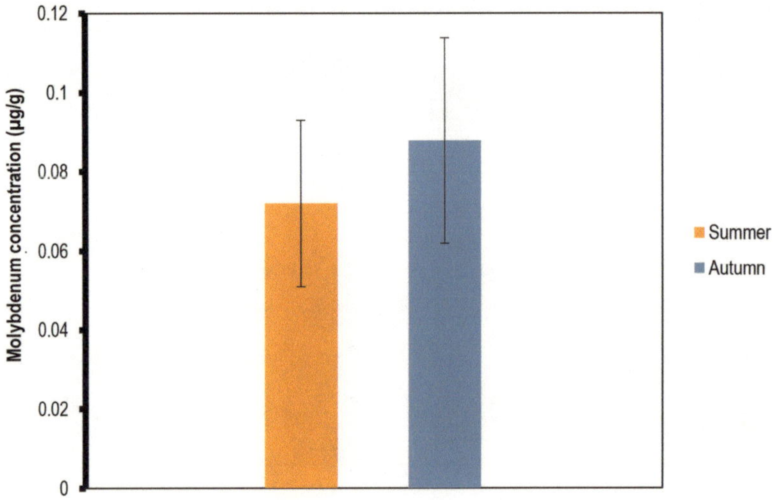

Figure 16.
Variation in the average Mo concentration accumulated by plant in summer and autumn.

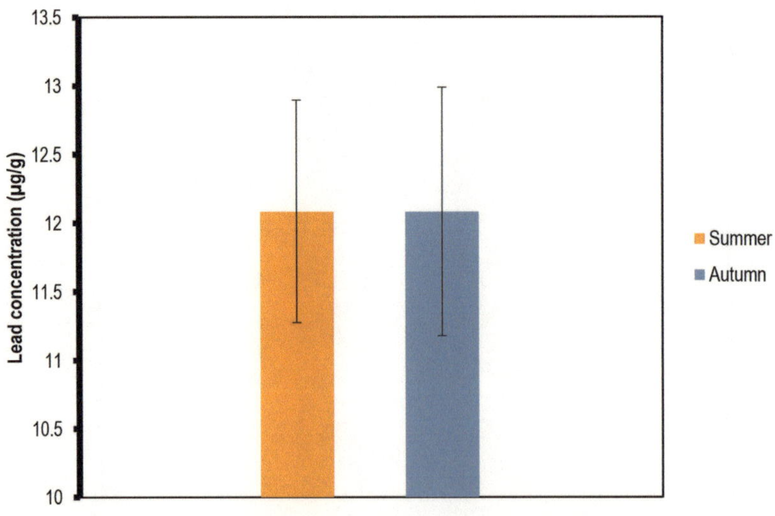

Figure 17.
Variation in the average Pb concentration accumulated by plant in summer and autumn.

2.13 Statistical significance of elements between water, sediments, and plants

The correlation coefficients were performed to determine the relationship between the levels at which the elements were absorbed by the three species of *Schoenoplectus, Phragmites,* and the *Typha spp*. The concentrations of elements in the water, sediments, and plants were compared by the use of the significant variance at p = 0.0500. When the significant correlations coefficient was performed, several significant correlations were found to take place between the plants, water, and sediment. A correlation was found between sediments and plants, where Magnesium was negatively correlating with Cr and Fe and positively correlating with P, S, Mn, Co, and Ni. Their associated r values were (Cr) r = −0.06143 and (P) r = −0.6232. The r values for the positive correlations were (P) r = 0.7598, (S) r = 0.6406, (Mn) r = 0.8714, (Co) r = 0.8021 and (Ni) r = 0.6704. The p values were significant at p = 0.0500 and all the p values were in arrange of p = 0.000 and 0.0015.

P was found to be having both negative and positive correlations with the elements in water, sediments, and plants. The negative correlations were found between the concentrations of Cr, Fe, and Pb while the positive correlations were obtained between the concentrations of Mn and Co. The r values of the elements with negative concentration correlations were (Cr) r = −0.6488, (Fe) r = −0.7973 and (Pb) r = −6581, while the correlation for the elements with positive concentrations of elements were (Mn) r = 0.07636 and (Co) r = 0.7841. Their p-values were significant at p = 0.0500 and their p values were in arrange of p = 0.000 and p = 0.009 in water, sediments, and also in plants. Cr was also found to have a negative correlation with Mn and positive correlations with Fe and Pb.

The correlations were significant at p = 0.0500 and the p values were in a range of p = 0.000 and p = 0.038. The correlations were (Mn) r = −0.5389, (Fe) r = 0.8005 and (Pb) r = 0.8584 respectively in water, sediments and plants respectively as shown in **Appendix 1.**

2.14 Translocation and bioconcentration factors of sediments and plants

Bioconcentration factor is defined as the ratio of metal concentration in plant aboveground part to the total metal concentration in the soil. The translocation factor is the ratio of metal concentration in the shoots to the metal concentration in the shoots. Bioconcentration of elements of the plants' species between the seasons. Since the amounts of elements enter the aquatic ecosystem after being washed from the mine dump, the elements (some of which are toxic) become accumulated in the water column, in the sediments, and also uptake by the plants which then pose some health threats when accumulated in higher amounts. Bioconcentration factor (BCF) is described as the measure of the amount of an element accumulated in the plants from their surrounding environment that is in contact with it [26]. It can be obtained by dividing the trace element concentration in plant tissues harvested by the initial concentration of the element in the external nutrient solution. Translocation factor (TF) on the other hand is defined as the ratio of element concentration in the root to shoot (**Table 4**) [27, 28].

This resulted in BCFs for the different types of plant organs (**Table 4**). The BCF for the shoots, rhizomes, and leaves was calculated from the elements accumulated by the plants in both the summer and autumn seasons.

Table 4 below shows the TF and BCF of quantified elements in the study. In both seasons, elements that were mostly taken with high BCF were S, Mg, Zn, Mn, P, Cu,

Ni, and Co, and elements that were mostly taken with the highest TF were P, Mg, S, Cr, Zn, Pb, Cu, and Mn. In autumn, the plants' organs that were found to have the highest BCF were the leaves of *T. capensis* and *P. communis* as well as the roots of *Scirpus corymbosus*. In summer, the highest BCF was obtained in the roots and rhizomes of *P. communis* and *S. corymbosus*. In summer, the plant species with the highest BCF and TF was *S. corymbosus;* and in autumn, *T. capensis* was the plant species with the highest BCF while *S. corymbosus* was the plant species with the highest TF. It was observed that both the TF and the BCF are affected by seasonality. The TF was higher in the autumn season than in the summer season, and the BCF was higher in the summer season than in the autumn season. It became evident that BCF active growth of the plants in summer as most of the elements are used by the plants during processes such as photosynthesis which is active in green leaves compared to when the plants' leaves start drying up and photosynthesis ceases and most element losses occur, as the leaves die off and become brittle.

When making comparisons of the results of the translocation, Bioconcentration factors for plants and sediments, it was found that P, S, Mn, Mo, and Pb were lower in all three plant species and higher in the sediments. On the other hand, elements such as Mg, Cr, Fe, Co, Ni, Cu, and Zn were accumulated in higher concentrations by the plants and lower in the plants.

The results for TF and BCF indicate that the investigated plants accumulate in higher concentrations of certain elements and some in smaller concentrations. The elements that were found to be accumulated in lower concentrations by the plants were on the other hand found to be accumulated at higher concentrations by the sediments. This could be the case where the plants release the elements back into the substratum when they die off. This was observed in the accumulation of Molybdenum, where the measured Mo concentration was below the detection level by the plants. The plants with high BCF were regarded as suitable to be used to decontaminate soils. Although the plants showed high BCF, they still do not meet the criteria of being hyperaccumulators. The plants accumulated levels of elements such as Cu, Zn, and Pb in amounts with higher BCF (92.40, 200.79, and 17.55 in summer and 30.65, 76.14, and 11.98 in autumn) but the concentration of these elements was not greater than 1000 mg/kg to be regarded as hyperaccumulators. The plants were regarded as moderate accumulators [26]. The plants were suitable to be applied in contaminated soils for phytoremediation processes [29].

2.15 Comparison of elements in water with the international organizations

Table 5 illustrate the current drinking water quality guidelines by international organizations, and for the basis of this study, the levels of elements in water were compared with the water quality guidelines to indicate whether the level of elements in water was either above or below the required or acceptable levels. The last column indicates the concentration levels of elements measured in the water sample sites of this study.

Zn concentration in water was within the acceptable range of 267 μg/l, and when compared with the international guidelines for water quality standards which were above 500 μg/l. Fe, Ni, Mn, and Cu were found to be highly concentrated above the acceptable levels in water when compared to the international water quality guidelines, with the concentration levels of 2230, 282, 5900, and 14,080 μg/l respectively. The units for the elements concentrated in water were illustrated in μg/l in this section to easily compare with international guidelines as the standards were expressed in μg/l

Heavy metal	WHO[a]	USEPA[b]	ECE[c]	FTP-CDW[d]	PCRWR[e]	ADWG[f]	NOM-127[g]	This study
Zinc		500		5000	5000	3000	5000	267
Iron		300	200	300		300	300	2230
Nickel	70		20		20	20		282
Manganese	100	50	50	50	500	500	150	5900
Copper	2000	1300						14,080

[a]*World Health Organization (WHO 2011).*
[b]*United States Environmental Protection Agency (USEPA, 2011).*
[c]*European Commission Environment (ECE, 1998).*
[d]*Federal-Provincial-Territorial Committee on Drinking water (CDW), Health Canada (FTP-CDW, 2010).*
[e]*Pakistan Council of Research in Water (PCRWR, 2008).*
[f]*Australian Drinking Water Guidelines (DDWG, 2011).*
[g]*Norma Official Mexicana NOM-127-SSA1–1994 (DOF, 1994).*

Table 5.
Drinking water quality guidelines ($\mu g/L^{-1}$) for elements in water in this study.

Variable	Descriptive statistics of elements accumulated by sediments n = 15							
	Mean	p-value	Median	Min.	Max.	Range	Variance	Std. Dev
Mg (μg/g)	0.1887	1.000	0.2027	0.039	0.321	0.282	0.0072	0.09
P (μg/g)	487	2.679	526.33	145.37	649	503.63	17973.47	147.39
S (μg/g)	1891.46	0.000	1813.67	519.67	3927	3407.33	629351.4	793.32
Cr (μg/g)	192.03	1.371	167.25	97.13	282.67	185.53	3450.42	58.74
Mn (μg/g)	1081.76	0.000	873.33	222.17	2807	2584.83	628712.2	792.91
Fe (μg/g)	7.23	0.9455	7.01	3.16	11.59	8.43	3.48	1.93
Co (μg/g)	29.51	1.6718	27.47	9.53	51.83	42.3	112.41	10.60
Ni (μg/g)	59.23	0.0001	60.33	20.67	75.67	55	175.61	13.25
Cu (μg/g)	38.37	0.1889	39.43	16.00	41.60	30.1	50.44	7.10
Zn (μg/g)	39.11	0.0022	40.23	10.5	50.27	39.77	94.29	9.71
Mo (μg/g)	1.73	0.6704	1.6	0.000	3.5	3.5	1.380	1.175
Pb (μg/g)	29.20	0.0210	29.67	8.53	41.3	32.77	55.71	7.464

Table 6.
Summary of descriptive statistics of elements accumulated by sediments.

rather than in mg/l as shown in **Table 6**. To evaluate the effectiveness of these macrophytes in the accumulation of selected elements, a study of the whole life cycle of the plants has to be conducted, to capture all the processes that the plants undergo to survive the heavy metal contaminated environment. Amongst the three macrophytes investigated, there is no single plant that accumulates more than four elements than the other plants. In this study, it was observed that there are elements that are highly accumulated by either the roots, and less by the rhizomes, and more by the leaves and less by the roots. The rate of metal accumulation between the plants as well as the plant parts varies between the two seasons of autumn and summer investigated. This indicates that this could be a result of the functionality of the elements during the period of maximum absorption.

Although the plants were found to accumulate the elements investigated, it is clear that some factors lead to the plants not uptake some elements in higher concentrations than the others. This creates a knowledge gap in this study, and as a result, further studies should be undertaken to investigate the reason the plants accumulate certain elements in lower concentrations. Another factor to be investigated is the overall performance of the plants in elements uptake throughout the whole growth cycle, and the results of such a study could be able to fill the gap in this study.

3. Conclusions

Rehabilitation of acid mine drainage contaminated sites requires an in-depth understanding of the appropriate macrophyte plants that could be used effectively in the accumulation of various elements from the sediments, and water column. Harvesting of the plants' biomass before the winter season when the plants' growth becomes inactive and dies off should be prioritized as more elements accumulated in the different plants' organs are released back into the soils and this adds more elements to the substratum. Harvesting of the plants' biomass for elements extraction for various metallurgical purposes could also be viewed as a value-adding initiative as this will be reducing the concentration amounts of these elements in the environment and also give more room for other less hardy plants to grow and thrive in those contaminated sites. The accumulation of elements in the sediment especially of most toxic elements enhances the release of other reactive ions into the environment, in addition, the reduction of such toxic elements also improved the pH conditions of the soil and water. In addition, there was no single plant amongst the three investigated species that was found to accumulate a higher amount of certain elements in all the plants' organs of leaves, rhizomes, and roots. The investigated macrophytes species have demonstrated merit results in decontaminating AMD in the soil and further portrayed the potential of being used for phytostabilisation and phytoextraction as they are fast and easily spreading on contaminated sites, and can withstand high metal toxicity.

Acknowledgements

I would like to express my gratitude to Prof Luke Chimuka of the University of the Witwatersrand, Isabel Weiersbye, Innocent Rabohale, Sashnee Raja, Ike Rampedi, Maxine Joubert, and the entire Ecological Engineering and Phytoremediation Programme, for the tireless effort efforts and contributions that they have made in this study. I also send my sincere gratitude to Ms. Isabel Weiersbye for assisting with funding from DTI-THRIP.

Acronyms and abbreviations

AMD	Acid Mine Drainage
ARD	Acid Rock Drainage
Zn	Zinc
Ni	Nickel
Fe	Iron
Mn	Manganese

Cu	Copper
Mg	Magnesium
Co	Cobalt
S	Sulfur
P	Phosphorus
Cr	Chromium
Mo	Molybdenum
Pb	Lead
Cd	Cadmium
Hg	Mercury
Se	Selenium
As	Arsenic
FeO	Iron oxide
MnO	Manganese oxide

Appendix 1.

Summary of correlation matrix of elements in water, sediments, leaves, rhizomes and rhizomes.

Variable	Correlating variables	Valid N	Correlation co-efficient	p value
Magnesium	Phosphorus	15	0.7598	0.001
	Sulfur	15	0.6406	0.010
	Chromium	15	−0.6143	0.015
	Manganese	15	0.8714	0.000
	Iron	15	−0.6232	0.013
	Cobalt	15	0.0821	0.006
	Nickel	15	0.6704	0.006
Phosphorus	Chromium	15	−0.6488	0.009
	Manganese	15	0.7636	0.001
	Iron	15	−0.7973	0.000
	Cobalt	15	0.7841	0.001
	Lead	15	−0.6581	0.008
Chromium	Manganese	15	−0.5389	0.038
	Iron	15	0.8005	0.000
	Lead	15	0.8584	0.000
Manganese	Iron	12	−0.7044	0.003
	Cobalt	15	0.8173	0.000
	Nickel	15	0.6145	0.015
	Lead	15	−0.5644	0.028
Iron	Cobalt	15	−0.5775	0.024
	Lead	15	0.8350	0.000

Variable	Correlating variables	Valid N	Correlation co-efficient	p value
Cobalt	Nickel	15	0.8303	0.000
	Molybdenum	15	−0.5262	0.044
Sulfur	Copper	15	0.6003	0.018
	Zinc	15	0.7279	0.002

Author details

Tinyiko Salome Mthombeni
University of the Witwatersrand, Johannesburg, South Africa

*Address all correspondence to: tsmthombeni@gmail.com

References

[1] Inter-Ministerial Committee. Mine water management in the Witwatersrand goldfields with a special emphasis on Acid Mine Drainage. 2010

[2] McCarthy TS. The impact of acid mine drainage in South Africa. South African Journal of Science. 2011;**107**: 5-6

[3] Johnson DB, Hallberg KB. Acid mine drainage remediation options: A review. Journal of Science of the Total Environment. 2005;**338**:3-14

[4] Naicker K, Cukrowska E, McCarthy TS. Acid mine drainage from gold mining activities in Johannesburg, South Africa. South Africa and Environs. Environmental Pollution. 2003;**122**: 29-40

[5] Sutton MW, Weiersbye IM, Galpin JS. Use of Remote Sensing and GIS in a Risk Assessment of Gold and Uranium Mine Residue Deposits and Identification of Vulnerable Land Use. Johannesburg, South Africa: University of the Witwatersrand; 2012

[6] Akcil A, Koldas S. Acid mine drainage (AMD): Causes, treatment, and case studies. Journal of Cleaner Production. 2006;**12–13**(14):1139-1145

[7] Oelefse SHH, Hobbs PJ, Rascher J, Cobbing JE. The pollution and destruction threat of gold mining waste on the Witwatersrand – A West Rand case study. In: Natural Resources, and the Environment, Pretoria, South Africa. CSIR; 2007. pp. 617-627

[8] Adler R, Rascher J. A strategy for the management of Acid Mine Drainage from Gold Mines in Gauteng. In: CSIR Pretoria. CSIR Report Number CSIR/NRE/PW/ER/2007/0053/C.

[9] Hu H. Human health and heavy metals exposure. In: McCally M, editor. Life Support: The Environment and Human Health. MIT Press; 2002

[10] Mohiuddin KM, Ogawa Y, Zakir HM, Otom K, Shikazono N. Heavy metals contamination in water and sediments of an urban river in a developing country. International Journal of Environmental Science Technology. 2011;**8**(4):723-736

[11] Lang D. Dynamics of Heavy Metals in Reedbeds along the Banks of the River Scheldt. Belgium: Ghent University; 2006. p. 284

[12] Van der Merwe CG, Schoonbee HJ, Pretorius J. Observations on concentrations of the heavy metals zinc, manganese, nickel, and iron in the water, in the sediments, and two aquatic macrophytes, *T. capensis* (Rohrb.) N. N. BR. and *Arundo donax L.*, of a stream affected by goldmine and industrial effluents. Journal of Water. 1990;**16**:2

[13] Mortimer DC. Freshwater aquatic macrophytes as heavy metal monitors-the Ottawa River experience. Environmental Monitoring and Assessment. 1985;**5**:311-323

[14] Tilahun G, Ashagre T. Heavy metals accumulation by aquatic macrophytes from Lake Hawassa, Ethiopia: Phytoremediation for water quality improvement. In: Proceedings of the 2nd National Workshop of 2012 on Challenges and Opportunities of Water Resource Management in Tena Basin. Ethiopia: Upper Blue Nile Basin; 2011

[15] Baker AJM. Accumulators and excluders: Strategies in the response of plants to heavy metals. Journal of Plant Nutrition. 1981;**3**:643-654

[16] Radulescu C, Stihi C, Popescu IV, Ionita Dulama ID, Chilian A, Bancuta OR, et al. Assessment of heavy metals level in some perennial medicinal plants by flame atomic absorption spectrometry. Romanian Reports in Physics, Environmental Physics. 2012; **65**(1):246-260

[17] Shafi N, Pandit AK, Kamili AN, Mushtaq B. Heavy metal accumulation by *Azolla pinnata* of dal Lake ecosystem, India. Journal of Environment Protection and Sustainable Development. 2015;**1**: 8-12

[18] Koorimannil K, Abdussalam AK, Chandra RP, Salim N. Bio-accumulation of heavy metals in *Bacopa monnieri (L.)* Pennell growing in different habitats. International Journal of Ecology & Development. 2010;**15**(10)

[19] Phukan P, Phukan R, Phukan SN. Heavy metal uptake capacity of Hydrilla verticillata: A commonly available aquatic plant. International Research Journal of Environment Sciences. 2015; **4**(3):35-40

[20] Cardwell AJ, Hawker DWM, Greenway M. Metal accumulation in aquatic macrophytes from Southeast Queensland. Australia. Journal of Chemosphere. 2012;**48**:653-663

[21] Ye ZH, Baker AJM, Wong MH. Zinc, lead and cadmium tolerance, uptake and accumulation by the common reed, Phragmites australis (Cav.). Annals of Botany. 1997;**80**:363-370

[22] Reisinger S, Schiavon M, Terry N, Pilon-Smits EAH. Heavy metal tolerance and accumulation in Indian mustard *(Brassica juncea l.)* expressing bacterial γ -glutamylcysteine synthetase or glutathione syntheses. International Journal of Phytoremediation. 2008;**10**: 1-15. DOI: 10.1080/15226510802100630

[23] Soudek P, Kotyza J, Lenikusová I, Petrová Š, Benešová D, Vaněk T. Accumulation of heavy metals in hydroponically cultivated garlic *(Allium sativum L.)*, onion *(Allium cepa L.)*, leek *(Allium porrum L.)* and chive *(Allium schoenoprasum L.)*. Journal of Food, Agriculture & Environment. 2009; 7(3&4):761-769

[24] Basic N, Keller C, Galland N. Ecological preferences and heavy metal hyperaccumulation of wild populations *of Thlaspi caerulescens* in Switzerland. In: Workshop "Phytoremediation of toxic metals" June 2003 Meeting in Stockholm, Sweden. 2003. pp. 12-15

[25] Salt DE, Smith RD, Raskin I. Phytoremediation. Annual Review of Plant Physiology and Plant Molecular Biology. 1998;**49**:643-668

[26] Sukumaran D. Phytoremediation of heavy metals from industrial effluent using constructed wetland technology. Journal of Applied Ecology and Environmental Science. 2013, 2013;**5**: 92-97

[27] Zayed A, Gowthaman S, Terry N. Phytoaccumulation of trace elements by wetland plants: I. aestivum Linn. Mutation Research/Generic Toxicology and Environmental Mutagenesis. 1998; **537**:29-41

[28] Lorestani B, Cheraghi M, Yousefi N. Accumulation of Pb, Fe, Mn, Cu, Zn in plants and choice of hyperaccumulation in the industrial town of Vian, Iran. Archives of Biological Sciences. 2011; **3**(63):739-745

[29] Nazir A, Malik RN, Ajaib M, Khan N, Siddiqui MF. Hyperaccumulators of heavy metals of industrial areas of Islamabad and Rawalpindi. Pakistan Journal of Botany. 2011;**43**(4):1925-1933

Chapter 27

Nutritional Potential of *Erythrina edulis* as a Forage Alternative for Supplementation in Feeding Ruminants

Oscar Giovanny Fuentes Quisaguano
and Santiago Alexander Guamán Rivera

Abstract

The main limiting factor in livestock production is fluctuation in the quantity and quality of forage resources. Therefore, it is necessary to determine the chemical composition and degradation kinetics of the feed that is used for ruminant feeding regime. *Erythrina edulis* (*Euphorbia edulis*) is a multipurpose legume plant with high nutritional quality and possibly the capacity to meet dairy ruminant requirements. The study showed that the two phenological stages leaves (SV) and sheath without seed (SF) had greater CP contents than other sources than are typically used for feeding ruminants. Nevertheless, the SF had lower fiber contents, so the highest DM and CP degradation parameters than SV, it was obtained. Consequently, *E. edulis* might be considered as a forage alternative for inclusion in ruminant feeding.

Keywords: shrubs, phenological stages, chemical composition, degradability, *Erythrina edulis*

1. Introduction

The sustainability of livestock farming systems plays a central role in addressing policies aimed at sustained and planned rural development [1]. In Latin America, the dairy sector has been more dynamic in the past 20 years than in the rest of the world with an average growth of 12.5% for its 3.15 million milk producers [2]. Dairy production in Ecuador is concentrated mostly in the Andean highlands, the Sierra Region [3]. In addition, FAO [4] stated that in Ecuador the cattle breeding systems tend to be extensive (5 million hectares dedicated to livestock with 4.1 million cattle) with low productivity (5.38 liters of milk per cow) and with poor use of pastures.

The crude protein (CP) and energy requirements for ruminants are the most important limitans in the livestock industry worldwide [5–8]. Soybean products are commonly fed concentrates in highly productive ruminants because of their high content of protein and good profile in essential amino acids [9–11], although the high global demand has resulted in price increases. In the Ecuadorian Highland Region approximately 57% of livestock farms are less than 10 hectares, and they are managed as smallholder production with low intensification levels and economic incomes [4], so in these production systems the use of soybean meal is too expensive. According to Camero et al. [12], feeding leguminous fodder that is high in protein, can improve rumen fermentation parameters leading to increased digestibility and intake of low-quality feeds, and hence improved animal production. For this reason, for the Ecuadorian Andean Region is necessary to research new forage alternatives to use in ruminant feeding. In this sense, trees and shrubs have had an increasing interest due to their high potential for supplying fodder which provides greater nutritive value and environmental services [1, 13–15].

Erythrina edulis is a leguminous plant with a wide range of uses from human (mainly seeds) to animal (forage) diets, as well as in the recovery of the soil nitrogen content [16–18]. Furthermore, this species has a higher protein content (ranging from 18 to 25%) than other legumes and is similar in terms of quality to egg protein [19–21]. South America countries such as Colombia, Peru, and Venezuela have already studied this species and its potential for use in animal production [12, 18, 22–24]. However, little available information on this multipurpose plant in the context of Ecuadorian conditions is available, despite that, *E. edulis* can be found as a wild plant, the lack of knowledge on its properties and potential for livestock nutrition, this species has gone unnoticed with a latent danger of extinction.

2. Materials and methods

The study was performed in (INIAP), Pichincha, Ecuador and all experimental procedures were approved by them.

2.1 Study area

The study was carried out in Valle de los Chillos, Pichincha province, located in the northern Highland region of Ecuador. The soil type in the area of study is Molisols, at 45.88%, followed by Andisols at 17.88% [25]. The climatic conditions predominant in this zone are, on average temperature (22.9°C), rainfall (1200 mm/year), relative humidity (78%), and an altitude of 2200 m.a.s.l.

2.2 Experimental design, collection, and preparation of samples

For the experimental procedure, 150 vegetative stakes of *E. edulis* were planted with a 6 × 6 m distance between them. After 2 years of establishment, the trees of *E. edulis* were randomly divided into two equilibrate groups to assess the vegetative stage (SV = leaf) and fructification stage (SV = previously seed was removed). The leaf samples were cut 50–60 cm from the tip of the second youngest branch at the top of each tree. Thereafter, the 60 samples of each phenological stage were pooled, according to SV and SF, giving 9 samples, and then frozen at −20°C for further analysis. Before analyses, samples were conditioned at 60°C for 48 h, and then milled and homogenized through a cyclone mill (Model 4 Wiley Mill, Thomas Scientific, Swedesboro, NJ, EUA) with a 1 mm mesh, and for *in situ* rumen incubation, samples were milled with a 2 mm mesh.

2.3 Chemical analysis

All determinations were performed, according to official reference methods [26]. Thus, Dry matter (DM) was determined at 103°C for 24 h, and ashes burnt at 550°C for 5 h. Whereas wall cell components such as crude fiber (CF), neutral detergent fiber (NDF), acid detergent fiber (ADF), and acid detergent lignin (ADL) using the Ankom200 Fiber Analyzer (Ankom Technology, Fairport, NY, USA). Furthermore, Crude protein (CP) was calculated as a percentage of N $\times$ 6.25 using the Kjeldahl method. Besides this, phenols and total tannins were determined using the Folin–Ciocalteu method, before and after the treatment of extracts with polyvinylpolypyrrolidone [27].

2.4 In situ rumen incubation

For this study, two 4-year-old Holstein non-lactating cows, (650 $\pm$ 5.0 kg BW) equipped with permanent rumen fistula were used. The cows, were subjected to a 15-d adaptation period, being fed (2% of BW) with only *Pennisetum clandestinum* (chemical composition, as % DM; ash, 7%; CP, 13.2; NDF, 38%; ADF, 26%; CF, 25.0%; and EE, 1.53). Throughout the experiment, the cows had free access to mineral and vitamin block (Na, 12 g; Ca, 20 g; P,10 g; Mg, 0.10 g; S, 0.29 g; Zn, 0.16 g; Mn, 0.12 g; Fe, 0.12 g; I, 0.020 g; Co, 0.002 g; Se, 0.003 g; Zinc, 0.16 g; and Cu, 0.002 g; Favetex, Favesal, Ecuador, milk production).

The DM and CP *in situ* degradability were carried out by incubating nylon bags (Ankom Technology Corporation, Fairport, NY, USA) in the rumen, which was 10 $\times$ 20 cm, 47 μm, pore size, containing 10 g of samples. Previously, zero-hour disappearance was estimated by washing duplicate bags containing feed samples in cold water (without passing through the rumen). After that, the samples were incubated in duplicate in the rumen for 3, 6, 12, 24, 36, 48, and 72, h, according to Aufrère [28] and NRC [29], before feeding at 0830 h. Once the bags were removed from rumen incubation, were immediately washed with clean water several times (three washing cycles of 5 min). They were also frozen at −20°C for 24 h to halt fermentative activity. After all this, the bags were dried at 60°C for 48 h, weighed, and so the residues were mixed for chemical analysis. Finally, using the equation of Orskov and Mcdonald [30], DM and CP degradation parameters were calculated:

$$D = \alpha + b(1 - e - \text{ct}). \tag{1}$$

where D is the fraction corresponding to the disappearance of either DM or CP at time t; α is an intercept representing the DM or CP soluble fraction; b is the fraction of insoluble but potentially degradable DM or CP; c is the rate of disappearance of fraction b; t is incubation time. The non-linear parameters α, b, and c were estimated using an interactive least-squares procedure of SAS (v. 9.4; SAS Institute Inc., Cary, NC). Therefore, the effective degradability (ED) of DM and CP was calculated using Equation:

$$\text{ED} = \alpha + [\text{bc}/(c + k)], \tag{2}$$

where α, b, and c are the same parameters as described earlier, and k is the estimated solid passage rate. In this study, we reported a k 6%/h for most lactation feeding conditions INRA [31], and according to low–middle–high level of intake (2, 5, and 8%/h, respectively).

2.5 Statistical analyses

Firstly, all data were checked with a normality test and then analyzed under a general linear model, using the GLM procedure of SAS v. 9.4 (SAS Institute Inc., Cary, NC). The means were determined using the PDIFF option of SAS, and Tukey's multiple range tests were used to compare means between SV and SF. Statistical differences were declared at $p < 0.05$, and tendencies at $p < 0.10$.

3. Chemical composition of leaves and sheath from *Erythrina edulis*

The chemical determinations of leaves (SV) or sheath without seed (SF) from *E. edulis*, are shown in **Table 1**. The SV had 61% greater DM content than SF (31.4 vs. 12.2 ± 5.0%; $p = 0.009$), but its ash content was 27% higher compared to SF (10.5 vs. 7.7 ± 1.40%; $p < 0.001$; **Table 1**). Besides this, the SV showed a greater CP content than the obtained in SF (28.7 vs. 20.3 ± 2.59; $p = 0.022$; **Table 1**), but no differences in CF (24 ± 3.15%, on average; $p = 0.50$) and E.E (1.3 ± 0.21%, on average, $p = 0.50$) contents between them, were observed (**Table 1**).

On the other hand, lower NDF (42.1 vs. 62.4 ± 8.60%; $p = 0.021$) and ADL (6.4 vs. 13.8 ± 2.07%; $p = 0.007$) contents in SF than SV were observed, with a trend in the ADL content (36.2 vs. 51.2 ± 7.98%; $p = 0.080$; **Table 1**). As for nutritive values, differences were observed in NE_L values between both studied stages (1.43 vs. 1.47 ± 0.06 Mcal kg/DM; $p = 0.010$; **Table 1**), while that MP content showed a tendency (1050 vs. 1227 ± 109 g/d; $p = 0.070$; **Table 1**).

Regarding antinutritional factors, the different metabolites did not differ between both studied stages ($p = 0.80–0.21$; **Table 1**), being their averages, for phenols (0.7 ± 0.11%), steroids (0.5 ± 0.10%), alkaloids (0.6 ± 0.11%) or saponins (0.7 ± 0.13%). The mineral contents observed in *E. edulis* differed according to the phenological stage, as shown in **Table 1**. Although the Na content did not vary between both studied stages (0.02 ± 0.01%, on average; $p = 0.62$), however, SF had greater P (0.31 vs. 0.19 ± 0.07%; $p = 0.04$) and K (3.0 vs. 1.3 ± 0.39%; $p = 0.05$) contents than SV, but with a 93% lower Ca content (0.1 vs. 1.4 ± 0.39%; $p = 0.05$; **Table 1**).

According to McDonald et al. [32], the chemical composition is highly correlated with feed digestibility. For this reason, in the grazing livestock systems, the forages should meet nutritional requirements for milk and meat production at a cheaper cost [33, 34]. This study showed that the vegetative stage (SV) had greater DM content than the fructification stage (SF), which was similar to the one found by Naranjo [35]. Consequently, the leaves of *E. edulis* will have greater DM content than the reported for *P. clandestinum* (12.0%), *Dactylis glomerata L.* (23.8%), and *Lolium perenne* (26.1.0%) which are the most important grasses in feeding ruminants in the livestock systems in the Andean region [36], although these values differ when compared for SF (12.2%). Despite huge differences in DM contents between both phenological stages, the SF showed a lower mean ash value than SV with clear differences in OM content. Additionally, as was expected, the EE values were low, as is typical for temperate climate forages [34, 37].

According to Schwab and Broderick [38] and Pfeffer and Hristov [39], CP content is essential for multiple organic functions and also serves as a substrate for rumen bacteria [40]. Additionality, McDonald et al. [32] and Gosselink et al. [41], revealed that there is a positive relationship between protein intake and the digestibility of

Item	*Erythrina edulis*		SEM	*p*-value
	SV[1]	SF[2]		SV vs. SF
DM at 60°C	31.4	12.2	5.0	0.009
Composition, % DM				
OM	89.5	92.3	0.89	0.020
Ash	10.5	7.7	1.40	0.001
CP	28.7	20.3	2.59	0.022
CF	24.9	23.1	3.15	0.50
NFE[3]	35.8	48.9	5.81	0.003
NDF	62.4	42.1	8.60	0.021
ADF	51.2	36.2	7.98	0.080
ADL	13.8	6.4	2.07	0.007
EE	1.5	1.1	0.21	0.50
NE⊠, Mcal kg/DM	1.43	1.47	0.06	0.010
MP[4], g/d	1053	1227	109	0.070
Antinutritional factors, % in DM				
Phenols	0.7	0.6	0.11	0.80
Steroids	0.5	0.5	0.10	0.53
Alkaloids	0.7	0.5	0.11	0.50
Saponins	0.7	0.7	0.13	0.21
Minerals, % in DM				
Ca	1.4	0.1	0.39	0.05
P	0.19	0.31	0.07	0.04
K	1.3	3.0	0.39	0.05
Na	0.02	0.02	0.01	0.62

[1]*SV, leaves.*
[2]*SF, sheath without seeds.*
[3]*NFE, N-free extract (or non-fiber carbohydrates) = OM − CP − CF.*
[4]*MP, was calculated based on 13 kg of DMI, according to NRC (2001); SEM, standard error of the mean.*

Table 1.
Chemical composition of Erythrina edulis *(leaves and sheath without seed).*

feed. Overall, the two studied phenological stages showed high CP contents (>20% on DM basis) in comparison to the common grasses used in the Andes region (*P. clandestinum*, 13.2%; *D. glomerata L.*, 18.4%; and *L. perenne*, 17%, respectively). In the same way, our CP contents obtained in *E. edulis* are similar to the reported for other legume plants *Medicago sativa* (22%) and *Trifolium pratense* (23%), respectively [36]. Additionality, our results were similar to those reported by Rosales (CP, 25%) [13], Bedoya et al. (CP, 18%) [16], Intiquilla et al. (CP, 22%) Perez et al. (CP, 20%) [22], and Pérez et al. (CP, 23%) [42], when they determined CP contents in leaves of *E. edulis*. Some studies have showed that, CP contents less than 7% in DM basis, are not adequate for feeding ruminants [43–45]. Based on our results, the *E. edulis* had enough CP contents to supply the requirements of protein and ammonia need by rumen microbial [40, 45, 46]. Possibly, this could explain the no correlations detected between CP contents and DM degradability for SV (p = 0.97) and SF (p = 0.95), as shown in **Table 2**.

With regard to the fiber contents, a study with leaves of *E. edulis* reported similar NDF contents (61 vs. 62%) to our work [13], and greater compared to those obtained by Naranjo [35] (62 vs. 50%). Whereas other researchers have reported in other varieties of *Erythrina (indica, subumbrans, and variegata)* lower NDF contents (54%, on average, on DM basis) than *E. edulis* [12, 18]. Positive correlations were detected between CP and NDF contents (r = 0.99; p < 0.001; **Table 2**), such as it has been seen in studies with pastures.

In the same way, Rosales [13] and Naranjo [35] reported lower ADF contents than our study (ranged from 45 to 26%), but this last author worked with other varieties, so cannot be comparable. Anyhow, in this work, SF showed lower ADF contents that above mentioned studies. The NDF stimulates the rumination and salivation, being important for remaining the normal rumen function [47, 48]. Nevertheless, negative correlations have been observed on DMI when NDF contents is above the limit require for ruminants (35–28%, in DM basis) [49, 50]. For this reason, the NDF content of forages is a good predictor to know its gut fill capacity, reflecting its nutritional value [51].

Differences on chemical composition between studied phenological stages, were reflected in the nutritive value, as shown in **Table 1**. The NE_L contents were greater in SF than SV (NE_L, 1.47 vs. 1.43 ± 0.06 Mcal kg/DM; p = 0.010) with a tendency in the MP contents (1227 vs. 1053 g/d; p = 0.070). Analyzing the correlations, the NE_L contents in SV were positively correlated with OM, CP, NFE and NDF (r = 0.99; p < 0.001) contents, but no correlated for SF (r =-0.99; p < 0.001). On the contrary, MP contents showed be positively correlated with CP, NFE and NDF contents in both phenological stages (r = 0.99; p < 0.001), as shown in **Table 2**.

Effects of condensed tannins on feeding efficiency, N losses and animal health are highly relevant, and a greater understanding of how grazing management might be refined to enable these potential benefits to be realized is critical [15, 52]. Several studies [43, 52–54] stated that temperate legumes contain moderate levels of secondary compounds, such as condensed tannins and flavonoid which could reduce environmental problems by increasing nitrogen use efficiency in protein utilization. In this sense, several studies have showed that antinutritional factors at low concentrations (20–40 g kg/DM) are nutritionally beneficial through decreased degradation of dietary protein in the rumen, and increase protein availability for digestion and absorption leading to good animal performance [15, 44]. Based on these evidence, *E. edulis* in both phenological stages showed lower antinutritional levels (<2% in DM basis) which was similar to the reported by Rosales [13]. Therefore, theoretically, the studied phenological stages (SV and SF, respectively) should not represent a nutritional problem in feeding ruminants according with Moore et al. [34] and Mehrez and Keely [55]. Although, we hypothesized that the latex or resins presents in the plant could inhibit the initial colonization of rumen microbes. Anyway, it must be confirmed with more studies.

4. Ruminal degradability data of DM and CP from *Erythrina edulis* according to different incubation times

Ruminal degradability data of DM and CP of *Erythrina edulis*, are shown in **Table 3**. Statistical differences in the DM degradability at 24/h between SV vs. SF

DOI: http://dx.doi.org/10.5772/intechopen.107496

Stages		DM	OM	Ash	CP	CF	NFE	NDF	ADF	ADL	EE	NE	MP
SV	DMd	0.017 0.94	−0.008 0.97	−0.008 0.97	−0.008 0.97	− 0.008 0.97	− 0.008 0.97	− 0.008 0.97	− 0.008 0.97	− 0.008 0.97	− 0.008 0.97	− 0.008 0.97	− 0.008 0.97
	CPd	− 0.05 0.81	0.11 0.64	0.11 0.64	0.11 0.64	0.11 0.64	0.11 0.64	0.11 0.64	0.11 0.64	0.11 0.64	0.11 0.64	0.11 0.64	0.11 0.64
	DM		− 0.50 0.02	−0.50 0.02	−0.50 0.02	−0.50 0.02	−0.50 0.02	−0.50 0.02	−0.50 0.02	−0.50 0.02	−0.50 0.02	−0.50 0.02	−0.50 0.02
	OM	− 0.50 0.02		0.99 0.001	0.99 0.001	0.99 0.001	0.99 0.001	0.99 0.001	0.99 0.001	0.99 0.001	0.99 0.001	0.99 0.001	0.99 0.001
	Ash	−0.50 0.02	0.99 0.001		0.99 0.001	0.99 0.001	0.99 0.001	0.99 0.001	0.99 0.001	0.99 0.001	0.99 0.001	0.99 0.001	0.99 0.001
	CP	−0.50 0.02	0.99 0.001	0.99 0.001		0.99 0.001	0.99 0.001	0.99 0.001	0.99 0.001	0.99 0.001	0.99 0.001	0.99 0.001	0.99 0.001
	CF	−0.50 0.02	0.99 0.001	0.99 0.001	0.99 0.001		0.99 0.001	0.99 0.001	0.99 0.001	0.99 0.001	0.99 0.001	0.99 0.001	0.99 0.001
	NFE	−0.50 0.02	0.99 0.001	0.99 0.001	0.99 0.001	0.99 0.001		0.99 0.001	0.99 0.001	0.99 0.001	0.99 0.001	0.99 0.001	0.99 0.001
	NDF	−0.50 0.02	0.99 0.001	0.99 0.001	0.99 0.001	0.99 0.001	0.99 0.001		0.99 0.001	0.99 0.001	0.99 0.001	0.99 0.001	0.99 0.001
	ADF	−0.50 0.02	0.99 0.001	0.99 0.001	0.99 0.001	0.99 0.001	0.99 0.001	0.99 0.001		0.99 0.001	0.99 0.001	0.99 0.001	0.99 0.001
	ADL	−0.50	0.99 0.001	0.99 0.001	0.99 0.001	0.99 0.001	0.99 0.001	0.99 0.001	0.99 0.001		0.99 0.001	0.99 0.001	0.99 0.001
	EE	−0.50 0.02	0.99 0.001	0.99 0.001	0.99 0.001	0.99 0.001	0.99 0.001	0.99 0.001	0.99 0.001	0.99 0.001		0.99 0.001	0.99 0.001
	NE	−0.50 0.02	0.99 0.001	0.99 0.001	0.99 0.001	0.99 0.001	0.99 0.001	0.99 0.001	0.99 0.001	0.99 0.001	0.99 0.001		0.99 0.001
	MP	−0.50 0.02	0.99 0.001	0.99 0.001	0.99 0.001	0.99 0.001	0.99 0.001	0.99 0.001	0.99 0.001	0.99 0.001	0.99 0.001	0.99 0.001	

Stages		DM	OM	Ash	CP	CF	NFE	NDF	ADF	ADL	EE	NE	MP
SF	DMd	− 0.024 0.91	− 0.012 0.95	− 0.012 0.95	− 0.012 0.95	− 0.012 0.95	− 0.012 0.95	− 0.012 0.95	− 0.012 0.95	− 0.012 0.95	− 0.012 0.95	0.012 0.95	− 0.012 0.95
	CPd	0.056 0.80	0.11 0.62	0.11 0.62	0.11 0.62	0.11 0.62	0.11 0.62	0.11 0.62	0.11 0.62	0.11 0.62	0.11 0.62	− 0.11 0.63	0.11 0.62
	DM		0.50 0.02	0.50 0.02	0.50 0.02	0.50 0.02	0.50 0.02	0.50 0.02	0.50 0.02	0.50 0.02	0.50 0.02	−0.50 0.02	0.50 0.02
	OM	0.50 0.02		0.99 0.001	0.99 0.001	0.99 0.001	0.99 0.001	0.99 0.001	0.99 0.001	0.99 0.001	0.99 0.001	−0.99 0.001	0.99 0.001
	Ash	0.50 0.02	0.99 0.001		0.99 0.001	0.99 0.001	0.99 0.001	0.99 0.001	0.99 0.001	0.99 0.001	0.99 0.001	−0.99 0.001	0.99 0.001
	CP	0.50 0.02	0.99 0.001	0.99 0.001		0.99 0.001	0.99 0.001	0.99 0.001	0.99 0.001	0.99 0.001	0.99 0.001	−0.99 0.001	0.99 0.001
	CF	0.50 0.02	0.99 0.001	0.99 0.001	0.99 0.001		0.99 0.001	0.99 0.001	0.99 0.001	0.99 0.001	0.99 0.001	−0.99 0.001	0.99 0.001
	NFE	0.50 0.02	0.99 0.001	0.99 0.001	0.99 0.001	0.99 0.001		0.99 0.001	0.99 0.001	0.99 0.001	0.99 0.001	−0.99 0.001	0.99 0.001
	NDF	0.50 0.02	0.99 0.001	0.99 0.001	0.99 0.001	0.99 0.001	0.99 0.001		0.99 0.001	0.99 0.001	0.99 0.001	−0.99 0.001	0.99 0.001
	ADF	0.50 0.02	0.99 0.001	0.99 0.001	0.99 0.001	0.99 0.001	0.99 0.001	0.99 0.001		0.99 0.001	0.99 0.001	−0.99 0.001	0.99 0.001
	ADL	0.50 0.02	0.99 0.001	0.99 0.001	0.99 0.001	0.99 0.001	0.99 0.001	0.99 0.001	0.99 0.001		0.99 0.001	−0.99 0.001	0.99 0.001
	EE	0.50 0.02	0.99 0.001	0.99 0.001	0.99 0.001	0.99 0.001	0.99 0.001	0.99 0.001	0.99 0.001	0.99 0.001		−0.99 0.001	0.99 0.001
	NE	−0.50 0.02	−0.99 0.001	−0.99 0.001	−0.99 0.001	−0.99 0.001	−0.99 0.001	−0.99 0.001	−0.99 0.001	−0.99 0.001	−0.99 0.001		0.99 0.001
	MP	0.50 0.02	0.99 0.001	0.99 0.001	0.99 0.001	0.99 0.001	0.99 0.001	0.99 0.001	0.99 0.001	0.99 0.001	0.99 0.001	−0.99 0.001	−0.99 0.001

Table 2.
Correlation matrix amongst chemical composition and digestibility of dry matter and crude protein from Erythrina edulis.

were observed (47.9 vs. 75.4 ± 0.68%/h; $p < 0.001$; **Table 3**). Whereas at 72/h rumen incubation the SF had 38% higher DM degradability than to the observed in SF (86.1 vs. 53.1 ± 0.68%/h, $p < 0.001$; **Table 3**). With regard to CP degradability, at 24/h incubation time the SV showed lower data than SF (53.4 vs. 61.3 ± 2.50%/h; $p < 0.001$), with marked differences at 72 h rumen incubation (53.4 vs. 77.9 ± 2.50%/h; $p < 0.001$; **Table 3**).

In situ procedures of the artificial synthetic fiber bag which have evolved since the earlier part of the 20th century, there is a standard means of estimating the extent of feed degradation [30, 56]. Consequently, degradation kinetics in the reticulorumen are one key tool for evaluating ruminant feedstuffs [5, 57, 58]. Naranjo [35] found a lower DM degradability value than our study (40 vs. 53%/h), although our value was similar to the obtained by Rosales [13]. In contrast, other reports have shown high degradability values at a 48 h incubation time (81.45%/h, on average) by Pedraza et al. [59] and Camero et al. [12] (68%/h, on average) although with other *Erythrina* varieties (*berteroana, variegate and poeppiginia*). Anyway, the SF showed high DM degradability at 48 h incubation time (84.1%/h). Unfortunately, there were not more reports on the rumen degradability kinetics with sheath of *E. edulis*, and our data are proposed as referential values. In summary, our work showed acceptable values compared with other forage used for livestock feeding [33, 60–62].

According to Jian et al. [63] and Buxton et al. [64] DM intake is essential for dairy ruminants to maintain and production performance [65]. However, the contents of NDF and ADF in forage are also considered the major factors affecting feed intake and feed conversion efficiency of ruminants, influencing animal performance [49, 66, 67]. Therefore, in this study, the lower DM and CP degradability observed in SV could be explained due to higher hemicellulose and the degree of lignification, which reflects the difficulty of degradation [68]. Despite that, no correlation was observed between ADL contents and DM degradability ($p = 0.95$; **Table 2**).

	Incubation time (h)								SEM	*p*-value
Item	**0**	**3**	**6**	**12**	**24**	**36**	**48**	**72**		
Dry matter										
SV¹	28.6^{bF}	32.8^{aE}	36.4^{bD}	41.7^{bC}	47.9^{bB}	50.8^{bA}	52.2^{bA}	53.1^{bA}	0.36	0.001
SF²	35.6^{aH}	44.5^{aG}	51.8^{aF}	62.8^{aE}	75.4^{aD}	81.6^{aBC}	84.1^{aA}	86.1^{aA}	0.49	0.001
SEM	0.68	0.68	0.68	0.68	0.68	0.77	0.68	0.68	-	-
p-value	<0.001	<0.001	<0.001	<0.001	<0.001	<0.001	<0.001	<0.001	-	-
Crude protein										
SV	35.0^{C}	48.7^{AB}	50.7^{A}	52.9^{A}	53.4^{A}	53.4^{bA}	53.4^{bA}	53.4^{bA}	2.5	0.001
SF	34.8^{B}	39.6^{B}	43.7^{B}	50.8^{B}	61.3^{AB}	68.2^{aA}	72.9^{aA}	77.9^{aA}	2.5	0.001
SEM	2.55	2.60	2.57	2.60	2.50	2.60	2.40	2.50	-	-
p-value	1.00	0.527	0.867	1.00	0.737	0.025	0.001	0.001	-	-

¹*SV, leaves.*
²*SF, sheath without seeds.*
$^{a-b}$*Mean values with different letter in the same column differ* (p < 0.05).
$^{A-G}$*Mean values with different letter in the same row differ* (p < 0.05); *SEM, standard error of the mean.*

Table 3.
Average of DM and CP degradability (%) after 0, 3, 6, 12, 24, 48, and 72 h incubation in situ technique.

5. Degradability parameters of *Erythrina edulis* obtained *in situ*

In situ degradability parameters of *E. edulis,* are shown in **Table 4**. The SF had highest disappearance of fast and slow fractions compared to SV (α, 35.6 vs. 28.6 ± 0.57%/h; p = 0.003; b, 51.1 vs. 24.8 ± 0.57%/h; $p < 0.001$, **Table 4**). But, the "c" degradation rate in SF was lower when compared to SV (0.55 vs. 0.62 ± 0.57%/h; $p < 0.001$; **Table 4**). Despite this, SF had greater effective degradability values in all passage rates than SV 2%/h (73.1 vs. 47.4 ± 0.57%/h; $p < 0.001$), 5%/h (62.4 vs. 42.3 ± 0.57/h; $p < 0.001$), 6%/h (60.0 vs. 41.2 ± 0.54%/h; $p < 0.001$) and 8%/h (56.4 vs. 39.4 ± 0.57%/h; $p < 0.001$), respectively, as shown in **Table 4**.

In contrast, no differences in the soluble fraction "α" for CP between both phenological stages (35 ± 2.03%, on average) were observed (**Table 4**). Nevertheless, the SF showed a highest insoluble but potentially "b" degradable fraction than SV (50.0 vs. 18.4 ± 2.03%/h; $p < 0.001$; **Table 4**). Whereas the ED only varied at passage rate of 2% between SF vs. SV (66.4 vs. 52.3 ± 2.0%/h; p = 0.011; **Table 4**).

The rumen degradability in SF was highest for all degradation parameters "α," "b," $\alpha + b$, and ED. These differences might be related with lower fiber contents, which increase of total degradable fraction ($\alpha + b$) [9]. Furthermore, with regard the lower ED for SV, we hypothesized that greater cell wall components especially lignin will be implicated. Naranjo [35] in leaves of *E. edulis* reported lower degradation parameter α (21.1 vs. 28.6%) than the obtained in this study, although with higher b (35.2 vs. 24.8%) and c (0.09 vs. 0.06/h) fractions. Differences that might be as a consequence of our greater ADF contents (51.2 vs. 32.2%). In this sense, forages with high wall cell components, have showed lower digestibly data, due to decreases the colonization of

Item	Degradation parameter[1]				Effective degradability			
					INRA	Passage rates[2]		
	α	b	$\alpha + b$	c	6%	2%	5%	8%
Dry matter								
SV[3]	28.6[b]	24.8[b]	53.4[b]	0.062[a]	41.2[b]	47.4[b]	42.3[b]	39.4[b]
SF[4]	35.6[a]	51.1[a]	86.7[a]	0.055[b]	60.0[a]	73.1[a]	62.4[a]	56.4[a]
SEM	0.57	0.57	0.57	0.57	0.54	0.57	0.57	0.57
p-value	0.003	0.001	0.001	0.001	0.001	0.001	0.001	0.001
Crude protein								
SV	35.0	18.4	53.4	0.038	51.0	52.3[b]	51.0	49.7
SF	35.0	50.0[a]	82.0[a]	0.034	53.0	66.4[a]	55.2	50.0
SEM	2.03	2.03	2.03	2.03	2.10	2.0	2.0	2.0
p-value	1.00	0.001	0.001	1.00	1.00	0.011	0.20	1.00

[1]α: *Soluble fraction;* b: *Insoluble but degradable fraction; kd: The rate (%/h) of disappearance of* b *fraction.*
[2]*Passage rates, according to Bhargava and Ørskov [69] for different intake level. INRA [31] with a fixed transit rate of 6%/h.*
[3]*SV, leaves.*
[4]*SF, sheath.*
[a–b]*Mean value with different letter in the same columns differ* (p < *0.05).*

Table 4.
Disappearance of fast and slow fractions and effective degradability values of DM an CP from Erythrina edulis.

rumen microbiota reducing its degradation [49, 66, 70]. Despite all differences above mentioned, the SF had higher DM degradability than SV and other forage sources [35, 71], which might be related to its lower NDF, ADF and ADL contents. To the best of our knowledge, there are no more reports with sheath of *E. edulis*, so our data could be proposed as referential.

In ruminants, the quantity of amino acids that reach the small intestine depends on the microbial protein synthesis and feed proteins which escape ruminal degradation. Therefore, the feed protein degradation can be affected by its nature [72–75]. Consequently, the CP content degraded in the rumen is determined by their fractional degradation and passage rates [38, 45, 76]. At present, the model of Ørskov et al. [30, 77] for describing protein degradation and escape, it is widely accepted and applied. How in other studies, the CP degradability increased with incubation time. According to Ibrahim et al. [47], the CP from lush pasture is highly soluble and rapidly and extensively degraded in the rumen. Possibly, this could explain our no differences in the α fraction rate between both studied stages (35%, on average). Nevertheless, the SF showed a higher insoluble but potentially degradable fraction "b" (50 vs. 18 $\pm$ 2.03%/h), although with a slower CP disappearance rate than to the observed in SV, as shown in **Table 4**. Regarding the lower "b" fraction observed for SV, it could be explained by its chemical composition: greater fiber contents, resulting in a lower CP degradability than SF. Additionality, Bhargava and Ørskov [50] and Camero et al. [12] stated that when protein degradability is high, the release rate of microbial substrates is high, as well as the period after feeding, during which the microbial substrates are released over a short period of time. Therefore, the CP content in the rumen of both phenological stages was mainly affected by the retention time agree with Ørskov and McDonald [30]. In this study, at 24 h, the CP degradability in SF was 13% higher than SV (61 vs. 53 $\pm$ 2.50%/h), which is opposite to the mentioned by Ma [63] that higher contents of CP had beneficial for CP degrading in the rumen. Similar trend was observed at 72 h incubation time, with clear differences between both studied stages (78 vs. 53 $\pm$ 2.50%/h). Despite this, the passage rate of 5%, 6% and 8% per hour did not vary between leaves or sheath, because of similar degradation rate (0.36%/h, on average), with exception for at 2% passage rate (66 vs. 52 $\pm$ 2.0%/h; **Table 4**). Based on these results, the "b" degradability rates should be taken in account for the ration in feeding ruminant programs. Referential values in leaves of *E. edulis* reported by Rosales [35] have shown a slight greater α fraction than our study (41 vs. 35%/h) but with a lower "b" potentially degradable fraction (38 vs. 50%) and no differences in disappearance rate (0.03%/h). Contrary to this, higher degradation parameters for "α," "b," and ED in other *Erythrina* varieties *(indica, subumbrans, variegata and berteroana)* than the obtained in *E. edulis* have been reported by Kongmanila et al. [18] and Pedraza et al. [59]. Anyhow, the chemical composition as well as degradability parameters obtained in this study, could be similar to the other multipurpose fodder trees and shrubs (e.g., *Leucaena leucocephala, Gliricidia sepium, T. tetraptera, L. diversifolia*, and *L. sericeus*) widely used in ruminant nutrition [13, 78, 79].

6. Conclusions

The *E. edulis* showed similar CP contents to other widely sources used in diets for livestock nutrition, but the leaves showed greater wall cell contents than the sheath. As a consequence, at 72 h incubation time the leaves had lower DM and CP disappearance than the sheath. For this reason, the ED of DM in SF was higher than SV, and only differences at outflow rate of 2%/h for CP was observed. Based on these findings, it necessary to perform an *in vivo* study to determine adequate levels of inclusion in ruminant feeding. Anyhow, *E. edulis* is an interesting feed to include as a supplementation to improve low quality forages in livestock systems.

Acknowledgements

All authors are very grateful with the small-livestock farmers of Ecuador, which were involved in this study. In the same way, to INIAP for help us and provides all experimental conditions.

Conflict of interest

The authors declare no conflict of interest.

Abbreviations

Acid detergent fiber (ADF), Acid detergent lignin (ADL), Crude fiber (CF), Crude protein (CP), Dry matter (DM), Dry matter intake (DMI), Effective degradability (ED), *Erythrina edulis* (*Euphorbia edulis*), Ether extract (E.E), Fructification stage (SF), Microbial protein (MP), Neutral detergent fiber (NDF), N-free extract (NFE), Next energy for lactation (NE_L), Organic matter (OM), Standard error of the mean (SEM), vegetative stage (SV)

Author details

Oscar Giovanny Fuentes Quisaguano[1] and Santiago Alexander Guamán Rivera[2*]

1 Veterinary Medicine, Central University of Ecuador, Quito, Ecuador

2 Research Group Causana Yachay (GICAY), Escuela Superior Politècnica de Chimborazo, Sede Orellana, El Coca, Ecuador

*Address all correspondence to: santiagoa.guaman@espoch.edu.ec

References

[1] Sturaro E, Marchiori E, Cocca G, Penasa M, Ramanzin M, Bittante G. Dairy systems in mountainous areas: Farm animal biodiversity, milk production and destination, and land use. Livestock Science 2013;**158**:157-168. http://dx.doi.org/10.1016/j.livsci.2013.09.011

[2] FAO. Good practices for the feed sector implementing the codex Alimentarius code of practice on Good Animal Feeding. 2020. Available from: www.fao.org. [Accessed March 3, 2022]

[3] Muñoz EC, Andriamandroso AL, Blaise Y, Ron L, Montufar C, Kinkela PM, et al. How do management practices and farm structure impact productive performances of dairy cattle in the province of Pichincha, Ecuador. Journal of Agriculture and Rural Development in the Tropics and Subtropics. 2020; **121**(121):233-241. https://doi.org/10.17170/kobra-202010191971

[4] FAO. Climate smart livestock production in Ecuador a strategic partnership between FAO and the private sector. 2019. Available from: www.fao.org. [Accessed January 12, 2021]

[5] Van der Walt J, Meyer JH. Protein digestion un ruminants. Arch für Tierernaehrung. Dec 1983;**8**(18):853-862 http://www.tandfonline.com/doi/abs/10.1080/17450398309426933

[6] Calsamiglia S, Busquet M, Cardozo PW, Castillejos L, Ferret A. Invited review : Essential oils as modifiers of rumen microbial fermentation. Journal of Dairy Science 2007;**90**:2580-2595. http://dx.doi.org/10.3168/jds.2006-2644

[7] Niderkorn V, Baumont R. Associative effects between forages on feed intake and digestion in ruminants. Animal. 2009;**3**:951-960

[8] González-Marcillo RL, Castro Guamàn WE, Guerrero Pincay AE, Vera Zambrano PA, Ortiz Naveda NR, Guamàn-Rivera SA. Assessment of guinea grass panicum maximum under silvopastoral systems in combination with two management systems in Orellana Province, Ecuador. Agriculture. 2021;**11**:117

[9] Salas H, Castillejos L, López-Suárez M, Ferret A. Rumen fermentation and N metabolism of camelina co-products for beef cattle studied with a dual flow. Animals. 2019;**9**:1079

[10] Yao KY, Gu FF, Liu JX. In vitro rumen fermentation characteristics of substrate mixtures with soybean meal partially replaced by microbially fermented yellow wine lees. Italian Journal of Animal Science. 2020;**19**:18-20

[11] González J, Andrés S, Rodríguez C, Remedios AM. In situ evaluation of the protein value of soybean meal and processed full fat soybeans for ruminants. Animal Research. 2002;**51**: 455-464

[12] Camero A, Ibrahim M, Kass M. Improving rumen fermentation and milk production with legume-tree fodder in the tropics. Agroforestry Systems. 2001; **51**:157-166

[13] Rosales M. In vitro assessment of the nutritive value of mixtures of leaves from tropical fodder trees [PhD thesis]. University of Oxford; 1996

[14] Phimphachanhvongsod V, Ledin I. Performance of growing goats fed panicum maximum and leaves of *Gliricidia sepium*. Asian-Australasian

Journal of Animal Sciences. Jan 2002;**1** (15):1585-1590. http://ajas.info/journal/view.php?doi=10.5713/ajas.2002.1585

[15] Oppong SK, Kemp PD, Douglas GB. Browse shrubs and trees as fodder for ruminants: A review on management and quality. Journal of Science and Technology. 2008;**28**:65-75

[16] Bedoya OA, Caicedo M, Guerrero Y. Obtención de un extracto proteico a partir de harina de chachafruto (*Erythrina edulis*). La Revista Universidad y Salud. 2012;**14**:161-167

[17] Escamilo CS. El Pajuro *(Erythrina edulis)* alimento andino en extinción. Investing in Society. 2012;**16**:16-20

[18] Kongmanila D, Bertilsson J, Ledin I. Degradability of leaves from three Erythrina species in Lao PDR. Livestock Science 2013;**155**:273–276. http://dx.doi.org/10.1016/j.livsci.2013.05.029

[19] Acero-Duarte L. Guía para el cultivo yaprovechamientodel Chachafruto o Balú *(Erythrina edulis)*. Spain: Convenio Andrés Bello; 2002

[20] Intiquilla A, Jiménez-Aliaga K, Zavaleta AI, Arnao I, Peña C, Chavez-Hidalgo EL, et al. *Erythrina Edulis* (Pajuro) seed protein: A new source of antioxidant peptides. Natural Product Communications. 2016;**11**:781-786 h ttp://journals.sagepub.com/doi/10.1177/1934578X1601100620

[21] Intiquilla A, Jiménez-aliaga K, Guzmán F, Alvarez CA, Zavaleta AI, Hernández-ledesma B. Novel antioxidant peptides obtained by alcalase hydrolysis of *Erythrina edulis* (pajuro) protein. Journal of the Science of Food and Agriculture. 2018;**99**:2420-2427

[22] Pérez G, de Martínez C, Díaz E. Evaluation of the protein quality of *Erythrina edulis* (balú). Archivos Latinoamericanos de Nutrición. 1979;**29**: 193-207. http://europepmc.org/abstract/MED/533329

[23] Bustamante J, Ibrahim M, Beer J. Evaluacion agronomica de ocho gramineas mejoradas en un sistema silvopastoril con poro *(Erythrina poeppigiana)* en el tropico humedo turriaba. Costa Rica: CATIE; 1998

[24] Guevara PJ, Díaz P, Bravo N, Vera M, Crisostomo O, Barbachán H, et al. Uso de harina de pajuro *(Erythrina edulis)* como suplemento en la alimentación de cuyes *(Cavia porcellus)*. Lima. Rev Per Quím Ing Quím. 2013;**16**: 21-28

[25] GADMCR. Plan de desarrollo y ordenamiento territorial actualización 2014–2019 Gobierno autónomo descentralizado municipal Cantòn Rumiñahui. 2014. https://ruminah ui.gob.ec/. Last accessed on 21 January 2022

[26] AOAC. Official Methods of Analysis. Virginia, USA: Association of Analytical Chemists; 2000

[27] Makkar P, Dawra R, Singh B. Determination of both tannin and protein in a tannin-protein complex. Journal of Agricultural and Food Chemistry. May 10 1988;**36**(3):523-525. https://pubs.acs.org/doi/abs/10.1021/jf 00081a600

[28] Aufrère J, Graviou D, Michalet-Doreau. Degradation in the rumen of proteins of 2 legumes: soybean meal and field pea. Reproduction, Nutrition, Development. 1994;**34**:483-490

[29] NRC. Requirements of Dairy Cattle. Seventh Revised ed. Washington, DC: National Academy Press; 2001. p. 405

[30] Ørskov ER, McDonald I. The estimation of protein degradability in the rumen from incubation measurements weighted according to rate of passage. The Journal of Agricultural Science. Apr 1979;**27**(92):499-503. https://www.cambridge.org/core/product/identifier/S0021859600063048/type/journal_article

[31] INRA. Alimentation Des Ruminants. Versailles, France: Éditions Quæ; 2018. p. 728

[32] McDonald P, Edwards RA, Greenhalgh JFD, Morgan CA, Sinclair LA, Wilkinson RG. The animal and its Food. Seventh ed. London, UK; 2010

[33] Hoffman PC, Sievert SJ, Shaver RD, Welch DA, Combs DK. In situ dry matter, protein, and fiber degradation of perennial forages. Journal of Dairy Science 1993 Sep; 76:2632–2643. http://dx.doi.org/10.3168/jds.S0022-0302(93)77599-2

[34] Moore KJ, Collins M, Jerry NC, Redfearn DD. Forages, the Science of Grassland Agriculture, II. Seventh ed. Chichester, UK: John Wiley & Sons Ltd; 2020

[35] Naranjo J. Nutritional characterization and ruminal degradation kinetics of some forages with potential for ruminants supplementation in the highland tropics of Colombia. Revista CES Medicina Veterinaria y Zootecnia. 2011;**6**:9-19

[36] León R, Bonifaz N, Francisco G. Pastos y Forrajes del Ecuador: Siembra y producción de pasturas. Vol. 148. Cuenca, Ecuador: Editorial Universitaria Abya-Yala; 2018

[37] Amrita P, Chairman N. Nutritive Value of Commonly Available Feeds and Fodders in India, Group Anand. Anand, India: National Dairy Development Board; 2012. pp. 1-128

[38] Schwab CG, Broderick GA. A 100-year review: Protein and amino acid nutrition in dairy cows 1. Journal of Dairy Science 2017;**100**:10094–10112. http://dx.doi.org/10.3168/jds.2017-13320

[39] Pfeffer E, Hristov A. Nitrogen and Phosphorus Nutrition of Cattle. ProQuest Ebook Central. Moscow, USA: CABI; ID 83844-2330; 2005. http://ebookcentral.proquest.com/lib/uab/detail.action?docID=289439.

[40] Fox DG, Sniffen CJ, O'Connor JD, Russell JB, Van Soest PJ. A net carbohydrate and protein system for evaluating cattle diets: Carbohydrate and protein availability. Journal of Animal Science. 1992;**70**:3578-3596

[41] Gosselink JMJ, Dulphy JP, Poncet C, Jailler M, Tamminga S, Cone JW. Prediction of forage digestibility in ruminants using in situ and in vitro techniques. Animal Feed Science and Technology. 2004;**115**:227-246

[42] Pérez A, Herández E, Sandoval C, Otárola F. Presencia del chachafruto *(Erythrina edulis Triana ex Micheli)* en el estado de Merida, Venezuela. Rev Electr'onica Conoc Libr y Licenciamiento. 2015;**6**:140-153

[43] Mertens DR. Regulation of forage intake. 2015. p. 450–493. Available from: https://onlinelibrary.wiley.com/doi/10.2134/1994.foragequality.c11

[44] Givens DI, Owen E, Axford RFE, Omed HM. Forage evaluation in ruminant nutrition. Wallingford, UK: CABI Publishing; 2000

[45] Fahey Jr GC, Collins M, Mertens DR, Moser LE. Forage Quality, Evaluation,

and Utilization. Wisconsin, USA: Madison; 1994

[46] Wallace RJ, Lahlou-Kassai A. Rumen ecology research planning. In: Proceedings of a Workshop Held at ILRI, Addis Ababa, Ethiopia; International Livestock Research Institute. 1995. p. 270 pp. Available from: http://linkinghub.elsevier.com/retrieve/pii/S0377840197888633

[47] Van Soest PJ. Nutritional Ecology of the Ruminant. Cornell University, ProQuest Ebook Central; 1994

[48] Preston NG. Effect of in vitro NDF digestibility of barley cultivars on ensiling, digestibility and lamb performance [MS thesis]. University of Saskatchewan, Saskatoon, SK; 2016.

[49] Hoover WH. Chemical factors involved in ruminal fiber digestion. Journal of Dairy Science 1986; 69,2755-2766. http://dx.doi.org/10.3168/jds.S0022-0302(86)80724-X

[50] Harper KJ, McNeill DM. The role iNDF in the regulation of feed intake and the importance of its assessment in subtropical ruminant systems (the role of iNDF in the regulation of forage intake). Agriculture. 2015;**5**:778-790

[51] Preston NG, Hünerberg M, Silva TM, Nair J, Yu P, Christensen DA, et al. Digestibility and performance of feeder lambs fed mixed barley grain-barley silage diets with varieties of barley silage selected on the basis of in vitro neutral detergent fibre degradability. Canadian Journal of Animal Science. 2017;**97**: 418-430

[52] Rochon JJ, Doyle CJ, Greef JM, Hopkins A, Molle G, Sitzia M, et al. Grazing legumes in Europe: A review of their status, management, benefits, research needs and future prospects. Grass and Forage Science. 2004 Sep;**59** (3):197-214 Available from: https://onlinelibrary.wiley.com/doi/10.1111/j.1365-2494.2004.00423.x

[53] Boval M, Dixon RM. The importance of grasslands for animal production and other functions: A review on management and methodological progress in the tropics. Animal. 2012;**6**: 748-762

[54] Broderick GA. Desirable characteristics of forage legumes for improving protein utilization in ruminants. Journal of Animal Science. 1995;**73**:2760-2773

[55] Mehrez R., Keely F. Feed Supplementation Blocks. FAO animal production and health, Rome, Italy: Food and Agriculture Organization of the United Nations, 248 p ISBN: 978-92-5-105438-3. 2009.

[56] Dhanoa MS, López S, Sanderson R, France J. Simplified estimation of forage degradability in the rumen assuming zero-order degradation kinetics. The Journal of Agricultural Science. 2009; **147**:225-240

[57] Hristov AN, Bannink A, Crompton LA, Huhtanen P, Kreuzer M, McGee M, et al. Invited review: Nitrogen in ruminant nutrition: A review of measurement techniques. Journal of Dairy Science. 2019;**102**:5811-5852

[58] Russell JB, O'Connor JD, Fox DG, Van Soest PJ, Sniffen CJ. A net carbohydrate and protein system for evaluating cattle diets: I Ruminal fermentation. Journal of Animal Science. 1992 Nov;**1**(70):3551-3561 https://academic.oup.com/jas/article/70/11/3551-3561/4705802

[59] Pedraza RM, La OO, Estévez J, Guevara G, Martínez S. Nota Técnica:

Degradabilidad ruminal efectiva y digestibilidad intestinal in vitro del nitrógeno. Pastos y forrajes. 2003;**26**: 237-241

[60] Taghizadeh A, Danesh Mesgaran M, Valizadeh R, Eftekhar Shahroodi F, Stanford K. Digestion of feed amino acids in the rumen and intestine of steers measured using a mobile nylon bag technique. Journal of Dairy Science 2005;**88**:1807-1814. http://dx.doi.org/10.3168/jds.S0022-0302(05)72855-1

[61] Ledea JL, La O, Ray JV. Characterization of in situ ruminal degradabilty of dry matter in new varieties of drought tolerant Cenchrus purpureus. Cuban Journal of Agricultural Sciences. 2016;**50**:421-433

[62] Lei YG, Li XY, Wang YY, Li ZZ, Chen YL, Yang YX. Determination of ruminal dry matter and crude protein degradability and degradation kinetics of several concentrate feed ingredients in cashmere goat. Journal of Applied Animal Research. 2017;**46**:134-140

[63] Ma J, Sun G, Shah AM, Fan X, Li S, Yu X. Effects of different growth stages of amaranth silage on the rumen degradation of dairy cows. Animals. 2019;**9**:793

[64] Buxton DR, Mertens DR, Fisher DS. Forage Quality and Ruminant Utilization, Cool-Season Forage Grasses. Madison Wisconsin, USA: American Society of Agriculture; 1996. pp. 229-266

[65] Moorby JM, Fraser MD. Review: New feeds and new feeding systems in intensive and semi-intensive forage-fed ruminant livestock systems. Animal 2021;**15**:100297. DOI: 10.1016/j.animal.2021.100297

[66] Jung HG, Vogel KP. Influence of lignin on digestibility of forage cell wall material. Journal of Animal Science 1986; **62**:1703-1712. https://doi.org/10.2527/jas1986.6261703x

[67] Ma Y, Zahoor-Khan M, Liu Y, Xiao J, Chen X, Ji S, et al. Analysis of nutrient composition, rumen degradation characteristics, and feeding value of Chinese Rye Grass, Barley Grass, and Naked Oat Straw. Animals. 2021;**11**:2486

[68] Van Soest PJ, Robertson JB, Lewis BA. Methods for dietary fiber, neutral detergent fiber, and nonstarch polysaccharides in relation to animal nutrition. Journal of Dairy Science. 1991; **74**:3583-3597

[69] Bhargava P, Ørskov E. Manual for Use of Nylon Bag Technique in the Evaluation Feedstuffs. Aberdeen, Scotland, UK: Rowett Research Institute; 1987

[70] Licitra G, Hernandez TM, Van Soest PJ. Feedbunk management evaluation techniques. Animal Feed Science and Technology. 1996;**57**: 347-358

[71] Rosales RB. The Use of Leguminous Shrubs and Trees as Forages in Tropical Ruminant Production Systems. Manejo de la proteína en la producción de ganado bovino: Seminario taller internacional; 2014

[72] Bach A, Calsamiglia S, Stern MD. Nitrogen metabolism in the rumen. Journal of Dairy Science 2005;**88**,E9-E21. http://dx.doi.org/10.3168/jds.S0022-0302(05)73133-7

[73] Dufreneix F, Faverdin P, Peyraud JL. Influence of particle size and density on mean retention time in the rumen of dairy cows. Journal of Dairy Science 2019;**102**:3010-3022. http://dx.doi.org/10.3168/jds.2018-15926

[74] Foley AE, Hristov AN, Melgar A, Ropp JK, Etter RP, Zaman S, Hunt WC, Huber K, Price WJ. Effect of barley and its amylopectin content on ruminal fermentation and nitrogen utilization in lactating dairy cows. Journal of Dairy Science 2006;**89**:4321-4335. http://dx.doi.org/10.3168/jds.S0022-0302(06)72479-1

[75] Sauvant D, Nozière P. Quantification of the main digestive processes in ruminants: The equations involved in the renewed energy and protein feed evaluation systems. Animal. 2016;**10**:755-770

[76] Broderick G. Determination of protein degradation rates using a rumen in vitro system containing inhibitors of microbial nitrogen metabolism. The British Journal of Nutrition. 1987;**58**:463-475

[77] Ørskov E, DeB Hovell F, Mould F. The use of the nylon bag technique for the evaluation of feedstuffs. Tropical Animal Production. 1980;**5**(3):195-213

[78] Ibrahim MNM, Tammingab S, Zemmelink G. Degradation of tropical roughages and concentrate feeds in the rumen. Animal Feed Science and Technology. 1995;**54**:81-92

[79] Larbi A, Smith J, Kurdi I, Adekunle A, Raji A, Ladipo D. Chemical composition, rumen degradation, and gas production characteristics of some multipurpose fodder trees and shrubs during wet and dry seasons in the humid tropics. Animal Feed Science and Technology. 1998;**72**:81-96